Nitrogen in the Environment: Sources, Problems, and Management

Nitrogen in the Environment: Sources, Problems, and Management

Edited by

R.F. Follett

USDA-ARS, Soil-Plant-Nutrient Research Unit, P.O. Box E, Fort Collins, CO 80522, USA

and

J.L. Hatfield

USDA-ARS, National Soil Tilth Laboratory, 2150 Pammel Drive, Ames, IA 50011, USA

2001

ELSEVIER

Amsterdam – London – New York – Oxford – Paris – Shannon – Tokyo

ELSEVIER SCIENCE B.V.
Sara Burgerhartstraat 25
P.O. Box 211, 1000 AE Amsterdam, The Netherlands

First edition 2001

Library of Congress Cataloging in Publication Data
A catalog record from the Library of Congress has been applied for.

British Library Cataloguing in Publication Data
A catalogue record from the British Library has been applied for.

ISBN: 0-444-50486-9

∞ The paper used in this publication meets the requirements of ANSI/NISO Z39.48-1992 (Permanence of Paper).
Printed in The Netherlands.

EXECUTIVE SUMMARY

R.F. Follett and J.L. Hatfield

Nitrogen (N) is among the most important elements required in agricultural systems to produce food and to supply protein for the increasingly larger World population. The amount of N required increases directly in response to the requirement for protein in the diets of growing population numbers. World population is projected to increase from current levels of about 6.1 to nearly 8 billion people during the next 25 years, and the requirement for N inputs will increase accordingly to meet these growing dietary protein requirements. Countries such as China are rapidly increasing the use of N fertilizer as a result of their large population and an increased demand for dietary protein. Such trends must be expected and planned for in many other parts of the World with growing populations, increased demands for dietary protein, and improved living standards. This book holistically describes the topic of N in the environment, the sources of N and the issues associated with its management. Twenty chapters were prepared by fifty authors who provide an extensive scientific and literature documentation of the issues.

Section 1: Background and Importance of Nitrogen

Within this Section, chapters have been prepared that provide a historical background and the basis whereby issues of N in the environment must be approached. To begin to address the potential needs and consequences of the increase in N inputs that will be required and their real and potential impacts upon the environment, it is necessary that policy makers, industry, the public, and producers be aware of the emerging issues. These issues will require a strategy that encompasses both policy and science over the next decades in order to develop viable solutions. It will be important that a multidisciplinary approach be used to address the wide array of issues. Policymakers, joined by experts from many disciplines that can address both the agricultural production as well as the environmental issues, and other interested stakeholders must become increasingly engaged in a sustained discussion to address the multifaceted issue that are involved.

Evaluation of the transport and fate of N through agricultural systems must be done within the concept of the N cycle. In order to develop strategies that improve the efficiency of N-fertilizer use, reduce the transport and flow of N through agricultural systems, improve N-use efficiency, and that benefit farming and livestock operations, the flow of N needs to be budgeted as it is followed through the pathways of the N cycle at farm, watershed, basin, and even larger scales. The Introductory Section includes chapters on the importance of N to crop quality and health, the role of major crops as a source of amino acids, human nutrition and health issues, and the necessity of N and dietary protein. Growing human populations are the driving force that requires increased production of dietary protein and more N inputs into crop production systems, but results in increased N losses into surface and groundwater resources, emissions of N compounds into the atmosphere, and livestock and human-excretory N contributions into the environment.

Section 2: Water Quality

The Section on water quality has five chapters beginning with the chapter on drainage and subsurface drains. Subsurface drainage is a common water management practice in highly productive agricultural areas with poorly drained soils that have seasonally perched water tables or shallow groundwater. This management practice increases crop productivity, reduces risk, and improves economic returns to crop producers. Such practices are also a source of N to surface waters. Information presented in this chapter clearly shows the value of long-term drainage research conducted across a wide range of soil, crop, and climatic conditions when attempting to improve N management, reduce nitrate losses, and influence policy making. The second chapter in this section (Chapter 6) discusses the occurrence of N in groundwater beneath agricultural systems and is presented by examining the factors influencing aquifer vulnerability to N contamination, and by characterizing the geographic distribution of groundwater contamination by N. Factors that influence aquifer vulnerability are presented in the context of exposure to N sources from general agricultural systems and hydrologic conditions that facilitate transfer of those sources to groundwater. This analysis focuses on the occurrence of nitrate in the United States because data are readily available on many variables needed for such an analysis. Chapter 7 extends the discussion of the transport of N to the role of watersheds. By considering N transport over a range of spatial and temporal scales, it is possible to improve understanding of the factors affecting the fate of N in watersheds, including the effects of land use and N sources (point, nonpoint, agricultural, urban, organic, and inorganic), N transformations (mineralization, nitrification, denitrification, and immobilization), and transport mechanisms (runoff, percolation to ground water, and ground water transport). Knowledge of the variability in N transport in relation to these factors is critical to developing and implementing effective strategies for mitigating unacceptably high N inputs to receiving waters. The scope is limited to technical issues of flow and chemistry of fixed (i.e., biologically reactive) organic and inorganic N forms in watersheds and the chapter concludes with a discussion of the results of empirical modeling methods that have been used to separate the effects of N supply and loss processes and estimate the fate of N sources in watersheds.

Chapter 8 provides an overview of the present state of N pollution in European streams, rivers and lakes from diffuse sources of N and measures that need to be developed to protect drinking water supplies and maintain the environmental quality of rivers, lakes and coastal waters. Elevated riverine N loading has been associated with increased primary production and nuisance algal growth in coastal zones and semi-enclosed and enclosed areas of European seas. Knowledge of the fate of the riverine N transport (mostly as nitrate) is important for allowing accurate estimations of N emissions and the resulting net escape to the open sea. Hence, quantification of the transformation of nitrate under anoxic conditions into nitrous oxide (N_2O) and N_2 gases in rivers, lakes, wetlands and estuaries is an important issue. Even though a high proportion of total anthropogenic N loading to the aquatic environment is from agriculture, reducing such inputs may be difficult for both technical and political reasons. Another approach, besides reduction of N loading at the source, that appears to hold considerable promise is to introduce measures to reduce the level of N by reinstating formerly drained lakes and wetlands in which the potential for nitrate removal through denitrification normally is high as high nitrate waters flow through them.

The final chapter in this section (Chapter 9) discusses N effects on coastal marine environments. Riverine and atmospheric enrichments, combined with burgeoning coastal zone populations of humans and their associated land use changes and wastewater releases, make coastal systems a frontline receiving system for much of the enormous global N enrichment that is occurring. This chapter emphasizes that we do know much about coastal eutrophication and the "proliferation of [autotrophic] cells." There has been strong progress made in connecting inputs, concentrations, and effects of N on coastal waters. But the author emphasizes that the stimulation of some of the most undesired cells (e.g. toxic dinoflagellates) or the precise point at which adverse secondary consequences of cell proliferation will occur in any given system cannot yet be predicted nor is there yet a good *generalized and quantitative* description of adverse effects of N loading, in part because of the wide variety of coastal systems. Increased loading of coastal systems over the past few decades indicate many individual system's limits have been passed and oxygen-depleted, mortality-inducing conditions are resulting. More subtle effects than fish kills occur, so there are related fundamental questions: when and where does N stimulate undesired species changes or wholesale food web shifts? The state of knowledge does not yet provide sufficient answers so that critical protective N limits can be set for many systems.

Section 3: Atmospheric Effects

The third section of this book on atmospheric effects has three chapters. Chapter 10 addresses livestock farming systems, which on a global scale contribute ~70% and ~30% to the total anthropogenic emissions into the atmosphere of ammonia and nitrous oxide (N_2O), respectively. Chapter 10 discusses the origin and controlling factors of gaseous N emissions from livestock farming, uncertainty in the estimates, and possible measures that may be taken to decrease these emissions. Chapter 11 addresses the role of industrial, automobile and food production systems as important contributors to local, regional and global NH_3, NO_x ($NO + NO_2$) and N_2O budgets through the emissions mainly of NO_x and NH_3 via the atmosphere and release of nitrate into surface and groundwaters. These reactive N species are highly mobile and may cross national boundaries and be deposited as NO_y and NH_X or nitrate far from their source. Emissions into the atmosphere serve to redistribute fixed N to local and regional aquatic and terrestrial ecosystems that otherwise may be disconnected from the sources of the N gases. Emission of NO_x contributes to local elevated ozone concentrations while emission of N_2O contributes to global greenhouse gas accumulation and to stratospheric ozone depletion. Gaseous redistribution of N has and is having profound impacts on the quality of terrestrial and aquatic ecosystems and on the atmosphere. Deposition of N can stimulate productivity in N-limited grasslands, forests and aquatic systems. Chapter 12 discusses the impact of N deposition on forest ecosystems. The author considers forest litter and soil systems that typically interact with vegetation to form "closed" N cycles in which annual rates of plant N uptake per unit area approximately balance annual N returns to the forest floors and soil as litter and root material. Increased N emission and deposition that occur and are projected to continue into the future across more of Earth's surface, serve to diminish the degree of N limitation of forest growth, alter forest community composition, change plant-soil interactions, and open forest N cycles. Thus N deposition has implications not only for forests, but also for the atmosphere and for ecosystems that receive nutrient outputs from forests. This chapter summarizes research addressing the influences of N deposition on forest N cycles and on forest ecosystem structure and function including the acidification and eutrophication of aquatic systems and to forest decline.

Section 4: Emerging Prediction and Management Technologies

Section 4 in this book has six chapters that describe various emerging, prediction, and management technologies to help minimize and control impacts of N in the environment. Opportunity for improvement at the farm level largely lies with technologies that enable timely, quick, and accurate measurement of the spatial variability of crop yield potential, soil N availability, and within season indication of crop N health. Soil N excess and deficiency often exist on the same field. Thus, it is variability in space and time of processes that regulate availability of N to plants and fate of N in soil that make predictive technologies and precision N management attractive. Chapter 13 considers on-farm technologies and practices to improve efficiency of N use in agricultural production systems and the authors review the tools currently available, or are becoming available, to help producers make better N management decisions. Soil and crop measurements are considered and compared. Nonetheless, "on-farm" implies that producers will be at the center of implementing these changes. Prerequisites for grower adoption requires that technologies and practices be reliable, incur minimal additional expense (time and equipment), and integrate with ease into current operations. When these criteria cannot be met, external incentives (e.g., regulation, private or government cost sharing programs, etc.) may be necessary. The next three chapters discuss the use of computer models as one of the tools to assist with on-farm management decisions to aid in improving efficiency of N use from livestock manures (Chapter 14), for water quality prediction and protection and use of best management practices of crop production (Chapter 15), and N prediction models for greenhouse gas emissions (Chapter 16). The authors of Chapter 14 identify that increasingly crop and livestock producers will be required to develop nutrient management plans to demonstrate that their operations have sufficient cropland area, seasonal land availability, manure storage capacity, and application equipment to manage animal manures, commercial fertilizers, and other land-applied nutrient resources in an environmentally responsible manner. Computer software has been and will continue to be used to help develop these plans. New nutrient management planning software needs to increasingly address the temporal and spatial nature of nutrient management, provide ways to accommodate regional areas and changing regulatory reporting requirements, utilize national databases and standards, and take advantage of modern software technologies. The authors of Chapter 15 discuss how potential scenarios for cropping system, N, and/or water management should be developed in cooperation with local producer, commodity, and action agency groups. Comparisons among simulations of various management scenarios should be done taking into account uncertainty in the results obtained from calibration and validation studies. Potential differences should be targeted when selecting management scenarios to test with models. Much progress is occurring to develop computer modeling to describe gaseous N_2O emissions. The authors of Chapter 16 describe how simulations are being used primarily to explore how N_2O gas emissions respond to changes in land management, soil texture, and precipitation; they describe how annual N_2O emissions can be reliably simulated for some native and managed systems. They conclude that mitigation of N gas losses from soils is strongly dependent on land management but that generalizations based solely on soil N and water inputs are likely to be limited because soil texture, SOM levels, and timing of management practices are also important.

Chapters 17 and 18 address the important issue of the remediation of high-nitrate drinking water supplies for rural and urban populations, respectively. Contamination can result from a number of activities. More intensive crop production has caused a continuing increase of nitrates derived from animal wastes or fertilizers deposited over many years on agricultural soils and which migrate slowly downward through the overlying soils to aquifers or by overland and interflow processes to surface waters that are used as drinking water supplies. Chapter 17 discusses how reverse osmosis, ion exchange, and distillation can be used in home systems to remove nitrate from raw water. Reverse osmosis and distillation are small point-of-use systems that provide water for cooking and drinking. In the future, "bio-barriers" might be used to protect a well from a contaminated aquifer or to protect an aquifer used for drinking water from a source of nitrate pollution. Above ground denitrification reactors might be used to provide nitrate-free water to rural communities in the not too distant future. Chapter 18 describes how, for most cities where high-nitrate drinking water supplies exist, establishing alternate supplies of source water is the most cost-effective alternative. Generally, nitrates are a problem in surface or shallow ground water supplies. Drilling a new, deeper well, may also be an alternative for some cities. The author of Chapter 18 describes the alternative used for the city of Des Moines Water Works to serve approximately 350,000 people. Nitrate concentrations from their traditional water supplies had risen to where the use of those supplies was no longer feasible. Feasibility studies found that ion exchange, using a strong anion resin, was the most desirable treatment process. The ion exchange process was implemented and is being successfully used.

Section 5: Economic and Policy Issues

Section 5 in this book has two chapters to address the economic and policy issues. These chapters strongly address issues related to agriculture. In Chapter 19 the authors point out that users of N (farmers) live not only in biophysical space but also in social space. It is pointed out that salient social and biophysical processes are not random in space or time. Inappropriate behaviors at vulnerable locations or times create many of the "problems" that policy and research agendas are oriented toward today. This situation of "disproportionality" provides a warning that there is not a universal policy or technical solution to these problems. The authors argue that patterns and processes must first be described and measured before seeking explanation and prediction. This chapter contributes toward that objective by describing how concepts such as scale and context apply to the management of N by farmers. Novel approaches are suggested and illustrated to measure these concepts in space because social and biophysical data occur along a continuum of differing scales, and are often incompatible because one or the other was measured at a finer or coarser scale.

Chapter 20 addresses new policy directions. The author contends that advances in linking the science of ecology with economics and the development of agro-environmental modeling systems will require they be coupled with new information technology, thus suggesting that new public policy approaches may well need to reward agricultural producers for providing ecological services to also advance the social and economic agendas of the public and of policy makers. Such a model to simulate complex relationships between agricultural land management practices, economic costs, and environmental impacts also allows for commoditization of services, and the possibility of new markets. Markets, however, require clarity regarding property rights to what is traded. Government needs to define what it is that is traded and with the use of these models, help establish who owns the product of the services (as well as who is liable for nonperformance of service and the extent of liability in many cases), and establish conditions for monitoring and enforcement where the primary beneficiary of the service is the public.

Finally, to those who are concerned with the environmental, policy, or economic issues, the editors hope that all will recognize that agriculture, as a business, is conducted outdoors and each year faces considerable risk and adversity as a result of the vagaries of climate. Depending upon location and year, climatic vagaries can include regional heat waves, droughts, severe storms, and/or flooding. From an economic perspective, the purpose for the use of N in agriculture is to increase crop yields, improve the economic return realized by the producer, and foremost to meet human needs for food and dietary protein. To address the issue of N in the environment, which is in fact increasingly becoming a complex societal (and global) problem, will require a sustained dialogue among policymakers, scientists, practitioners, and affected stakeholders. The interaction of scientists and policymakers, in particular, often does not occur in a regular or orderly manner to help benefit policymaking. In general, the more complex the problem is, which may require legislative solutions, the longer and more deliberative must be the process to craft the legislation. In the case of issues with a complex scientific underpinning, such as the issues associated with N in the environment, the deliberative process will undoubtedly require a thorough, thoughtful interface between scientists, who tend to speak in terms of 'uncertainties', and policymakers, who seek the bottom-line 'certainties' necessary to create good policy and prudent laws.

About the Editors

Dr. Ronald F. Follett is Supervisory Soil Scientist with the Agricultural Research Service (ARS) of the United States Department of Agriculture (USDA) with 35 years of research experience. For the past 15 years he has been Research Leader with the ARS Soil-Plant-Nutrient Research Unit in Fort Collins, Colorado. He previously served 10 years as a National Program Leader with ARS headquarters in Beltsville, Maryland. He has also been a Research Soil Scientist with ARS in Mandan, North Dakota and Ithaca, New York. Dr. Follett is a Fellow of the Soil Science Society of America, American Society of Agronomy, and the Soil and Water Conservation Society. He has received USDA's highest award twice, the USDA Distinguished Service Award, and in June of 2000 he received an Individual USDA Superior Service Award "For promoting sensible management of natural, soil, and water resources for an environmentally friendly and sustainable agriculture." Dr. Follett organized and wrote the ARS Strategic Plans for both Groundwater Quality Protection – Nitrates' and Global Climate Change – Biogeochemical Dynamics'. Dr. Follett has edited eleven books including a recent book entitled "The Potential of U.S. Grazing Lands to Sequester Carbon and Mitigate the Greenhouse Effect", coauthored the book entitled "The Potential of U.S. Cropland to Sequester Carbon and Mitigate the Greenhouse Effect", and served as a guest editor for the Journal of Contaminant Hydrology. His numerous scientific publications include topics about nutrient management for forage production, soil-N and C-cycling, groundwater quality protection, global climate change, agroecosytems, soil and crop management systems, soil erosion and crop productivity, plant mineral nutrition, animal nutrition, irrigation, and drainage.

Dr. Jerry Hatfield received his Ph.D. from Iowa State University in 1975 in the area of Agricultural Climatology and Statistics. From 1975 through 1983 he was the Biometeorologist at the University of California-Davis and from 1983 through 1989 Research Leader at the USDA-ARS Plant Stress and Water Conservation Laboratory in Lubbock, Texas. Since 1989 he has served as the Laboratory Director of the USDA-ARS National Soil Tilth Laboratory in Ames, Iowa. Dr. Hatfield has been responsible for the development of the scientific program in the Tilth Laboratory and the management of a multi-agency, multi-location environmental quality program to assess the impact of farming practices on water quality. He has developed several watershed scale projects to address concerns about the spatial and temporal impacts of farming practices on environmental quality. His research interests focus on the interaction of water, nutrients, carbon, and light in the response of crops to management systems across varying landscapes. He serves as the Editor-in-Chief of the American Society of Agronomy and is a Fellow of the American Society of Agronomy, Crop Science Society of America, and Soil Science Society of America and a 1997 recipient of the A.S. Flemming Award for Outstanding Federal service and an ARS Outstanding Scientist of the Year in 1999. He is the author over 280 referred publications and the editor of seven monographs.

List of Contributors

R. B. Alexander. *U.S. Geological Survey, 413 National Center 12201 Sunrise Valley Drive, Reston, VA 20192 USA*

David D. Baltensperger. *Panhandle Research & Extension Center, University of Nebraska, 4502 Avenue I, Scottsbluff, NE, 69361 USA*

André Bannink. *Institute for Animal Science and Health (ID), Wageningen University and Research Center P.O. Box 65, 8200 AB Lelystad, The Netherlands*

Jürg M. Blumenthal. *Panhandle Research & Extension Center, University of Nebraska, 4502 Avenue I, Scottsbluff, NE, 69361 USA*

P. Boers. *Institute for Inland Water Management and Wastewater Treatment, P.O. Box 17, 8200 AA, Lelystad, The Netherlands*

J. V. Brahana. *University of Arkansas, Department of Geosciences 114 Ozark Hall, Fayetteville, AR 72701 USA*

Michael R. Burkart. *USDA/ARS, National Soil Tilth Laboratory, 2150 Pammel Drive, Ames, IA 50011-4420 USA*

Perry E. Cabot. *Institute of Environmental Studies, Land Resources Program, 1450 Linden Dr., Ag. Hall Rm. 420, University of Wisconsin, Madison, WI 53706 USA*

Kenneth G. Cassman. *Department of Agronomy and Horticulture, University of Nebraska, 279 Plant Science, Lincoln, NE 68583-0915 USA*

Laurence B. Cutforth. *Institute of Environmental Studies, Land Resources Program, 1450 Linden Dr., Ag. Hall Rm. 420, University of Wisconsin, Madison, WI 53706 USA*

Jorge A. Delgado. *USDA-ARS, Soil Plant Nutrient Research, P.O. Box E, Fort Collins, CO 80522 USA*

S.J. Del Grosso. *Natural Resource Ecology Laboratory, Colorado State University, Fort Collins, CO 80523 USA*

Jennifer R. Follett. *Nutrition Department, 3402 Meyer Hall, 1 Shields Avenue, University of California, Davis, CA 95616 USA*

Ronald F. Follett. *USDA/ARS, Soil-Plant-Nutrient Research Unit, PO Box E, Fort Collins, CO,80522 USA*

M. J. Goss. *Land Resource Science, 140 Richards Bldg., University of Guelph, Guelph, Ontario N1G 2W1, CANADA*

K.W.T. Goulding. *Soil Science Department, IACR-Rothamsted, Harpenden, Herts. AL5 2JQ, UK.*

M.D. Hartman. *Natural Resource Ecology Laboratory, Colorado State University, Fort Collins, CO 80523 USA*

J.L. Hatfield. *USDA-ARS-National Soil Tilth Laboratory, 2150 Pammel Drive, Ames, IA 50011 USA*

Philip J. Hess. *Agronomy Dept., Purdue University, 1150 Lilly Hall of Life Sciences, West Lafayette, IN 47907 USA*

C.C. Hoffmann. *National Environmental Research Institute, Department of Streams and Riparian Areas Vejlsoevej 25, DK-8600, Silkeborg, Denmark*

W. J. Hunter. *USDA/ARS, Soil-Plant Nutrient Research Unit, P. O. Box E, Fort Collins, CO 80522 USA*

J.P. Jensen. *National Environmental Research Institute, Department of Lake and Estuarine Ecology Vejlsoevej 25, DK-8600, Silkeborg, Denmark*

Brad C. Joern. *Agronomy Dept., Purdue University, 1150 Lilly Hall of Life Sciences, West Lafayette, IN 47907 USA*

Bruce M. Kahn. *Institute of Environmental Studies, Land Resources Program, 1450 Linden Dr., Ag. Hall, Room 420, University of Wisconsin, Madison, WI 53706 USA*

D.R. Keeney. *Professor Emeritus, Department of Agronomy, Iowa State University, 3402 Eisenhower Ave. Ames, IA 50010 USA*

John R. Kelly. *U.S. Environmental Protection Agency (US EPA), National Health and Environmental Effects Research Laboratory, Mid-Continent Ecology Division, 6201 Congdon Blvd., Duluth MN 55804 USA*

C.A. Keough. *Natural Resource Ecology Laboratory, Colorado State University, Fort Collins, CO 80523 USA*

N.R. Kitchen. *USDA/ARS, 240 Agricultural Engineering Bldg., University of Missouri, Columbia, MO 65211 USA*

B. Kronvang. *National Environmental Research Institute, Department of Streams and Riparian Areas Vejlsoevej 25, DK-8600, Silkeborg, Denmark*

John A. Lory. *Agronomy Dept., University of Missouri, 210 Waters Hall, Columbia, MO 65211 USA*

L.D. McMullen. *CEO and General Manager Des Moines Water Works, 2201 Valley Drive, Des Moines, IA 50321 USA*

Andrew P. Manale. *US Environmental Protection Agency, Office of Policy, Economics, and Innovation, 1200 Pennsylvania Ave., NW, Ariel Rios Bldg., Washington, D.C. 20460 USA*

Stephen C. Mason. *Department of Agronomy and Horticulture, University of Nebraska, 279 Plant Science, Lincoln, NE 68583-0915 USA*

A. R. Mosier. *USDA/ARS, Soil-Plant-Nutrient Research Unit, P.O. Box E Fort Collins, CO 80522 USA*

K. J. Nadelhoffer. *The Ecosystems Center, Marine Biological Laboratory, 7 MBL Street, Woods Hole, MA 02543-1015 USA*

Pete J. Nowak. *University of Wisconsin, Department of Rural Sociology, 346D Agriculture Hall, 1450 Linden Drive, Madison, WI 53706 USA*

Oene Oenema. *Alterra, Wageningen University and Research Center, P.O. Box 47, NL-6700 AA Wageningen, The Netherlands*

D.S. Ojima. *Natural Resource Ecology Laboratory, Colorado State University, Fort Collins, CO 80523 USA*

W.J. Parton. *Natural Resource Ecology Laboratory, Colorado State University, Fort Collins, CO 80523 USA*

Alexander D. Pavlista. *Panhandle Research & Extension Center, University of Nebraska, 4502 Avenue I, Scottsbluff, NE 69361 USA*

G.A. Peterson. *Soil and Crop Sciences Department, Colorado State University, Fort Collins, CO 80523 USA*

G. W. Randall. *Southern Research and Outreach Center, University of Minnesota, 35838 120th St., Waseca, MN 56093-4521 USA*

T. J. Sauer. *USDA/ARS, National Soil Tilth Laboratory 2150 Pammel Drive, Ames, IA 50011-4420 USA*

D.S. Schimel. *Max Planck Institut für Biogeochemie, Carl-Zeiss-Promenade 10, Room ECO 004, 07745 Jena, Germany.*

Marvin J. Shaffer. *USDA/ARS, Great Plains System Research, P.O. Box E, Fort Collins, CO 80522 USA*

R. A. Smith. *U.S. Geological Survey, 413 National Center 12201 Sunrise Valley Drive, Reston, VA 20192 USA*

Sven G. Sommer. *Danish Institute of Agricultural Sciences (DIAS), P.O. Box 536, 8700 Horsens, Denmark*

Jeffrey D. Stoner. *U.S. Geological Survey, National Water Quality Assessment Program, Nutrient Synthesis Office, Bldg. 53, Denver Federal Center, Mail Stop 425, Box 25046, Lakewood, CO 80225 USA*

Gerard L. Velthof. *Alterra, Wageningen University and Research Center, P.O. Box 47, NL-6700 AA Wageningen, The Netherlands*

Contents

Section 1:

Background and Importance of Nitrogen

Nitrogen in the Environment: Sources, Problems, and Management
R.F. Follett and J.L. Hatfield (Eds)

Chapter 1. The Nitrogen Cycle, Historical Perspective, and Current and Potential Future Concerns

D.R. Keeney[a] and J.L. Hatfield[b]

[a]Professor Emeritus, Department of Agronomy, Iowa State University
3402 Eisenhower Ave. Ames, Iowa 50010 drkeeney@iastate.edu

[b]Director, USDA-ARS-National Soil Tilth Laboratory, 2150 Pammel Drive, Ames, Iowa 50011 hatfield@nstl.gov

Nitrogen (N) along with carbon is the most complex and crucial of the elements essential for life. Supplementing grain and grass forage crops with organic and inorganic fertilizers has long been recognized as a key to improving crop yields and economic returns. Nitrogen is the most widely utilized plant nutrient in fertilizers and is a major component of animal manures. Nitrogen compounds also have been recognized for their many potential adverse impacts on the environment and health.

From 1850 to 1980, biological scientists concentrated on unraveling the biological and physical-chemical intricacies of N. We now know the paths of its comings and goings, the route it takes as it moves, at rates varying from milliseconds to centuries, through nature's compartments (atmosphere, soil, water, and living matter) and the interactions of N with various elements. We know as well as its oxidation/reduction status under varying environmental conditions. But nature, in its clever way, has kept science from tracking precisely the actual ledger of this whimsical element and curiously of predicting the impact of N on the environment when it accumulates at levels far above that for which stable ecosystems have adapted.

Many ecological problems occur when N is separated from its most common partner, carbon (Asner et al. 1997). Nitrification, denitrification, nitrous oxide formation, leaching of nitrate, and volatilization of ammonia are fates of the "mobile" N atom. Effects within the environment vary with the N form. The atmosphere might receive more nitrous oxide than it can assimilate, resulting in stratospheric ozone destruction. Combined N in the atmosphere and precipitation fertilizes natural ecosystems resulting in lowered biodiversity, stress, and N leakage, while acidity from nitric oxide and ammonia oxidation depletes ecosystems of bases and results in acid lakes and declining health of forests.

Lakes, coastal waters, and estuaries overloaded with biologically available N, produce organic materials in abundance. The N atom gets connected to carbon, but the unwanted effects of excess growth and subsequent decay create anaerobic conditions. Nitrogen is widely regarded as responsible for the hypoxia (low oxygen) zone in the Gulf of Mexico that is currently concerning ecologists and conservationists as well as those financially dependent on fish and shrimp catches. Decaying organic matter removes oxygen, changing the ecology and productivity of the bottom waters in a large area of the Gulf. Productive agricultural regions of the Central U. S. are the major source of the nitrate to the Gulf.

Can the N cycle be managed to avoid some of the problems N generates? Given the world's needs for food, the great ability of annual grains to produce the needed food (and animal feed), and the relatively cheapness of N fertilizer due to undervalued energy, change on the larger scale does not seem to be forthcoming. The UK and Western Europe have adopted strict manure and fertilizer application regulations with stiff fines for failing to adhere to the regulations. Other countries, including the U. S. and Canada, have relied on education and demonstration programs to lessen environmental effects from excess fertilizer N use. Iowa and some other states in the U. S. have had some modest success at decreasing N fertilizer use through research and education projects. However, ground water quality measurements in Iowa have shown little long term change in nitrate concentrations illustrating again the problems of second guessing the N cycle.

The “solutions” to the issues on environmental effects of N will involve looking beyond the edge effects to redesigning agriculture in ways that will "tighten up" the N cycle and that will provide for N sinks such as grasslands and wetlands. To do this, policies will need to be developed that assure farmer and the public that such measures will not cost productivity, and that a redesigned agriculture can provide for future food needs. Turning back is not possible. The road ahead will demand a level of innovation of agricultural research and development of new agriculture systems.

1. THE NITROGEN CYCLE

AZOT, the German word for N, was the subject of ancient philosophers. AZOT is believed to be formed from the ancient scientific alphabets, A (the beginning of scientific Latin Greek and Hebrew) and zet, omega and tov, the last letters of these alphabets. The term “saltpeter” came from the association of nitrate salts with the salt of the earth or the salt of fertility. Potassium nitrate was manufactured for gunpowder in the 14th century. By the 1650's Johannes Rudolph Glauber spoke of ‘nitrum’ as the “soul” or “embryo” of saltpeter. He states, “It is like a wingless bird that flies day and night without rest; it penetrated between all the elements and carries with it the spirit of life --from nitrum are originated minerals, plants and animals. (It) never perishes; it only changes its form; it enters the bodies of animals in the form of food and then is excreted. It is thus returned to the soil, from which part of it again rises into the air with vapors, and hence it is again among the elements.” The N cycle was never better described even though this was 3 1/2 centuries ago. (Much of the material for this paragraph originated from Vorhees and Lipman, (1907); Waksman, (1952); and Harmsen and Kolenbrander, (1965).

Nitrium was the subject of numerous other early authors. In the 1780's Cavendish discovered that the inert gas of air would combine with oxygen to give oxides. The stage was set for the linkage of the lifeless AZOT and saltpeter. The French scientist, Boussingault, who many regard as the founder of modern agrochemistry, did this with the sand culture research during the 1830’s to 1860’s. He deduced that the fertilizing properties of manures came from the ammonium formed in the soil and that ammonium was taken up by the plant root (Waksman, 1952; Burns and Hardy, 1975).

Research during the 1880 to 1910 period revealed many basic reactions of the N cycle and set the stage for five decades of vigorous and detailed N cycle studies. Schloessing first demonstrated denitrification in 1868. Gayon and Dupetite reported their research on denitrification in 1882 and coined the term at that time (see Payne, 1981). Davy, in 1813, first attributed the beneficial effects of legumes to soil AZOT. While Boussingault quantified this benefit, Liebig was not convinced and hence the classical experiments of Lawes, Gilbert, and Pugh were established at Rothamsted in 1857. Unfortunately their sterile sand experiments destroyed the Rhizobium population and it was not until the late 1880's and early 1890's that biological N fixation was confirmed by Hellriegel and Wilfarath. Beijerinck isolated Rhizobia in 1888 and Azotobacter in 1901. Winogradsky identified Clostridium in 1890. Burns and Hardy, (1975) reported much of this history of N research.

Nitrification received much study during the early 1900's on the belief that nitrate was the dominant form of N used by plants. King and Whitson, (1902) at Wisconsin conducted some excellent research on the effects of environmental variables on the rate of nitrification. Their research also examined the effects of cropping on profile nitrate concentrations and leaching of nitrate. The use of a nitrification test to measure soil fertility was proposed, tested, and abandoned. Heterotrophic nitrification was identified and following the acceptance in 1926 of two-electron shifts during sequential oxidation, research began on determining nitrification intermediates. Allison identified the first nitrification inhibitors, the cyanamides, in the 1920's. During this time denitrification received little attention. In 1910, Beijerinck and Minkman, and Suzuki in 1912, concluded that nitric oxide and nitrous oxide were obligatory intermediates in denitrification and that organic matter was the major electron donor (see Alexander, 1955 and Stevenson, 1965). Waksman and Starkey, (1931) dismissed denitrification as of little economic importance. However, (Allison, 1955) in a seminal review, pointed out that nitrogen balances are never obtained in field and lysimeter experiments

Mineralization and immobilization were recognized as important reactions but most scientists looked at these as separate rather than coupled transformations. Much time was spent evaluating the fertilizer values of manures and compost (e.g., Blair, 1917). Methods for identification of organic forms of N were developed and extracellular enzymes were recognized. Conrad (1940) identified extracellular soil urease. Soil tests to estimate N availability by the rate of ammonium formation in incubated samples were first studied at about the turn of the century. The first modern "N Cycle" was probably the one published in 1913 by Lohnis (see Lohnis, 1926). The concepts he proposed are valid today. Blair, in 1917, presented a more ecosystem-oriented N cycle, including abiotic reactions. Science has spent the last 84 years on the details of the N cycle in various soils and cropping systems.

1.1 The N Fertilizer Era

By 1921 a manufacturing plant using the Haber-Bosch process for synthesis of ammonia was operating in the U. S. Synthetic ammonia plants were not widely used until the WW II munitions plants were converted to ammonium nitrate fertilizer plants. Most important was the Tennessee Valley Authority (TVA) complex at Muscle Shoals, Alabama that was completed just as WW II was ending. Until 1945, Chilean nitrate, ammonium nitrate, and ammonium sulfate were the major inorganic fertilizers.

1.2 Historical and Current Trends in N Fertilizer Use

The availability of relatively cheap ammonia-based fertilizers marked a significant change in the way N was supplied in agriculture. However, replacement of traditional N sources for crops by fertilizer N proceeded slowly until the early 1960's. The TVA began a demonstration program in the late 1940's to facilitate information on proper N fertilizer use and established a state of the art research facility at its Muscle Shoals, Alabama facility.

Cooperative research programs in key U. S. agricultural colleges also helped forward the TVA research program and enabled scientists to fund research and graduate students in the areas of N fertilizer use and N cycle reactions. This cadre of soil chemists and biochemists made up the bulk of the research community in N cycle reactions during the 1950-1970 era. The senior author was privileged to share in this particular period. It was an era never to be repeated, one full of excitement, enthusiasm, and accomplishments in understanding the N cycle. Annual cooperators meetings at Muscle Shoals were events to be treasured because of the sharing of research ideas, results, and philosophy. This program accomplished the goal of increasing N fertilizer use. Nitrogen fertilizer use became the mainstay of modern World agriculture. Some now feel that the overemphasis on fertilizer to increase crop yields came at the expense of sound ecological farming approaches (e.g., Kjaergaard, 1995; Moffat, 1999; Keeney and Muller, 2000). By the 1960's, fertilizer use in agricultural regions such as the Midwest Corn Belt began a marked increase. An example is the sales values for Iowa (Figure 1).

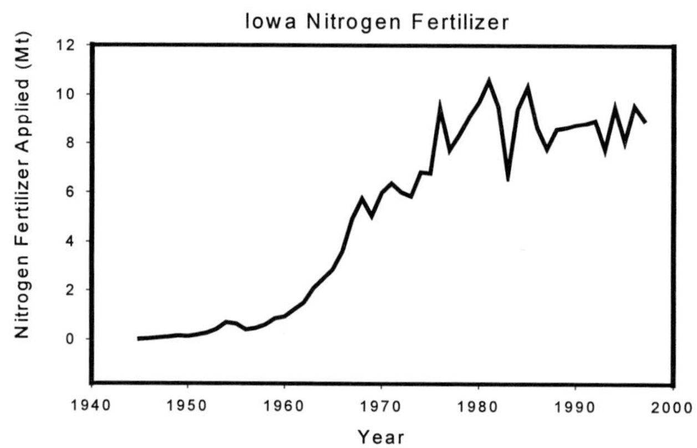

Figure 1. Nitrogen fertilizer applied in Iowa per year from 1945 through 1998.

Although N fertilizer consumption rapidly increased, N from animal manures did not show a concurrent reduction. For example, in the U. S. annual production of N from all animal sources has ranged from 5.7 Mt (Million Metric tons) in 1982, 5.6 Mt in 1987 and 1992 and 5.9 Mt in 1997 (Kellog et al., 2000). Total N from manures are relatively small compared to fertilizer sources, the move to concentrated animal feeding operations has resulted in high N outputs in local areas.

2. MODERN N CYCLE RESEARCH

The 1945-1980 period was marked by a spectacular increase in research activity on all facets of N in agriculture. The mass spectrometer developed for the Manhattan project was subsequently used for innovative ^{15}N research. The application of ^{15}N isotope methods by Bremner, Burris, Broadbent, and Norman, in the late 1940's and 1950's demonstrated the tremendous power of stable isotope research. The National Fertilizer Development Center at TVA was established in the mid-1950's. This center aided greatly in development and application of ^{15}N methods in agricultural research. Many other analytical methods were improved, some were automated and others were developed. Sensitive gas chromatographic methods for identification of gaseous intermediates of nitrification and denitrification facilitated research on these reactions. Computer technology was sufficient by the early 1970's to permit construction of sophisticated, mathematical models of the N cycle.

Table 1. Nitrogen fertilizer consumption (Mt) N in the World, United States, and China, 1961-1998.

Year	World	United States	China
1961	11.588	3.057	544
1965	19.097	4.832	1.832
1970	31.756	7.363	3.369
1975	44.420	9.445	5.039
1980	60.776	10.817	12.112
1985	69.830	9.457	13.650
1990	77.246	10.047	19.450
1995	77.905	11.161	23.382
1998	82.421	11.281	22.446

FAO, 2000

The use of ^{15}N permitted renewed emphasis on mineralization-immobilization research. Researchers included Bartholomew, Broadbent, Bremner, Harmeson, Hauck, Jansson, Jenkinson, Paul, Perrson, and Van Schreven. Bremner clarified the denitrification process, and developed many methods for analysis of ^{15}N. Jannson, (1958) identified the central role of ammonium in mineralization-immobilization. Major reviews by Harmesen and Van Schreven, (1955), Bartholomew and Clark, (1965) and Jansson and Persson, (1982) and Stevenson, (1982) set the stage for current concepts of mineralization-immobilization. Jansson and Persson's review solidified their concepts of the universal N cycle. Major advances in nitrification pathway research were the establishment of nitrous oxide as a byproduct of ammonium oxidation and the development of commercial nitrification and urease inhibitors. Denitrification research was expanded with discovery that acetylene blocked nitrification as well as nitrous oxide reduction. The interest in nitrous oxide in ozone destruction and as a greenhouse gas gave impetus to studies to quantify its output from various agricultural and natural ecosystems.

By the early 1980s breakthrough research on N was largely complete. Nitrogen research moved out of the public eye. Nitrogen fertility research continued on a site and crop specific basis, but less attention was paid to environmental issues. In a recycling of issues, N is now gaining new attention as new environmental problems come to the forefront and old issues resurface. And world food problems remain.

3. THE ISSUES

Nitrogen from anthropogenic sources, including fertilizers, biological N fixation, ammonia volatilization, combustion, and activities that bring N from long term storage pools such as forests have been estimated by several groups to be close to the same order of magnitude as the N from natural (pre industrial) sources (Vitousek et al., 1997), (Table 2). This doubling of the available N pool worldwide has many implications. While most N issues are local and thus the global N cycle would not seem applicable, many issues have regional or global implications.

Excess N in rivers, lakes and groundwater can be toxic to humans and causes water quality problems in natural water systems (Hallberg and Keeney, 1993). Excess N in the estuaries of the oceans enhances growth of aquatic organisms to the point that they affect water quality and lower dissolved oxygen levels (Rabalais et al., 1996; Downing, 1999; Howarth, 2000). This affects the metabolism and growth of oxygen requiring species, causing a condition referred to as hypoxia (Rabalais et al., 1996; Downing, 1999). Nitrogen in the atmosphere comes from emission of ammonia from human activities such as feedlots (Jackson et al., 2000) and from combustion sources. This N contributes, as nitric acid, to acid rain, damaging lakes, rivers, and forests. In land ecosystems, excess atmospheric N may enhance growth of exotic species or accelerate growth of trees, causing disruption of ecosystem functions (Vitousek et al., 1997; Jordan and Weller, 1996) such as over-fertilizing natural grasslands and lakes (Keeney, 1997).

Table 2. Estimates of global nitrogen fixation (MMt of N)
(extrapolated from Vitousek et al., 1997).

Source	1960	1990
Legume crops	30	40
Fossil fuel emissions	10	15
Fertilizer	20	80
Total	60	145
Natural N fixation	80-130	80-130

3.1 Water Quality

Numerous studies, summarized by Hallberg (1989) and Keeney (1989) have documented the large increase in nitrate in ground waters in the U.S. relative to pre-industrial levels. Natural background levels commonly are less than 2 mg nitrate-N/L while agricultural areas often exhibit seasonal concentrations greater than 10 mg/L especially in tile drained regions (Randall and Mulla, 1998). The numerous sources and sinks of nitrate make evaluation and control of sources difficult and hence establishment of policies and groundwater protection goals are controversial and often unproductive. Hallberg's (1989) review points out the interaction of ground and surface water systems, particularly those that impact shallow ground waters. Vitousek et al. (1997) summarized accumulation of N in surface waters, particularly riverine systems, but also estuaries. For example, Howarth et al. (2000) estimated that riverine fluxes from lands surrounding the North Atlantic Ocean have increased from pre-industrial times by 2- to 20-fold.

3.2 Health Issues

Methemoglobinemia was first recognized by Comley (1945) who related infant illnesses to nitrate-contaminated private wells in Iowa. Nitrate can be reduced to nitrite in the digestive tract and the nitrite interferes with oxygen transport in the blood. The main health effect is with infants. The U. S. health standard (maximum contaminant level, MCL) of 10 mg/L of nitrate-N was established by the United States Environmental Protection Agency (U.S. EPA) in 1977 as a safeguard against infantile methemoglobinemia (Kross et al., 1992). Ground water is the primary water source of concern and many potable ground water supplies, especially those in rural areas of productive agriculture, are above the MCL (Hallberg and Keeney, 1993). For example, a recent study showed that in the early 1990's 18% of Iowa private water supply wells had nitrate-N levels above 10 mg/L (Kross et al., 1992).

Several infant illnesses and deaths from consumption of high nitrate waters have been documented and while the numbers are small, there is also concern about sub clinical exposure and under reporting of deaths and illnesses because the problems often occur in rural areas (Kross et al., 1992). Often rural water supplies are also high in bacterial pathogens and the ensuing gastric illnesses may complicate the effects of nitrate. Few government agencies enforce remedial actions for private water supplies, but most European countries and the U.S. EPA require actions if municipal supplies are above the MCL. This has forced some municipalities such as Des Moines, Iowa to install expensive nitrate removal plants that operate only when the MCL gets close to the maximum concentration. New health risk policies set by the U.S. EPA in 1995 that evaluate risk to infants and children separately from adults may raise new concerns about the health risks of nitrate in potable water (Meyer, 1995).

Recently a researcher for the National Institutes of Health (Avery, 1999) suggested that nitrate toxicity was not a problem because no cases of methemoglobinemia had occurred recently. He also suggested that nitrate was beneficial to human health because of internal formation of nitrous oxide. He proposed increasing the MCL to 20 mg/L. Recently, Knobeloch et al. (2000) reported occurrence of blue baby syndrome in two Wisconsin households served by private water wells. They stated that their findings could not support Avery's conclusions. A review of risks of nitrate to humans (Wilson et al., 1999) points out that nitrate exposure could cause not only methemoglobinemia but also diabetes and cancer. Yet Addiscott (1999) in summarizing the series of papers in the Wilson et al., (1999) book provides some support for Avery's view that nitrite in stomach would be beneficial for control of pathogens and may be a natural defense mechanism in humans.

4. ENVIRONMENTAL ISSUES

4.1 Surface water quality and ecology

Phytoplankton and vascular plant growth in surface waters is limited by numerous factors, just as is growth of land-based plants. However, complications in assessing limiting factors, especially nutrients, arise because of the dynamic nature of surface waters relative to inputs, seasonal changes, turbidity that limits light penetration and differing nutrient and environmental requirements of phytoplankton and algae. In most instances, a moderate amount of growth will support a healthy food chain including fish, shrimp, etc. Undesirable effects of excessive growth, most often a result of the decay of the excessive plant growth that consumes oxygen at a higher rate than it is replenished will result in low to zero oxygen levels, especially in bottom waters of lakes and estuaries. High levels of algae or aquatic plants also impair use of water for recreational or industrial purposes.

Fresh water bodies differ from estuaries in their requirements for N and phosphorus (P) relative to excessive growth. Fresh waters almost always are limited by P (Correll 1998) while coastal zones and estuaries, which have the bulk of the biological activity in the oceans, usually are N and perhaps Si (for diatom growth) limited (Rabalias et al., 1996). Hence, much attention has been paid to N management to improve water quality in important estuaries such as the Gulf of Mexico and the Chesapeake Bay on the Atlantic coast of the US.

Addressing control of nitrate sources at the estuary level is much more difficult than smaller watersheds because of the large geographical area involved. The number of point and nonpoint sources of nitrate is many, control of nonpoint sources cannot be done through regulations, and N, being so pervasive, is very hard to manage. An example of the difficulty is an evaluation of nitrate concentrations in the Des Moines River from 1945, before N fertilizers were used in the highly productive agriculture watershed, through 1980's, when the watershed was predominantly row crops (corn and soybeans) and N fertilizer use was high (Keeney and DeLuca 1993). The results indicated that the yearly flow-weighted average nitrate-N (about 7 mg/L) changed insignificantly over the 40 years of monitoring. This in spite of large increases in fertilizer N use and in row crops. Nitrate concentrations did relate directly to yearly precipitation; however, being highest during years of high flow. These results indicate that highly productive, tile drained watersheds such as the Des Moines River, have large reservoirs of N in the soil organic matter and that the mineralization-immobilization processes dominate the output. Denitrification, the most likely control mechanism for nitrate removal, has been minimized by tile drainage (that shunts the nitrate directly to drainage ditches rather than allowing it to flow overland) and by removal of almost all of the wetlands, the primary N sink aside from the annual corn crop.

4.2 Hypoxia in the Gulf of Mexico

Many papers and opinion pieces have discussed the current state of agriculture in the Upper Mississippi River Basin (drainage above Cairo, IL-refer to Brezonik et al, 1999 for a more complete description). Mollisols (soils formed under prairie in till and loess and high in organic matter and clay but with poor internal drainage) dominate but there are also Alfisols (soils formed under mixed prairie and forest) that have high clay subhorizons and lower amounts of organic matter and clay in the surface horizons. These soils are highly productive but have poor internal drainage. Subsurface tile drainage of soils in this region is necessary for crop production. Corn and soybean require a well-drained warm soil for optimum growth. But water moves so slowly downward to the water table that soils in the spring are often too wet and cold to be planted to corn or soybean in a timely manner. If a porous tile (farmers now use perforated plastic pipe) is placed below the seasonal water table, water flows to the tile by the force of gravity. Tiles are placed so that water flows to larger collector tile and finally to open ditches. Tile drainage short circuits the natural drainage pattern and effectively flushes nitrate out of the soil before it is either denitrified or leached to the water table. The tile drainage systems thus become a major source of nitrate to surface water (Randall and Mulla, 1998).

Before habitation, this nitrate would have been denitrified in wetlands and ponds, or taken up by prairie grasses. Effectively, human intervention has allowed the agricultural N cycle to change markedly from its natural state.

Five states (Illinois, Indiana, Iowa, Ohio, and Minnesota) comprise the heart of the Corn Belt and will be referred to as the Upper Mississippi River Basin (UMRB). They have the greatest amount of artificially drained soil, the highest percentage of total land in agriculture (corn and soybean) and the highest use of N fertilizers in the nation. The region has abundant precipitation most years for crop growth and only rarely suffers from major yield declines because of drought.

Data analyzed by Goolsby et al (1999) showed that the UMRB generates about 19% of the flow but 43% of the nitrate load to the Mississippi River basin. Two states, Iowa and Illinois, provide 16% and 19% of the nitrate, respectively. These two states have the most intensive corn-soybean cropping systems, the most productive soils, and the highest total N fertilizer use.

The UMRB basin is the most productive agricultural regions in the world. Total N output to the Gulf of Mexico has increased 3 to 7-fold compared to outputs before settlement (Downing 1999). The tributary rivers have been straightened and dams have been installed on the Mississippi River and many of its major tributaries. Industrialization at the mouth of the river has diminished wetlands and added to the pollutant load.

The apparent result of the dramatic increase in N input to the Gulf of Mexico has been a major change in the ecology of the Gulf (Rabalais et al., 1996, Downing, 1999). Higher productivity of phytoplankton because of increased nutrient input has provided more organic residue from dead cells. This has led to increased oxygen consumption during decomposition of the material. The result has been the development of an extensive region of oxygen deficiency (less than 2 mg/liter of dissolved oxygen, commonly referred to as hypoxia). This level of dissolved oxygen is below the threshold for survival of most aquatic organisms, hence the popular term, “dead zone”. The zone runs roughly directly west from Louisiana to Texas and is the third largest hypoxia zone in the world (Downing, 1999). The area varies between 12,000 to 18,000 square kilometers in mid-summer during normal rainfall to high years, but is smaller during drought years (Downing, 1999). For example, it was only about 5,000 square kilometers in June of 2000 because of low rainfall in the basin.

Goolsby et al., (1999) recently examined the nitrate loads to the Gulf of Mexico from the Mississippi and Atchafalaya Rivers. Their results indicated that since 1985, the amount of N released to the Gulf has been more or less constant, varying directly in proportion to the streamflow. Streamflow is related to land drainage. In normal and wet years, much of the excess nitrate (that is nitrate that has not been denitrified or used by plants) is leached from the soil profile. In dry years, it is retained, but in wet years profile drainage leaches nitrate to the tile systems before it can be used by the corn crop (Randall and Mulla, 1998; Brezonik et al., 1999).

Nitrogen fertilizer use and manure production has been approximately constant in the basin over the past 15 years. The increasing yields of corn and soybean without additional fertilizer N implies a more efficient N use by the cropping system. Frequently the soil system is sufficiently high in available N that nitrogen-fixing legumes are not active. Quantification of N cycling at the basin level is not a simple balance sheet process.

5. ECOLOGICAL ISSUES

The large increase in mobile N worldwide has had many other significant ecological effects, many so subtle that they are not noticed at the public or policy maker level (Vitousek et al., 1996). Modern day activities ranging from industry to agriculture to land clearing has increased the rate of release of several N gases in trace amounts, including nitrous oxide, nitric oxide, and ammonia. Fossil fuel combustion is the major source of nitric oxide, which is a causative effect of photochemical smog and high levels of troposphere ozone. Further oxidation of nitric oxide gives rise to nitric acid, a major component of acid rain now that sulfur emissions have been lowered. Nitrous oxide has been implicated in stratospheric ozone destruction, leading to increased ultraviolet light at the Earth's surface, and as a major contributor to greenhouse gases. Ammonia has a fairly short retention time in the lower atmosphere, but will cause significant fertilization effects on N-limited ecosystems such as prairies, forests, and waters, increasing biological productivity and lowering biodiversity (Vitousek et al., 1996).

The concept of N saturation (Aber, 1992) has been introduced to explain ecosystem changes that occur in forests. A fully N saturated system will be one that has a net zero retention of N, that is, carbon storage through primary productivity is nil. These systems leak N to the environment rather than being net sinks as they were in unaltered ecosystems. The same concept can be applied to other natural ecosystems.

Nitrogen contributes significantly to ecosystem acidity by direct deposition of nitric acid, by oxidation of ammonium, and by leaching of cations, especially calcium, from soils. Landscapes that are poorly buffered, that is with soils that are already acid or have low exchange capacities, and whose ecosystems are N saturated will lose cations rapidly. Plant growth, species diversity, and water quality are adversely affected.

6. NITROGEN AND SUSTAINABLE HUMAN ACTIVITY

By 2020, the world will have added another 2.3 billion people (equivalent to another China) (Rosegrant and Livernash, 1996). Population control can do little to stop this trend, only slow its steady rise. Food production, particularly in Asia but also in Africa, must intensify. Green revolution strategies which worked well to offset earlier food stresses likely will be hard to repeat for several reasons: 1) They work well in countries with an established infrastructure, e.g., roads, educated workers, and credit, and these countries are among the list of those now largely self sufficient in food; 2) The international research centers no longer have the funding available or the political support to help develop and transfer new technologies; and 3) New germplasm to take advantage of current technologies, including fertilizer and pesticides, is difficult to obtain. Grain production for meat production continues to take precedence over that grown for food. A related issue is the declining land under irrigation due to urbanization, salinization, and other demands for water.

There are at least three distinct schools of thought on how to meet world food needs in the next century. One is to press for high yield, high input agriculture (Avery, 1995). Rosegrant and Livernash (1996) echo the opinions of others that agricultural intensification has imposed heavy environmental costs in many developing countries. Lester Brown and the WorldWatch Institute (Brown, 1995) present a pessimistic scenario, somewhat like that of

Kendall and Pimentel (1994). These groups feel that rapidly expanding populations, poor land use decisions, environmental degradation, and demands for a changing lifestyle will move the world to a bidding war for available food within the next 2 to 3 decades. Somewhere in the middle are calls for moderation in our views of the future policies and practices. The challenges for equitable production and availability of food even the next 25 years will require concentrated large scale efforts in both developed and developing countries to upgrade agricultural research and extension, improve the policy environment, and to remain committed to further improvements in agricultural technologies.

7. FOOD PRODUCTION AND NITROGEN FERTILIZERS

The above discussion on world food needs perhaps digresses from the main topic of the paper, but yet is critical to our decisions as a society and as scientists regarding where N fits in the overall scheme of world food production, global environmental issues, and our collective futures. Fertilizer N use in the developed world is stable or declining on a per hectare basis. This is not reflected in yields, showing clearly that earlier inefficiencies in fertilizer N use can be overcome even without clear economic or political pressures. Some of our efforts must concentrate on increasing fertilizer N efficiency (e.g., use of nitrification inhibitors, precision farming, other efficiency technologies). Even more bold approaches involving plant as well as management alternatives are needed. But it is critically important to spend more of our intellectual capital developing new systems and redesigning the ones we have. What are the "more sustainable technologies?" It is up to us to ask these questions and to find the answers.

Market forces will dominate land use and agricultural management decisions in developed countries in the foreseeable future barring unforeseen catastrophes. Grain will continue to be the major farm commodity, and N fertilizers will be essential. Nitrogen fertilizer efficiency improvements will be widely heralded, particularly if the price of N fertilizers rises as might be expected in an energy-short future. Further plant breeding efforts should be applied to developing highly efficient grains. System science can be applied even in grain growing regions of the world to develop approaches that tighten up the N cycle through combinations of legumes, perennials, and annuals. Efficient use of animal manure must increase. Land use that provides N sinks, such as including wetlands and overflow regions for runoff rather than short-circuiting the hydrologic cycle through tile drainage will be major policy decisions of the future. Property rights issues will need to be solved to advance land use planning to minimize N pollution. In the long term, N fixing legume perennial grains might be one solution.

8. REGULATORY AND EDUCATIONAL APPROACHES

The use of regulatory approaches to modify human behavior can have some success, particularly when the activities being regulated are outside the bounds of society's desires. Europe for example has had some success using rules and stiff fines to modify fertilizer and animal waste use. Educational programs are important to change the way farmers manage N in their farms. A concerted effort by Iowa educators has helped lower N fertilizer use by about 20% over the past decade (Hallberg, 1996). However, these approaches work best when profits are also shown to increase.

9. OTHER HUMAN ACTIVITIES

This has been addressed to some extent earlier and is included again to emphasize that more than agriculture is involved in solving the recurring issues of N in the environment. Emissions from autos and industry must be addressed as part of the ozone, smog, and acid rain issue. Land use is critical. Development now takes our best agricultural lands as well as lands critical to environmental stability. The way we build and populate landscapes must be reassessed.

10. ARE THERE SOLUTIONS?

This review attempts to broaden the perspective on N in the environment beyond that of agriculture and crop production. The issues are major, and are recurring. Some are recycled and some are new, brought on by advances in science and monitoring that increased our awareness (e.g., hypoxia) and some are new because ecosystems are now showing stress that they were able to overcome earlier (global climate change, acid rain, ecosystem degradation). The issues may ebb and flow, but they will not go away. Perhaps as a society wide spread dissemination of knowledge, open and informed discussion at world forums, and consensus on appropriate actions is called for. Technical solutions are the domain of the scientist, but such solutions must fit the world needs for a sustainable future.

REFERENCES

Aber, J.D. 1992. Nitrogen cycling and nitrogen saturation in temperate forest ecosystems. Trends in Ecology and Evolution 7:220-223.

Addiscott, T.M. 1999. Nitrate and health. Introductory comments. p. 247-249. *In* Managing Risks of Nitrates to Humans and the Environment. Royal Society of Chemistry. Cambridge, UK.

Alexander, M. 1965. Nitrification. *In* W.V. Bartholomew and F.E. Clark, (eds.) Soil Nitrogen. p 307-343. Agronomy 10. American Society of Agronomy. Madison, WI.

Allison, F.E. 1955. The enigma of soil nitrogen balance sheets. Adv. Agron. 7:213-250.

Asner, G.P., T.R. Seastedt, and A.R. Townsend. 1997. The decoupling of terrestrial carbon and nitrogen cycles. BioScience. 47:226-234.

Avery, A.A. 1999. Infantile metheglobenemia Reexaming the role of drinking water nitrates. Environ. Health Perspectives 107:583-586.

Avery, D.T. 1995. Saving the planet with pesticides and plastic: The environmental triumph of high-yield farming. Hudson Institute. Indianapolis, Indiana.

Bartholomew, W.V., and F.E. Clark. 1965. Soil Nitrogen. Agronomy 10. American Society of Agronomy. Madison, WI

Blair, A.W. 1917. Maintaining the nitrogen supply of the soil. New Jersey Agr. Exp. Stn. Bull. 305.

Brezonik, P.L. et al. 1999. Effects of reducing nutrient loads to surface waters within the Mississippi River Basin and Gulf of Mexico. NOAA Coastal Ocean Program Analysis Series. Report 4.

Brown, L.R. 1995. Who will feed China? A wake up call for a small planet. The WorldWatch Environmental Alert Series.

Burns, R.C., and R.W.F. Hardy. 1975. Nitrogen fixation in bacteria and higher plants. Springer Verlag, N. Y.

Comly, H.H. 1945. Cyanosis in infants caused by nitrate in well water. J. Amer. Med. Assoc. 129:112-116.

Correll, D.L. 1998. The role of phosphorus in the eutrophication of receiving waters: A review. J. Environ Qual. 27:261-266.

Downing, J.A. (chair) et al. 1999. Gulf of Mexico Hypoxia: Land and Sea Interactions. CAST Task Force. Report 134.

Frink, C.R., P.E. Waggoner, and J.H. Ausubel. Nitrogen fertilizer: Retrospect and prospect. Proc. Natl. Acad. Sci. 96:1175-1180.

Goolsby, D.A. et al. 1999. Fluxes and sources of nutrients in the Mississippi-Atchafalaya River Basin. NOAA Coastal Ocean Program Decision Analysis Series.

Hallberg, G.R. 1989. Nitrate in ground water in the United States. p. 35-74. *In* R.F. Follett (ed.) Nitrogen management and ground water protection.

Hallberg, G.R. 1996. Water quality and watersheds: An Iowa perspective. p. 1-5-23. *In* Agriculture and Environment Conference Building Local Partnerships. Iowa State University Extension. Ames, IA.

Hallberg, G.R., and D.R. Keeney. 1993. Nitrate. p. 297-332. *In* W. M. Alley (ed.) Regional Ground-water Quality. Van Norstrand Reinhold, N. Y.

Harmsen, G.W., and G.J. Kolenbrander. 1965. Soil inorganic nitrogen. *In* W.V. Bartholomew and F.E. Clark (eds.) Soil Nitrogen. p. 43-92. Agronomy 10. American Society of Agronomy. Madison, WI.

Harmesen, G.W., and D.A. Van Schreven. 1955. Mineralization of organic nitrogen in soil. Adv. Agron. 7:299-398.

Howarth, R.W. 2000. Clean Coastal Waters: Understanding and reducing the effects of nutrient pollution. National Academy of Sciences. Washington, DC.

Jackson, L.L., D.R. Keeney, and E.M. Gilbert. 2000. Swine manure management plans in North-Central Iowa: Nutrient loading and policy implications. J. Soil Water Consv. 55:205-212.

Jansson, S.L. 1958. Tracer studies on nitrogen transformations in soil. Ann. Roy. Agric. Coll. Sweden 24: 01-361.

Jordan, T.E., and D.E. Weller. 1996. Human contributions to terrestrial nitrogen flux. Bioscience 46:655-664.

Keeney, D.R. 1989. Sources of nitrate to groundwater. p. 23-34. *In* R.F. Follett (ed.) Nitrogen management and groundwater protection. Elsevier Press, N. Y.

Keeney, D.R. 1997. What goes around comes around-the nitrogen issues cycle. Dahlia Greidinger Symposium on fertilizer and the environment. Technion Israel Institute of Technology. Haifa, Israel.

Keeney, D.R., and T.H. DeLuca. 1992. Des Moines River nitrate in relation to watershed practices: 1945 versus 1980s. J. Env. Qual. 22:267-272.

Keeney, D.R., and M. Muller. 2000. Nitrogen and the Upper Mississippi River. Institute for Agriculture and Trade Policy. Minneapolis, MN. www.iatp.org.

Kellogg, R.L., C.H. Lander, D.C. Moffitt, and N. Gollehon. 2000. Manure nutrients relative to the capacity of cropland and pastureland to assimilate nutrients: Spatial and temporal trends for the U.S. NRCS Report. Washington, DC. (In Press).

Kendall, H.W, and D. Pimentel. 1994. Constraints on the expansion of the global food supply. Ambio. 23:198-205.

King, F.H., and A.R. Whitson. 1902. Development and distribution of nitrates in cultivated soil-second paper. Univ. Wis. Agr. Exp. Stn. Bull. 93.

Kjaergaard, T. 1995. Agricultural development and nitrogen supply from an historical point of view. p. 3-14. Nitrogen Leaching in Ecological Agriculture. A.B. Academic, Great Britain.

Knobeloch, L., B. Salina, A. Hogan, J. Postle and H. Anderson. 2000. Blue babies and nitrate-contaminated well water. Env. Health Perspectives 108: 675-678.

Kross, B.C., A.D. Ayebo, and L J. Fuortes. 1992. Methemoglobinemia: Nitrate toxicity in rural America. America Family Physician. p. 183-188.

Lipman, J.G., and A.W. Blair. 1916. Cylinder experiments relative to the utilization and accumulation of nitrogen. N.J. Agr. Exp. Stn Bull. 289.

Lohnis, F. 1926. Nitrogen availability of green manures. Soil Sci. 22:253-290.

Moffat, A.S. 1998. Global nitrogen overload problem grows critical. Science. 279:988-999.

Meyer, A. 1996. Nonpoint Source News Notes. April-May 1996, Issue no. 44. p. 3-4.

Rabalias, N. N., R.E.Turner, G. Dortch, W.J. Wiseman, J.R., and B.K. Sen Gupta. 1996. Nutrient Changes in the Mississippi River Basin and system responses on the adjacent continental shelf. Estuaries. 19:396-407.

Randall, G.W., and D.J. Mulla. 1998. Nitrate-N in surface waters as influenced by climatic conditions and agricultural practices. *In* Hypoxia around the world: Causes and solutions. Soil Sci Soc. Amer. (in press).

Rosegrant, M.W. and R. Livernash. 1996.Growing more food, doing less damage. Environment 38:6-11, 30-31.

Stevenson, F.J. (ed.). 1982. Nitrogen in agricultural soils. Agronomy 22. Amer. Soc. Agron. Madison, WI.

United Nations Food Agricultural Organization (FAO). 2000. http://apps.fao.org/page/form?collection=Fertilizers&Domain

Vitousek, P.M., J.D. Aber, R.M. Howarth, G.E. Likens, P.A. Watson, D.W. Schindler, W.H. Schlesinger, and D.W. Tilman. 1997. Human alterations of the global nitrogen cycle: sources and consequences. Ecol. Applic. 7:737-750.

Vorhees, E., and J.G. Lipman. 1907. A review of investigations in soil bacteriology. U.S. Dept. Agr. Bull. 194.

Waksman, S.A. 1952. Soil Microbiology. p 1-28. Wiley, N.Y.

Waksman, S.A. and R.L. Starkey. 1931. The soil and the microbe. Wiley, N.Y.

Williams, W.S., A.S. Ball, and R.H. Hinton. 1999. Managing Risks of Nitrates to Humans and the Environment. Royal Society of Chemistry. Cambridge, UK.

Nitrogen in the Environment: Sources, Problems, and Management
R.F. Follett and J.L. Hatfield (Eds)
Published by Elsevier Science B.V.

Chapter 2. Nitrogen Transformation and Transport Processes

Ronald F. Follett[a]

[a]USDA/ARS, Soil-Plant-Nutrient Research Unit, PO Box E, Fort Collins, CO, 80522, United States.

Nitrogen (N) is ubiquitous in the environment. It is one of the most important nutrients and is required for the survival of all living things. It is also central to the production of all crop plants. Nitrogen accounts for 78% of the atmosphere as elemental dinitrogen (N_2) gas. Elemental N_2 gas is inert, does not impact environmental quality, and is not directly available for plant uptake and metabolism. The purpose of this chapter is to review the fate and transport processes for N in agricultural systems. Some of the most mobile substances found in the soil-plant-atmosphere system contain N and the need to understand N transport and transformations in the environment has been the subject of several reviews (Follett, 1989; Follett et al., 1991; Hallberg, 1987, 1989; Keeney, 1982, 1989; Laegreid, et al., 1999; Mosier, et al., 1998; and Power and Schepers, 1989). Nitrogen represents the nutrient most applied to agricultural land. This is because available soil-N supplies are often inadequate for optimum crop production and because commercial fertilizer, manures, and other sources of N are generally easily and economically applied. An important consideration is to keep applied and residual sources of N within the soil-crop system by curtailing transport processes (leaching, runoff, erosion, and gaseous losses) that carry N into the surrounding environment. The objective is to lower the rate and duration of the loss processes themselves. Practices and concepts that lessen the opportunity for loss processes to occur help decrease the amount of N that may be lost. In some cases improved efficiency is achieved by using less N and in other cases it can be achieved by increasing plant growth while using the same amount of N input. The fate and transport of N from any of the various sources from which it may enter the environment must always be considered in the context of the N cycle. Often a N budget, or mass balance, approach is needed to understand the options to minimize and/or mitigate the environmental impacts of N that may occur and to improve N management in farming and livestock systems.

1. NITROGEN TRANSFORMATIONS

1.1. Biological nitrogen fixation

Through the process of biological nitrogen fixation (BNF), symbiotic and nonsymbiotic organisms can fix atmospheric N_2 gas into organic N forms (Figure 1). A few living organisms are able to utilize molecular N_2 gas from the atmosphere. The best known of these are the symbiotic Rhizobia (legume bacteria), nonsymbiotic free-living bacteria such as Azotobacter and Clostridium, and blue-green algae. Generally, in a symbiotic relationship, one organism contains chlorophyll and uses light energy to produce carbohydrates. The other organism receives some of the carbohydrates and uses them as an energy source to enzymatically fix atmospheric N_2 into the ammonia (NH_3) form of N and thence into amino acids and other nitrogenous compounds that are nutritionally useful to the chlorophyll containing organism. To agriculture, the most important type of BNF is symbiotic fixation by legumes (i.e. alfalfa, clovers, peas, beans, etc.). Follett et al. (1987) estimated that leguminous crops were returning about 700 Gg N yr^{-1} of symbiotically fixed N to cropland soils in the U.S. and the amount may now reasonably be more than 1000 Gg N yr^{-1}. Global N fixation estimates from the use of legumes in agriculture were recently reported as 40,000 Gg N yr^{-1} (Laegreid et al., 1999; from various sources). Even though fixed N resulting from BNF is initially within the non-symbiotic or symbiotic organism/plant system; the fate, transport and entry of this N into the environment must also be considered in the context of the N cycle.

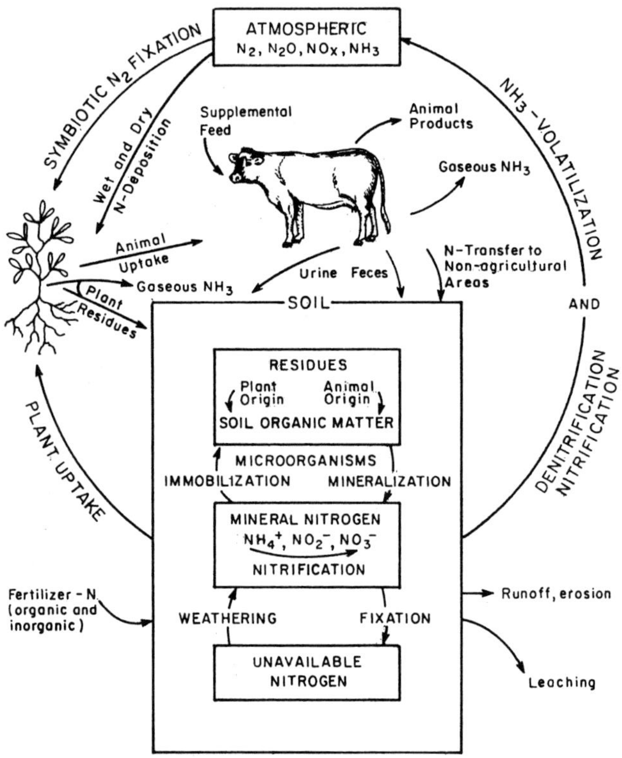

Figure 1. The Nitrogen cycle

1.2. Immobilization and mobilization of soil nitrogen

Nitrogen taken up by plants from the soil originates from indigenous organic and inorganic forms. Organic N occurs naturally as part of the soil's organic matter fraction; it can also be added to the soil from manure, symbiotic and nonsymbiotic biological N fixation, plant residues, and from other sources. Soil microorganisms and their activities are an integral part of immobilization and mineralization processes in soil (Figure 1); soil-organic N can be transformed to ammonium (NH_4^+) by the process of ammonification. Inorganic (mineral) forms of N include NH_4^+ or nitrate (NO_3^-), both readily taken up by crops, and nitrite (NO_2^-) that occurs as an intermediate form during mineralization of NH_4^+ to NO_3^-. Even though NH_4^+ is the preferred form, microbes in soil can convert either NH_4^+ or NO_3^- to satisfy their need for N, a process called immobilization. Immobilization of NO_2^- and NO_3^- back to organic forms of N can also occur through enzymatic activities associated with plant or microbial N uptake and N utilization processes. Microbes and soil animals use organic matter in soil as food and excrete nutrients in excess of their own needs. When NH_4^+ is released, it is called mineralization. When oxygen is present, microbes in the soil can readily transform NH_4^+ to NO_3^- with NO_2^- as an intermediate form, a process called nitrification. This is a fairly rapid process that, under aerobic conditions, can be completed in a few days. Although NO_2^- can potentially accumulate in soils under some conditions, it usually does not because it is rapidly transformed to NO_3^- as part of the nitrification process or else it is denitrified.

1.3. Gaseous transformations

1.3.1. Ammonia volatilization

Ammonium ions in the soil solution enter into an equilibrium reaction with NH_3 in the soil solution. The soil solution NH_3 is, in turn, subject to gaseous loss to the atmosphere. Soil pH and concentration of NH_4^+ in the soil solution are important factors affecting the amount of NH_3 loss to the atmosphere. As soil pH increases, the fraction of soil-solution NH_4^+ plus soil-solution NH_3 in the NH_3 form increases by an order of magnitude for every unit of pH above 6.0; thus, increasing the loss of soil-solution NH_3 to the atmosphere. As summarized by Stevenson (1986), NH_3 volatilization:

1. Is of most importance on calcareous soils, especially as soil pH exceeds 7;
2. Losses increase with temperature and can be appreciable for neutral or alkaline soil as they dry out;
3. Is greater in soils of low CEC, such as sands;
4. Losses can be high when high N organic wastes, such as manure, are permitted to decompose on the soil surface;
5. Losses are high from urea applied to grass or pasture as a result of hydrolysis of the urea to NH_3 by indigenous urease enzyme; and
6. Losses of soil and fertilizer N are decreased by growing plants.

Anhydrous, or gaseous, NH_3 is a very important direct-application N fertilizer. Gaseous NH_3, when in contact with moist soil, dissolves in, and reacts with, soil water to form NH_4^+ and OH^- ions. The pH is increased dramatically immediately around the application zone of anhydrous NH_3. Therefore, depending upon buffering capacity of the soil and the resulting soil pH, an equilibrium is approached between soil solution NH_4^+ and NH_3 in the soil solution and gaseous NH_3. If anhydrous NH_3 is placed in dry soil or at too shallow a depth, the NH_3 is also subject to volatilization. However, the N that is in NH_4^+ form is readily sorbed to the CEC of the soil.

1.3.2. Denitrification

As organic matter in soil decomposes, first NH_4^+, then NO_2^- and finally NO_3^- ions are formed by the process of nitrification (Figure 1). Nitrite usually does not accumulate in soils because it is rapidly transformed to NO_3^- or is denitrified to N_2 gas, nitrous oxide (N_2O), nitric oxide (NO), or one of the other gaseous N oxide (NO_x) compounds. Nitrate can also be lost to the atmosphere through the denitrification processes. Nitrous oxide is a product of incomplete denitrification, is a greenhouse gas that contributes to global climate change, and potentially to thinning of the ozone layer. Not only denitrification (a reductive process), but also the oxidative process of nitrification causes emission of a small amount of N_2O (Tortoso and Hutchinson, 1990). However, denitrification is the route for most losses of gaseous N compounds to the atmosphere. The potential for denitrification is increased as oxygen levels in the soil decreases. Under favorable environmental conditions, *Nitrosomonas* spp. bacteria in the soil readily transform NH_4^+ to NO_2^- that in turn is transformed by *Nitrobacter* spp. bacteria to NO_3^- (Figure 1). The small quantity of N_2O produced during nitrification of NH_4^+ in aerobic soils is a direct metabolic product of chemoautotrophic NH_4^+-oxidizing bacteria or results from other soil processes dependent upon these organisms as a source of NO_2^- (Tortoso and Hutchinson, 1990).

Table 1
Factors affecting the proportion of N_2O and N_2 produced during denitrification

Factor	Will increase N_2O/N_2
$[NO_3^-]$ or $[NO_2^-]$	Increasing oxidant
$[O_2]$	Increasing O_2
Carbon	Decreasing C availability
PH	Decreasing pH
$[H_2S]$	Increasing sulfide
Temperature	Decreasing temperature
Enzyme status	Low N_2O reductase activity

Biogenic production in soil is the principle source of atmospheric N_2O. In addition, several factors affect the ratio of N2O to N2 during denitrification (Table 1). Anaerobic soil processes, rather than nitrification (an aerobic process) are the principle biogenic sources of atmospheric N_2O (Freney et al., 1979; Goodroad and Keeney, 1984; Klemedtsson et al., 1988). Dinitrification is a bacterial process, during which NO_3^- or NO_2^- are reduced to gaseous N species NO, N_2O or N_2, and is capable of producing and consuming N_2O and NO. Nitrate is reduced first to NO_2^-, then to NO, next to N_2O and finally to N_2 (Eq. 1).

$$NO_3^- \rightarrow NO_2^- \rightarrow \overset{\displaystyle NO}{\overset{\uparrow}{[X]}} \rightarrow N_2O \rightarrow N_2 \qquad \text{(Equation 1)}$$

The general conditions required for denitrification to occur include: (a) presence of bacteria possessing the metabolic capacity; (b) availability of suitable reductants such as organic C; (c) restriction of O_2 availability; (d) availability of N oxides, NO_3^-, NO, or N_2O (Firestone and Davidson, 1989; Klemedtsson et al., 1988; Mosier, this volume). Either the NH_4^+ or NO_3^- form can potentially contribute to the release of N_2O to the atmosphere, especially where excess NO_3^- accumulates in the soil profile and is available for denitrification. Because N_2O is the greenhouse gas of concern, the proportion of N_2O produced relative to N2 under denitrifying conditions becomes of special concern. A number of factors affect the proportion of N_2O to N2. A model by Betlach and Tiedje (1981) predicts accumulation of N_2O whenever one of the factors shown in Table 1 slows the rate of overall reduction.

2. TERRESTRIAL TRANSPORT AND RELATED PROCESSES

2.1. Fertilizer and manure

For highly-water soluble compounds with NH_4 as part of their chemical formula (Table 2), the NH_4^+ cation can be sorbed to the CEC, incorporated (fixed) into clay and other complexes within the soil, released by weathering back into the available mineral pool, or immobilized into organic form by soil microbial processes. Ammonium that is associated with soil colloids can be transported into surface water during water or wind erosion of soil or, under certain conditions, can volatilize into the atmosphere as NH_3 gas and be aerially transported across the landscape, including into surface water. Gaseous NH_3 often is returned to the soil-plant system by direct uptake into plant leaves or dissolved in precipitation. Urea and calcium cyanamide (Table 2), are forms of N that, when applied to soil, are acted upon by enzymes in the soil to mineralize the N in them to NH_4^+ ions. Once in the NH_4^+ form and until nitrified to the NO_3^- ion form, the N in these two fertilizers is also sorbed to the CEC of the soil and is subject to the soil-erosion transport process described above. The N in organic materials such as crop residues is also first mineralized to NH_4^+, again being subject to sorption to the CEC of the soil until nitrified to the NO_3^-. The NO_3^- ion, when it is part of the chemical formula in compounds shown in Table 2, does not sorb to the CEC of the soil. Nitrate, a water-soluble anion, is very mobile, and moves readily with percolating water (leaching). It is not sorbed to the negatively charged sites on soil colloids, the cation exchange capacity (CEC) of the soil. The primary transport mechanism for NO_3^- ions is with percolating water by leaching or surface runoff (including return flow). Nitrate that is leached below the crop root zone often ends up as a pollutant in ground water supplies. Nitrate can also be dissolved in surface runoff water or in return-flow water that returns to the surface to become part of the runoff. Nitrate and NO_2^- ions can also be denitrified and lost to the atmosphere (Eq. 1) as NO, N_2O, or NO_x.

Table 2
Nitrogen fertilizer materials, their formulas and chemical analysis

Material	Chemical formula	Chemical analysis
		(%N)
Anhydrous ammonia	NH_3	83
Ammonium nitrate	NH_4NO_3	33.5
Ammonium sulfate	NH_4SO_4	21
Diammonium phosphate	$(NH_4)2H_2PO_4$	18-21
Monoammonium phosphate	$NH_4H_2PO_4$	11
Calcium nitrate	$Ca(NO_3)_2$	15
Calcium cyanamide	$CaCN_2$	20-22
Potassium nitrate	KNO_3	13
Sodium nitrate	$NaNO_3$	16
Urea	$CO(NH_2)_2$	45
Urea-ammonium nitrate	$CO(NH_2)_2+NH_4NO_3$	32

2.2. Runoff

Amount and timing of rainfall and soil properties are key factors that influence loss of dissolved N in runoff. Soils with low runoff potential usually have high infiltration rates, even when wet. They often consist of deep, well to excessively-drained sands or gravels. In contrast, soils with high runoff potential have one or more of the following characteristics: very slow infiltration rates when thoroughly wetted and containing high clay content possibly of high swelling potential, high water tables, a claypan or clay layer at or near the surface, or are shallow over a nearly impervious subsurface layer. A combination of soil conditions of high runoff potential and high precipitation amount are especially conducive to surface runoff losses. Steeper slope gradients increase amount and velocity of runoff, while depressions, soil roughness, and presence of vegetative cover or crop residue decrease runoff by improving the infiltration.

The dissolved concentration of N in surface runoff from soils under conservation or no-tillage often is higher than from soil under conventional tillage (McDowell and McGregor, 1984; Romkens, 1973). Reasons may include incomplete incorporation of surface-crop residues, and higher dissolved N concentration in the surface soil because of residue accumulation and decomposition. In addition, high concentrations of soluble N can occur when there is a soil horizon barrier (e.g. Fragipan) present in the soil profile that results in return flow of leached N back to the soil surface (Lehman and Ahuja, 1985; Smith et al., 1988).

Some of the effects on dissolved nutrients in surface and subsurface water discharges that are associated with agricultural nutrient management for crop production and the use of conservation tillage for erosion control, are illustrated (Figure 2) by the work of Alberts and Spomer (1985). Their study site, for this ten year study, was in the deep loess hills in western

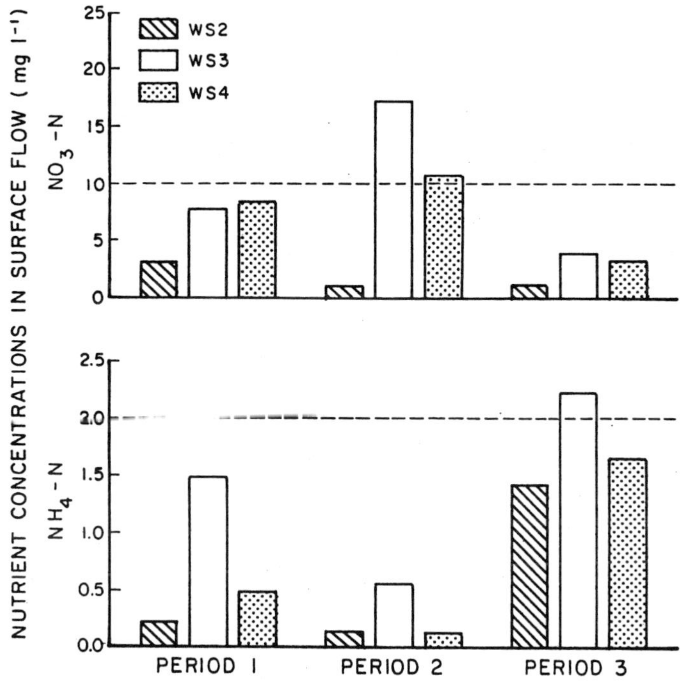

Figure 2. Runoff-weighted concentrations of NO_3^--N and NH_4^+-N in surface flow by seasonal period. Dashed lines represent current water quality standards (from Alberts and Spomer, 1985).

Iowa. The loess is underlain by nearly impervious glacial till at depths of 4.6 to 24.4 m. Lateral water movement occurs in a saturated soil zone that exists at the loess-till interface. Water from both surface runoff and subsurface flow was sampled. In their study, watershed 2 (WS2) was conventionally tilled (33.5 ha) while watersheds 3 and 4 (WS3 and WS4) were contour-till planted (43.3 ha) and terrace-till planted (60.8 ha), respectively. About sixty-five head of cattle gleaned the corn stalks from WS3 and WS4 from mid-November to March each year. Figure 2 shows the ten year, runoff-weighted concentrations of NO_3^- and NH_4^+ for three time periods; April through June (fertilization, seedbed preparation, and crop establishment period); July through November (crop reproduction and maturation period); and December through March (crop residue period) or periods 1, 2, and 3 respectively. Water quality criteria for NO_3^- and NH_4^+ are shown by dashed lines (Fletcher, 1991; U.S. EPA, 1982) in Figure 2 as 10.0 and 2.0 mg L^{-1}; respectively. Highest NO_3^- concentrations from the till-planted watersheds (WS3 and WS4) occurred during July through November (period 2), perhaps as a result of evaporative drying moving previously applied fertilizer salts to the soil surface. Preplant applications of fertilizer for the conventionally tilled watershed (WS2) had been incorporated with a disk. Ammonium N concentrations were generally from cattle manure and leaching of NH_4^+ from partially decomposed corn stalks. Issues illustrated by this study include the need to place fertilizer below the soil surface while still maintaining residue cover for soil erosion control. Fall and winter livestock grazing of crop residues likely contributes to N runoff since the manure and urine may be deposited on frozen ground.

2.3. Erosion

Detachment of sediments and nutrients from the parent soil is selective for soluble nutrients (such as NO_3^-) and for the fine soil fractions with which nutrients (such as NH_4^+ and the soil organic matter N) are associated. Therefore, N contained in runoff and/or associated with sediments is present in higher concentrations than in the parent soil. This difference is termed the enrichment ratio. Enrichment of sediment loads is a two-step process: enrichment during particle suspension and enrichment due to redeposition of coarser particles during overland and channel flows. In order for management practices to decrease the effect of water erosion processes on the production and transport of sediment associated N, they must directly influence the processes involved. Such practices need to protect against soil particle detachment, slow sediment transport, and enhance sediment deposition within the landscape rather than allowing the sediments to move into surface water.

Soil erosion is important to the movement of N into surface water which primarily occurs with soil erosion by water, rather than by wind. Briefly, soil erosion by water includes the processes of detachment, transport, and deposition of the soil particles by raindrops or surface flow (Foster et al., 1985). Some sediments may travel only a few millimeters while others may be transported long distances before either being deposited or reaching a surface water resource (i.e. a lake, reservoir, or stream). The movement of NH_4^+ results because it is sorbed to the surfaces of clays and finer sediments. The NO_3^- is completely water soluble and thus moves with the water until it re-enters the available soil pool, is utilized by microbes or plants, becomes denitrified, is possibly deposited and buried, or enters and possibly degrades surface- and/or ground-waters. A major source of the N that degrades surface water is that which is transported in soil organic matter. A large part of the soil organic matter and soil organic N (SON) contained in it are concentrated near the soil surface and are therefore vulnerable to erosional and oxidative (mineralization/nitrification) processes. Also, within the U.S., about 400 million m^3 of sediment are dredged each year in the maintenance and establishment of waterways and harbors (Sopper, 1993). Loss of topsoil and the SON contained in dredged sediments is a primary environmental impact of accelerated soil erosion. Two independent methods of estimating the amount of eroded SON are to utilize information about river sediment loads or to use estimates of amounts of eroded sediments themselves. To use the sediment load approach, data by Leeden et al. (1991) showed that the suspended load in 12 major U.S. rivers during 1991 was 336,000 Gg yr^{-1}. Assuming that 75% of the suspended load (mostly due to soil erosion) was contributed by cropland, then the sediment transport attributed to cropland is about 250,000 Gg yr^{-1}. Assuming a delivery ratio of 10% and SON content of sediment of 0.25% (Lal, 1995; Follett et al. 1987), the total SON displaced by soil erosion from cropland was about 6.25 Tg yr^{-1}. Alternatively, (Lal, et al., 1998) used an estimate of the amount of eroded sediments to calculate soil organic carbon (SOC) losses. By assuming a SOC:SON ratio of 110:9 in sediment (Follett et al. 1987) the total SON displaced by soil erosion would be about 9.6 Tg yr^{-1}. Thus considering only the U.S., soil erosion serves as an environmental source of 6 to 9 Tg N yr^{-1} as SON.

Much still needs to be learned about managing cropland soil erosion. For example, Follett et al. (1987) assessed effects of tillage practices and slope on amount of organic N in eroded sediments from cultivated land surfaces in Minnesota (USA) for major land resource areas (MLRAs) 102, 103, 104, and 105. Their estimates, using the Universal Soil Loss Equation (USLE), average organic matter in topsoil by slope category, and dominating slope gradient and soil series indicates that conservation tillage compared to conventional tillage decreases the amount of organic N associated with eroded sediments by about half with some additional decrease resulting from the use of no-tillage. One can assume that added fertilizer N responds similarly to organic N when it is sorbed to clay surfaces, finer sediments, or to soil organic matter.

2.4. Leaching

Nitrate is a negatively charged ion that is repelled by, rather than attracted to, negatively-charged clay mineral surfaces in soil (i.e. the CEC). It is the primary form of N leached into ground water, is totally soluble at the concentrations found in soil, and moves freely through most soils. As described by Jury and Nielson (1989), movement of NO_3^- through soil is governed by convection, or mass-flow, with the moving soil solution and by diffusion within the soil solution. The widespread appearance of NO_3^- in ground water is a consequence of its high solubility, mobility, and easy displacement by water. An extensive literature concerning N-management, leaching, and ground water quality exists including that assembled by CAST (1985), Follett (1989), Follett et al. (1991), and Follett and Wierenga (1995).

Juergens-Gschwind (1989) reported on leaching losses observed under widely varying conditions (lysimeters, drainage water measurements in field trials, catchment areas, profile and groundwater research in field trials) (Figure 3). The results were made comparable by referencing the N-losses at each site to an ~ 300 mm drainage level per year. The leaching risk was distinctly higher on arable land than on grassland, and on lighter textured soils than on heavy textured soils. An upward shift in the data was observed when going from lower nutrition rates obtained by normal fertilization practices to the very high rates that can result from excessive N-fertilization and animal manure disposal (rates in excess of the plant nutrient requirements) on agricultural lands. As shown in Figure 3, soil texture influences how rapidly NO_3^- leaching occurs. This influence of soil texture, in sandy soils is further documented by

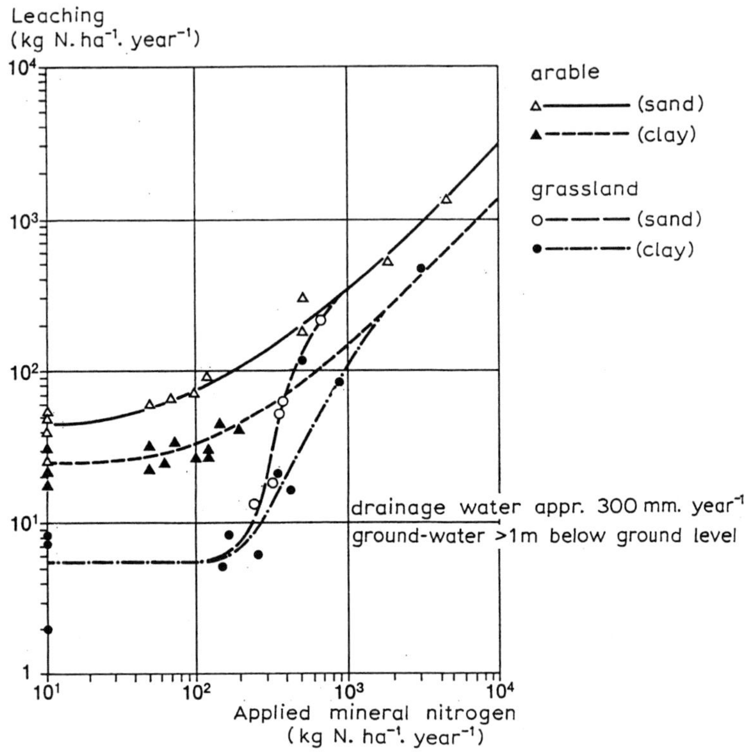

Figure 3. Leaching of nitrogen from arable and grassland systems (adapted from Juergens-Gschwind, 1989).

Delgado, et al. (1999) in which more NO_3^- leaching was observed on a loamy sand than on a sandy loam. Also, unless the soil is anaerobic, excess amounts of NO_3^- also leach on heavy-textured soils, as illustrated in an N-rate study with irrigated corn (*Zea mays L.*) by Godin (1999). Godin used ^{15}N-labeled fertilizer on a clay loam soil, he observed that the recommended fertilizer rate (135 kg N ha^{-1}) adequately satisfied the crop N requirement and resulted in higher percent recovery of N than did the excess N rate (200 kg N ha^{-1}). At the excess N rate, fertilizer ^{15}N had leached below the crop root zone (0.9 m) by harvest of the first year and to a depth of over 1.5 m by harvest of the second year.

3. NITROGEN CYCLING IN PASTURE SYSTEMS

Inputs are from fertilizers and manures, biological N fixation, wet and dry deposition, supplemental feed to livestock, and mineralization of soil organic matter (Figure 1). Losses may occur through harvest of animal or plant products, transfer of N within the pasture with animal excreta, fixation of N in the soil, soil erosion, surface runoff, leaching, volatilization, and denitrification. The soil compartment includes a pool of available N (NO_3^- and NH_4^+) for plant uptake that is in equilibrium with N in residues (organic N) and, especially for some soils, with fixed NH_4^+ which is held between mineral layers of the clay. Plant N uptake is from the available soil pool. The N in the herbage is then either harvested and removed from the field, returned to the soil as crop residue and root material, and/or eaten by grazing animals and either utilized by the animal or excreted as feces or urine and returned to the soil.

3.1. Role of soil organisms

Soil microfauna and microflora have a major role in N cycling. Release of N from plant and animal residue depends on microbial activity. Soil bacteria utilize the more readily available, soluble, or degradable organic fractions. Fungi and actinomycetes decompose the resistant cellulose, hemicellulose, and lignin. Dung beetles, earthworms, and other soil fauna increase the decomposition rates of feces and plant litter by mixing them with soil. Rhizobia and vesicular arbuscular mycorrhizae (VAM) associate with plant roots to fix N and increase nutrient and water scavenging ability, respectively. VAM infection of roots is considered more helpful for tap rooted pasture legume species than for fibrous rooted grasses. At any time, soil-microbial biomass contains much of the actively-cycling N of the soil and represents a relatively available N pool, capable of rapid turnover (Bristow and Jarvis, 1991). The energy flux through the soil microbial biomass (SMB) drives the decomposition of organic residues (Smith and Paul, 1990) and soil organic matter. Plant root biomass and soil microbial processes are intimately linked in grassland systems as described by Reeder et al. (2000). If decomposition exceeds C inputs, the soil organic matter will decline. The resulting mineralization of N (and other nutrients) will result in their becoming vulnerable to possible losses into the environment by leaching, denitrification, or other mechanisms (Follett, et al., 1995). Because its levels are relatively stable for a particular soil/land-use system, even though the SMB pool is very active for nutrient cycling, SMB can serve as a measure (index) of the effects of agricultural management practices on soil quality. In their study, Follett et al. (1995) utilized ^{15}N labeled fertilizer and followed the N in the SMB fraction under no-till in a 4 yr (winter wheat-sorghum-fallow-winter wheat) cropping sequence. Their conclusion was that, under no-till, biological processes conserved the N by accumulation of crop residue carbon (C) and N near the soil surface, recycling of N through the crop-SMB system, and maintenance of N in organic forms.

3.2. Role of the grazing animal

Grazing animals affect plant growth by defoliation, traffic patterns, herbage fouling, partitioning of ingested N to body weight, feces, and urine, redistribution of herbage N in excreta, and N turnover rate. Defoliation by grazing animals prevents senescence of plant tissue, removes N in animal products, changes the N pathway from internal plant recycling or leaf fall to return as feces and urine, increases light penetration into the canopy and, through selective grazing, may alter botanical composition by promoting one species over another. Animal traffic compacts soil, sometimes making soil characteristics for plant growth less desirable. Herbage fouling by feces reduces its acceptability for grazing, thereby increasing maturity and reducing forage quality and/or consumption by grazers. Urine does not cause herbage to be unacceptable for grazing. Livestock recycle much of the N that they consume from forage back to the soil. The N retention of forage N by livestock, as a percentage of dietary intake, ranges from about 8+ % of live weight gain (LWG) (e.g. in steers) to 20 % (Follett and Wilkinson, 1995) in high producing animals (e.g. milk cows). For example, a 250 kg steer that ingests 6 kg forage d^{-1} (containing 3% N in the forage) and gaining 0.8 kg d^{-1} may ingest 180 g N d^{-1}, retain about 20 g in LWG (12 % retention) and excrete the remainder, about 160 g N d^{-1}. Excretion as feces and urine both result in volatile losses of NH_3. About 74% of the total N excreted is in the urine (Follett and Wilkinson, 1995) and a single urine spot can have an N concentration corresponding to more than 600 kg N ha^{-1} (Whitehead, 1995). Some of the N is released to the atmosphere as volatile NH_3, while the N remaining in the excreta and its associated plant residues return to available nutrient pools in the soil.

Animals on range may utilize more of the forage near watering points. Greater density of dung and increased levels of soil-profile NO_3^- are frequently observed in areas near watering and shade points (Haynes and Williams, 1993; Wilkinson et al., 1989). Even without transfer of N to unproductive areas such as woods, shade, watering points, fence lines, and paths, consumption and excretion of N by ruminants results in gathering of N from large areas of the pasture, and deposition of the N to smaller areas. This concentrating effect frequently means that N cycled through livestock cannot be used efficiently by forage plants. On an annual basis, less than 35% of pasture areas receives excretal N; some areas receive one or more applications (overlapping of excreta). This uneven distribution means some of the pasture will be under fertilized and some over fertilized.

4. PRIMARY AND SECONDARY FLOWS OF NITROGEN

Primary and secondary flows of N are very much a part of the animal/plant N cycling ecosystem as discussed above. The following discussion is focused upon cropland and surrounding ecosystems but also relates to a livestock system. Figure 4 from Duxbury et al. (1993) illustrates some of the flows of N following input of 100 kg of fertilizer N. The primary flows are shown by dashed lines. In this example, fifty of the 100 kg are harvested in the crop and fifty are lost by the combination of leaching (25 kg), surface runoff (5 kg), and gaseous loss (20 kg, primarily denitrification). If 10% of the gaseous N loss is N_2O, then 2 kg N_2O-N would be generated in the primary cycle.

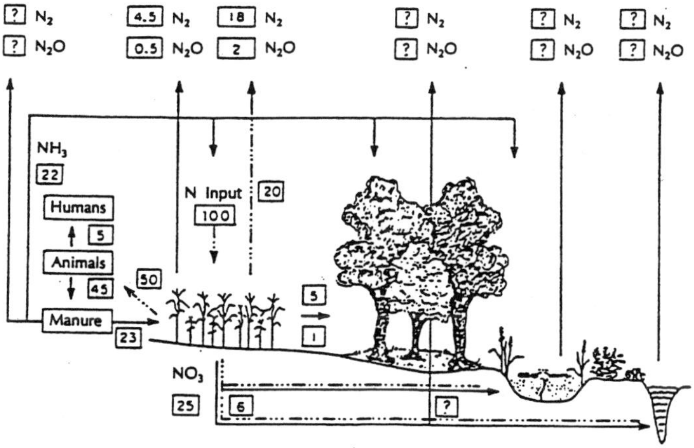

Figure 4. A simplified flow of nitrogen fertilizer through the environment (from Duxbury et al., 1993).

Mosier et al. (1998) evaluated the International Panel on Climate Change (IPCC) methodolgy (IPCC 1997) as part of an effort to provide a more comprehensive N_2O emission calculation methodology. Using mid-point values, they recommended that the emission factor relating N_2O directly from soil to fertilizer-N application should be 1.25 $\pm$ 1% N_2O-N of the applied fertilizer N. If both direct and indirect emissions are considered about 2.0% of N input into agricultural system would be emitted as N_2O-N annually.

Secondary flows, shown by the solid lines in Figure 4, include feeding of the 50 kg of harvested N to animals, which might generate about 45 kg of manure N. The manure is returned to cropland to create a secondary flow of the original fertilizer N. Part of this secondary flow of applied fertilizer N is again removed from the field by the harvested crop; through gaseous losses as NH_3, N_2O, NO_x, and as N_2 gas, surface runoff, and NO_3^- leaching. However, about half of the manure N is volatilized as NH_3 prior to or during manure application. Volatilized NH_3 is aerially dispersed to eventually be returned to and cycled through both natural ecosystems and cropland (Duxbury et al., 1993). Estimates are that, over the course of about fifty years, more than 80% of the N applied to a field will eventually return to the atmosphere through denitrification (Cole et al., 1993). Generally, greater than 95% of this N returns to the atmosphere as N_2 gas but some unknown amount is released as N_2O.

5. GROUND AND SURFACE WATER

5.1. Ground water

Nitrate that moves below the crop root zone is totally soluble and can potentially leach into ground water. Ground water flows within permeable geologic formations called aquifers.

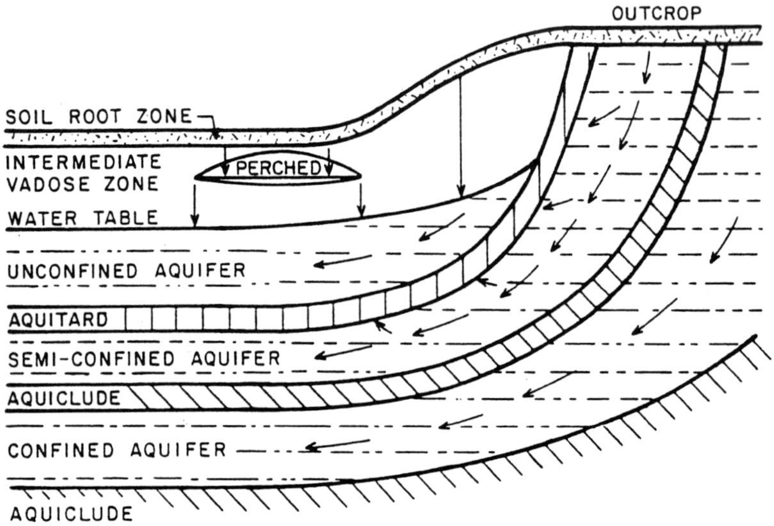

Figure 5. Schematic of vadose zone, aquifer system, and flow direction (from Pionke and Lowrance, 1991).

Aquifers are natural zones beneath the earth's surface that often yield economically important amounts of water. In a very simple system, water and dissolved NO_3^- percolate below the root zone and through the intermediate vadose zone to an aquifer. From there, these waters can recharge deeper aquifers or discharge to streams or water bodies.

Aquifers are subdivided based upon geology. A meaningful division, from the perspective of ground water quality, is between confined and unconfined aquifers. Confined aquifers are separated from the earth's surface by flow-impeding layers that, depending upon the degree of flow impedance, are referred to as aquicludes or aquitards (Figure 5). Unconfined aquifers are not separated from the earth's surface by a flow-impeding layer, and are therefore in contact with the atmosphere through the unsaturated zone. Aquifer systems are often complex. To minimize the amount of NO_3^- that may enter ground water, it is necessary to understand the aquifer system and then to identify and apply improved N management practices to the recharge area of the aquifer. Structure of the aquifer system and subsequent flow patterns affect NO_3^- dilution, transport, and removal.

Ground water can rejoin the ground surface downslope and adjacent to a perennial stream, often along a riparian zone similar to that shown in Figure 6. In a riparian zone, that water table moves progressively toward the land surface and the intermediate vadose zone is lost as the stream channel is approached. During storms or wet periods, the water table can rise rapidly to intersect the land surface at some distance from the stream - discharge of ground water to the soil surface results. The system can be dynamic, with water table levels, extent of the saturated zone, and flow directions changing substantially and rapidly with precipitation (Pionke and Lowrance, 1989). As the ground water and its dissolved NO_3^- move into the more biologically and chemically active soil zones, the NO_3^- becomes available for uptake by riparian vegetation. Also, if oxygen levels become limited, activation of soil biological and chemical regimes result in denitrification.

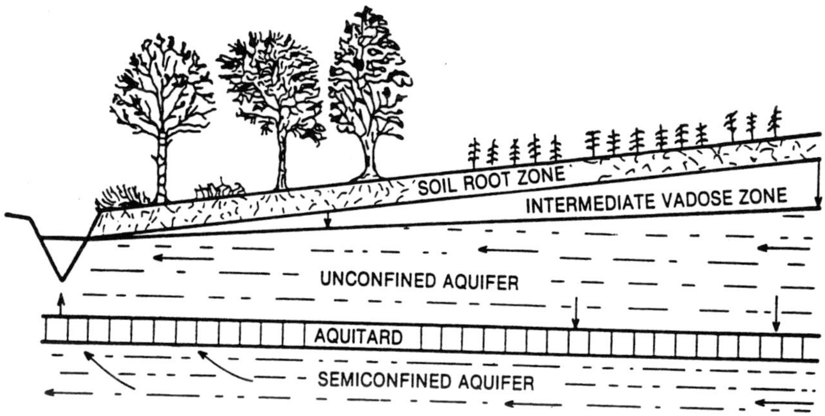

Figure 6. Schematic of the vadose zone, aquifers, and flow directions in a typical riparian zone subject to a humid climate (from Pionke and Lowrance, 1991).

Nitrogen is the nutrient of most concern in the contamination of ground water, primarily resulting from NO_3^- leaching. Leaching of NH_4^+ is generally not important since it is strongly adsorbed by soil, except for sands and soils having low retention (cation exchange) capacities. However, NO_3^- is readily leached deeper into the soil profile, below the bottom of the root zone, and may eventually leach into ground water supplies.

Water quality impact zones for N are wells, ground water supplies, streams, and surface water bodies. Because 95% of rural inhabitants and substantial livestock populations consume ground water, NO_3^- concentration is most important and can cause both human and animal health effects (Chapter this book, by Follett and Follett). Those factors that control NO_3^- concentration in groundwater, such as dilution and well position relative to the primary source areas for NO_3^-, can greatly affect their impact on ground water quality. In contrast, stream flow tends to mix ground water discharge and surface runoff from different land uses and time periods, thus causing generally much lower and more stable NO_3^- concentrations. Although elevated concentrations of NO_3^- are most often observed at shallow water table depths, long term increases in deeper wells are possible where deep aquifers are recharged by NO_3^--rich waters. Movement of NO_3^- with percolating water, through the unsaturated zone, can be very slow and time required for present-day inputs of NO_3^- to reach the ground water reservoir may be many years. Shuman et al. (1975) observed an average rate of NO_3^- movement through silt soils (loess) of about one meter per year for Iowa. Where 168 kg N ha^{-1} (the recommended N rate) was applied, N did not accumulate beneath the crop-root zone. Ground water flows from areas of high pressure toward areas of low pressure (hydraulic head). Generally movement is slow and there is little mixing of contaminated with uncontaminated ground water as they flow through the saturated zone, contaminants tend to remain concentrated in zones. However, because of the slow rate of movement and lack of dilution, contamination may persist for decades and centuries, even if input sources of NO_3^- are decreased or eliminated. Unfortunately, reclamation is technically and economically impossible in most cases (Keeney, 1982).

Many sites of excessive NO_3^- accumulation are recognized. Viets and Hageman (1971) conducted a comprehensive review of studies in the U.S. Substantial accumulations of NO_3^- were found in deep profiles of irrigated Colorado soils, except where alfalfa was the crop (Stewart et al., 1967). Muir et al. (1973) conducted a study of factors influencing NO_3^- content of ground water in Nebraska. Their data indicated that the quality of Nebraska water was not being materially influenced by agricultural use of commercial fertilizers previous to that time except on sites of intensively irrigated sandy soils and in valley positions with a shallow underlying water table.

There are numerous sources of N in the environment. Keeney (1986) identified intense land-use activities (e.g. irrigation farming of high value crops, high density of animal operations, or septic tank systems) as causes of excessive NO_3^- in ground water. Irrigation of cropland is widely practiced in the U.S., particularly in the more arid west and in the southeast where economic returns are high. The review by Pratt (1984) shows that in situations where roots have access to the entire soil solution, NO_3^- is not leached unless excess fertilizer N is added or the soils are over-irrigated.

Because the subsurface system is generally large and not uniform in structure, function, or efficiency, it is much easier to focus on source areas rather than on the whole system. The source area is a bounded area or volume within which one or a set a of related processes dominate to provide excessive production (source), permanent removal (sink), detention (storage), or dilution of NO_3^-. Source area effects, by definition, are disproportionately large relative to the area or volume occupied. If the source area(s) can be identified, then positioned relative to the generalized flow pattern within the system, a basis is possible for estimating effects on an impact zone.

Systematic data on production practices, input use, and management systems are insufficient to do many of the assessments that are needed. However, quantity and quality of soil survey data, climate data, and assessments of NO_3^- concentrations in various aquifers are increasing. Statistical techniques and simulation models used in conjunction with Geographical Information Systems (GIS) technology show promise in identifying and assessing NO_3^- leaching across regions (Christy, 1992; Wylie et al., 1994). Models such as the Nitrate Leaching and Economic Analysis Package (NLEAP) (Shaffer et al., 1991; Delgado et al., 2000) use farm management, soil, and climate information to estimate NO_3^- leaching at a farm or even the soil series level. Thus, allowing determination of potential landscape NO_3^--leaching hot spots when sufficient information is available. As technology continues to improve, the targeting of improved practice to those areas, farm enterprises, fields within a farm, or even locations (hot spots) within a field that cause the most damage should become possible for decreasing losses of N to the environment.

Two general approaches to minimize NO_3^- leaching into ground water are: 1) optimum use of the crop's ability to compete with processes whereby plant available N is lost from the soil-plant processes themselves. Key elements of the first approach are to assure vigorous crop growth and N assimilation capacity, and to apply N in phase with crop demand; 2) The second approach might include use of nitrification inhibitors or delayed release forms of N to directly lower potential leaching losses. In addition, realistic crop-yield goals must be selected. Olson (1985) emphasizes that a realistic yield goal would be no more than 10% above recent average yield for a given field or farm. Such a yield goal will still likely be difficult to achieve because of limitations imposed by environmental factors and/or the farmers own operational skills.

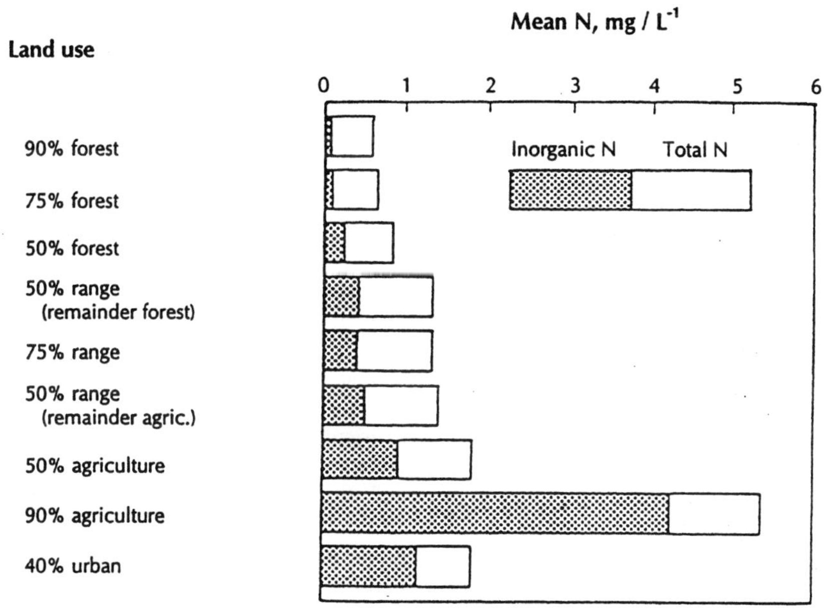

Figure 7. Land use and mean inorganic and total nitrogen concentrations from stream data from 904 nonpoint source-type watersheds (from Omernik, 1977).

The high NO_3^- flux that often occurs in streams draining agricultural land, does not come from the overland runoff, but primarily from the ground water contributions (including tile-drainage effluent) to stream flow. During discharge events, the ground water and its NO_3^- load will include shallow interflow (sometimes referred to as subsurface runoff). However, during the majority of time, deeper baseflow that rejoins surface water provides the major contribution of NO_3^- (Hallberg, 1989).

Stream water quality data from 904 nonpoint source-type watersheds across the U.S. were summarized by Omernik (1977). The watersheds ranged in character from forested areas, to urbanized regions, to areas dominated by row-crop agriculture. The data were compared to landuse and, as shown in Figure 7, especially the inorganic N concentrations are directly related to the amount of the watershed used for agriculture.

The data in Figure 7 are over two decades old now; however, reviews of temporal trends since then also show significant increases in NO_3^- (Hallberg, 1989). Referring to Figure 7, long-term environmental concern about the impact may not only need to be the increasing loads of soluble N, but also the dramatic change in the proportion of the particulate and soluble N concentrations. In forest and range systems the major N load was as organic N, much of it in the particulate fraction (related to organic matter); but now the major load in agricultural areas is as soluble NO_3^-.

5.2. Surface water

Agricultural production has been identified as a major nonpoint source of pollution in U.S. lakes and rivers that do not meet water quality goals. Nitrogen can be transported into aquatic systems from airborne, surface, underground, and in situ sources (Table 3). Sediment is the largest single type of pollutant followed by nutrients (NRC, 1993). As discussed above, much of the N that enters lakes and rivers is associated with eroding sediments (NH_4^+), eroding soil organic matter (organic forms of N and NH_4^+), and dissolved in surface runoff (primarily NO_3^-). The water that runs over the soil surface during a rainfall or snowmelt event, by rill or sheet flow, or even high-order channelized flow, may have a relatively high concentration of organic N related to suspended particulate matter, but it is typically quite low in NO_3^- concentration.

Agricultural sources of N can arrive in surface water via airborne dust from wind erosion, through gaseous transport of NH_3 volatilized from livestock manure or from some fertilizer materials. Surface sources of N from agriculture are perhaps the best understood, and N delivered with eroded soil sediments is a major source.

Ground water delivery of NO_3^- to lakes and streams is no doubt very important but difficult to gauge. In situ sources include biological N fixation, such as by blue-green algae and the leaching of N from lake sediments. An additional source of N and other nutrients is from wild aquatic birds; however, their role in the nutrient regime of a water body may be more that of cycling agents than of direct sources.

Table 3
Sources and sinks for the nitrogen budgets of aquatic systems

Sources	Sinks
Airborne	Effluent loss
Rainwater	
Aerosols and dust	Ground water recharge
Leaves and miscellaneous debris	
	Fish harvest
Surface	
Agricultural drainage, including tile drainage	Weed harvest
Water erosion of sediment from agricultural land	
Animal waste runoff	Insect emergence
Marsh drainage	
Runoff and erosion from forest and rangeland	NH_3 volatilization
Urban storm water runoff	
	Evaporation (aerosol formation from surface foam)
Domestic waste effluent	
Industrial waste effluent	
Wastes from boating activities	Denitrification
Underground	Sediment deposition of detritus
Natural ground water	
Subsurface agricultural and urban drainage	
Subsurface drainage from septic tanks	Sorption of ammonia onto sediments
In situ	
Nitrogen fixation	
Sediment leaching	

When waters become too enriched by nutrients, the aquatic environment can become eutrophic - a result of the ensuing luxuriant growth of algae and macrophyte growth to levels that can choke navigable waterways, increase turbidity, and depress dissolved oxygen concentrations. Rapid growth of algae is the greatest and most widespread eutrophication problem. When a large mass of algae dies and begins to decay, the oxygen dissolved in water is depleted and certain toxins are produced, both of which can kill fish. The complexities of eutrophication are that nutrient status of various species of algae can vary from lake to lake or even from different areas and depths of the same lake on the same day. Excess algal growth can create obnoxious conditions in ponded waters, increase water treatment costs by clogging screens and requiring more chemicals, and cause serious taste and odor problems.

Sawyer (1947) was the first to propose quantitative guidelines for lakes. He suggested that 0.3 mg L^{-1} of inorganic N and 0.015 mg L^{-1} of inorganic phosphorus are critical levels above which algal blooms can normally be expected in lakes. However, development of nutrient criteria or recommended methodologies for protecting waterbodies from excessive nutrient loading are very much needed. National criteria that are available for NO_3^-, NO_2^-, and NH_3 are generally established to protect human health and aquatic life from toxic eutrophication, or impairments to recreational uses such as swimming, fishing, and boating (Tetra Tech, Inc., 1994).

Under natural conditions, NO_3^- and NO_2^- occur in moderate concentrations and have little toxicological significance for aquatic life. Because the levels that are toxic to aquatic life are much higher than those expected to occur naturally in surface waters, restrictive water quality criteria for these elements have not been recommended. Two of the main concerns about the impacts of NO_3^- and NO_2^- on the environment are the primary water quality concern about their potential health effects on humans and ruminant animals associated with contaminated drinking water.

On the other hand, NH_3 is highly toxic to aquatic organisms. Acute toxicity in fish causes loss of equilibrium, hyperexcitability, increased breathing, cardiac output, convulsions, coma, and death, if concentrations are extreme. Chronic toxic effects include reduced hatching success, growth rates, and developmental or pathological changes in gill, liver, and kidney tissues (U.S. EPA, 1982).

6. WITHIN AGRICULTURAL SYSTEMS

6.1. Accounting for all Nitrogen Sources

Nitrogen budgets provide a valuable framework to quantify and examine N inputs and losses for agricultural production systems (also see Figure 1). Accounting for the major sources of N to cropping systems and into the envionment, in general, is especially important. The following are some of the sources that should be considered.

1. Fertilizer N inputs and amounts are easily determined and can be managed.
2. Organic wastes are an important N source. Organic wastes available for use on cropland in the U.S. include livestock wastes, crop residues, sewage sludge and septage, food processing wastes, industrial organic wastes, logging and wood manufacturing wastes, and municipal refuse. Animal manures and crop residues account for the majority of organic wastes applied to agricultural land.
3. Manure N inputs are uncertain because the N content is related not only to livestock type, age, and health, but also to variations in N content. Once excreted, the N content can change considerably depending on type and amount of bedding, type and time of manure storage, and manure management and placement when being applied. The best way to overcome these uncertainties is through the use of manure analysis and calibration of application equipment. Manure credits are often used to try to account for N that becomes available from applied manure.
4. Biological N fixation (BNF), especially by legumes, can be an especially important source of N. Although the importance of BNF has been known for centuries, there are

few quantitative methods for estimation of BNF. Currently, the method most used is that of recognizing BNF by legumes with legume credits.

5. Nitrate contained in irrigation water is available to the crop and should be considered when making fertilizer recommendations. Crop utilization of NO_3^- from irrigation water is greatest when plant-N requirement is greatest and other N sources are not excessive.

6. Atmospheric additions, including volatilized NH_3 from livestock operations, are another source of N to agricultural systems and to the environment. The mechanisms of additions that are identified include N dissolved in precipitation, dry deposition, and direct plant absorption of gaseous NH_3.
7. Contributions of residual soil N require soil testing for NO_3^- and NH_4^+ within the root zone and will be discussed below.
8. Nitrogen mineralization is the term given to biological decomposition of organic material in soils and their conversion and contribution to inorganic forms is significant.

6.1.1. Soil nitrogen availability tests

Available soil N represents residual N in the soil profile, plus N mineralized from the soil organic matter during the growing season. While residual N has proven to be a useful index in certain regions of the U.S., no generally accepted index exists for N mineralization. Obviously, such a development would represent a major advance for avoidance of excessive fertilizer N applications. A complement to a soil N test may be a plant tissue N test. An attractive feature of tissue tests is that the plant root system tends to integrate spatial variability of soil N supplying power over a relatively large field volume.

6.1.2. Soil organic nitrogen availability

A significant portion of plant-N requirements are supplied by mineralization of soil organic matter during the growing season. Various N availability indexes exist, but they typically provide qualitative rather than quantitative measures of SON availability. Early concepts of a N availability index have been modified; but to date, no soil organic N availability procedure has received general acceptance from a soil test standpoint. Ultimately, a systems-type, mass balance N approach may be the best alternative. The present recommendation is to follow pertinent N fertilizer guides which have been developed locally for specific crop needs and soil areas.

6.2. Agricultural practices

6.2.1. Nitrification inhibitors

The NH_4^+ ion is sorbed to the CEC of the soil; whereas, NO_3^- ion is not and can be readily leached or denitrified. Both forms can be readily utilized by crops. Nitrification inhibitors include chemicals added to soils to stabilize fertilizer applied as NH_3 or in the NH_4^+ form by inhibiting the activity of the *Nitrosomonas* bacteria in the first step of the nitrification process.

6.2.2. Control/slow release fertilizer

Methods of altering the release of N from soluble materials has been to coat water-soluble N fertilizer with less water-soluble materials in order to retard entry of water into the particle and the movement of N out. Coatings applied to soluble N materials generally have been of three types: 1) Impermeable coatings with small pores that allow slow entrance of water and slow passage of solubilized N out of the encapsulated area; 2) Impermeable coatings that require breakage by physical, chemical, or biological action before the N is dissolved; and 3) Semipermeable coatings through which water diffuses and creates internal pressures sufficient to disrupt the coating. Sulfur-coated urea (SCU) has been developed for a number of years as a product with characteristics of slow-N release. Elemental sulfur (S) was chosen because of its relatively low cost and ease of handling. Newer control-release N fertilizer materials are also being developed and marketed (Shaji and Gandeza, 1992). These newer materials have polyolefin resin coatings. The coatings can be tailored to provide a range of N release rates that are suitable for a variety of cropping systems. However, further field research is needed to insure the utility of these newer materials for cropping systems.

6.2.3. Conservation tillage

Use of conservation or reduced tillage (including no-till) continues to increase as an alternative for nearly all forms of crop production. Management systems which maintain crop residues at or near the soil surface have several attractive features, including less on-farm fuel use and its associated carbon dioxide emissions (Follett, 2001), more available soil water, and reduced soil erosion. However, adoption of conservation tillage practices may result in some N moving from the soil-plant systems into the environment under certain conditions.

There is no question that conservation tillage is effective in decreasing particulate N losses associated with soil erosion and surface water runoff as discussed above. However, effects of conservation tillage on leachable N are not so well delineated as are surface losses. Generally, conservation tillage provides a wetter, cooler, more acidic, less oxidative soil environment. Under such conditions, processes of ammonification and denitrification may be favored over nitrification. Conversely, for NO_3^- that is already present, the leaching potential may be greater under conservation tillage. This is because more undisturbed soil-macropores exist for NO_3^- and water movement. Increased water flow, into and through the root zone, has been observed under no-till compared to conventional-tillage soils. This higher flow has been attributed to decreased water evaporation because of surface residues and increased numbers of undisturbed channels (e.g. earthworm and old roots) continuous to the soil surface. The surface mulch enhances the environment for earthworms and the lack of tillage preserves existing channels for several years.

6.2.4. Rotations, cover crops, and nitrogen-scavenging crops

Rotations and cover crops, historically used as a means of conserving soil and/or providing an organic N source, have received renewed interest as an aid in avoiding excessive N losses to the environment. Whereas monocultures of grain crops (e.g. corn and wheat) require high inputs of fertilizer N, such inputs can be decreased with crop rotations which require less, or fix atmospheric N. Because less excess profile N may be expected with a rotation, there should be less potential for N leaching. An exception may be under certain rotation-fallow conditions designed to conserve water in drier areas.

Winter cover crops can be effective in absorbing both NO_3^- and available water during the fall, winter, and spring, thereby decreasing the N leaching potential. When the cover crop is returned to the soil, some of the absorbed N is then available to the following crop. Both legumes and non-legumes are used from a strictly N leaching standpoint. While an annual crop such as rye can be effective in scavenging excess available N from within crop rooting zones, deep-rooted perennials should be considered for NO_3^- accumulation below normal rooting depths. Alfalfa, with a potential rooting depth in excess of fifteen feet, is a crop which merits particular attention.

6.2.5. Filter strips

Vegetative filter strips, also referred to as buffer strips and riparian zones, remove sediment, organic matter, and other pollutants from runoff and waste waters. Under field conditions, excess runoff from terraces is frequently diverted to a strip. Upon entering the strip, both the flow velocity and transport capacity of the runoff are reduced. The sediment and its associated pollutants are then removed from the runoff by filtration, deposition, infiltration sorption, decomposition, and volatilization processes. The effectiveness of filter strips in removing sediment and particulate N is well established. Less certain is the effectiveness of filter strips for removing soluble N in runoff. Uptake by filter strip vegetation of mineral N transported by runoff water may occur during times of active growth but less during other times of the year. Also, some denitrification may be occurring. Scavenging of N from underground water and the vertical horizon by riparian vegetation, especially by deeper rooted plants, also may be important for removing dissolved N in surface and subsurface flows before the N is transported into streams and lakes.

7. SUMMARY

Nitrogen (N) is ubiquitous in the environment. It is also one of the most important nutrients and is central to the growth of all crops and other plants. However, N also forms some of the most mobile compounds in the soil-plant-atmosphere system; and there is mounting concern about agriculture's role in N delivery into the environment. Nitrogen represents the mineral fertilizer most applied to agricultural land. This is because available soil-N supplies are often inadequate for optimum crop production. This manuscript reviews the fate and transport of N from the various sources used to supply the N-requirements of crops in the context of the N cycle. Use of N budgets or a mass balance approach is needed to understand the options for improving management of N in farming and livestock systems and for mitigating the environmental impacts of N. Fertilizing crops for crop N uptake that will be near the point of maximum yield generally is an economically and environmentally acceptable practice. The objective is to lower the rate and duration of the loss processes themselves. Practices and concepts that lessen the opportunity for loss processes to occur and that help decrease the amount of N that may be lost to the environment are considered. In some cases improved efficiency is achieved by using less nutrients and in other cases it can be achieved by increasing the yield while using the same amount of N-input. In either case, the goal is to decrease the total residual mass of N in the soil. Another approach is to keep the residual N in the soil-crop system by curtailing the transport processes (leaching, runoff, erosion, and gaseous losses) that carry pollutants out of the soil crop.

REFERENCES

Alberts, E.E. and R.G. Spomer. 1985. Dissolved nitrogen and phosphorus in runoff from watersheds in conservation and conventional tillage. J. Soils and Water Cons. 40:153-7.

Betlach, M.R. and J.M. Tiedje. 1981. Kinetic explanation for accumulation of nitrite, nitric oxide, and nitrous oxide during bacterial denitrification. Appl. Environ. Micrbiol. 42:1074-84.

Bock, B.R. and G.W. Hergert. 1991. Fertilizer nitrogen management. pp 139-64. *In* R.F.

Follett, D.R. Keeney, and R.M. Cruse (eds). Managing Nitrogen for Ground water Quality and Farm Profitability. SSSA. Madison, WI.

Bristow, A.W. and S.C. Jarvis. 1991. Effects of grazing and nitrogen fertilizer on the soil microbial biomass under permanent pastures. J Sci Food Agric 54:9-21.

CAST. 1985. Agriculture and ground water quality. Council for Agricultural Science and Technology Report 103, 62 pp.

Christy, A.D. 1992. Managing agricultural chemical use to protect ground water. Geo Info Systems. 2:34-9.

Cole, C.V., K. Flach, J. Lee, D. Sauerbeck, and B. Stewart. 1993. Agricultural sources and sinks of carbon. Water, Air, and Soil Pollution 79:111-22. *In* J. Wisniewski and R. Neil Sampson (eds). Terrestrial Biospheric Carbon Fluxes: Quantification of Sinks and Sources of CO_2. Klewer Academic Publishers. Dordrecht, Netherlands.

Delgado, J.A., R.F. Follett, and M.J. Shaffer. 2000. Simulation of nitrate-nitrogen dynamics for cropping systems with different root depths. Soil Sci. Soc. Amer J. 64:1050-1054.

Delgado, J.A., R.T. Sparks, R.F. Follett, J.L. Sharkoff, and R.R. Riggenbach. 1999. Use of winter cover crops to conserve water and water quality in the San Luis Valley of south central Colorado. pp 125_142. *In* R. Lal (ed) Soil Quality and Soil Erosion. CRC Press, Boca Raton, Fl.

Duxbury, J.M., L.A. Harper, and A.R. Mosier. 1993. Contributions of agroecosystems to global climate change. pp 1-18. *In* L.A. Harper, A.R. Mosier, J.M. Duxbury, and D.E. Rolston (eds). Agroecosystem Effects on Radiatively Important Trace Gases and Global Climate Change. Spec Pub no 55. ASA. Madison, WI.

Firestone, M.K. and E.A. Davidson. 1989. Microbiological basis of NO and N_2O production and consumption in soil. pp 7-21. *In* M.O. Andreae and D.S. Schimel (eds). Exchange Trace Gases Between Terrestrial Ecosystems and the Atmosphere. John Wiley and Sons, Ltd.

Fletcher, D.A. 1991. A national perspective. pp 10-7. *In* R.F. Follett, D.R. Keeney, and R.M. Cruse (eds). Managing Nitrogen for Groundwater Quality and Farm Profitability. SSSA. Madison, WI.

Follett, R.F. (ed). 1989. Nitrogen Mangement and Ground Water Protection. Elsevier Sci. Publishers, Amsterdam. 395 p.

Follett, R.F. 2001. Soil management concepts and carbon sequestration in cropland soils. Soil and Tillage Res. (In Press).

Follett, R.F., S.C. Gupta and P.G. Hunt. 1987. Conservation practices: relation to the management of plant nutrients for crop production. pp 19-51. *In* R.F. Follett, J.W.B. Stewart and C.V. Cole (eds). Soil Fertility and Organic Matter as Critical Components of Production Systems. Spec Pub no 19. SSSA. Madison, WI.

Follett, R.F., D.R. Keeney and R.M. Cruse (eds). 1991. Managing Nitrogen for Ground water Quality and Farm Profitability. SSSA. Madison, WI. 357 p.

Follett, R.F., L.K. Porter, and A.D. Halvorson. 1995. Nitrogen-15 labelled fertilizer dynamics in soil in a 4 year, no till cropping sequence. pp. 165-174. *In* Nuclear Techniques in Soil-Plant Studies for Sustainable Agriculture and Environmental Preservation. IAEA, Vienna, Austria.

Follett, R. F. and S.R. Wilkinson. 1995. Nutrient management of forages. pp. 55-82. *In* R.F. Barnes, D.A. Miller, and C.J. Nelson (eds.). Forages: The Science of Grassland Agriculture. Vol II. IA State Univ. Press. Ames, IA.

Follett, R.F. and P.J. Wierenga (Guest Eds). 1995. Integrated Nitrogen Management in Relation to Leaching and Groundwater Quality. J. Contam. Hydrol. (Special Issue) 20(3&4):i-iv, 168-350.

Foster, G.R., R.A. Young, M.J.M. Romkens and C.A. Onstad. 1985. Processes of soil erosion by water. pp 137-62. *In* R.F. Follett and B.A. Stewart (eds). Soil Erosion and Crop Productivity. ASA, CSSA, and SSSA. Madison, WI.

Freney, J.R., O.T. Denmead and J.R. Simpson. 1979. Nitrous oxide emissions from soil at low moisture content. Soil Biol. Biochem. 16:167-73.

Godin, R.E. 1999. Effects of Irrigation and Nitrogen Management on Water and Nitrogen Use Efficiency of Irrigated Corn. PhD Thesis. Colorado State Univ. 148p

Goodroad, L.L. and D.R. Keeney. 1984. Nitrous oxide production in aerobic soils under varying pH, temperature, and water content. Soil Biol. Biochem. 16:39-43.

Hallberg, G.R. 1987. Nitrates in ground water in Iowa. pp 23-68. *In* F.M. D'Itri and L.G. Wolfson (eds). Rural Ground water Contamination. Lewis Publish. Chelsea, MI.

Hallberg, G.R. 1989. Nitrate in ground water in the United States. pp 35-74. *In* R.F. Follett (ed). Nitrogen Management and Ground Water Protection. Elsevier Sci. Publishers, Amsterdam.

Haynes, R.J. and P.H. Williams. 1993. Nutrient cycling and soil fertility in grazed pasture ecosystems. Advances in Agron. 46:119-99.

IPCC., 1997. Revised 1996 IPCC Guidelines for National Greenhouse Gas Inventories: Vols (2) and (3)., IPCC, Geneva, Switzerland.

Juergens-Gschwind, S. 1989. Ground water nitrates in other developed countries (Europe) - Relationships to land use patters. pp. 75-138. *In* Follett (ed). Nitrogen Management and Ground Water Protection. Elsevier Sci. Publishers, Amsterdam.

Jury, W.A. and D.R. Nielson. 1989. Nitrate transport and leaching mechanisms. pp 139-57. *In* R.F. Follett (ed). Nitrogen Management and Ground Water Protection. Elsevier Sci. Pubs., Amsterdam.

Keeney, D.R. 1982. Nitrogen management for maximum efficiency and minimum pollution. pp. 605-649. *In* F.J. Stevenson (ed). Nitrogen in Agricultural Soils, Agron. Monograph 22. ASA. Madison, WI.

Keeney, D.R. 1989. Sources of nitrate to ground water. pp. 23-34. *In* R.F. Follett (ed). Nitrogen Management and Ground Water Protection. Elsevier Sci. Pubs. Amsterdam, The Netherlands.

Klemedtsson, L., B.H. Svensson and T. Rosswall. 1988. Relationships between soil moisture content and nitrous oxide production during nitrification and denitrification. Biol. Fert. Soils. 6:106-11.

Laegreid, M., O.C. Bockman, and O. Kaarstad. 1999. Agriculture Fertilizers and the Environment. CABI Publishing in association with Norsk Hydro ASA. Wallingford Oxon UK. 294 p.

Lal, R., J.M. Kimble, R.F. Follett, and C.V. Cole. 1998. The Potential of U.S. Cropland to Sequester Carbon and Mitigate the Greenhouse Effect. Ann Arbor Press. Chelsea, MI. 128 pp.

Leeden, F., F.L. Van der Troise, and D.K. Todd. 1991. The Water Encyclopedia, Second Edition. Lewis Pubs. Chelsea, MI.

Lehman, O.R. and L.R. Ahuja. 1985. Interflow of water and tracer chemical on sloping field plots with exposed seepage faces. J. Hydrology. 76:307-17.

McDowell, L.L. and K.C. McGregor. 1984. Plant nutrient runoff from conservation tillage corn. Soil and Tillage Res. 4:79-91.

Mosier, A., Kroeze, C., Nevison, C., Oenema, O., Seitzinger, S., Van Cleemput, O., 1998. Closing the global N20 budget: nitrous oxide emissions through the agricultural nitrogen cycle. Nutrient Cycling in Agroecosystems 52:225-248.

Muir, J., E.C. Seim and R.A. Olson. 1973. A study of factors influencing the nitrogen and phosphorus contents of Nebraska waters. J. Environ. Quality. 2:466-70.

National Research Council (NRC). 1993. Soil and Water Quality: An Agenda for Agriculture. Committee on Long-Range Soil and Water Conservation - Board on Agriculture. National Academy Press. Washington, DC.

Olson, R.A. 1985. Nitrogen Problems. pp 115-37. *In* Proceedings: Plant Nutrient Use and The Environment. Kansas City. The Fertilizer Institute. Washington, DC.

Omernik, J.M. 1977. Nonpoint source-stream nutrient level relationships: a nationwide study. EPA-600/3-77-105, U.S. Environmental Protection Agency. U.S. Govt. Print. Off. Washington, DC.

Pionke, H.B. and R.R. Lowrance. 1991. Fate of nitrate in subsurface drainage waters. pp 237-57. *In* R.F. Follett, D.R. Keeney and R.M. Cruse (eds). Managing Nitrogen for Ground water Quality and Farm Profitability. SSSA. Madison, WI.

Power, J.F. and J.S. Schepers. 1989. Nitrate contamination of ground water in North America. Agriculture, Ecosystems, and Environment. 26:165-88.

Pratt, P.F. 1984. Nitrogen use and nitrate leaching in irrigated agriculture. CH 21. *In* R.D. Hauck (ed). Nitrogen in Crop Production. ASA. Madison, WI.

Reeder, J.D., C.D. Franks, and D.G. Milchunas. 2000. Root biomass and microbial processes. pp. 139-166. *In* R.F. Follett, J.M. Kimble, and R.Lal (eds.) The Potential of U.S. Grazing Lands to Sequester Carbon and Mitigate the Greenhouse Effect. Lewis Publishers, Boca Raton, FL.

Romkens, M.J.M. 1973. Nitrogen and phosphorus composition of surface runoff as affected by tillage method. J. Environ. Qual. 2:292-5.

Sawyer, C.N. 1947. Fertilization of lakes by agricultural and urban drainage. J. New Eng. Water Works Assn. 61:109-27.

Schuman, G.E., T.M. McCalla, K.E. Saxton and H.T. Knox. 1975. Nitrate movement and its distribution in the soil profile of differentially fertilized corn watersheds. Soil Sci. Soc. Amer. Proc. 39:1192-7.

Shaffer, M.J., A.D. Halvorson and F.J. Pierce. 1991. Nitrate leaching and economic analysis package (NLEAP): Model description and application. pp 285-322. *In* R.F. Follett, D.R. Keeney and R.M. Cruse (eds). Managing Nitrogen for Ground water Quality and Farm Profitability. Soil Sci. Soc. Amer. Madison, WI.

Shaji, S. and A.T. Gandeza (eds). 1992. Controlled release fertilizers with polyolefin resin coating. Konno Printing Co., Ltd. Sendai, Japan. 92 p.

Smith, J.L. and E.A. Paul. 1990. The significance of soil microbial biomass estimation. pp. 357-396. *In* J. Bollag and G. Stotzky (eds.). Soil Biochemistry (Vol 6). Marcel, Dekker, Inc. NY.

Smith, S.J., J.S. Schepers and L.K. Porter. 1990. Assessing and managing agricultural nitrogen losses to the environment. Adv. Soil Sci. 14:1-43.

Stevenson, F.J. 1982. Cycles of Soil: Carbon, Nitrogen, Phosphorus, Sulfur, Micronutrients. John Wiley & Sons, NY. 380 p. 210:347-50.

Stewart, B.A., F.G. Viets, G.L. Hutchinson, W.D. Kemper, F.E. Clark, M.L. Fairbourn and F. Strauch. 1967. Distribution of nitrates and other water pollutants under fields and corrals in the middle South Platte Valley of Colorado. USDA-ARS 41-134. 206 p. U.S. Govt. Printing Off. Washington, DC.

Tetra Tech, Inc. 1994. Summary of State of Water Quality Criteria and Standards for Nutrients. Draft Report. Prepared for the U.S. Environmental Protection Agency. Office of Science and Technology. Washington, DC.

Tortoso, A.C. and G.L. Hutchinson. 1990. Contributions of autotrophic and heterotrophic nitrifiers to soil NO and N_2O emissions. Applied and Environ. Micrbiol. 56:1799-805.

U.S. EPA. 1982. Water Quality Criteria Manual (Gold Book). U.S. Environmental Protection Agency, Office of Water. Washington, DC.

Viets, F.G., Jr. and R.H. Hageman. 1971. Factors affecting the accumulation of nitrate in soil, water and plants. U.S. Department of Agriculture. Agricultural Handbook 413, 63 p.

Whitehead, D.C. 1995. Grassland Nitrogen. CAB International. Wallingford, U.K.

Wilkinson, S.R., J.A. Stuedemann and D.P. Belesky. 1989. Distribution of soil potassium in grazed K-31 tall fescue pastures as affected by fertilization and endophytic fungus infection. Agron J. 81:508-12.

Wylie, B.K., M.J. Shaffer, M.K. Brodahl, D. Dubois and D.G. Wagner. 1994. Predicting spatial distribution of nitrate leaching in northeastern Colorado. J. Soil and Water Conserv. 49:288-93.

Nitrogen in the Environment: Sources, Problems, and Management
R.F. Follett and J.L. Hatfield (Eds)

Chapter 3. Importance and Effect of Nitrogen on Crop Quality and Health

Jürg M. Blumenthal[1], David D. Baltensperger[1], Kenneth G. Cassman[2], Stephen C. Mason[2], and Alexander D. Pavlista[1].

[1] Panhandle Research & Extension Center, University of Nebraska, 4502 Avenue I, Scottsbluff, NE, 69361, United States. [2] Department of Agronomy and Horticulture, University of Nebraska, 279 Plant Science, Lincoln, NE 68583-0915, United States.

Nitrogen (N) is often the most limiting factor in crop production. Hence, application of fertilizer N results in higher biomass yields and protein yield and concentration in plant tissue is commonly increased by. Nitrogen often affects amino acid composition of protein and in turn its nutritional quality. In cereals, abundant supply of N decreases the relative proportion of lysine and threonine, thus, reducing the biological value of the protein. Increasing N supply generally improves kernel integrity and strength, resulting in better milling properties of the grain. In oil seed crops, protein levels are increased upon N fertilization, whereas oil concentration is decreased. Effects of N fertilization on oil composition and quality are inconsistent. In sugarbeet production, abundant supply of N results in a reduction of sucrose concentration per unit fresh matter and to an increase in impurities (alpha-amino-nitrogen, invert sugars, and lime salts), which negatively affect efficiency of sucrose extraction. Nitrogen supply to potatoes primarily influences tuber size, dry matter, and sugar contents. Nitrogen supply is managed according to market classes (table stock, French fries, and potato chips), which require different quality parameters.

1. CEREALS

1.1 Corn

Corn (*Zea mays* L.) is the third most important crop worldwide following rice (*Oryza sativa* L.) and wheat (*Triticum aestivum* L.). The corn kernel is composed of approximately 72% starch, 10% protein, 5% oil, 2% sugar, and 1% ash with the remainder being water (Perry, 1988). The corn protein biological value is low due to the low concentration of the essential amino acids, lysine, and tryptophan, although opaque-2 corn has been developed with higher levels of these two amino acids (Mertz et al. 1964). Recent breeding efforts by CYMMYT have improved the hardness of opaque-2 corn cultivars (Vasal et al., 1980), thus improving the agronomic characteristics of this corn type. The oil in corn is an important energy source for livestock feed, and due to a high degree of unsaturation, is widely used for human consumption (Perry, 1988).
Increasing nitrogen supply to corn generally resulted in increased grain and protein yields and increased grain protein concentration (Anderson et al., 1984; Cromwell et al., 1983; Kniep and Mason, 1991; Oikeh et al; 1998; Olsen et al., 1976; Pierre et al., 1977; Sabata and Mason, 1992; Tsai et al., 1983; Tsai et al., 1992). Research reported by Tsai et al. (1983) suggested that protein concentration of corn grain increases with nitrogen supply due to preferential deposition of zein

over the other endosperm proteins. It is apparent that the amount of fertilizer nitrogen required to maximize grain yields is not the same as the amount that will produce maximum grain protein concentrations (Sander et al., 1987). As the protein concentration of corn grain increases, zein makes up an increasing proportion of the protein (Frey et al., 1949; Frey, 1951; Tsai et al., 1992). Rendig and Broadbent (1977) reported that concentration of the protein fraction zein in corn grain was closely associated with the level of soil nitrogen, with each added increment of nitrogen increasing the percentage of zein. Tsai et al. (1983) reported that as nitrogen levels increased, zein accumulated preferentially in normal corn grain, but not in grain of opaque-2 hybrids. Tsai et al. (1992) reported that protein yield increase from nitrogen application was accompanied by an increase in the amount of zein present in the endosperm, creating harder, less brittle and more translucent grain. Since zein contains lower amounts of the most limiting essential amino acids lysine and tryptophan, increased grain yields change the amino acid balance by reducing the lysine and typtophan concentrations, thus reducing the biological value of the grain protein. However, this may be compensated for in some cases since nitrogen fertilizer application increases the size of the germ, which has a better amino acid balance than the endosperm (Bhatia and Rabson, 1987). In contrast, studies with opaque-2 corn hybrids indicate that increased nitrogen supply maintains or increases the lysine concentration of grain (Cromwell et al., 1983; Tsai et al., 1983).

Kniep and Mason (1991) found that irrigation increased grain yield, reduced protein concentration, had no effect on percent lysine per sample and increased percent lysine of protein of normal corn. Nitrogen application increased grain yield, protein concentration and percent lysine of sample, but decreased percent lysine of protein. Irrigation decreased percent lysine per sample for opaque-2 hybrids. The grain from the above study was used in a rat feeding experiment, and found that rats fed grain produced with irrigation had greater and more efficient rates of gain, while those fed grain from plots with nitrogen application had lower and less efficient rates of gain (Hancock et al., 1988). These studies and the one previously reported by Bullock et al. (1989) clearly indicate that irrigation increases grain and protein yields, and lowers protein concentration, but improves the biological value of the protein. In contrast, nitrogen application increases grain and protein yields, and protein concentration, but reduces the biological value of the protein. Similar results for nitrogen fertilizer application, except the adverse effect of nitrogen fertilizer application on percent lysine and tryptophan was less for opaque-2 hybrids than for normal hybrids were reported by Breteler (1976). MacGregor et al. (1961) found that amino acid concentrations of grain did not increase uniformly to nitrogen fertilizer application, and that the concentrations of lysine, methionine and phenyalanine did not increase. Rendig and Broadbent (1979) found that nitrogen fertilizer application decreased the concentrations of tryptophan, lysine, glycine, arginine and threonine in protein, while concentrations of alanine, phenylalanine, tyrosine, glutamic acid and leucine were increased. It is apparent that irrigation has a positive effect on corn grain amino acid balance, while nitrogen fertilizer application has a negative effect.

Large uniform kernel size is desired for dry milling (Paulsen and Hill, 1985), wet milling (Watson, 1987), alkaline cooked products (Shumway et al., 1992) and livestock feed when processed by rolling or cracking. Most agronomic studies have focused on kernel weight which often is highly associated with both kernel density and/or size (Watson, 1987). Personal communications from livestock feeders and dry millers indicate that uniformity of kernel size is important, but has not been studied scientifically. Kernel weight has been shown to increase with increasing nitrogen application (Bauer and Carter, 1986; Kniep and Mason, 1989; Cromwell et al., 1983; Rendig and Broadbent, 1979).

Physical quality of grain is usually measured by kernel hardness (density), kernel breakage susceptibility (brittleness) and stress cracking. Physical quality is a primary concern of the corn dry milling industry to optimize the production of the highest value end-product of uniform, large flaking grits (Paulsen and Hill, 1985). Bauer and Carter (1986) and Kniep and Mason (1989) found that kernel breakage decreased and kernel density increased with nitrogen fertilizer application.

There are few studies on agronomic practice influencing the oil concentration of corn grain, although the development of the specialty high-oil corn for livestock feed (Alexander, 1988) has spurred much recent interest. Earle (1977) showed variations in oil concentration by year from 1917 to 1972, but no correlations were found between oil concentration and variations in temperature, rainfall or fertilization. Welch (1969) reported that nitrogen, phosphorus and potassium applications increased the oil concentration of corn grain slightly, but more important was that the increased grain yield resulted in greater oil production per unit of land area. In contrast, Jellum et al. (1973) found that increasing nitrogen application rate had no influence on the oil concentration of corn grain.

1.2 Wheat

Wheat is grown across a wide range of environments around the world. A broad range of food products stem from wheat. The protein of wheat is unique among cereals. The glutein protein fraction of wheat can trap and retain CO_2 generated during the fermentation of dough. The most common products are leavened and unleavened baked bread. Other important wheat products are pastries, crackers, biscuits, and pasta. Production factors that increase grain yield also increase the starch concentration of grain while reducing the grain protein concentration (McDermitt and Loomis, 1981). This inverse relationship has been shown in barley (*Hordeum vulgare* L.), corn, oats (*Avena sativa* L.), rice, sorghum [*Sorghum bicolor* (L.) Moench], and wheat (Frey, 1977). The negative relationship between protein concentration and grain yield is partly associated with the higher glucose costs for synthesis of protein than carbohydrates (Penning de Vries et al., 1974); thus the higher cost of protein synthesis is logically inversely related to the grain yield (Bhatia and Rabson, 1976).

Benzian and Lane (1979) analyzed the relationship between nitrogen supply, grain yield and grain protein concentration for wheat. They found that a greater nitrogen supply increased grain protein concentration linearly while grain yield response to added nitrogen had a diminishing return relationship. They also found that when nitrogen was very limiting, small nitrogen additions resulted in greater grain yield with decreased protein concentration caused by dilution of the plant nitrogen. However, at higher levels of nitrogen, which are far more common, grain and protein yields usually increased while the grain protein concentration increased as well.

Goos et al. (1982) determined protein threshold values of 120 g protein per kg grain for winter wheat and later 140 g protein per kg grain in spring wheat (Goos, 1984). If wheat protein concentration was below these threshold values, the yield of the crop was limited by nitrogen. Higher protein concentrations than the threshold values could be obtained by increasing nitrogen fertilization. At this level of fertilization, however, only protein concentration increased, whereas no significant effects on crop yield were observed (Christensen and Killhorn, 1981; Fowler and de la Roche, 1984). Foliar applications of nitrogen and fertilization late in the growing season tended to have a greater effect on increasing grain protein concentration than on increasing grain yield (Altman et al., 1983; Gallagher et al., 1973).

Wheat protein is high in glutamic acid and proline, whereas lysine, threonine, methionine, and cystine concentrations are lower than those recommended by the World Health Organization (WHO) (Simmonds, 1978). Kies and Fox (1974) determined that lysine was the most limiting amino acid for human nutrition in wheat protein. The composition of amino acids in wheat protein are affected by nitrogen fertilization. Poulsson (1973) and Timms et al. (1981) found that nitrogen fertilization increased the proportion of glutamine, proline, and phenylalanine in wheat protein, while threonine, serine, glycine, alanine, valine, and sulfur amino acids decreased. The balance between nitrogen and sulfur nutrition of the wheat crop also had an effect on grain amino acid composition. Byers and Bolton (1979) and Wrigley et al. (1980) observed that high levels of nitrogen supply and marginal levels of sulfur supply resulted in marked reductions of the concentration cysteine and methionine. Timms et al. (1981) observed the similar reductions caused by a nitrogen to sulfur imbalance induced by late-season fertilization of wheat with urea. These changes in amino acid composition affected the quality of the gluten. Timms et al. (1981) found their study that bread baked with flour with the highest protein concentration had a lower loaf volume and a lower texture score.

Almost 80% of the kernel protein is found in the endosperm (Vogel et al., 1976). Eighty-five percent of the protein in the endosperm is storage protein. The gliandin (alcohol soluble) fraction and the glutenin (alkali or acid soluble) fraction are the two forms of storage protein. When protein increases in a given cultivar, the gliandin fraction increases, while the glutenin fraction decreases (Ulmer, 1973). Shorter baking mixing times due to this shift in storage protein composition are a consequence of increasing protein levels in wheat (Finney et al., 1987). In a study conducted in Nebraska, Blumenthal et al. (2000) found that increased levels of protein caused by nitrogen fertilization resulted in shorter baking mixing times of flour of 12 different winter wheat varieties (Table 1).

1.3 Rice

Rice is the primary source of dietary energy and protein for nearly three billion people in Asia. Head rice is more valuable than brokens and consumers prefer translucent, white grains. Nitrogen nutrition has a large influence on rice grain quality through effects on milling characteristics, translucence and color.

Most rice is consumed as “white rice”. White rice is obtained after dehulling the rough rice, as harvested from the field, to produce brown rice, which undergoes abrasive milling to remove the outer layers of pericarp, seed-coat and nucellus, the germ or embryo, and aleurone layers (Juliano, 1993). After milling, white rice represents about 70% of the original yield of rough rice.

Table 1. Effect of fertilizer nitrogen on grain protein and dough mixing time (mixograph) of winter wheat (Blumenthal et al., 2000).

Nitrogen fertilizer treatment† (kg N ha^{-1})	Grain protein concentration (g kg^{-1})	Time to peak resistance (min.)
0	104††	4.73††
15	107	4.64
30	110	4.44
45	113	4.39
60	117	4.23
75	121	4.11
$LSD_{0.05}$	4	0.21

† Fertilizer treatments were ammonium nitrate applied topdress in early spring.
†† Values are means of 12 winter wheat varieties, 4 years, and 4 locations each year.

Crop nitrogen supply and plant nitrogen status have a marked effect on final protein content of both the brown and white rice. Early work with improved rice varieties conducted in the 1960s and 1970s demonstrated that grain protein could be significantly increased by ensuring adequate nitrogen fertilizer application up to panicle initiation stage (Nangju and De Datta, 1970; De Datta et al., 1972). Subsequent field research has found that improved fertilizer-N use efficiency to achieve both high yields and high grain quality, requires careful attention to the rate and timing of N fertilizer applications such that the total available nitrogen supply from soil and fertilizer is congruent with crop nitrogen demand. In high-yield production systems, improved congruence between nitrogen supply and crop demand sometimes requires several split applications—including a final nitrogen topdressing at flowering stage (Perez et al., 1996). In one field study with transplanted rice in the Philippines, an additional nitrogen topdressing at flowering resulted in a 6% increase in rough rice yield and a 25% increase in grain protein compared to a treatment that received all of the nitrogen fertilizer by panicle initiation stage (Table 2). Although the rate and timing nitrogen treatments are confounded in this study, additional studies have consistently demonstrated the effectiveness of nitrogen application at flowering to optimize both yield and grain protein in high yield systems where soil nitrogen is not sufficient to meet crop nitrogen demand during grainfilling.

An increase in rice grain protein is primarily reflected in greater amounts of the storage proteins glutelin and prolamin (Cagampang et al., 1966). These proteins are located in protein bodies within the starchy endosperm. When total grain protein is less than 10%, an increase in grain protein causes little reduction in lysine content. Most commercial rice has grain protein below 10%. Therefore, protein quality should not be significantly reduced when N management increases grain protein up to this 10% threshold.

The influence of nitrogen management on rice milling properties are as important as the effects of plant N nutrition on grain protein. Increased grain protein makes brown rice more resistant to cracking and breakage during abrasive milling than low protein rice of the same variety. Typically there is a tight, positive correlation between grain protein content and head rice yield. For example, in the previously cited study, head rice percentage was 58% of the rough rice for high-protein rice that was obtained when nitrogen fertilizer was applied at flowering versus only 47% head rice when the last nitrogen topdressing was applied at panicle initiation (Table 2). Taken together with the 6% increase in yield of rough rice, with an nitrogen application at flowering, head rice yield increased by 30% and head rice protein yield increased by 63% compared to the treatment without nitrogen application after panicle initiation.

Grain protein also has a large impact on other quality traits. Translucency is often positively correlated with protein content of milled rice, while grain whiteness can decrease as grain protein increases (Cagampang et al., 1966). Through appropriate nitrogen fertilizer management, however, it is possible to increase protein content and translucency while maintaining whiteness within acceptable limits. Such was the case in the Perez et al. (1996) study in which translucency increased from 76.4 to 85.5% with an N application at flowering stage (Table 2). While whiteness remained above 40% (data not shown), which is comparable to milled rice whiteness in most Asian and European markets. Although grain protein content also affects other quality traits related to palatability, such as stickiness and chewiness (Hamaker, 1994), preferences for these traits vary widely in different regions and countries.

Table 2. Effect of rate and timing of nitrogen fertilizer application on rice yield and quality characteristics. Modified from Perez *et. al.* 1996.

N Fertilizer Treatment † (kg ha^{-1})					Rough Rice	Head Rice	Head Rice	Protein	Translucency
PP	MT	PI	FL	Total	Yield (kg ha^{-1})	(%)	Yield (kg ha^{-1})	(%)	(%)
0	0	0	0	0	5.3c	37.5c	2.0c	5.6c	58.2c
120	0	60	0	180	9.3b	47.1b	4.4b	7.6b	76.4b
60	60	60	45	225	9.9a	57.7a	5.7a	9.6a	85.5a

† Nitrogen fertilizer timing designations: PP = preplant, MT = maximum tillering, PI = panicle initiation, FL = flowering.

†† Means in the same column followed by different letters are statistically different at $P < 0.01$ by Duncan's multiple range test.

2. OIL SEED CROPS

Oil seed crops are recognized as those whose oil is the most valuable component of the seed, being utilized for both edible and industrial purposes. There is also considerable vegetable oil produced as a byproduct of extraction for other components as is the case with corn oil. Oil serves primarily as a source of energy and carbon precursors in germinating seed. Synthesis of storage lipids occurs in the seed and, thus, oil composition is genetically determined by the embryo, and relative weight of the embryo to endosperm and seed coat determines oil content. It is generally accepted that there is a negative relationship between protein and oil content. Oil and protein constituents are synthesized at different rates and times during oilseed development. Variation in nitrogen fertility during seed development and maturation affect the synthesis of fatty acids and, therefore, their final proportions in the oils of mature seeds. Since not only oil composition, but oil content as well is affected by nitrogen fertility, this can affect oil utilization and the value of specific oilseed crops. While nitrogen limiting situations generally reduce total oilseed production and, hence, oil yield per acre, there are few instances where the crop quality is actually reduced by inadequate nitrogen availability.

2.1 Sunflower

Sunflower (*Helianthus annuus* L.) is capable of considerable translocation of N from leaves to maturing seeds (Cheng and Zubriski, 1978; Mathers and Stewart, 1982; Hocking and Steer, 1983). Numerous trials have been conducted on response to N fertility, but the response estimate proposed by Danke et al. (1981) seems to be rigorous enough to fit most other reports. It is based on the equation: N fertilizer applied = 0.05 * Yield goal - (Nitrate nitrogen in a 150-cm soil sample). Higher levels of N fertility can affect seed quality by increasing protein and decreasing oil concentrations (Steer et al., 1984, 1986). The increase in yield response to N fertilizer, however, resulted in an overall increase in oil yield. Geleta et al. (1997) found both a significant decrease in oil content and return per acre as N rates increased from 40 to 115 kg/ha. They broke the 37 site years into three groups (high, medium and low yielding) and found a similar response for all three groups. Besides N fertilizer effects on oil concentration, Steer and Seiler (1990) found that the application of N fertilizer before floret initiation increased the concentrations of palmitic and linoleic acids, but decreased those of stearic and oleic acids. On the contrary, Bahl et. al. (1997) found a decrease in oil content with additional N, but an increase in the ratio of oleic acid:linoleic acid. The discrepancy may have to do with N P interactions, since the studies varied in their P treatments. This information may be helpful in managing mid-oleic sunflowers that are currently increasing in market share.

2.2 Rapeseed

Rapeseed (*Brassica napus* L.) also tends to have reduced oil content with high N fertility (Krogman and Hobbs, 1975; Holmes and Ainsley, 1979). As glucosinolates are synthesized from sulfur amino acids, glucosinolate content is affected by both nitrogen and sulfur availability and the nitrogen:sulfur ratio should be taken into account when assessing the effect of nitrogen on glucosinolate content. A plentiful supply of both elements can result in high levels of glucosinolates and, according to Grant and Bailey (1993), the optimal ratio of nitrogen to sulfur is 12 at flowering time. High glucosinolate content has been recorded after restricted nitrogen fertilizer application: the effect of nitrogen on seed glucosinolate content varied from one year (Bilsborrow et al., 1993). Forster (1978) found that in pot studies N rate lowered seed oil content and increasing K content, at the high N rate, increased oil content. He also found an increase in glucosinolate and protein content with increasing N rates.

A negative correlation between oil and protein content in the traditional determination of seed quality is well documented. High nitrogen applications reduce oil content and increase protein content. The economic value of the oil has led to reduced nitrogen usage and protein content of the meal has been decreasing.

2.3 Soybean and Peanuts

Soybeans (*Glycine Max.* (L.) Merr.) and peanuts (*Arachis hypogaea* L.), both nodulating legumes, are still responsive under many conditions to increasing N levels in terms of yield. However, the oil content appears to be less negatively impacted by N rates (Pawar et al., 1982; Bishnoi and Dutt, 1980; Abdel et al., 1988; Nagre et al., 1991). However, there are still several reports of decreasing oil concentration at higher N rates (Hassan et al., 1985; Jadhav et al., 1994). Ohyama et al. (1994) report on variation in amino acid ratios in soybeans under N limiting and adequate conditions, with aspartic acid being higher under the N limiting conditions.

3. ROOT AND TUBER CROPS

3.1 Sugarbeet

For thousands of years, sugar, as sucrose is commonly called, has been a valuable part of human diet. Sucrose is an important component or additive to a broad range of foods, beverages, and pharmaceuticals. Two-thirds of consumed sugar stems from sugarcane (*Saccharum officinarum* L.) and one-third of consumed sugar stems from sugarbeet (*Beta vulgaris* L.). Sugarcane is produced in tropical and subtropical climates, whereas sugarbeet is produced in temperate climates.

For the last fifty years, refined nitrogen nutrition has been a major focus of agronomic practices in growing sugarbeet. There are several reasons for this. Firstly, soil supply of mineral nitrogen forms is often not sufficient for optimal crop growth (biomass) and a producer has, through addition of nitrogen fertilizers, immediate control over the nitrogen economy of the crop; secondly, correct nitrogen nutrition is crucial for the quality of the crop; and thirdly, losses of nitrogen from cropping systems have been implicated in the contamination of surface and groundwater

Once harvested, sugarbeet will undergo a series of processing steps until almost pure sugar (>99.9% sucrose) is obtained. Beets are stored for a period of up to several months. At the factory, beets are flumed, washed, and sliced in thin strips. Raw juice is obtained from these strips by counter-extraction (diffusion) with hot water. Raw juice is submitted to several purification and concentration processes. From concentrated juice sucrose is harvested, crystallized under vacuum, in several steps involving repeated dissolution of non-extactable sugars in order to gain as much sucrose as possible. In the end of these processes, there are basically three components left of the sugarbeet crop: sucrose, molasses, and beet pulp. Molasses is the syrup that contains all the components from which no more sugar can be extracted economically and beet pulp is the insoluble beet tissue left over from the process of counter-extraction. Both, molasses and pulp are commonly sold as animal feed.

While adequate nitrogen supply to the crop must be insured to obtain optimal root yield, excessive nitrogen supply can have several detrimental effects on sugarbeet quality. Excessive nitrogen supply, especially late in the growing season, has in general two main effects on the quality of the harvested beets: 1) it decreases the concentration of sucrose in the beets; and 2) it increases the impurities of the juice from which sugar is extracted.

The decrease in sucrose concentration is mainly caused by dilution, i.e. the roots retain more water, which in turn reduces the concentration of sucrose per unit fresh matter (Wieninger and Kubadinow, 1973). Abundant nitrogen supply also reduces the amount of marc (Wieninger and Kubadinow, 1973). Marc is the insoluble part of the beet root. It is the main component of beet pulp. As such, it is the second most valuable product of sugarbeet. A secondary component of the reduction of marc by oversupply of nitrogen, is that under such conditions beet roots tend to be heavier and contain relatively less marc (Beiss, 1988). Under high nitrogen supply, the physical strength of the beet tissue is lower (Bürcky et al., 1978; Drath et al., 1984). This is undesirable for two reasons. Firstly, lower tissue strength results in higher losses during harvesting and handling of beets due to breakage and associated losses by invasions of pathogen through wounds. Secondly, lower physical strength of the root tissue also causes problems during slicing of beets. Strips obtained from such roots often are not optimally conducive to sucrose extraction process.

Juice purity is reduced by excessive nitrogen through increases in alpha-amino-nitrogen (Wiklicky, 1971; Burba et al., 1984), invert sugars, lime salts (Reinefeld and Baumgarten, 1975), and raffinose (Burba and Nitzschke, 1980). Sugarbeets take up most of the nitrogen (150-170 kg N / ha) before row closure of the canopy. After row closure, they have only modest demand for additional nitrogen (about 1 kg N / ha / day). When nitrogen uptake exceeds demand, the concentration of nitrogenous impurities, especially alpha-amino acids, increases. At high levels, they hurt the crystallization process of sugar, thereby, rendering sugar extraction less efficient (Armstrong and Milford, 1985). Invert sugars (glucose and fructose) are obtained by acid or enzyme (invertase) hydrolysis of sucrose. High levels of invert sugars means losses of sucrose. Additionally, high levels of invert sugars cause discoloration of white sucrose. Invert sugars are also the primary cause of increased lime salts (K, Na) (Oldfield et al., 1971). Lime salts in turn increase the amount of sucrose lost to molasses. High levels of raffinose can decrease the crystallization rate of sucrose and impair sucrose crystal morphology (Vaccari et al., 1986).

3.2 Potatoes

Many production variables influence the quality of potatoes (*Solanum tuberosum* L.) for processing into chips and fries, and for the count-carton fresh market (Pavlista, 1995). Beside fertilization, cultivar selection, seed quality, planting density, soil moisture (irrigation), timeliness of farming operations, crop rotations, vine desiccation, handling, and storage are controllable production variables that effect quality (Pavlista and Ojala, 1997). Processors require high quality for chips and fries with desirable color, flavor, texture, and appearance (Gould and Plimpton, 1985). Nitrogen fertilization has a key impact on potato quality for processing, as well as influencing potato size for the fresh market. Not only is tuber size affected by nitrogen but also starch content, estimated by specific gravity, and sugar content are greatly affected (Salunkhe et al., 1991; Talbert and Smith, 1975).

Soils used for potato production nearly always require nitrogen to be added for optimum yields and quality. Nitrogen is typically added at the beginning and the middle of the season. It is often applied before planting or during early vegetative stages of growth. Nitrogen may be additionally supplied through irrigation systems during early tuber growth. Petiole nitrate-N is often monitored during vine and early tuber growth, and guidelines are available for many cultivars. Adequate nitrogen will usually improve tuber quality, size, maturity, dry matter and sugar content (Table 1).

Tubers from plants with low nitrogen are usually small with low dry matter content and high reducing sugar levels (Iritani and Weller, 1980; Sowokinos and Preston, 1988). Excessive nitrogen delays tuber initiation, while promoting excessive vine growth. Too much nitrogen results in tubers with a lower dry matter content and immature skin prone to bruising and susceptibility to shatter bruise (Dean and Thornton, 1992). Proper nitrogen fertilization of potato crops is important to achieve optimum quality for processing.

Table 3. Influence of nitrogen levels of potato crops on tuber quality.

Deficient	Adequate	Excess
Very Small Tubers	Optimal-sized Tubers	Slightly Small Tubers
High Sugar Levels	Low Sugar Levels	High Sugar levels
Low Dry Matter	High Dry Matter	Medium Dry Matter *
Few Useables	Most Useable	Many Useables
Over-mature Tubers	Mature Tubers	Immature Tubers *
Disease Susceptible	Disease Resistant	Disease and Bruise Susceptible

* Phosphorus fertilizer application will help to improve quality (e.g., skin maturity and dry matter content) of tubers at harvest when potato crops have excessive nitrogen fertility levels.

Quality processed potato chips and French fries can only be made from high-quality raw product. The major attributes important for potato chip and French fry processing include optimum size and shape, minimal external and internal defects, and desirable dry matter and sugar contents. These attributes determine finished product quality, e.g., size or length, color, texture, uniformity, and desired appearance (Gould and Plimpton, 1985). Two key quality characteristics affected by nitrogen are specific gravity (dry matter content) and reducing sugar (glucose) content.

Dry matter content varies from 15 to 24% between potato types and cultivars (Pavlista, 1997). Potatoes used for potato chips and French fries require a narrow range in dry matter content. Higher dry matter content (higher specific gravity) in raw product improves recovery rate during processing and directly influences texture and appearance, and indirectly influences the color of potato chips and French fries. Since specific gravity has a near-linear relation with dry matter and starch content in tubers, it is the common way for measuring these quality characters.

For potato chips, specific gravity is critical because it affects the thickness, color, crispness, oiliness, and taste of the product. It also affects the recovery rate of chips from raw potato tubers. The preferred specific gravity for chipping potato tubers is between 1.090 and 1.099. Higher shear forces, resulting from high specific gravity, increase the mealiness and dryness giving the chip its characteristic "snap" or brittleness. The principle component of texture is starch content. The lower the specific gravity of the tuber, the more fragile are the final potato chips and the more likely they are to break. This breakage is called "hash" in packages.

For French fries, specific gravity has more influence on the quality of the finished fries and the production efficiency of the processing plant than any other single tuber quality. Specific gravity determines the French fry's texture, appearance, recovery rate, oil usage, and energy usage by the processing plant. Potatoes processed into French fries should be within a range of specific gravity, 1.080 to 1.089, to produce fries with excellent texture and appearance. High specific gravity does not guarantee the best texture, but low specific gravity does guarantee that the product will have an inferior consistency or texture.

North American consumers prefer potato chips and French fries to have a uniform light color. Light brown, or darker color, chips are usually not well received. Finished product color in potato chip and French fry processing is determined particularly by glucose content. For low glucose content, proper maturation before harvest is essential for allowing reducing sugars to reach low levels in tubers. During frying, the critical chemical reactions that causes darkening of potato chips and French fries is non-enzymatic browning (Maillard reaction) during frying.

The brown color caused by the Maillard reaction is due to polymerized cyclamen. Glucose in potato tissues reacts with asparagine, an amino acid, to produce cyclamen which then polymerizes during frying to form a discoloration. Asparagine is usually present at a sufficiently high concentration for this reaction, so, glucose is the limiting substrate. Glucose levels higher than about 0.35 mg/gram of fresh potato will result in brown product.

4. FORAGES

Forages are an important component of agricultural systems. In the United States more than 50% of agricultural land is kept in grassland pasture and cropland used only for pasture (USDA, 1992). The forage resource is the backbone of animal meat production.

Nitrogen is the most limiting factor for grassland productivity. Hence, application of nitrogen fertilizers profoundly affects grassland systems in many ways. At low rates, nitrogen fertilization increases forage yield with little effect on forage nitrogen. It stimulates tiller development, increases leaf size, and lengthens the period of green leaves (Rhykerd and Noller, 1974). At a higher level of nitrogen fertilization, yield and nitrogen concentration in the forage are increased. Increased water concentration and decreased soluble carbohydrates are commonly observed after nitrogen fertilization of grasses (Messman at al., 1991; Brink and Fairbrother, 1992).

Most of the nitrogen taken up by the forages is incorporated into protein. However, 10-30% of the absorbed nitrogen exists in the plant as non-protein nitrogen, mainly amino acids, nitrate, and secondary plant metabolites (Dougherty and Rhykerd, 1985). Accumulation of nitrates within forages can be a problem under conditions of high nitrogen supply and impaired photosynthesis, e.g. low level of solar radiation or drought stress (Stritzke and McMurphy, 1982). Animal health problems are a consequence when nitrate is reduced to nitrite in the rumen and absorbed in the bloodstream. A wide range in critical forage nitrate concentration has been observed. Dougherty and Rhykerd (1985) reported that a nitrate concentration as low as 0.1% can cause problems, whereas Kemp and Geurink (1978) observed problems only at the much higher concentration of 6%. Low levels of soluble carbohydrates may cause problems with silage fermentation, resulting in poor silage quality.

Nitrogen application will also increase the concentration of total organic acids and alkaloids in forage. Wedin (1974) and Odom et al. (1980) suggested that increased alkaloid concentration in reed canary grass (*Phalaris arundinacea* L.) and tall fescue (*Festuca arundinacea* Scherb.) may be responsible for lower forage palatability, resulting in lower forage intake. Bush et al. (1979) observed that accumulation of perloline in tall fescue was a contributing factor to summer toxicosis syndrome.

Nitrogen fertilization also has an impact on mineral composition of the forage. If nitrogen is supplied in the ammonium form, uptake of cations such as calcium, potassium and magnesium is reduced (George and Thill, 1979). By lowering magnesium concentration, ammonium fertilization may induce grass tetany. Other effects of nitrogen fertilization on mineral composition of forages are inconsistent. Whitehead et al. (1986) found lower concentrations of macro and micronutrients, caused mainly by dilution, whereas Reid and Jung (1974) reported little effect of nitrogen fertilization on mineral composition of forage.

Nitrogen fertilization of forages often had little impact on forage digestibility. Van Soest (1982) reported that as nitrogen fertilization increases lignin production, digestibility may be reduced. Others (Messman et al., 1991; Puoli et al., 1991) found that the rate of neutral detergent fiber digestion and with it animal feed intake was increased upon nitrogen fertilization. Nitrogen fertilization of forages was more likely to increase digestibility of warm season grasses because of their lower nitrogen content as compared to cool season grasses (George and Hall, 1983). The main beneficial effect of nitrogen fertilization of forages, on meat production per unit land, is increases in yield resulting in increased carrying capacity (Wedin, 1974). In warm season grasses, many studies found that increased nitrogen fertilization caused higher beef gains in kg/ha (Rhem et al., 1977; Perry and Baltensperger, 1979).

REFERENCES:

Abdel-Wahab, A.M., G.M. Yakout, A.A. Mohamed, and H.M. Abd-El-Motaleb. 1988. Effect of different Levels of Nitrogen, Phosphorus, Potassium and Calcium on Peanut. Egyptian J. Agron. 11:79-92.

Alexander, D.E. 1988. Breeding special nutritional and industrial types. p. 869-880. In: C.F. Sprague and J.W Dudley (eds.) Corn and Corn Improvement, 3rd Edition. American Society of Agronomy, Madison, WI.

Altman, D.W., W.L. McCuistion, and W.E. Kronstad. 1983. Grain protein percentage, kernel hardness, and grain yield of winter wheat with foliar applied urea. Agron. J. 75:87-91.

Anderson, E.L., E.J. Kamprath, and R.H. Moll. 1984. Nitrogen fertility effects on accumulation, remobilization, and partitioning of N and dry matter in corn genotypes differing in prolificacy. Agron. J. 76:397-404.

Armstrong, M., and G. Milford. 1985. The nitrogen nutrition of sugar beet: the background to the requirement of sugar yield and amino-N accumulation. British Sugar Beet Rev. 53 (4): 42-44.

Bahl, G.S., N.S. Pasricha, and K.L Ahuja. 1997. Effect of Fertilizer Nitrogen and Phosphorus on the Grain Yield, Nutrient Uptake and Oil Quality of Sunflower. J. Indian Soc. Soil Sci. 45:292-296.

Bauer, P.J. and P.R. Carter. 1986. Effect of seeding date, plant density, moisture availability, and soil nitrogen fertility on maize kernel breakage susceptibility. Crop Sci. 17:362- 366.

Beiss, U. 1988. Influence of some factors on marc content of sugar beet. Zuckerindustrie. 113:1041-1048.

Benzian, B. and P. Lane. 1979. Some relationships between grain yield and grain protein of wheat experiments in south-east England and comparisons with such relationships elsewhere. J. Sci. Food Agric. 30:59-70.

Bhatia, C.R. and R. Rabson. 1976. Bioenergetic considerations in cereal breeding for protein improvement. Science 194:1418-1421.

Bhatia, C.R. and R. Rabson. 1987. Relationship of grain yield and nutritional quality. p. 11-44. In: R.A. Olson and K.I. Frey (eds.) Nutritional Quality of Cereal Grains: Genetics and Agronomic Management. American Society of Agronomy, Madison, WI.

Bilsborrow, P.E., E.J. Evans, and F.J. Zhao. 1993. The Influence of Spring Nitrogen on Yield, Yield Components and Glucosinolate Content of Autumn-sown Oilseed Rape (*Brassica napus*). J. Agric. Sci. 120:219-224.

Bishnoi, K.C., and R. Dutt. 1980. Effect of Rhizobium Isolates Inoculation Methods and Nitrogen Levels on Nodulation and Quality of Soybean. Indian J. Agron. 25:544-545.

Blumenthal, J.M., D.D. Baltensperger, D.R. Shelton, L.A. Nelson, G.D. Binford, and S. Geleta. 2000. Nitrogen and variety effects in winter wheat. Abstracts of the third International Crop Science Congress, Hamburg, Germany.

Breteler, H. 1976. Nitrogen fertilisation, yield and protein quality of a normal and high-lysine maize variety. J. Sci. Food Agric. 27:978-982.

Brink, G.B., and T.E. Fairbrother. 1992. Bermudagrass-subterranean clover response to nitrogen application. J. Prod. Agric. 5:591-595.

Bullock, D.G., P.L. Raymer, and S. Savage. 1989. Variation of protein and fat concentration among commercial corn hybrids grown in the southeastern USA. J. Prod. Agric. 12:157-160.

Burba, M., and U. Nitzschke. 1980. Nachweis und Eigenschaften der Saccharose Spaltenden Enzyme der Zuckerrübe. Zuckerindustrie. 105:149-155.
Burba, M., U. Nitzschke, and R. Ritterbusch. 1984. Die N-Assimilation der Pflanze unter besonderer Berücksichtigung der Zuckerrübe. Zuckerindustrie. 109:613-628.
Bush, L.P., J.A. Boling, and S. Yates. 1979. In: Tall Fescue. Agron Monogr 20. American Society of Agronomy, Madison, WI.
Bürcky, K. U. Beiss, C. Winner, L. Drath, and H. Schiweck. 1978. Versuch zur Bedeutung des Nährstoffangebotes für die Qualität der Zuckerrübe. II: Stickstoff und Kalium. Zuckerindustrie. 103:190-200.
Byers, M., and J. Bolton. 1979. Effects of nitrogen and sulfur fertilizers on yield, N and S content, and amino acid composition of the grain of spring wheat. J. Sci. Food Agric. 30:250-261.
Cagampang, G.B., L.J. Cruz, S.G. Espiritu, R.G. Santiago, and B.O. Juliano. 1966. Studiens on the extraction and composition of rice proteins. Cereal Chem. 43:145-155.
Cheng, S.F., and J.C. Zubriski. 1978. Effects of nitrogen fertilizer on production of irrigated sunflower, plant uptake of nitrogen, and on water use. p. 400-409. In: Proc. 8th Int. Sunflower Conf., Minneapolis, MN 23-27 July. International Sunflower Association, Paris, France.
Christensen, N.W., and R.J. Killhorn.1981. Wheat and barley growth and N fertilizer utilization under sprinkler irrigation. Agron. J. 74:840-844.
Cromwell, G.L., M.J. Bitzer, T.S. Stahly, and T.H. Johnson. 1983. Effects of soil nitrogen fertility on the protein and lysine content and nutritional value of normal and opaque-2 corn. J. Anim. Sci. 57:1345-1351.
Dahnke, W.C., J.C. Zubriski, and E.H. Vasey. 1981. Fertilizing sunflowers. North Dakota State Univ. Coop. Ext. Serv. Circ. SF-713, Fargo, ND.
De Datta, S.K., W.N. Obcemea, and R.K. Jana. 1972. Protein content of rice grain as affected by nitrogen fertilizer and some triazines and substituted ureas. Agron. J. 64:785-788.
Dean, B.B. and R.E. Thornton. 1992. The specific gravity of potatoes. Washington State Univ. Cooperative Extension Bulletin #1541, Pullman, WA.
Dougherty, C.T., and C.L. Rhykerd. 1985. The role of nitrogen in forage-animal production. p. 318-326. In: ME Health, RF Barnes, and DS Metcalfe (eds.) Forages: The Science of Grassland Agriculture, 4th ed. Iowa State Univ. Press, Ames IO.
Drath, L., R. Strauss, and H. Schiweck. 1984. Untersuchungen über die mechanischen Eigenschaften von Zuckerrüben. II: Einflussfaktoren auf die Bruchfestigkeit von Rüben. Zuckerindustrie. 109:993-1007.
Earle, F.R. 1977. Protein and oil in corn: Variation by crop years from 1907 to 1972. Cereal Chem. 54:70-79.
Finney, K.F., W.T. Yamazaki, V.L. Youngs, and G.L. Rubenthaler. 1987. Quality of hard, soft, and durum wheats. p. 677-748. In: E.G. Heyne (ed.) Wheat and wheat improvement, second ed. American Society of Agronomy, Madison, WI.

Forster, H. 1978. Influence of N and K fertilizers on the Quality and Yield of Oil from Old and New Varieties of Rapeseed (*Brassica napus ssp. oleifera*). p. 305-310. In Fertilizer use and production of carbohydrates and lipids.

Fowler, D.B., and I.A. de la Roche.1984. Winter wheat production on the north central Canadian prairies: protein quality classes. Crop Sci. 24:873-876.

Frey, K.J. 1951. The inter-relationships of protein and amino acids in corn. Cereal Chem. 28:123-132.

Frey, K.J. 1977. Proteins of oats. Zeitschrift fur Pflanzenzüchtung 78:185-215.

Frey, K.J., B. Brimhall, and G.F. Sprague. 1949. The effects of selection upon protein quality in the corn kernel. Agron. J. 41:399.

Gallagher, P.J., L.S. Murphy, C.L. Harms, and W.A. Moore. 1973. Comparisons of effects of nitrogen carriers, rates, and time of application on yield and quality of winter wheat. Kansas Fert. Res. Rep. Prog. 202:59-60.

Geleta, S., D.D. Baltensperger, G.D. Binford, and J.F. Miller. 1997. Sunflower Response to Nitrogen and Phosphorus in Wheat-Fallow Cropping Systems. J. Prod. Agric. 10:466-472.

George, J.R., and K.E. Hall. 1983. Herbage quality of three warm-season grasses with nitrogen fertilization. Iowa State J. Res. 58:247-259.

George, J.R., and J.L. Thill. 1979. Cation concentration of N- and K-fertilized smooth bromegrass during the spring grass tetany season. Agron. J. 71:431-436.

Goos, R.J., D.G. Westfall, A.E. Ludwick, and J.E. Goris. 1982. Grain protein content as an indicator of N sufficiency for winter wheat. Agron. J. 74:130-133.

Goos, R.J. 1984. Post-harvest evaluation of nitrogen management - a new approach for "selling" soil testing to wheat farmers. J. Agron. Educ. 13:103-106.

Gould, W.A. and S. Plimpton. 1985. Quality evaluation of potato cultivars for processing. North Central Regional Research Publication #305.

Grant, C.A. and L.D. Bailey, 1993. Fertility management in canola production. Canadian J. Plant Sci. 73:651-670.

Hamaker, B.R. 1994. The influence of rice protein on rice quality. p. 177-193. In: W.E. Marshall and J.I. Wadeworth (eds.) Rice Science and Technology. Marcel Dekker, New York.

Hancock, J.D., E.R. Peo Jr., A.J. Lewis, K.R. Kniep, and S.C. Mason. 1988. Effects of irrigation and nitrogen fertilization of normal and high lysine corn on protein utilization by the growing rat. Nutr. Rep. Intern. 38:413-422.

Hassan, R.A., E.M. El-Hadidi, and S.M. Hassan. 1985. Oil, Protein, Amino Acid Content and Yield of Soybean seeds as affected by N and P Fertilization. J. Agric. Sci. 10:1, 53-57.

Hocking, P.J., and B.T. Steer. 1983. Uptake and partitioning of selected mineral elements in sunflower (*Helianthus annuus* L.) during growth. Field Crops Res. 6:93-107

Holmes, M.R.J. and A.M. Ainsley. 1979. Nitrogen Top-dressing Requirements of Winter Oilseed Rape. J. Sci. Food Agric. 30:119-128.

Iritani, W.M. and L.D. Weller. 1980. Sugar development in potatoes. Washington State Univ. Cooperative Extension Bulletin #0717, Pullman, WA.

Jadhav, P.J., A.S. Jadhav, and S.M. Bachchhav. 1994. Effects of N, Row Spacing and Plant Densities on the yield and Quality of Soybean. J. Maharashtra Agric. Univ. 19:75-77.

Jellum, M.D., F.C. Boswell, and C.T. Young. 1973. Nitrogen and boron effects on protein and oil of corn grain. Agron. J. 65:330-331.

Juliano, B.O. 1993. Rice in human nutrition. FAO Food Nutr. Ser. No. 26. International Rice Research Institute, Manila, Philippines.

Kemp, A., and J.H. Geuring. 1978. Grassland farming and minerals in cattle. Netherlands J. Agric. Sci. 26:161-169.

Kies, C., and H.M. Fox. 1974. Triticale and human nutrition. p. 202-211. In C.C. Tsen (ed.) Triticale: first man-made cereal. American Association of Cereal Chemists, St. Paul, MN.

Kniep, K.R. and S.C. Mason. 1991. Lysine and protein content of normal and opaque-2 maize grain as influenced by irrigation and nitrogen. Crop Sci. 31:177-181.

Kniep, K.R. and S.C. Mason. 1989. Kernel breakage and density of normal and opaque-2 maize grain as influenced by irrigation and nitrogen. Crop Sci. 29:158-163.

Krogman, K.K., and E.H. Hobbs. 1975. Yield and Morphological Response of Rape (*Brassica campestris L.Cv. Span*) to Irrigation and Fertilizer Treatments. Can. J. Plant Sci. 55:903-909.

MacGregor, J.M., L.T. Taskovitch, and W.P. Martin. 1961. Effect of nitrogen fertilizer and soil type on the amino acid content of corn grain. Agron. J. 53:211-214.

Mathers, A.C., and B.A. Stewart. 1982. Sunflower nutrient uptake, growth, and yield as affected by nitrogen or manure, and plant population. Agron. J. 74:911-915.

McDermitt, D.K. and R.S. Loomis. 1981. Elemental composition of biomass and its relation to energy content, growth efficiency, and growth yield. Ann. Bot. 48:275-290.

Mertz, E.T., L.S. Bates, and O.E. Nelson. 1964. Mutant gene that changes protein composition and increases lysine content of maize endosperm. Science 145: 279.

Messman, M.A., W.P. Weiss, and D.O. Erickson. 1991. Effects of nitrogen fertilization and maturity of bromegrass on in situ ruminal digestion kinetics of fiber. J. Anim. Sci. 69:1151-1161.

Nagre, K.T., S.B. Deshmukh, P.D. Bhalerao, and P.V. Thorve. 1991. Effect of Sowing Dates and Fertilizers on Growth, Yield and Quality of Soybean Varieties. PKV Res. J. 15:82-84.

Nangju, D. and S.K. De Datta. 1970. Effect of time of harvest and nitrogen level on yield and grain breakage in transplanted rice. Agron. J. 62:468-474.

Odom, O.J., R.L. Haaland, C.S. Hoveland, and W.B. Anthony. 1980. Forage quality response of tall fescue, orchardgrass, and phalaris to soil fertility level. Agron. J. 72:401-402.

Ohyama, T., N. Ohtake, T. Nishiwaki, K. Mizukoshi, R. Minagawa, K. Kobayashi, T. Chinushi, and Y. Takahashi. 1994. Effect of N Fertilization on Seed Quality and Xylam Transport Forms in Nodulating and Non-nodulating Soybean Isolines. Bul. Fac. Agric. Niigata Univ. 46:57-70.

Oikeh, S.O., J.G. Kling, and A.E. Okoruwa. 1998. Nitrogen management effects on maize grain quality in the West Africa moist savanna. Crop Sci. 38:1056-1061.

Oldfield, J.F.T., J.V. Dutton, and H.J. Teague. 1971. The significance of invert and gum formation in deteriorated beet. Internat. Sugar J. 73:3-8, 35-40:66-68.

Olsen, R.A., K.D. Frank, E.J. Deibert, A.F. Dreier, D.H. Sander, and V.A. Johnson. 1976. Impact of residual mineral N in soil on grain protein yields of winter wheat and corn. Agron. J. 68:769-772.

Paulsen, M.R. and L.D. Hill. 1985. Corn quality factors affecting dry milling performance. J. Agric. Eng. Res. 31:255-263.

Pavlista, A.D. 1995. Potato production stages: Scheduling key practices. Univ. Nebraska Cooperative Extension Circular # 95-1249, Lincoln, NE.

Pavlista, A.D. 1997. Potato types, characteristics and their uses. Amer. Biology Teacher 59:26-29.

Pavlista, A.D. and J.C. Ojala. 1997. Potatoes: chip and fry processing. Chapter p. 237-284. In: Smith, Cash, Nip, and Hui (eds.) Processing Vegetables: Science and Technology. Technomics Publ. Co. Inc., Lancaster, PA.

Pawar, V.P., N.T. Sarnaik, and V.D. Sondge. 1982. Effect of Nitrogen and Phosphate on yield and quality of soybean (*Glycine max* (L.) Merrill) Varieties. J. Maharashtra Agric. Univ. 7:90.

Penning de Vries, F.W.T., A.H.M. Brunsting and H.H. van Laar. 1974. Products, requirements and efficiency of biosynthesis: A quantitative approach. J. Theor. Biol. 45:339-377.

Perez, Consuelo M, B. Juliano, S. Liboon, J.M. Alcantara, Jovencio, and K.G. Cassman. 1996. Effects of late nitrogen fertilizer application on head rice yield, protein content, and grain quality of rice. Cereal Chem. 73: 556-560.

Perry, W.P. 1988. Corn as a livestock feed. P. 941-963. In: C.F. Sprague and J.W Dudley (eds.) Corn and Corn Improvement, 3rd Edition. American Society of Agronomy, Madison, WI.

Perry, L.J., Jr., and D.D. Baltensperger. 1979. Leaf and stem yields and forage quality of three N-fertilized warm-season grasses. Agron. J. 71:355-358.

Pierre, W.H., L Dumenil, V.D. Jolley, J.R. Webb, and W.D. Shrader. 1977. Relationship of corn yield, expressed as a percentage of maximum and the N percentage in grain: I. Various N-rate experiments. Agron. J. 69.215-220.

Poulsson, E. 1973. Amino acid analyses of wheat grain and gain fractions. Sci. Rep. Agric. Univ. Norway. 52:1-18.

Puoli, J.R., G.A. Jung, and R.L. Reid. 1991. Effects of nitrogen and sulfur on digestion and nutritive quality of warm-season grass hays for cattle and sheep. J. Anim. Sci. 69:843-852.

Reid, R.L., and G.A. Jung. 1974. Effects of elements other than nitrogen on the nutritive value of forage p. 395-435. In: D.A. Mays (ed.) Forage fertilization. American Society of Agronomy, Madison, WI.

Reinefeld, E., and G. Baumgarten. 1975. Verarbeittungseigenschaften der Zuckerrübe in Abhängigkeit vom Stickstoffangebot. Zucker. 28:61-65.

Rendig, V.V. and F.E. Broadbent. 1979. Proteins and amino acids in grain of maize grown with various levels of N. Agron. J. 71:509-512.

Rhem, G.W., J.T. Nichols, R.C. Sorensen, and W.J. Moline. 1975. Yield and botanical composition of and irrigated grass-legume pasture as influenced by fertilization. Agron. J. 67:64-68.

Rhykerd, C.L., and C.H. Noller. 1974. The role of nitrogen in forage production. p. 416-424. In: D.A. Mays (ed.) Forage fertilization. American Society of Agronomy, Madison, WI.

Sabata, R.J. and S.C. Mason. 1992. Corn hybrid interactions with soil nitrogen level and water regime. J. Prod. Agric. 5:137-142.

Salunkhe, D.K., S.S. Kadam and S.J. Jadhav. 1991. Potato: Production, processing, and products. CRC Press.

Sander, D.H., W.H. Allaway, and R.A. Olson. 1987. Modification of nutritional quality by environment and production practices. p. 45-82. In: R.A. Olson and K.I. Frey (eds.) Nutritional Quality of Cereal Grains: Genetics and Agronomic Management. American Society of Agronomy, Madison, WI.

Shumway, C.R., J.T. Cothren, S.O. Serna-Saldivar, and L.W. Rooney. 1992. Planting date and moisture effects on yield, quality and alkaline-processing characteristics of food-grade maize. Agron. J. 32:1265-1269.

Simmonds, D.H. 1978. Structure, composition, and biochemistry of cereal grains. p. 105-137. In Y. Pomeranz (ed.) Cereals '78: better nutrition for the world's millions. American Association of Cereal Chemists, St. Paul, MN.

Sowokinos, J.R. and D.A. Preston. 1988. Maintenance of potato processing quality by chemical maturity monitoring (CMM). Minnesota Agricultural Experiment Station Bulletin #586-1988, St. Paul, MN.

Steer, B.T., P.D. Coaldrake, C.J. Pearson, and C.P. Canty. 1986. Effects of Nitrogen Supply and Population Density on Plant Development and Yield Components of Irrigated Sunflower (*Helianthus annuus L.*). Field Crops Res. 13:99-115.

Steer, B.T., P.J. Hocking, A.A. Kortt, and C.M. Roxburgh. 1984. Nitrogen Nutrition of Sunflower (*Helianthus annuus L.*): Yield Components, the Timing of Their Establishment and Seed Characteristics in Response to Nitrogen Supply. Field Crops Res. 9:219-236.

Steer, B.T., and F.J. Seiler. 1990. Changes in Fatty Acid Composition of Sunflower (*Helianthus annuus*) Seeds in Response to Time of Nitrogen Application, Supply Rates and Defoliation. J. Sci. Food Agric. 51:11-26.

Stritzke, J.F., and W.E. McMurphy. 1982. Shade and N effects on tall fescue production and quality. Agron. J. 74:5-8.

Talbert, W.F. and O. Smith. 1975. Potato processing. AVI Publications Co.

Timms, M.F., R.C. Bottomley, J.R.S. Ellis, and J.D. Schofield. 1981. The baking quality and protein characteristics of a winter wheat grown at different levels of nitrogen fertilization. J. Sci. Food Agric. 32:684-698.

Tsai, C.Y., H.L. Warren, D.M. Huber, and R.A. Bressan. 1983. Interactions between the kernel N sink, grain yield and protein nutritional quality of maize. J. Sci. Food Agric. 34:255-263.

Tsai, C.Y., I. Dweikat, D.M. Huber, and H.L. Warren. 1992. Interrelationship of nitrogen nutrition with maize (Zea mays) grain yield, nitrogen use efficiency and grain quality. J. Sci. Food Agric. 58:1-8.

Ulmer, R.L. 1973. Relationships of endosperm protein fractions from normal and high protein wheats to protein content, lysine content, quality, and nitrogen fertilization. Ph.D. diss. Univ. of Nebraska, Lincoln, NE (Diss. Abstr. 34:1835B)

USDA. 1992. Agricultural Statistics. US Gov. Print. Off., Washington, D.C.

Vaccari, G., G. Mantovani, G. Squaldino, D. Aquilano, and M. Rubbo. 1986. The raffinose effect on sucrose morphology and kinetics. Sugar Tech. Rev. 13: 133-178.

Van Soest, P.J. 1982. Nutritional ecology of the ruminants. Oregon O. and B. Books, Corvallis, OR.

Vasal, S.K., E. Villegas, M.Bjarnason, B. Gelaw, and P. Goertz. 1980. Genetic modifers and breeding strategies in developing hardendosperm opaque-2 materials. p37-71. In: W.G. Pollmer and R.H. Phipps (eds.) Improvement of quality traits for grain and silage use. Martinus Nijhoff Publishers, The Hague, Netherlands.

Vogel, K.P., P.J. Mattern, and G.W. Lenser. 1976. Modified milling procedure for separating endosperm and nonendosperm of wheat kernel for protein and lysine analysis. Agron. J. 68:686-690.

Watson, S.A. 1987. Measurement and maintenance of quality. p. 125-183. In: S.A. Watson and P.E. Ramstad (eds.) Corn: Chemistry and Technology. American Association of Cereal Chemists, St. Paul, MN.

Wedin, WF. 1974. Fertilization of cool-season grasses. p. 95-118. In: D.A. Mays (ed.) Forage fertilization. American Society of Agronomy, Madison, WI.

Welch, L.F. 1969. Effect of N, P, and K on the percent and yield of oil in corn. Agron. J. 61:890-891.

Whitehead, D.C., K.M. Goulden, and R.D. Hartley. 1986. Fractions of nitrogen, sulphur, phosphorus, calcium, and magnesium in the herbage of perennial ryegrass as influenced by fertilizer nitrogen. Anim. Feed Sci. Tech. 14:231-242.

Wieninger, L., and N. Kubadinow. 1973. Die Stickstoffdüngung und ihre Auswirkung auf technologische Qualitätsemerkmale der Zuckerrübe. Zucker. 26:65-70.

Wiklicky, L. 1971. The processing quality of sugar beet. Zucker. 21:667-672.

Wrigley, C.W., D.L. ducros, M.J. Archer, P.G. Downie, and C.M. Roxburgh. 1980. The sulfur content of wheat endosperm and its relevance to grain quality. Aust. J. Plant Physiol. 7:755-766.

Nitrogen in the Environment: Sources, Problems, and Management
R.F. Follett and J.L. Hatfield (Eds)
Published by Elsevier Science B.V.

Chapter 4. Utilization and Metabolism of Nitrogen by Humans

Jennifer R. Follett[a] and Ronald F. Follett[b]

[a] Nutrition Department, 3402 Meyer Hall, 1 Shields Avenue, Univ of California, Davis, CA, 95616, USA
[b] USDA/ARS, Soil-Plant-Nutrient Research Unit, PO Box E, Fort Collins, CO, 80522, USA

The need by humans to produce and consume food and other agricultural products is increasing. This need is directly related to increasing World population, demands for goods and services, and expectations. Nitrogen (N) is contained in all of the amino acids and proteins in the foods consumed by humans. The use of N to produce food and other products is generally increasing as human needs increase. Proteins are an essential component of the human diet because, unlike plants, humans are unable to utilize more simple forms of N and rely on food sources for protein which can then be digested to amino acids and used for protein synthesis in the body. Estimation of protein requirements for humans depends both on the content of essential amino acids and digestibility of the protein.

Although humans may not be able to utilize more simple forms of N for normal physiological functions such as energy and synthesis of proteins, they are exposed to and capable of absorbing other forms of N such as nitrate (NO_3^-) and nitrite (NO_2^-). Nitrate can be obtained from plant sources as well as from contaminated drinking water and is easily absorbed by the intestine. Nitrate itself is not generally considered as a health risk but may become a concern due to its conversion to NO_2^-. The main health risk associated with NO_3^- consumption is methemoglobinemia due to the conversion of NO_3^- to NO_2^-, which in turn can interact with hemoglobin leading to formation of methemoglobin, leading to oxygen deprivation of the cells. Secondary and related deleterious effects of exposure to NO_3^- include increased respiratory infections, inhibition of iodine uptake by the thyroid, and possible reproductive problems. Although concerns about NO_3^- and NO_2^- have been focused mainly on harmful effects there is growing acceptance for the beneficial effects of the related compound nitric oxide (NO). Nitric oxide is a free radical gas that acts as a messenger molecule for regulation of several systems including blood vessel dilation, hormonal and neurotransmission functions.

Another group of N-containing substances are N-Nitroso Compounds (NOC) which include nitrosamines and nitrosamindes. These compounds are important to consider in the human diet since they are capable of participating in DNA alkylation and appear to be among the most potent and broad acting carcinogens known. In addition to the effect of N-containing compounds on humans it is important to consider the impact of humans to contribute to N loading into the environment. We estimate the total intake and loss of N by humans in the US to be in the range of 0.8-1.0 million MT/year indicating that N excretion by humans has a large potential to impact the environment. Such impacts on the environment by N from humans are just as important to consider as those from livestock wastes, inefficient fertilizer-N use, or from other N sources. It is especially important to recognize the potentially serious environmental effects that may occur as the human population continues to grow.

1. INTRODUCTION

The need by humans to produce and consume food and other agricultural products is increasing. This need is directly related to increasing World populations, demands for goods and services, and expectations. With increased human demands, the need and use of nitrogen (N) to produce food and other products is increased because N is contained in all of the amino acids and proteins in the foods consumed and is essential for human survival and health. Diets in many parts of the World are protein deficient, often having an imbalance of essential amino acids. In other cases, the consumption of certain forms of N or the endogenous formation of harmful N-containing compounds can be harmful to human health. Finally, the N consumed by humans is also excreted and must be absorbed back into the environmental where it may have harmful effects. Too often, problems associated with N in the environment are dismissed as primarily resulting from agricultural production systems and inefficiencies in properly managing N source and their use. Numerous efforts continue to improve N-use efficiency, minimize losses and leakage of N into important water bodies, prevent natural resources degradation, and to understand the environmental impacts of N. However, it is also important to understand the important roles that the consumption of N-containing food products, the various forms of N in the human diet, and the potential for environmental impacts that can result from larger and larger amounts of excretory N as human populations grow. Amounts of N consumed and excreted by humans need to be quantified and acknowledged for their potential to impact the environment. Human numbers are increasing rapidly. The World's population reached a total of one billion people in 1900 and is currently projected at 6.1 billion (U.S. Census Bureau, 2001). Within the United States, the population increased from 76 million in 1900 to over 275 million people in 2000. As will be discussed later in this chapter the recommended dietary allowance (RDA, 1989) in the diet of a human adult weighing 70 kg is about 56 g of protein, or about 9 g of N/day or about 3.25 kg N/year (Wildman and Medeiros, 2000; RDA, 1989) The demand for protein, and thus the N it contains, is increasing dramatically. It is highly important to understand N intake and metabolism, beneficial and harmful effects on N in the human body, and the implications that can be derived about human dietary needs for N. The objective of this chapter is to discuss sources and forms of N in the diet, human requirements for N, metabolism, and potential harmful and/or beneficial effects of certain forms of N and N metabolites in the body on human health. In addition, estimates will be made of the amount of N consumed and excreted from food consumed by humans in the US and perspectives provided about World dietary N needs.

2. PROTEIN IN THE HUMAN DIET

Plants can obtain N from the soil through uptake of nitrate (NO_3^-) and/or ammonium (NH_4^+) ions by the root system. The NO_3^- can then be reduced to the NH_4^+ form and utilized by the plant to synthesize N-containing compounds (Fowden, 1981). Unlike plants, animals and humans are for the most part incapable of utilizing more simple forms of nitrogen and rely on food sources for pre-formed nitrogen compounds in the form of amino acids (Fowden, 1981). Amino acids are synthesized by plants and formed into proteins, which can then be ingested directly by animals and humans or indirectly through consumption of animal products such as milk, eggs, and meat.

2.1 Protein Digestion

Digestion of protein begins in the stomach first through denaturation of proteins by stomach acid. The denatured proteins can then be partially digested by a gastric enzyme called pepsin. Proteins are in most cases made up of more than one type of amino acid and the amino acids are connected by peptide bonds. Pepsin does not fully break proteins into single amino acids, as it is only capable of hydrolyzing peptide bonds involving aromatic amino acids such as phenylalanine, tryptophan, and tyrosine (Wildman and Medeiros, 2000; Wardlaw and Insel, 1996). The partially digested proteins or peptones enter the small intestine where they are acted upon by proteases secreted by the pancreas. The major pancreatic peptidases involved in protein digestion in the intestine are trypsin, chymotrypsin, and carboxypeptidase and result in formation of short peptides and free amino acids (Wardlaw and Insel, 1996). At the surface of the intestinal mucosal cells aminopeptidase will act on the peptides yielding individual amino acids and oligopeptides of 2-4 amino acids. The oligopeptides and amino acids are absorbed by the intestinal cells and finally broken down into free amino acids, which are transported through the portal vein to the liver where they can be used in protein synthesis.

Twenty different amino acids are required for synthesis of needed proteins in the body. Humans are capable of synthesizing amino acids, but are unable to synthesize 8 to 9 of the amino acids needed in the body, either at all or in sufficient levels for growth and maintenance. Cells in the human body can produce carbon skeletons to which amino groups from other amino acids can be added. The first step in catabolism of amino acids is the removal of the alpha-amino group. This amino group can then be incorporated into other compounds or excreted from the body. The removal of this amino group from most amino acids involves transfer to α-ketoglutarate. This results in the formation of an alpha keto acid (from the original amino acids) and the formation of glutamate. So alpha keto glutarate accepts the amino group and become glutamate. Glutamate can undergo transamination resulting in deamination for disposal of amino groups or it can donate its amino group to another carbon skeleton for the formation of non-essential amino acids in the body. Transamination is catalyzed by several enzymes classified as aminotransferases. Transamination is also important in excretion of amino acids brought into the urea cycle as glutamate. There are eight to nine essential amino acids that the body must get through the diet because the body either cannot make the carbon skeleton necessary for that specific amino acid, has no pathway for the addition of the amino group, or cannot process the amino acid in high enough concentration to meet the needs of the body (Wardlaw and Insel, 1996). These amino acids, referred to as essential amino acids, must be obtained from food sources and include isoleucine, leucine, lysine, methionine, phenylalanine, threonine, tryptophan, and valine. Infants additionally require histidine as an essential amino acid for growth and development (Wildman and Medeiros, 2000).

Protein synthesis is inhibited when there is a lack of only one amino acid in the diet. If a proper balance of the amino acids is not received protein synthesis will cease when the amino acid in the lowest concentration (limiting amino acid) is exhausted. This is known as the "all or none" law. If the limiting amino acid is depleted, then the remaining essential amino acids obtained from the diet will be broken down and be unavailable as a source for protein synthesis. Different types of food have different amounts of the amino acids needed for growth and maintenance. The biological value of food is the amount of nitrogen digested, absorbed, and used by the body but not excreted and reflects the distribution of amino acids found in the food and how well it meets the amino acid requirements of the individual (Wildman and Medeiros, 2000).

2.2 Sources of dietary protein

There is no one perfect food for protein requirements and just because a food source contains high concentrations of protein does not mean it has a high biological value. A protein with a high biological value, or a complete protein, is normally obtained from an animal source, such as meat and milk and would contain all essential amino acids in a distribution similar to what is required for growth and maintenance in the human (Wildman and Medeiros, 2000). Eggs show distribution of amino acids at mixtures most similar to what is required by the body and are used as the standard by which all other protein sources are measured. A protein source with a low biological value has all nine essential amino acids but not in a distribution necessary for normal growth and maintenance in humans. Protein sources with low biological values are also called incomplete proteins and most plants fall in this category (Wildman and Medeiros, 2000)

2.3 Protein quality

Unlike animal protein sources, plant protein sources are generally deficient in lysine, tryptophan, or methionine (Garrison and Somer, 1995). However, animal proteins do not need to be consumed to obtain a balance of all the essential amino acids. Incomplete protein sources can be combined to balance the limiting amino acids in another incomplete protein source. These two protein sources would be referred to as complementary proteins. For instance beans (low in methionine) can be combined with rice (low in lysine) so that a balance of all essential amino acids are consumed.

Healthy adults require approximately 15 % of the source of protein in their diets to be composed of the essential amino acids and a typical western diet supplies around 50 %. When obtaining essential amino acids through incomplete proteins, healthy adults do not need to consume both complementary proteins in the same meal but can consume them throughout the day. Infants and children require a higher percentage of essential amino acids (around 35 % of total protein) due to growth and development. If incomplete proteins are being consumed as the protein source in children, consumption of both complementary proteins in one meal is suggested (Wardlaw and Insel, 1996).

Protein is the second most plentiful substance in our bodies (with water being first), comprising one-fifth of our total body weight (Garrison and Somer, 1995). Proteins are a major constituent of all living cells and are important components of muscle, body organs, skin, hair, and nails as well as enzymes and immune system compounds

(Garrison and Somer, 1995). Proteins also participate in regulation of fluid balance and regulation of blood pH and can serve as an energy source.

The liver is the main site of protein amino acid metabolism. In a normal adult human 75 % of amino acids are utilized in protein synthesis due to the continuous catabolism and synthesis of proteins (Garrison and Somer, 1995). An average adult male can synthesize 300 g of new protein a day (Garrison and Somer, 1995) Amino acids not used for protein synthesis can be converted to other important nitrogen containing compounds including purines, pyrimidines, choline, creatine, niacin and porphyrins (Garrison and Somer, 1995).

Proteins supply about 2-5 % of the energy needs of the body. In comparing energy sources, fat generates 9 calories per g of fat and carbohydrates generate 4 calories per g of fat. As an energy source proteins are similar to carbohydrates in that they can generate 4 calories per g of protein, but considerable processing by the liver and kidneys is needed for utilization of energy from protein (Wardlaw and Insel, 1996). Proteins are first broken down into free amino acids and the amino group is removed for utilization as an energy source. Removal of nitrogen from amino acids results in the formation of carbon skeletons, which can be metabolized to acetyl Coenzyme A (acetyl CoA) and pyruvate and used in energy production through the citric acid cycle.

Catabolism (i.e. the breakdown), of amino acids results in the formation of amino (NH_2) groups which are converted to the ammonia (NH_3) form. Ammonia must be excreted as increased concentrations are toxic to the cells in the body. Excretion of NH_3 in humans occurs through the synthesis of urea. Urea is synthesized by liver and secreted into the bloodstream where it is then taken up by the kidneys for excretion in the urine. An NH_3 and an NH_2 group react with carbon dioxide through a series of steps known as the urea cycle to form urea and water. Urea can then be excreted through the urine.

2.4 Protein requirements by humans

When humans become adults and are in a healthy state there is no net gain in body N stores (Fowden, 1981). In other words, protein intake beyond that required for maintenance and excretion in a normal healthy adult equals protein loss. This is known as "nitrogen balance" and can be determined by measuring the difference between protein-N intake and excretion of N in urine or urea-N excretion (Zeman, 1991). During growth or recovery from illness, protein requirements are increased due to protein needed to build or repair new tissue. In this case a positive N balance would occur due to the intake of protein-N being higher than the output. If not enough protein is consumed and proteins are being broken down faster than they are being synthesized than a negative N balance will occur. In other words the catabolism and excretion of protein is higher than is the intake of protein (Wildman and Medeiros, 2000).

If the amount of protein-N consumed is equal to the amount lost then this is known as N balance. One equation for estimating the amount of protein that would need to be consumed to maintain N balance is:

$$\text{Nitrogen balance} = \frac{\text{24 hour protein intake}}{6.25} - (\text{UUN} + 4)$$

In this equation grams of protein ingested, which are made up of about 16 % nitrogen is converted to grams of N by dividing by 6.25. Excretion of N is equal to excreted urea over 24 hours in the urine (UUN) + 4 g/d non-urea N loss (this includes non-urea N in urine, fecal N loss from digestive enzymes and sloughed intestinal cells, and integumental loss (Zeman, 1991).

A normal adult male (70 kg) loses approximately 54 $\pm$ 2 mg of N/ kg body weight per day, meaning the estimated N requirement to maintain N balance would be 70 mg N/kg body weight per day, which can be converted to 0.45 g of protein/ kg body weight per day. (Wildman and Medeiros, 2000). However it is important to consider that as ingestion of protein increases to levels approaching the requirement of the individual there is a decrease in the utilization of dietary protein which can increase requirements from 0.45 to 0.57 g protein/kg body weight/day. In addition the source and quality of the protein can affect utilization and increase amounts that must be ingested. It has been determined that protein quality from Western diets is about 75 % of protein quality from eggs (Wildman and Medeiros, 2000). With this in mind dietary protein intake needed to maintain N balance would be 0.57 g protein/kg body weight per day x 100/75 = 0.8 g protein/kg body weight/day (Wildman and Medeiros, 2000). It is of interest to note differences between adults and infants. As stated above dietary intake needed to maintain N balance in adults would be 0.8 g protein/kg body weight/ day. However due to increased needs for growth, requirements for very young infants can be as high as 1.98 g of protein/ kg body weight/day during the first month of life and decreases to that of other age groups by about 4 months. (Fomon, 1993).

Estimated protein recommendations for most healthy adults are based on grams of protein needed/kg of desired body weight. However considerations for a recommended value that need to be taken into account are the quality of the protein, age and state of the individual as well as energy needs (Garrison and Somer, 1995). The recommended protein allowance in the US for maintenance of a normal, healthy individual is 0.8 g/kg body weight per day. This value can vary due to growth and illness due to increased requirements for growth and repair. Children under two may need 1.2 g/kg body weight for increased growth. Pregnant and lactating women can require up to 6 g of protein/kg body weight per day (Garrison and Somer, 1995).

Another method for estimation of N requirements is based on energy needs. If energy intake from carbohydrates and fats is too low then the body will use protein as an energy source, which can result in degradation of muscle. Carbohydrates can act as a N-sparing agent by prevention of amino acid catabolism for use as an energy source. The elderly population may have a higher protein need due to decreased energy intake leading to muscle loss and some studies have suggested an increase to 1.25 g/ kg body weight per day in this population (Wardlaw and Insel, 1996). For maintenance in a normal healthy adult the ratio of kilocalories to N (from protein) suggested is around 300:1 or 300 kcal for every 1 g of N (or 6.25 g of protein) (Zeman, 1991). For anabolism the suggested ratio of calories to N is 150 kcal/g N. A group that can be at risk of inadequate protein consumption related to caloric intake are the elderly.

In developing countries diets may be low in energy and protein leading to protein-energy malnutrition. It has been estimated that around 500 million children worldwide suffer from protein-energy malnutrition (Wardlaw and Insel, 1996). There are two major types of protein-energy malnutrition; marasmus and kwashiorkor. In marasmus or general starvation there is insufficient intake of both energy and protein resulting in overall wasting. There is little fat stored or muscle mass and death occurs frequently from infections. Kwashiorkor is a word in Ghana meaning disease of the first child. This is due to the mother having another child and stopping breastfeeding of the first child thus changing the child's diet from the nutrient rich milk to roots and gruels significantly decreasing protein availability in the diet. In kwashiorkor, there is marginal but adequate energy consumption and extremely low protein intake. This results in fat stores being preserved as well as decreased catabolism of muscle protein. Turn over of amino acids in muscle is a vital function for synthesis of essential proteins such as albumin and immunoglobulins (Brody, 1994). In kwashiorkor infection, edema and poor growth are common with continuing decreases in proteins essential for proper function.

Most American adults meet or exceed the recommended allowance and on average consume two to three times the recommended levels (Garrison and Somer, 1995). Increased intake in healthy adults above the RDA is not generally required as there are no identified benefits in exceeding amounts of 1.5 g protein consumption per kg body weight per day.

3. EFFECTS OF NITRATE IN THE DIET

Although humans may not be able to utilize more simple forms of N for normal physiologic functions such as energy and synthesis of proteins, they are exposed and capable of absorbing other forms of nitrogen in the diet such as NO_3^-, which must be considered for its effect on human health. The natural presence of NO_3^- in plants is due to the nitrogen cycle in which organic forms of N are converted to the mineral forms of N; i.e. NO_3^-, NH_4^+, and nitrite (NO_2^-). by microorganisms and taken up by the plants. Because NO_3^- can be accumulated in the tissues of a number of edible plants, vegetables can be a major dietary source of NO_3^-. Humans can also be exposed to NO_3^- that is present in rural drinking water supplies due to contamination from agricultural or other sources. Under normal conditions NO_3^- is found only in small amounts in drinking water and is mainly ingested in the diet from cured meats and green vegetables such as spinach and lettuce and roots such as beets and carrots. Nitrate is easily absorbed by the intestine and in general rapidly metabolized and excreted in the urine. In general NO_3^- is not usually considered a health risk but may become a concern due to its conversion to NO_2^- and N-nitroso compounds (NOC). Due to these concerns NO_3^- metabolism in humans must be examined.

3.1. Methemoglobinemia

The main health risk associated with NO_3^- consumption in humans is development of methemoglobinemia due to NO_2^- derived from NO_3^-. The iron contained within hemoglobin, in red blood cells is normally in the ferrous (reduced) state. Through oxidation, ferrous iron can be converted to ferric iron and this conversion of the iron to the oxidized state prevents transport of oxygen throughout the body. When the ferrous iron in heme is converted to ferric iron the resulting hemoglobin is referred to as methemoglobin (MHb). Methemoglobin doesn't have the capacity to transport enough oxygen to supply the cells with an adequate amount of oxygen.

Under normal conditions the presence of a small amount of MHb is normal, making up approximately 1 % of the total hemoglobin (Hb) in adults and around 2 % in infants (Wright et al, 1999). Red blood cells do contain mechanisms to protect against oxidation and are capable of converting MHb back to hemoglobin. However, the red blood cell has a finite life span and over time loses its ability to protect against oxidation. Continued oxidative stress resulting in the formation of MHb is considered to be a normal mechanism for the identification of aging cells that the body needs to remove from circulation (Coleman and Coleman; 1996; Faivre-Fiorina et al, 1998; and Wright et al, 1999). Although MHb is a normal product of metabolism in the body, methemoglobinemia occurs under abnormal conditions when levels of MHb become too high for adequate transport of oxygen to cells in the body.

One of the most apparent manifestations of methemoglobinemia is cyanosis in which the skin can appear blue in color, because of the lack of oxygen in the blood. Consequently, methemoglobinemia in infants is referred to as 'blue baby syndrome'. Cyanosis has been shown to be evident even at levels of 5-10 % of Hb in the form of MHb especially in the appearance of the bluish color in the lips and nails (Bruning-Fann and Kaneene, 1993a). Blood taken from an individual with methemoglobinemia exhibits a characteristic chocolate brown color due to oxidation. Other manifestations include drowsiness and lethargy as well as diarrhea and vomiting and, due to oxygen deprivation to body cells, can lead to death.

3.1.1. Contribution of dietary nitrate

Many compounds can contribute to the formation of methemoglobinemia but the most commonly reported is NO_3^-. Nitrate presents a health risk for humans and animals when they are converted to NO_2^- and absorbed by the body. Absorbed NO_2^- can interact with hemoglobin and oxidize the ferrous iron, inhibiting oxygen transport. Development of methemoglobinemia due to ingestion of water containing high NO_3^- is well known with the most common cases being due to contaminated well water. However, some municipal water contamination has been reported as well (Bruning-Fann and Kaneene, 1993a). Most cases of methemoglobinemia result from a water source with NO_3^- levels above 45 ppm (i.e. 10 ppm NO_3^- -N) however, some reports indicate increased concentrations of NO_3^- in the water boiled for sterilization due to evaporation (Bruning-Fann and Kaneene, 1993a). The ingested NO_3^- can be reduced to NO_2^- in the gut by nitrate-reducing bacteria. In addition, bacterial contamination of water high in NO_3^- is common and contributes to increased toxicity by reducing NO_3^- before consumption.

In humans, NO_3^- in food does not in general contribute significantly to methemoglobinemia. This may be due to the presence of compounds in foods such as ascorbic acid or other chelates, which can inhibit NO_2^- formation in the gut. The most common food related toxicity is the ingestion by infants of infant formula mixed with contaminated well water (Bruning-Fann and Kaneene, 1993a). Increased concentration of NO_3^- in breast milk has not been substantiated by the literature suggesting that it is not concentrated in human milk and reflects the plasma levels of the mother (Dusdieker et al., 1996; Bruning-Fann and Kaneene, 1993a)

There is some evidence to support improper storage or handling of foods as contributors to methemoglobinemia. Many vegetables such as lettuce, spinach, and root vegetables such as carrots contain high levels of NO_3^-. Plant NO_3^- converted to NO_2^- prior to ingestion have been implicated as a cause of contamination in home-prepared spinach, stored at room temperature (Fomon, 1993). Similar conditions were seen in the preparation and improper storage of carrots.

3.1.2. Factors affecting susceptibility to methemoglobinemia

Several factors contribute to increased sensitivity to methemoglobinemia. Age is a considerable contributor to increased sensitivity for several reasons. Infants under six months are more sensitive to methemoglobinemia than children and adults due to the stomach pH being greater than 4. This higher pH decreases the effectiveness of the stomach in prevention of the growth of nitrate-reducing bacteria. Older adults can also experience higher stomach pH and be at risk for methemoglobinemia for this same reason.

Infants also have an increased risk because they consume more water per unit of body weight than an adult and as mentioned above infants who's only food source is home prepared formula are more at risk than breast fed infants if the formula is made with tap water high in NO_3^- (Bruning-Fann and Kaneene, 1993a). Pre-natal methemoglobinemia has also been shown to occur in pregnant women consuming water containing high concentrations of NO_3^-. Nitrite has clearly been shown to cross the placenta and cause methemoglobinemia in the developing fetus (Wright *et al*, 1999). In addition, pregnancy in rats has long been associated with naturally higher levels of MHb which may exacerbate any contribution from nitrate ingestion and may need to be considered in humans as well (Tarburton et al, 1985). Another factor which might play a role in the sensitivity of the fetus and in the newborn infant, is the presence of fetal hemoglobin. Fetal red blood cells have higher oxygen affinity than adult red blood cells due to differences in the amino acid sequence of the globin chains. This causes fetal hemoglobin to be more easily oxidized than the predominant form of hemoglobin in adults (Wright et al, 1999). Although fetal hemoglobin decreases after birth, it is still present in the newborn infant and may contribute to sensitivity to MHb formation during the first months of life.

Under conditions of diarrhea and dehydration metabolic acidosis can occur. Although the mechanism is not completely understood, condition of acidosis can lead to methemoglobinemia even without known exposure to NO_3^- and may contribute to increased sensitivity in the presence of NO_3^-.

Over 140,000 people receive dialysis treatment each year. Dialysis patients are also at risk and must be careful to use water known to contain less than 2ppm N in the NO_3^- form (i.e. NO_3^--N) (Fan et al, 1987).

Although not related to increased MHb formation anemic patients might experience more deleterious effects due to less hemoglobin bioavailability. A normal person with a hemoglobin concentration of 15 g/dL with 20 % being composed of MHb would still have 12 g/dL of normal hemoglobin whereas an anemic individual with a hemoglobin value of 8 g/dL would only have 6.4 g/dL of normal hemoglobin available (Wright et al., 1999).

3.1.3. Metabolic pathways and protective mechanisms

Since the red blood cell is constantly exposed to high concentrations of oxygen and is thus exposed to oxygen free radicals, the body has developed protective mechanisms for reduction of iron in MHb back to the ferrous state and if nitrates are removed minor cases of methemoglobinemia can resolve on their own. In a normal adult human approximately 15 % of the MHb can be reduced back to hemoglobin per hour (Wright et al., 1999). The cytochrome-b_5-MHb reductase pathway is the main form of reduction protection in the body with 99 % of MHb reduction occurring through this pathway (Wright et al., 1999). The two enzymes involved in this pathway are cytosolic cytochrome-b_5 and cytochrome-b_5 reductase. NADH produced through the glycolytic pathway is also required for the reaction to occur. This pathway helps to control normal endogenous formation of MHb however factors related to decreases in enzymatic activity can contribute to formation of higher concentrations of MHb.

Under normal conditions infants under 6 months have reduced levels of cytochrome-b_5 reductase with concentrations at birth only being half that of an adult (Wright et al., 1999; Hjelt et al., 1995). In addition some congenital factors related to this pathway may aggravate conditions brought on by nitrate ingestion in adults. Cytochrome-b_5 reductase and cytochrome b_5 deficiencies can occur. Both deficiencies are autosomal recessive and result in increased levels of MHb due to decreased reduction capacity of this main pathway.

Another pathway to consider in the treatment of MHb is Nicotinamide adenine dinucleotide phosphate (NADPH)-MHb reductase. This enzyme is not specific for the reduction of MHb but instead functions in the metabolism of xenobiotics.. Methylene blue is commonly used in the treatment of methemoglobinemia. However, methylene blue itself is an oxidizing agent. NADPH-MHb reductase has an affinity for methylene blue and reduces the dye to leukomethylene blue, which has an affinity for MHb and can act as a reducing agent to reduce the iron to the ferrous state. The reaction occurs quickly and significant amounts of MHb can be reduced within 30 minutes. In a person with a congenital NADPH-MHb reductase deficiency the methylene blue will not be converted as efficiently leading to more methylene blue acting as an oxidizing agent than a reducing agent which can in turn cause hemolysis of the red blood cells (Wright et al., 1999).

Glucose-6-phosphate dehydrogenase (G6PDH) deficiency may also contribute to methemoglobinemia. This enzyme is part of the hexose monophosphate shunt and is therefore involved in the synthesis of NADPH, which is utilized in the reduction of methylene blue and may therefore decrease the effectiveness of the treatment (Wright et al., 1999). Although cytochrome-b_5 reductase is the main pathway for reduction of MHb and methylene blue is the most common clinical treatment used, other compounds can play minor roles. Both ascorbic acid and glutathione can act indirectly as cellular antioxidants and in some individuals with congenital deficiencies up-regulation of reduction is thought to occur through these compounds. However, deficiencies in these compounds do not lead to methemoglobinemia (Wright et al., 1999).

Dextrose can also act indirectly by contributing glucose for glycolysis and for the hexose monophosphate. The reduced nicotinamide adenine dinucleotide (NADH) created through glycolysis is utilized in the cytochrome-b_5 reductase pathway and the NADPH produced in the hexose monophosphate shunt is involved in the reduction of methylene blue. Therefore high enough concentrations of glucose for these pathways should be considered for increased efficiency (Wright et al., 1999).

3.2. Secondary and related effects

3.2.1 Acute respiratory infection

A possible secondary effect of methemoglobinemia from exposure to NO_3^- in water is the increased risk of respiratory tract infections. Animal studies have suggested changes in bronchi and lung cells along with increased presence of lymphocyte in the lung (suggesting increased infection) in animals consuming high NO_3^- diets. The damage to the tissue increased with higher NO_3^- levels in the ingested water. A study in 8 year old children in India demonstrated a strong positive correlation between methemoglobinemia due to well water NO_3^- and the incidence of respiratory tract infections. It is thought that the high NO_3^- leads first to methemoglobinemia, causing hypoxia and increased free radicals of nitric oxide (NO) and oxygen. Since NO can act as a vasodilator this may cause changes in pulmonary circulation and alveoli providing high- risk conditions for respiratory tract infections (Gupta et al., 2000).

3.2.2. Thyroid

Although methemoglobinemia has been identified as a major human health risk from exposure to NO_3^-, there are other health risks from NO_3^- in drinking water including deleterious interactions with the thyroid gland. The thyroid gland synthesizes two iodoamino acid hormones which play a role in general metabolism and developmental regulation, as well as regulation of tissue differentiation. Thyroid hormones are distinctive in that they require iodine to be active.

Nitrate appears to inhibit both uptake and retention of iodine by the thyroid. Nitrate is of similar in size and charge to the iodide ion and appears to compete with the iodide binding site in the thyroid (Zraly et al., 1997). The effect on the thyroid is similar to that seen with the administration of thiocyanate and perchlorate anions, which also inhibit accumulation of iodide in the thyroid (Jahreis, et al., 1986).

The effect of NO_3^- on the thyroid was first observed in rats by Wyngarden et al, (1952) (Bruning-Fann and Kaneene, 1993a). Further studies in rats and similar studies in chickens confirmed altered thyroid metabolism and decreased iodide uptake along with increased thyroid size, which is indicative of an attempt of the thyroid to compensate for decreased hormone synthesis (Bruning-Fann and Kaneene, 1993a).

Ruminant animals such as sheep also demonstrate decreased uptake of iodine by the thyroid. However there is some evidence to support an increased ability of ruminants to adapt to increased NO_3^- consumption over time (Bruning-Fann and Kaneene, 1993a). On the other hand administration of potassium nitrate to bulls resulted in decreased thyroxin levels, indicative of depressed thyroid gland activity. It is thought that this decrease in hormone level lead to an observed effect on the libido and delayed onset of erection and mounting suggesting that the effect of NO_3^- on ruminant animals should not be overlooked (Zraly et al., 1997).

It is also apparent that human populations exposed to high NO_3^- levels in drinking water show a similar increase in thyroid volume and decreased levels of thyroid stimulating hormone. The effect is dose dependent with differences in thyroid volume occurring above 50 mg/L (van Maanenen et al., 1994). Guidelines for concentrations of NO_3^- in water have been developed for the prevention of methemoglobinemia As summarized by Fraser and Chilvers, (1981) current WHO European Standards for drinking water recommend levels of nitrate of less than 50 mg/L. while the standard in the U. S. is 45 ppm, (USEPA, 1973, 2001,) Yet is important to note around 50 mg/L alteration in thyroid metabolism might manifest in humans (van Maaenen et al., 1994).

Alteration in thyroid metabolism must be seriously evaluated in both the monogastric and ruminant animal in that it may partially explain some other effects seen from NO_3^- consumption including immune function, reproduction, and fetal developmental problems.

3.2.3. Birth defects and reproduction

The potential effects of NO_3^- on reproduction and normal fetal development in humans remains a topic of controversy. It has been clearly shown that NO_2^- can cross the placental barrier in animals (Bruning-Fann and Kaneene, 1993b) and research suggests that transfer may occur in humans as well due to the presence of fetal methemoglobinemia when high NO_3^- was consumed by the mother during pregnancy (Bruning-Fann and Kaneene, 1993b). Due to placental transfer of NO_3^- the risk of spontaneous abortion associated with fetal methemoglobinemia has been studied. In ruminants a clear relationship between NO_3^- intake and abortion have been established. In cows, fetal death and abortion have been linked to ingestion of high NO_3^- by the mother. (Bruning-Fann and Kaneene, 1993a; R.F. Follett, unpublished observations). Similar results are seen in sheep (Bruning-Fann and Kaneene, 1993a). The cause of the abortion is thought to be due to fetal methemoglobinemia, which results in death due to hypoxia and is supported by demonstration of decreased oxygen saturation of umbilical cord blood and elevated nitrate in the dead calves (Bruning-Fann and Kaneene, 1993a).

The effect of NO_3^- ingestion in monogastric animals during pregnancy is less clear and in general is less likely to occur in the presence of NO_3^- and requires higher levels of NO_2^- than in ruminants for detrimental outcomes to occur. (Bruning-Fann and Kaneene, 1993a; Fan and Steinberg, 1996). Ground water NO_3^- and NO_2^- as a cause of fetal methemoglobinemia have also been implemented as a risk for spontaneous abortions in humans (Bruning-Fann and Kaneene, 1993b). However, other studies have shown no association between increased water NO_3^- consumption and abortion at levels higher than 40 ppm NO_3^- (Bruning-Fann and Kaneene, 1993b).

Nitrosating substances may influence DNA alkylation and transcription. Therefore, the effects of NO_3^- and NO_2^- during pregnancy need to be considered in the proper development of the fetus (van Maanen et al., 1996a;). A review by Fan et al. suggests that studies in animals such as mice rats and guinea pigs do not appear to fully support an increase in congenital malformations from ingestion of NO_3^- and NO_2^- at levels that might be encountered in drinking water (Bruining-Fann and Kaneene, 1993b; Fan et al., 1987).

Although links with NO_3^- and NO_2^- consumption and birth defects of the central nervous and musculoskeletal system have been alluded to in humans, epidemiological studies do not support a clear connection. However due to recognized detrimental effects of NO_3^- in the diet this area should not be ignored and further research is suggested (Bruning-Fann and Kaneene, 1993b).

4. NITRIC OXIDE

The role of NO_3^- and NO_2^- in the diet has mainly focused on their deleterious effects. However, there is growing acceptance for the beneficial importance of these compounds in the synthesis of nitric oxide. Nitric oxide (NO) is a free radical gas important in normal physiological function where it acts as a messenger molecule for regulation of several systems including blood vessel dilation, hormonal and neurotransmission functions. Nitric oxide can be produced endogenously with the main source being from the amino acid arginine. The enzyme nitric oxide synthase (NOS) catalyzes a five-electron oxidation of an amidine nitrogen of arginine resulting in the synthesis of citrulline and NO (Murray et al., 1994; Ellis et al., 1998). The NOS enzyme contains a tightly bound heme and is similar in structure to the cytochrome P-450 reductase enzyme (Pufahl and Marletta, 1993). Nitric oxide is a small molecule and is lipophilic allowing for rapid diffusion through cell membranes for interaction with intracellular target compounds (Ignarro et al., 1999). Nitric oxide has a biological half-life of approximately 5 seconds allowing it to act exceptionally well as a local mediator of physiological function (Moncada and Higgs, 1991). In healthy adults, NO reacts in the body with water to form NO_2-, which is an unstable compound in blood and quickly converts to NO_3^- (Ellis et al., 1998). Both NO_3^- and NO_2^- can then be excreted in the urine.

The role of NO in the body can be divided into two groups. First is its action as a messenger molecule where it plays a role in vascular tone, platelet activation, immune response and as a neurotransmitter. The second category is related to its function as a cytotoxic molecule important in host defense but also as a harmful compound related to autoimmune diseases (Moilanen and Vapaatalo, 1995).

4.1. Target compounds

Nitric oxide has five main molecular targets in the body; heme proteins, enzymes, DNA, thiols, and superoxide (Radomski, 1995). One of the most important heme proteins that NO binds is guanylate cyclase (GC). Binding of NO results in the activation of GC leading to increased synthesis of cyclic guanosine monophosphate (cGMP) a compound involved in the mediation of ion flux, modulation of cyclic adenosine monophosphate (cAMP)-mediated responses and increases in protein kinase phosphorylation reactions (Radomski, 1995). Nitric oxide also interacts with hemoglobin and myoglobin resulting in the degredation of NO and thus a decrease in the biological activity of NO (Radomski, 1995). Nitric oxide can act on other enzymes such as complex I and II of the mitochondrial electron transport chain and aconitase in the TCA cycle. Interaction with these enzymes has been linked to some of the cytotoxic effect of NO (Radomski, 1995). Inhibition of DNA synthesis can occur from macrophage NO through the inhibition of ribonucleotide reductase; the rate-limiting enzyme in DNA synthesis (Lepoivre et al., 1991). Interaction with thiols through the nitrosylation of sulfhydryl groups may offer a storage mechansim for readily available NO (Stamler et al., 1992; Radomski, 1995). Interaction of NO with superoxide results in the formation of peroxynitrite a highly reactive molecule, which can participate in the oxidation of many compounds and may relate to some of the more detrimental effects of NO.

4.2. Relaxation of smooth muscle

One of the most well-documented functions of NO in the body is its action as a relaxing factor on smooth muscle of blood vessel walls leading to vasodilatation and a decrease in blood pressure (Ellis et al., 1998). The interaction of NO_3^- with hemoglobin in methemoglobinemia allows for an understanding of the role of NO in vasodilatation. Nitric oxide first diffuses out of the cell it was synthesized in and acts on neighboring cells. The intracellular enzyme guanyl cyclase contains a heme prosthetic group to which the NO can bind resulting in a conformational change causing activation of the enzyme. This results in the production of cGMP, which causes relaxation of the vessel walls, which leads to vasodilatation and hypotension or a decrease in blood pressure. In the heart, cGMP will act to relax the muscle and decrease the force of the contractions by stimulating ion pumps that maintain low cytosolic Ca^{2+} concentrations.

Although the primary synthesis of NO in the body occurs through arginine metabolism, it has been suggested that dietary NO_3- and NO_2- can be converted to NO and contribute to endogenous NO synthesis as well (Gruetter et al., 1981). Vasodilator therapy is commonly used for management of congestive heart failure. Nitrate compounds such as nitroglycerin (glyceryl trinitrate), are among the oldest and most utilized compounds in vasodilator therapy with a well-documented decrease in intromyocardial pressure leading to improved perfusion of the heart. Although NO_3^- is generally considered an inorganic molecule, within the medical field it is common to refer to a number of compounds, including nitroglycerin, as "organic nitrates". While it has been known for over 100 years that ingestion of nitroglycerin resulted in dilation of veins and arteries, it was not until 20 years ago that a connection with NO was truly recognized (Ignarro, 1999). Nitroglycerin interacts with thiols such as cysteine and glutathione in the cell such that chemically unstable nitrosothiols are formed, upon which NO is released (Ignarro, 1999). The formation of NO is slow for nitroglycerin allowing for the effects of nitroglycerin to be long-lasting. Pharmacological administration of nitroglycerin along with many other organic nitrates has been clearly shown to result in the formation of NO. The formation of NO from inorganic NO_3- and NO_2- is important in considering the role diet might play in regulation of NO formation.

The role of orally administered sodium nitrite is well documented and is known to increase guanylate cyclase activity leading to vasodilatation (Classen et al, 1990). Inorganic NO_2- is a weak activator on its own but similar to nitroglycerin is slightly enhanced in the presence of thiols which decrease its chemical stability resulting in increased release of NO (Ignarro, 1999). Although a strong connection has not yet been established, some discussion has centered around the contribution of dietary NO_3- to blood pressure. Hypertension appears to be lower in vegetarians and the major contribution of NO_3- in the diet is from vegetables. Therefore the potential contribution of dietary NO_3- to decreased blood pressure has been considered but not fully evaluated (Classen et al., 1990). However research suggests "inorganic nitrates" may have little consequence on vasodilatation unless converted to NO_2- (Classen et al., 1990).

Ingestion of NO_3- and NO_2- from water has also been studied in relation to hypertension but the studies are conflicting (Bruning-Fann and Kaneene, 1993b). With higher incidences of hypertension in communities with nitrate-free water than in communities with water levels averaging 45 ppm and reported increases in hypertension in communities consuming water with nitrate when compared with communities consuming nitrate-free drinking water suggesting the need for further investigation on the effect of dietary nitrates and well water nitrates on hypertension (Bruning-Fann and Kaneene, 1993b).

The importance of NO in relaxation of smooth muscle also appears to contribute to regulation of other physiological functions. Nitric oxide has been identified in exhaled air and is thought to contribute to the moderation of normal respiration through bronchial dilation (Ward et al., 1993). Nitric oxide mediated relaxation of smooth muscle is also of benefit to the gastrointestinal tract serving as a regulator of gut motility (Calignano et al., 1992). Interestingly NO may be extremely important in the development of the infant gut as well. Infants appear to show high production of NO shortly after birth (Honold et al, 2000). In addition, NO_3^- and NO_2^- present in human breastmilk may also result in the formation of NO. As discussed earlier, NO_3^- and NO_2^- are present in breastmilk at levels reflecting maternal plasma. At normal physiological levels NO_3^- and NO_2^- may be converted to NO and play a critical role in regulating the infant's developing mucosal blood flow and gastric motility. Nitric oxide may also play a role in the gut that is similar to adults through development of bacteriostasis (Iizuka et al, 1999).

Nitric oxide has been shown to increase during pregnancy with a concomitant rise in cGMP and is important in the prevention of pre-eclampsia. Pre-eclampsia is a multisystem disorder seen in 4-5 % of all pregnancies and is a leading cause of both maternal and neonatal death. Symptoms include increased blood pressure and hypertension. Long-term inhibition of NO synthesis results in manifestation of the symptoms and an NO deficiency has been linked to women exhibiting pregnancy-induced hypertension (Weiner et al., 1994).

4.3. Platelet aggregation

Endothelial cell release of NO not only results in vaso-relaxation but also is a potent inhibitor of platelet adhesion and aggregation (Radomski et al.,1993). Nitric oxide is an inhibitor of platelet aggregation by stimulation of platelet cGMP, which appears to interfere with binding of platelets to endothelial cells.

4.4. Neurotransmission

Increased cGMP synthesis due to the presence of NO may also be important in neuronal transmission and may be involved in regulation of memory, learning, and pain perception (Radomski, 1995). Although the function is not completely elucidated, it is clear that NO synthase if found in high concentrations in the myenteric plexus of nerves and that inhibitors of NO synthesis impair proper neurotransmission (Ignarro, 1996). Cyclic GMP may act as an intracellular messenger in the target cell and elicit a response such as increased synaptic connections (Ignarro, 1996). Nitric oxide may also act on non-nueronal cells in a vasodilatory function leading to increased blood flow.

4.5. Cytotoxic effects

4.5.1. Food treatment

Nitrate and NO_2^- salts are commonly added to cured meats to give color and to increase product stability through anti-oxidant and cytoxic mechanisms. The anti-oxidant and cytotoxic effects are in fact due to the interaction of iron with NO formed from the reduction of the added NO_3^- and NO_2^-. The first recognition of NO in the curing of meat was in 1908 when it was determined that the red pigment of cooked cured meats such as ham was due to the presence of a heat denatured NO-myoglobin complex. The flavor of rancid meat is due to the oxidation of unsaturated fats and is catalyzed by iron containing compounds such as heme. Addition of NO_2- stabilizes heme compounds by formation of a NO-heme complex. In addition NO inhibits oxidation through conversion of heme to a ferrous heme-NO radical, which can act as an anti-oxidant, as well as the binding of NO to free ionic iron to form such compounds as low molecular weight cysteine-iron-NO radicals which also have anti-oxidant properties (Cornforth, 1996).

4.5.2. Prevention of infection

The anti-microbial action of NO is varied. Nitric oxide can inhibit and form complexes with heme, iron-sulfur and copper proteins thus impairing function as well as interfering with incorporation into key enzymes. Inhibition of clostridia appears to be due in part to interaction of NO with pyruvate-ferredoxin oxidoreductase an iron-sulfur enzyme involved in regeneration of adenosine triphosphate (ATP) (Cornforth, 1996). Inactivation of enzymes containing no redox metals such as glyceraldehyde-3-phosphate dehydrogenase, part of the glycolytic pathway, can occur as well. Nitric oxide can also inhibit ribonucleotide reductase, a crucial enzyme in the formation of DNA through quenching of a tyrosine radical. Interference with DNA also occurs through NO by promoting deamination of N-terminal and other amino groups of proteins and through this process can be mutagenic such as seen in *Salmonella typhimurium* and *P. stutzeri* where mutations in the nucleotide sequence occur that prove lethal to the bacteria. The common practice of addition of ascorbate accelerates the formation of NO and decreases formation of nitrosamines thus aiding in the anitbacterial effects NO_3- and NO_2- salts (Cornforth, 1996).

The body is capable of utilizing dietary NO_3- and NO_2- as a source of NO for use as an anti-microbial agent. Dietary NO_3- mainly from foods high in NO_3^- such as green leafy vegetables are absorbed from the stomach and the small intestine into the plasma. Nitrate is then concentrated in saliva such that the more dietary NO_3- consumed the more NO_3- and NO_2- found in the saliva (McKnight et al., 1999). Approximately 25% of NO_3- ingested in the diet is re-secreted into the saliva. Salivary NO_3- can be converted to NO_2- by the action of lactoperoxidase, lysozymes and lactoferrin which are involved in cleaning the oral mucosal cells. Conversion of NO_3- to NO_2- in the saliva can also occur due to the presence of NO_3- reducing bacteria on the tongue (McKnight et al., 1999; van Maanen et al., 1996b). Salivary NO_2- is then swallowed, where in the acidic conditions of the stomach, the NO_2- can be protonated to form nitrous acid. Nitrous acid can then decompose into nitrogen oxides including NO which is believed to act in the stomach as an anti-microbial agent for swallowed bacteria (McKnight et al., 1999). McKnight et al showed large concentrations of stomach NO formed after intake of oral NO_3- (potassium nitrate). The importance of NO formation due to protonation under acidic conditions is supported by increased bacterial colonization in patients with achlorhydria, in which conversion of NO_2- to NO would be impaired.

4.6. Immune system effects

Another location of NO in relation to infection is macrophage from the immune system. (Ellis et al., 1998). The cytotoxic activity of macrophage is partially due to the endogenous production of NO from argininc which is catalyzed by cytokine inducible nitric oxide synthase. Cytotoxicity is due to at least a 100-fold increase in local NO concentrations in comparison to levels generated by other cells in the body such as endothelial cells. If macrophage are arginine deficient or if inhibition of the enzyme is induced by administration of arginine analogs there is a decrease in bactericidal effectiveness. Nitric oxide production relates to their cytotoxic effect and has been shown to act on bacteria, tumor cells, viruses, fungi, protozoans and helminths (Moilan and Vapaatalo, 1995). Although NO is the main oxide produced, synthesis of peroxynitrite (ONOO-) occurs in macrophage due to the spontaneous reactions of NO with O_2- and H_2O_2 also produced by the macrophage. This more reactive oxide leads to formation of a highly reactive hydroxide radical OH^- and nitrogen dioxide (NO_2) which kill bacteria through oxidative damage.

4.7. Autoimmune disease

Nitric oxide production is not always beneficial in its role in free radical production as it can be cytotoxic to the hosts cells as well. Nitric oxide can act as a cyotoxic molecule in autoimmune disease such as seen in diabetes. Increased NO synthesis due to an autoimmune response results in decreased insulin secretion and damage to the islet cells. Similarly tissue damage during arthritis has been linked to toxic levels of NO due to an autoimmune response (Moilan and Vapaatalo, 1995). Nitrate concentrations are higher in the sinovial fluid of patients with rheumatoid arthritis and levels are increased in urine as well suggesting increased endogenous formation of NO (Moilan and Vapaatalo, 1995). The origin of NO in the joints is not completely known but stimulation of chondrocytes by IL-2 appears to increase NO production by the chondrocytes (Moilan and Vapaatalo, 1995). Nitric oxide plays more of a harmful effect in arthritis appearing to contribute to inflammation and destruction of tissue.

The destructive effect is supported by evidence showing alleviation of symptoms when inbitors of NO synthesis are given (Moilan and Vapaatalo, 1995). The role of NO in inflammation has not been fully elucidated with both pro-inflammatory and anti-inflammatory effects. However dual function of mediators is common in the inflammatory response (Moilan and Vapaatalo, 1995).

Although this review touches upon only a few established roles of NO, research continues to expand as the significance of NO in many other physiological functions becomes more apparent and more in depth reviews of the literature are available (McKnight et al., 1999 ; Radomski, 1995; Moilanen and Vapaatalo, 1995).

5. CANCER

Many epidemiological studies have demonstrated a relationship between diet and cancer risk and it has been suggested that diet is associated with as much as 35 % of all deaths from cancer in the United States (Doll and Peto, 1981; Ferguson, 1999; Howe et al, 1986). Nitrate itself is not considered to be carcinogenic, however, NO_3^- can be reduced under a variety of conditions to NO_2^-. Nitrite is a more reactive compound and can participate in the nitrosation of many substrates in the diet including secondary and tertiary amines and amides resulting in the formation of N-nitroso compounds (NOC)

5.1. The role of N-nitroso compounds

In general NOC compounds include both nitrosamines and nitrosamides and can participate in DNA alkylation which can lead to tumor formation. Animal experiments have demonstrated N-nitroso compounds to be the most potent and broadly acting carcinogens known (Ferguson, 1999). Several hundred N-nitroso compounds have been tested as carcinogens and over 80 % have been found to be carcinogenic in at least 40 animal models (Eichholzer and Guzwiller, 1998).

Of the NOC that have been tested in the laboratory for carcinogenic activity, humans are exposed to only a small percentage and almost all are **nitrosamines** (Lijinsky, 1999). In animal models nitrosamines have been linked to bladder, esophagus, kidney, liver, lung, nasopharyngeal, and thyroid cancer (Mirvish, 1995). In general, nitrosamines are stable compounds and require metabolic activation to have any carcinogenic effect. The activated compounds are unstable and have a relatively short half-life. Therefore, it is thought that the sensitivity of the organ to the nitrosamine might be influenced by the nitrosamine activating systems within the specific organ capable of producing reactive alkylating compounds (Magee, 1989). Metabolic activation of nitrosamines is thought to occur due to the presence of a class of enzymes with overlapping substrate specificity known as cytochromes P-450; more specifically a sub-family of cytochromes P-450 involved in ethanol detoxification (Magee, 1989).

Humans are also exposed to **nitrosamides.** Nitrosamides do not need to be activated and if applied directly through ingestion in animal models will cause tumors in the stomach and duodenum. Nitrosamides can also act distantly and are primarily linked in animal models to cancer of the lymphatic and nervous systems (Mirvish, 1995). The sensitivity of the brain and possibly some of the other target organs has been hypothesized to be due to a deficiency or lack of the enzyme alkyltransferase, which can facilitate the repair of alklylated DNA (Mirvish, 1995).

5.1.1. Dietary exposure

Dietary related exposure to NOC can occur through both exogenous and endogenous sources. Few 'western style' foods contain detectable amounts of pre-formed NOC (Hotchkiss, 1989). Asian-style foods show a slightly higher content of NOC due to preparations involving smoking (Hotchkiss, 1989). A main source of pre-formed NOC in western diets that has been of concern is from NO_2^- containing foods such as cured meats, especially bacon (Hotchkiss, 1989). Foods exposed to nitrogen oxides such as beer are another substantial source of pre formed NOC in the western diet (Ferguson, 1999). Malt based beverages such as beers and scotch whiskey contain NOC due to reaction of amines in the barley with nitrogen oxides produced during drying of the malted barley in natural gas kilns (Mirvish, 1995). An average beer drinker is exposed to twice the amount of nitrosamines as an average bacon eater (Whitney et al, 1994). Concentrations of NOC have been reduced by decreasing the kiln temperature or indirect heating of the barley (Mirvish, 1995). Nitrate containing foods may also contribute to exogenous NOC if contaminated with nitrate-reducing bacteria, resulting in the formation of NO_2^- which can then interact with amines and amides found in the the food (McKnight et al, 1999).

In the past, fried bacon was found to contain up to 100 ppb of nitrosamines (mainly in the form of N-nitrosopyrrolidine and dimethylnitrosamine). After addition of NO_2^- was lowered and the inhibitory effect of ascorbic acid on nitrosamine formation was discovered levels have decreased significantly (17ppb N-nitrosopyrrolidine, 9 ppb N-nitrosothiazolidine, 4 ppb dimethylnitrosamine, 0.7 ppb N-nitrosopiperidine). (Mirvish, 1995). Presently, the average intake of exogenously produced NOC from a western style diet has dropped significantly and has been estimated to be on average 0.5-1.0 ug/day (Hotchkiss, 1989).

In addition to exogenous formation there is a concern as to the contribution of endogenous NOC formation. Nitrites and secondary amines are present in a variety of foods and may be able to contribute to the endogenous formation of NOC. NOC synthesis in the body is thought to take place through acid catalyzed and bacterial nitrosation in the stomach and by NO formation and involves the presence of NOC precursors (Mirvish, 1995). Nitrite, participating in acid catalyzed nitrosation in the stomach, can be supplied through food, water and salivary re-secretion (Bruning-Fann and Kaneene, 1993a). It has been suggested that approximately 80 % of NO_2^- found in the stomach is due to the reduction of ingested or endogenous NO_2^- re-secreted in saliva and 20 % is due to the ingested NO_2^- from preserved meats and other foods as well as water (Mirvish, 1995). To participate in nitrosation nitrite must first be converted to nitrous acid (HNO_2) through acidification in the stomach (Mirvish, 1995). It is thought (HNO_2) can then be protonated to form H_2NO_2. which may interact with amines and amides from food to form NOC (Mirvish, 1995).(McKnight et al, 1999).

$NaNO_2 + H^+ \text{-->} HNO_2 + Na^+$
$2HNO_2 \Leftrightarrow N_2O_3 + H_2O$
$2N_2O_3 + O_2 \rightarrow 2N_2O_4$
$HNO_2 + H^+ \Leftrightarrow H_2NO_2^+$
H_2NO_2+ RNHCOR-> RN (NO)COR (nitrosamide) + $H_2O + H^+$

Or
$N_2O_3 + R_2NH \rightarrow R_2N.NO$ (nitrosamine) + HNO_2

Nitrosation can still occur in the stomach at higher pH such as seen in achlorhydria due to the increased presence of bacteria. Bacteria such as *Escherichia coli* are capable of reducing NO_2^- to NO and in the presence of oxygen can form N_2O_3 which can participate in the formation of nitrosamines (Mirvish, 1995). This reaction occurs at neutral pH.

$2NO + O_2 \rightarrow 2NO_2$
$NO + NO_2 \Leftrightarrow N_2O_3$
$N_2O_3 + R_2NH \rightarrow R_2N.NO$ (nitrosamine) + HNO_2

5.1.2. Inflammatory and immune responses

Conditions in the stomach and other organs may also contribute to NO and NOC formation. Ulcerative colitis, parasitic infection of the bladder and liver, hepatitis B, and colonization of the stomach by *Helicobacter pylori* are all examples of conditions which may result in localized formation of NO during the inflammatory response due to immune stimulation of macrophage and neutrophils utilizing arginine for the synthesis of NO (Leaf et al, 1989) . It has been hypothesized that NO can react with oxygen in other locations in the body in a similar reaction to that seen in the stomach during achlorhydria. This results in the formation of N_2O_3 which is capable of interacting with endogenous amines to form NOC (Mirvish, 1995) (Leaf et al, 1989) .

In animal experiments NOC have been shown to be the most broadly acting and potent group of carcinogens known (Eichholzer and Gutzwiller, 1998; Lijinsky, 1999; McKnight et al, 1999). N-nitroso compounds have been shown to induce cancer in over 40 animal species including monkeys (Magee, 1989; Dayal and Ertel, 1997). In addition, no animal species has been found to be resistant to NOC induced carcinogenesis (Magee, 1989). Humans are exposed to a variety of exogenous and endogenous NOC as well and exposure to NOC in humans has been linked to several cancers including increased risk of stomach, esophageal, nasopharyngeal, bladder, and colon cancer (Bartsch et al, 1990). Research in humans however is fairly limited to epidemiological case-control studies and the compiled data remains inconclusive (Eichholzer and Gutzwiller, 1998). Although the epidemiological data may be unclear, the substantial evidence from animal studies suggests serious consideration be taken as to the potential risk of these compounds in humans.

5.2. Stomach and gastric cancer

Incidences of stomach cancer vary geographically with the highest occurrence seen in Japan and China and the lowest in North America and Greece (Forman, 1989). Stomach cancer is the second cause of death from cancer worldwide and in the United States ranks seventh (Yamaguchi and Abe, 1999). If dietary NO_3^- is to be considered a risk for stomach cancer through endogenous formation of NOC then vegetarians should be at a greater risk in that they can consume three times the concentration of NO_3^- when compared to omnivores (McKnight et al, 1999). In fact diets high in vegetables, and thus also high in NO_3^- have been shown to be protective against some types of cancer including gastric cancer (McKnight et al, 1999). This is thought to be due to compounds present in vegetables that can inhibit nitrosation such as ascorbic acid, alpha tocopheral, and polyphenols (Mirvish et al, 1995;Yamaguchi and Abe, 1999).. Polyphenols are also present in green and black tea both of which have been shown to inhibit nitrosation (Mirvish et al, 1995; Sobala et al, 1989).

Most human studies on gastric cancer risk and dietary NO_2^- intake showed no significant relationship. However a potential association was found in the presence of high amounts of the free amino acid methionine and high amounts of NO_2^- suggesting the occurrence of endogenous NOC formation (Eichholzer and Gutzwiller, 1998). However, substantial endogenous nitrosamine formation in the stomach from NO_2^- when consuming an average diet might be inhibited by availability of free amino acids because they are not generally present in the stomach in high concentrations. Pepsin, an enzyme in the stomach involved in proteolytic digestion cannot cut all peptide bonds into free amino acids as it only cleaves peptides bonds involving aromatic amino acids such as tyrosine, tryptophan or phenylanine. Therefore after gastric peptide digestion less than 5 % of amino acids found in the upper small intestine are in the free state (Nixon and Mawer, 1970). In fact even at the intestinal level, absorption of amino acids by the enterocytes occurs mainly through absorption of dipeptides.

Another consideration in the endogenous formation of NOC is related to pathological conditions of the stomach. Patients with gastric achlorhydria have been shown to be at higher risk for gastric cancer (Yamaguchi and Abe, 1999). Chronic atrophic gastritis results in decreased gastric juice and achlorhydria (Yamaguchi and Abe, 1999). It is thought that achlorhydria leads to increased stomach pH and more optimal conditions for colonization by bacteria. These bacteria might participate in NO_3^- reduction or induce NO production through the immune system leading to formation of NOC. This hypothesis is supported by the strong link between *Helicobacter pylori* infection and increased gastric cancer (Eichholzer and Gutzwiller, 1998). *Helicobacter pylori* is a gram negative bacteria most commonly located in the stomach and detection has been associated with the presence of atrophic gastritis (Yamaguchi and Abe, 1999). While this is considered to be a major hypothesis linking increased cancer rates seen in patients with achlorhydria and in patients with *H. pylori* infection, research still needs to be performed in order to validate the model.

Although NOC formation from dietary NO_3^- and NO_2^- might not show a strong link to gastric cancer, the contribution of pre-formed NOC must be considered. Induction of gastric tumors by pre-formed NOC has been well established in animal models (Eichholzer and Gutzwiller, 1998). It is also clear that pre-formed NOC can occur due to the interaction of NO_2^- with secondary amines in food and that certain foods and methods for food preservation increase the formation of free amines and amides. For example foods high in proteins that are dried and stored for an extended period such as fish, show increased concentration of amines and amides (Hotchkiss 1989 Cancer Surveys 8:295). In addition extended storage of uncured meat prior to processing has been demonstrated to increase NOC content after the meat had been cured and fried (Hotchkiss, 1989). Human studies provide support for increased gastric cancer risk with increased consumption of cured and smoked meats and fish and salted fish; foods potentially containing substantial amounts of NOC (Yamaguchi and Abe, 1999). A review by Eichholzer and Gutzwiller, 1998) of human epidemiological studies on the risk of gastric cancer from ingestion of pre-formed NOC suggest that that although the epidemiological evidence does lean towards the contribution of pre-formed NOC in foods to gastric cancer, the evidence may not be strong enough yet to fully support the hypothesis.

5.3. Esophageal and nasopharyngeal cancer

The strongest connection between NOC and human cancer can be seen in its potential role in esophageal and nasopharyngeal cancer (Eichholzer and Gutzwiller, 1998). Esophageal cancer incidences are found to be highest along a geographical area that has been termed the 'Asian esophageal cancer belt' and extends from Russia and Turkey to Eastern China (Eichholzer and Gutzwiller, 1998). Many possible reasons have been suggested for the prevalence of esophageal cancer and one connection may be due to nutrient deficiencies such as low consumption of ascorbic acid combined with the consumption of nitrosamines from mold-infected cereals eaten in some of the areas along the Asian esophageal cancer belt (Mirvish, 1995). Nitrosamine formation may occur in corn due to nitrosation of methylalkylamines present in the mold *Fusarium moniliforme* (Mirvish, 1995). Other types of food suspect due to possible NOC content are pickled foods and beer (Eichholzer and Gutzwiller, 1998; Hothckiss, 1989; Mirvish, 1995).

Nasopharyngeal cancer is rare in most countries but is high in areas of China, Greenland, and Tunisia (Mirvish, 1995). Dietary causes are thought to be associated with consumption of salted dried products such as fish, which contain volatile nitrosamines such as dimethylnitrosamine, clearly shown to induce nasopharyngeal cancer in rats (Mirvish, 1995; Eichholzer and Gutzwiller, 1998). The salted and dried fish also contains amines which might participate in endogenous nitrosation (Magee, 1989).

Studies have also suggested a link between increased consumption of foods containing NOC and childhood cancers; specifically brain tumors. Nitrosoureas which may be produced in bacon and salted dried fish have been shown to cause brain tumors and leukemia in rats (Mirvish, 1995). Consumption of foods containing pre-formed NOC such as hot dogs, by pregnant mothers or by the child have been studied as an increased risk factor for childhood brain tumors (Mirvish, 1995; Kuijten et al, 1990; Sarasua and Savitz, 1994). While the potential contribution of NOC to childhood cancers must be carefully examined due to the evidence from animal studies, the compiled literature in humans does not demonstrate a definitive connection (Eichholzer and Gutzwiller, 1998)). In summary, even though solid confirmation of a connection between cancer and NOC may not be apparent from human epidemiological studies, animal models offer strong support for cancer risk from NOC and therefore the link between human NOC exposure and cancer should continue to be investigated.

6. HUMAN USE AND CONTRIBUTION OF NITROGEN TO THE ENVIRONMENT

The requirement for protein consumption in human diets was discussed above (Section 2.4). From that information an estimate of N intake was calculated based upon the Recommended Dietary Allowance (RDA) for protein consumption (RDA, 1989). Nitrogen intake for US males and females (Table 1) was calculated by assuming that protein contains 16 percent N and that the RDA recommendation was the same per unit of body mass for all age, gender, and social and cultural groups. Average weights for US age groups were adapted from those from the US RDA for protein (RDA, 1989; Wildman and Medieros, 2000). Population statistics are from those reported by the U.S.Census Bureau (2001). The assumptions used for these calculations are expected to result in conservative estimates of the actual dietary N intake by the US population. Besides N intake, estimates of N losses were made based upon obligatory losses of protein for a carbohydrate-containing protein-free diet (Table 1). Such losses are reported as 54 mg N/kg body weight for a 70 kg male (Wildman and Medieros, 2000). Most of this loss is as N in urine (37 mg N/kg body weight) and feces (12 mg N/kg body weight), with the remaining 5 mg lost as a result of cutaneous and other miscellaneous N routes.

We consider our estimates for both total N intake and N loss (excretion) conservative. In the case of N intake, US diets in many sectors of the population exceed the RDA for protein intake and thus the N-intake estimate of about 775,000 MT N shown in Table 1 is also likely exceeded. It could be readily visualized that N intake approaches 1 million MT (MMT) of dietary N in the human population and that dietary N intake ranges from 0.8 to 1 MMT/year. Just as estimates of intake are likely low in Table 1, so too the estimate of obligatory N-loss is less than the actual N loss. The obligatory N-loss is measured on individuals on a carbohydrate containing protein-free diet. The base amount of N loss (54 mg/kg body weight) will be exceeded once protein is added back to the diet. Factors affecting the increased amount of N loss are protein quality, amount of protein consumed, including that in excess of the RDA, and because for adults N-loss should about equal N-intake. Consequently the estimate of N-loss for the US population is likely be near the estimate of N intake, i.e. somewhere in the range of 0.8 to 1.0 MMT N.

Table 1. Population, average weight, estimated N intake per person and annual US N intake, and estimated N loss (obligatory excretion) per person and annual US N loss as a function of age ranges for males and females in the US.

Male Age Range	Population (1000s)	Mean Weight Kg	Estimated Annual N-intake kg N/person	Estimated Annual US N-intake MT N	Estimated Annual N-loss kg N/person	Estimated Annual N-loss MT N
<5	9683	12	0.56	5429	0.24	2290
5-13	18303	32	1.50	27465	0.63	11587
14-17	8094	54	2.52	20392	1.06	8603
18-24	13579	72	3.36	45678	1.42	19270
25-49	50578	79	3.69	186677	1.56	78754
50+	34398	77	3.60	123745	1.52	52205
Total	134635					
Total				409385		172709

Female Age Range	Population (1000s)	Mean Weight Kg	Estimated Annual N-intake kg N/person	Estimated Annual US N-intake MT N	Estimated Annual N-loss kg N/person	Estimated Annual N-loss MT N
<5	9264	12	0.56	5194	0.24	2191
5-13	17458	31	1.45	25285	0.61	10667
14-17	7636	57	2.67	20369	1.13	8593
18-24	13015	60	2.80	36484	1.18	15392
25-49	51535	63	2.94	151686	1.24	63993
50+	41834	65	3.04	127041	1.28	53596
Total	140742					
Total				366059		154431
US Total	275377			775444		327140

In the above discussion of N intake and N loss, the US has been used as a case study because a base of data was readily available. As a developed country, it is likely that the data from the US can be extrapolated to other developed countries with similar age, weight, and dietary N intake patterns. We considered it inadvisable to extrapolate this data to less developed countries or to the World, because of the unlike dietary, age and weight distribution, income, and cultural differences that exist. Irrespective, it can be noted that the US population (about 275 million) is less than five percent of the World population of about 6.1 billion people. It can certainly be stated that humans excrete huge amounts of N in both the US and in the World. The amount can only increase when it is considered that the World population is projected to increase to about 8 billion people during the next 25 years. The N excreted by humans has just as much potential to impact the environment as does N from livestock wastes, inefficient fertilizer-N use or from any other N source. As a comparison in the US, where approximately 1 MMT N are excreted by humans, the amount of commercial fertilizer N consumed in 1998 was 11.3 MMT (FAO, 2001), while N excreted in animal manure was estimated as 4.1 and 7.8 MMT N for collectable and all animal manure in 1997, respectively (Chapter by Hess, et al., this book). Consequently, the need exists to recognize the highly significance role of N in the human diet, as protein, and also the very significant potential impact of increasing human excretion on N as it enters the environment and that this impact will increase as the human population grows.

7. SUMMARY AND CONCLUSIONS

The need by humans to produce and consume food and other agricultural products is increasing with a growing population leading to increased human demands for N to produce food and other products. Therefore it is important to consider the impact human intake, metabolism and excretion of N-containing compounds will have on both humans and the environment. Protein was shown to be an essential source of nitrogen for humans. Other N containing compounds were shown to impact human health as well in both potentially deleterious and beneficial ways. Therefore it is important to continue to analyze and recognize the impact that N containing compounds have on humans and the impact that will occur on the environment as the human population continues to grow.

8. LITERATURE CITED

Bartsch, H, H Ohshima, D Shuker, B Pignatelli, and S Calmels. 1990. Exposure of humans to endogenous N-nitroso compounds: implications in cancer etiology. Mutation Research 238:255-267.

Brody, T. 1994. pp. 295-352. Protein. In: Nutritional Biochemistry. Academic Press, New York, NY.

Bruning-Fann, C and JB Kaneene, 1993a The effects of nitrate, nitrite, and N-Nitroso compounds on human health: A review. Vet Human Toxicol. 35:521-538.

Bruning-Fann, C and JB Kaneene, 1993b. The effects of nitrate, nitrite, and N-nitroso compounds on animal health. Vet Hum Toxicol. 35:237-253.

Calignano, A; BJ Whittle, M; Di Rosa, and S Moncada. 1992. Involvement of endogenous nitric oxide in the regulation of rat intestinal motility in vivo. European Journal of Pharmacology. 229:273-6.

Classen, HG; C Stein-Hammer, and H Thöni. 1990. Hypothesis: the effect of oral nitrite on blood pressure in the spontaneously hypertensive rat. Does dietary nitrate mitigate hypertension after conversion to nitrite? Journal of the American College of Nutrition. 9:500-2.

Coleman, MD and NA Coleman. 1996. Drug-induced methaemoglobinaemia. Treatment issues. Drug Safety.14:394-405.

Cornforth, DP. 1996. pp 259-288. Role of nitric oxide in treatment of foods. In: (J Lancaster, ed) Nitric oxide : principles and actions. Academic Press. San Diego, CA:

Dayal, B and NH Ertel. 1997. Studies on N-nitroso bile acid amides in relation to their possible role in gastrointestinal cancer. Lipids 32:1331.

Doll, R. and R Peto. 1981. The cause of cancer: quantitative estimates of avoidable risks of cancer in the United States today. JNCI. 66:1192-1308.

Dusdieker, L.B.; P.J. Stumbo, B. C. Kross, and C.I. Dungy. 1996. Does increased nitrate ingestion elevate nitrate levels in human milk?. Archives of Pediatrics and Adolescent Medicine. 150:311-4.

Eichholzer M and F Gutzwiller. 1998. Dietary nitrates, nitrites, and N-nitroso compounds and cancer risk: a review of the epidemiologic evidence Nutrition Reviews 56:95-105.

Ellis, A; CG Li, and MJ; Rand. 1998. Effect of xanthine oxidase inhibition on endothelium-dependent and nitrergic relaxations. European Journal of Pharmacology. 356:41-7.

Faivre-Fiorina, B; A. Caron, P. Labrude, C. Vigneron.1998. Erythrocyte, plasma and substitute hemoglobins facing physiological oxidizing and reducing agents. Annales de Biologie Clinique. 56:545-56.

Fan, A.M; C.C. Willhite, and S.A. Book. 1987. Evaluation of the nitrate drinking water standard with reference to infant methemoglobinemia and potential reproductive toxicity. Regulatory Toxicology and Pharmacology. 7:135-48.

Fan, AM and VE Steinberg. 1996. Health implications of nitrate and nitrite in drinking water: an update on methemoglobinemia occurrence and reproductive and developmental toxicity. Regulatory Toxicology and Pharmacology. 23:35-43.

FAO, 2001. Food and Agricultural Organization of the United Nations, FAOSTAT Agricultural Data. <http://apps.fao.org/page/collections?subset=agriculture>.

Ferguson L. (1999) Natural and man-made mutagens and carcinogens in the human diet. Mut Res. 443:1-10.

Fomon, S. J. 1993. Nutrition of Normal Infants. Mosby. Baltimore, MD.

Forman, D. 1989. Are nitrates a significant risk factor in human cancer. Cancer Surveys 8:443-58.

Fowden, L. 1981. pp. 87-95. Contrasts in nitrogen metabolism between animals and plants. In(J.C. Waterlow and J.M.L. Stephen, eds.) Nitrogen Metabolism in Man. Applied Science Publishers. Englewood, NJ.

Fraser, P and C Chilvers,. 1981. Health aspects of nitrate in drinking water. Science of the Total Environment. 18:103-16.

Garrison, RH and E Somer. 1995. The Nutrition Desk Reference. Keats Pub. New Canaan, Conn.

Gruetter, CA; PJ Kadowitz, and LJ; Ignarro. 1981. Methylene blue inhibits coronary arterial relaxation and guanylate cyclase activation by nitroglycerin, sodium nitrite, and amyl nitrite. Canadian Journal of Physiology and Pharmacology. 59:150-6.

Gupta, SK; RC Gupta, AB Gupta; AK Seth, JK Bassin, and A Gupta. 2000. Recurrent acute respiratory tract infections in areas with high nitrate concentrations in drinking water. Environmental Health Perspectives. 108:363-6.

Hjelt, K; JT Lund, B. Scherling, S Bendixen, K Lundstrøm, S Støvring, P Voldsgaard, and K Linnet. 1995. Methaemoglobinaemia among neonates in a neonatal intensive care unit. Acta Paediatrica. 84:365-70.

Honold, J; N Pusser; L Nathan; G Chaudhuri,; LJ Ignarro, and; MP Sherman. 2000. Production and excretion of nitrate by human newborn infants: Neonates are not little adults. Nitric Oxide 4:35-46.

Hotchkiss J. 1989. Preformed N-nitroso compounds in foods and beverages. Cancer Surveys 8:295-321.

Howe, G, L Harrison, and M Main. 1986. A short diet history for assessing dietary exposure to N-nitrosamines in epidemiologic studies. American Journal of Epidemiology. 124:595-602.

Ignarro, LJ.1996. Physiology and pathophysiology of nitric oxide. Kidney International. Supplement. 55:S2-5.

Ignarro, LJ. 1999. Nitric oxide: a unique endogenous signaling molecule in vascular biology. Bioscience Reports. 19:51-71.

Iizuka, T; M Sasaki; K Oishi; S Uemura, M Koike, and M Shinozaki. 1999. Non-enzymatic nitric oxide generation in the stomachs of breastfed neonates. Acta Paediatrica 88:1053-1055.

Jahreis, GV, F Hesse , A Schöne, K Hennig, K Gruhn. 1986. Effect of chronic dietary nitrate and different iodine supply on porcine thyroid function, somatomedin-C-level and growth. Experimental and Clinical Endocrinology. 88:242-8.

Kuijten R, GR Bunin, CC Nass, and AT Meadows. 1990. Gestational and familial risk factors for childhood astrocytoma: results of a case-control study. Cancer Res 50:2608-2612.

Leaf, CD, JS Wishnok, and SR Tannenbaum. 1989. Mechanism of endogenous nitrosation. Cancer Surveys 8:323-34.

Lepoivre, M; F Fieschi, J Coves, L Thelander, and; M Fontecave, M. 1991 Inactivation of ribonucleotide reductase by nitric oxide. Biochemical and Biophysical Research Communications. 179:442-8.

Lijinsky W. 1999. N-nitroso compounds in the diet. Mutation Research 443:129-38.

Magee, PW.1989. The experimental basis for the role of nitroso compounds in human cancer. Cancer Surveys 8:207-239.

McKnight, GM; CW Duncan, C; Leifert, and MH Golden. 1999 Dietary nitrate in man: friend or foe? British Journal of Nutrition. 81:349-58.

Mirvish, S. Role of N-nitroso compounds (NOC) and N-nitrosation in etiology of gastric, esophageal, nasopharyngeal and bladder cancer and contribution to cancer of known exposure to NOC. 1995. Cancer Letters. 93:17-48.

Moilanen, E; and H Vapaatalo. 1995. Nitric oxide in inflammation and immune response. Annals of Medicine. 27:359-67.

Moncada, S; and EA Higgs, 1991. Endogenous nitric oxide: physiology, pathology and clinical relevance. European Journal of Clinical Investigation. 21:361-74.

Murray, JA; and ED Clark. 1994. Characterization of nitric oxide synthase in the opossum esophagus. Gastroenterology. 106:1444-50.

Nixon SE and GE Mawer. 1970. The digestion and absorption of protein in man. The site of absorption. Br J of Nutrition 24:227-40.

Pufahl, RA and MA Marletta.. 1993. Oxidation of NG-hydroxy-L-arginine by nitric oxide synthase: evidence for the involvement of the heme in catalysis. Biochemical and Biophysical Research Communications, 193:963-70.

Radomski, MW; P Vallance, G Whitley, N Foxwell, S Moncada. 1993. Platelet adhesion to human vascular endothelium is modulated by constitutive and cytokine induced nitric oxide. Cardiovascular Research. 27:1380-2.

Radomski, MW. 1995. Nitric oxide: biological mediator, modulator and effector. Annals of Medicine. 27:321-9.

RDA: 1989. Protein and amino acids. pp 52-77. In:Recommended Dietary Allowances. National Academy Press, Washington D. C.

Sarasua, S and D Savitz. 1994. Cured and broiled meat consumption in relation to childhood cancer. Cancer Causes and Control 5:141-8.

Sobala, GM, CJ Schorah, M Sanderson, MF Dixon, DS Tomkins, P Godwin, and AT Axon. 1989. Gastroenterology. 97:357-63.

Stamler, JS; O Jaraki, J Osborne, DI Simon, J Keaney, J Vita, D Singel, CR Valeri, and J Loscalzo. 1992. Nitric oxide circulates in mammalian plasma primarily as an S-nitroso adduct of serum albumin. Proceedings of the National Academy of Sciences of the United States of America. 89:7674-7.

Tarburton, J P; N. F. Metcalf, and W. K. Metcalf. 1985. Methemoglobinization of oxyhemoglobin in pregnant mice by sodium nitrite. Journal of Clinical Hematology and Oncology. 15:77-82.

U.S. Census Bureau, 2001. U.S. Department of Commerce <http://www.census.gov/>and <http://www.census.gov/population/estimates/nation/intfile2-1.txt>.

USEPA. 1973. Nitrogenous compounds in the environment. U.S. Government Printing Office, Washington, D.C. EPA-SAB-73-001.

USEPA. 2001. National Primary Drinking Water Regulations. Consumer Factsheet on: nitrates/nitrites <http://www.epa.gov/safewater/dwh/c-ioc/nitrates.html>

van Maanen, JM; A van Dijk, K Mulder, MH de Baets, PC Menheere, D van der Heide, PL Mertens, and JC Kleinjans. 1994. Consumption of drinking water with high nitrate levels causes hypertrophy of the thyroid. Toxicology Letters. 72:365-74.

van Maanen, JM; IJ Welle, G; Hageman, JW Dallinga, PL Mertens, and JC Kleinjans, JC. 1996a. Nitrate contamination of drinking water: relationship with HPRT variant frequency in lymphocyte DNA and urinary excretion of N-nitrosamines. Environmental Health Perspectives. 104:522-8.

van Maanen, JM; AA van Geel, and; JC Kleinjans. 1996b. Modulation of nitrate-nitrite conversion in the oral cavity. Cancer Detection and Prevention. 20:590-6.

Ward, JK; MG Belvisi, AJ Fox, M; Miura, S Tadjkarimi, MH Yacoub, and PJ; Barnes. 1993. Modulation of cholinergic neural bronchoconstriction by endogenous nitric oxide and vasoactive intestinal peptide in human airways in vitro. Journal of Clinical Investigation. 92:736-42.

Wardlaw, G. M. and P. M. Insel, 1996 Perspectives in nutrition. Mosby, St. Louis, MO.

Weiner, CP; RG Knowles, and S Moncada. 1994. Induction of nitric oxide synthases early in pregnancy. American Journal of Obstetrics and Gynecology, 171:838-43.

Whitney E, C Cataldo, and S Rolfes. 1994 Understanding normal and clinical nutrition. West Publishing Company, New York, NY. 478- 508

Wildman, REC and DM Medeiros. 2000. pp. 123-150. Protein. In: Advanced Human Nutrition. CRC Press New York, NY.

Wright, R, W Lewander, and A Woolf A 1999 Methemoglobinemia: Etiology Pharmacology and Clinical management. Annals of Emergency Medicine 34(646-656).

Yamaguchi N and K Abe. pp 477-487. 1999. Gastric Cancer In: (D Heber, GL Blackburn, and VLW Go, eds) Nutritional Oncology. Academic Press, San Diego, CA.

Zeman, F. J. 1991. Clinical nutrition and dietetics. Macmillan Pub. Co. New York, NY.

Zralý, Z; J. Bendová, D. Svecová, L. Faldíková, Z. Vezník, and A. Zajícová, A. 1997. Effects of oral intake of nitrates on reproductive functions of bulls. Veterinarni Medicina, 42:345-54.

Section 2:

Water Quality

Nitrogen in the Environment: Sources, Problems, and Management
R.F. Follett and J.L. Hatfield (Eds)

Chapter 5. Nitrate Losses to Surface Water through Subsurface, Tile Drainage

G. W. Randall[a] and M. J. Goss[b]

[a] Southern Research and Outreach Center, Univ. of Minnesota, 35838 120th St., Waseca, Minnesota, USA 56093-4521.

[b] Richards Building, Univ. of Guelph, Guelph, Ontario N1G 2W1, CANADA

Wet, poorly drained soils throughout North America and Europe are often artificially drained with subsurface tile systems to remove excess (gravitational) water from the upper 1 to 1.2-m soil profile. Crop production is improved due to better physical conditions for field operations and a deeper unrestricted root zone for greater crop rooting and yields. However, subsurface, tile drainage in agricultural production systems has been identified as a major source of nitrate entering surface waters.

Factors influencing the nitrate content in subsurface waters draining from agricultural production landscapes can be divided into two categories - - uncontrollable and controllable. Precipitation, including variations in annual amount, temporal distribution within a year, or extreme daily events, is an uncontrollable factor having the greatest impact on nitrate loss. Mineralization of soil organic N, an uncontrollable process, but one that can be manipulated, produces nitrate which can be lost in drainage if crop uptake is not synchronized with release of the nitrate.

Controllable factors are those management practices that crop producers use to improve yield and profitability of their enterprise. Cropping system is the controllable factor which has the greatest effect on nitrate losses. Annual row crops such as corn (*Zea mays* L.) and soybean (*Glycine max* L. Merr.) can lose 30 to 50X more nitrate compared to perennial grass and alfalfa (*Medicago sativa*) crops. Nitrate losses from fall and spring sown cereal crops can be 10X greater than for unfertilized grass. Rate of N application is the nutrient management practice that most influences nitrate losses. Time of N application and nitrification inhibitors play a significant role in minimizing nitrate loss, especially under wetter and warmer fall, winter, and spring conditions. Tillage systems generally have little effect on nitrate losses except when fall tillage is succeeded by warm and wet winter and early spring conditions promoting mineralization and tile flow. Limited field studies have shown decreased tile spacing to increase losses, although modeling efforts suggest a minimal effect of spacing and tile depth. Water table management through controlled drainage systems significantly affects nitrate losses.

Information presented in this chapter clearly shows the value of long-term drainage research conducted across a wide range of soil, crop, and climatic conditions when attempting to improve N management, reduce nitrate losses, and influence policy making.

1. INTRODUCTION

Subsurface drainage is a common water management practice in highly productive agricultural areas with poorly drained soils that have seasonally perched water tables or shallow groundwater. This management practice increases crop productivity, reduces risk, and improves economic returns to crop producers. In the North Central Region of the USA, eight states (Illinois, Indiana, Iowa, Michigan, Minnesota, Missouri, Ohio, and Wisconsin) are some of the most highly drained states in the nation (Zucker and Brown, 1998). All together these eight states account for over 20.6 million hectares or 37% of the total cropland drained by surface and subsurface drainage (Fausey et al., 1995). Substantial amounts of nutrients can be contained in subsurface tile drainage water and, depending on the amount of water flow, large losses may occur (Baker and Johnson, 1981; Baker et al., 1975; Randall, 1998).

Society's concerns about the quality of river water and its ecological impact on receiving bodies, i.e., lakes and coastal marine estuaries on the continental shelf, have escalated since the late 1980s. Hypoxia, a condition where the concentration of dissolved oxygen is <2 mg L^{-1}, has been known to exist in portions of the world's oceans and shallow seas for several decades (Diaz and Rozenberg, 1995), but the development of seasonal hypoxia in the Gulf of Mexico off the coast of Louisiana has recently been reported (Rabalais et al., 1994). The cause of hypoxia in this salt water system has been linked indirectly to the load of nutrients, primarily nitrogen, delivered to the Gulf via the Mississippi River drainage basin (Turner and Rabalais, 1994; Rabalais et al., 1996). Nitrate concentrations in the Mississippi River are generally highest in the tributaries emanating from Illinois, Iowa, and Minnesota and vary seasonably, usually being higher in winter, spring and early summer and lower in late summer and early autumn (Antweiler et al., 1995). Burkhart and James (1999) determined the hydrologic units with the largest residual N contributions available to steams and largest total N loss rates are located in the Upper Mississippi River and the Ohio River basins where row crops, particularly corn and soybean, dominate the landscape.

This chapter will focus on (i) the linkage between subsurface tile drainage of agricultural lands and nitrate in surface waters and (ii) the effect of uncontrollable factors (precipitation and soil mineralization) and controllable factors (cropping system, rate and time of N application, nitrification inhibitors, tillage, and drain tile spacing and depth) on nitrate losses to subsurface drainage.

2. ROLE OF AGRICULTURE

Agricultural drainage water has been identified as a major contributor of nitrate-N to receiving waters. Research conducted at widely different scales of watershed basins point to the fact that agricultural systems do affect nitrate levels in river waters. Similarly, long-term field plot research studies have demonstrated the effects of crop and nutrient management practices on edge-of-field losses of nitrate to subsurface drainage water.

2.1 Watershed basins

Omernik (1977) reported that total N concentrations were nearly nine times greater downstream from agricultural lands than downstream from forested areas with the highest concentrations being found in the Corn Belt States. Nitrate-N concentrations in stream-water collected from water years 1984 through 1993 for a portion of the Upper Mississippi River Basin were significantly greater (2 to 6 mg L^{-1}) from those rivers which drain a large percentage of agricultural land compared to those which drain a larger percentage of forested land (0.1 to 0.5 mg L^{-1}) (Kroening, 1996). In the Mississippi River, mean concentrations were significantly greater (1.8 to 2.5 mg L^{-1}) down stream of the confluence with a major river draining an agricultural watershed than upstream (0.2 to 0.9 mg L^{-1}).

Subsurface drains were suggested by Martin et al. (1996) as the cause of high levels of nitrate in the stream water of the White River Basin of Indiana. Subsurface tile drainage directly affected water quality in a small central Indiana stream (Sugar Creek) by transporting soil pore water and shallow groundwater containing high concentrations of nitrate to the creek (Fenelon and Moore, 1998). When tile drains were flowing (typically December through July), elevated nitrate-N concentrations (2-10 mg L^{-1}) in the creek correlated with high nitrate-N concentrations (2-23 mg L^{-1}) in tile drains discharging to the creek. When tile drains went dry, nitrate concentrations in the creek were low, indicating most groundwater discharge to the creek consisted of old or denitrified water. Although agrichemicals in Sugar Creek and many similar small streams throughout the Midwest are not considered a problem to many people, in part because stream water is not used to supply drinking water, elevated nitrate concentrations may be undesirable because of (i) contamination of downstream public water supplies such as reservoirs (Scribner et al., 1996), (ii) economic loss suffered by farmers who lose fertilizer to streams, (iii) potential stresses on fish communities, and (iv) eutrophication of downstream surface waters, e.g., the Gulf of Mexico, receiving discharge water in part from many small drainage basins (Fenelon and Moore, 1998).

In Illinois, elevated concentrations of nitrate-N (5 to 49 mg L^{-1}) were found in four monitored drainage tiles in the Embarras River watershed during a 6-yr period, and these concentrations were synchronous with nitrate concentrations in the Embarras River (David et al., 1997). These authors later estimated that Illinois contributed 15% of the annual total N load of the Mississippi River, and that any reduction strategy in Illinois must address agricultural sources (David and Gentry, 2000).

Generally, in the USA, rivers with higher concentrations of nitrate seem to be surrounded by landscapes and watershed basins with similar general characteristics. They are: 1) humid/high rainfall conditions, 2) soils high in organic matter, 3) poorly drained, fine-textured soils needing artificial subsurface drainage for optimum crop production, and 4) domination by intensive corn and soybean agricultural systems.

2.2 Classical field studies

Monitoring of nitrate losses in subsurface drainage water from agricultural fields began in England in the 1850's (Lawes et al., 1882). About 100 years elapsed before additional drainage studies were installed in both Europe and North America. These drainage facilities have been used to monitor tile effluents for assessing the impact of agricultural management practices on surface and groundwater quality (Addiscott et al., 1991; Hallberg et al., 1986; Kanwar et al., 1988). Subsurface drains integrate the effects of spatial variability on a field scale and may be a better tool for studying agricultural impacts on water quality than measurement methods such as suction cups and soil cores (Richard and Steenhuis, 1988). However, solute concentrations in subsurface drain flow have been shown to not respond immediately to changes in chemical application rates or residual levels in the soil (Jury, 1975a,b; Gast et al., 1978; Baker and Johnson, 1981). Some time lag is exhibited due to traveltime depending on drain spacing, soil hydraulic properties, and precipitation.

3. FACTORS AFFECTING NITRATES IN SUBSURFACE DRAINAGE

The primary factors that influence the nitrate content of surface waters draining agricultural production landscapes can be divided into two categories - - uncontrollable and controllable. Uncontrollable factors include (1) precipitation, and (2) soil organic matter mineralization, which to some degree can be manipulated, e.g., tillage, etc. Controllable factors include those agricultural management practices that are used by crop producers to best fit the needs of their own enterprise. The controllable factors include: cropping system used, rate of N applied, time of N application, placement method, use of a nitrification inhibitor, tillage system, and drain tile spacing and depth.

3.1 Precipitation

Loading of nitrate-N into surface water is a function of transport volume (amount of water) and nitrate-N concentration in the transported water. Since nitrate concentrations of drainage water within a year are normally fairly stable, with characteristic levels for each cropping system, the annual variation in nitrate-N flux lost through subsurface drains depends primarily on the volume of water drained (Bolton et al., 1970; Letey et al., 1977; Goss et al., 1988). Exceptions to this premise exist in particular climate/cropping scenarios as was shown for winter cereal grain production on clay soils in the UK. Nitrate-N concentrations in November often exceeded 50 mg L^{-1} but decreased to between 1 and 5 mg L^{-1} in late February and March (Harris et al., 1984). The amount of drainage water leaving the landscape is largely a function of climate and soil properties, i.e., precipitation, texture, infiltration rate, etc. Drainage is further influenced by the temporal distribution of precipitation within a year and the amount of annual or growing season precipitation that occurs. For instance, an 80-mm rainfall in the spring, when evapotranspiration (ET) losses are low and moisture in the soil above the drains is likely near field capacity, will have a much greater effect on drainage volume than the same rainfall during the middle of the summer when daily ET losses are high and soil moisture is far short of field capacity. In the former scenario, storage capacity is minimal and drainage water carrying nitrates is plentiful. A significant storage reservoir can exist in the soil in the latter scenario, and subsurface drainage may or may not occur.

The effect of climate on subsurface drainage volume is clear in the following tile drainage studies. Annual tile drainage in an 11-yr Minnesota study conducted with continuous corn ranged from 26 to 618 mm yr^{-1} with an average of 297 mm (Randall and Iragavarapu, 1995). Drainage was least in 1989 when growing season precipitation was 35% below normal and greatest in 1991 when growing season precipitation was 51% above normal (Table 1). In addition, drainage in the 3-yr dry period (1987-89) averaged only 43 mm compared to the following 3-yr wet period (1990-92) when drainage averaged 549 mm. A 6-yr study also conducted in Minnesota showed no tile drainage in 1988 and 1989 when annual precipitation was 69 and 76% of normal, respectively (Randall et al., 1997). Drainage from the corn and soybean row-crop systems averaged 22 mm in 1990, 223 mm in 1991, 143 mm in 1992, and 469 mm in 1993 (Table 2). Annual precipitation in those four years was 95, 125, 117, and 160% of normal, respectively. Data from these studies clearly indicate the strong relationship between annual or growing season precipitation and volume of subsurface, tile drainage.

Table 1
Influence of growing season precipitation on drainage volume and annual nitrate-N losses. (adapted from Randall and Iragavarapu, 1995).

Year	April-October rainfall [1/]	Drainage	Nitrate-N	
			Conc.[2/]	Lost
	mm	Mm	mg L^{-1}	kg ha^{-1}
1986	796	402	14	55
1987	586	42	9	4
1988	426	46	15	6
1989	414	26	12	2
1990	789	486	24	112
1991	961	618	24	139
1992	726	417	14	55

[1/] 1961-90 Normal = 639 mm
[2/] Annual flow-weighted concentration.

Drainage volume and nitrate losses have been shown to be affected also by the temporal distribution of annual precipitation within a year coupled with ET and wintertime temperatures. Subsurface drainage occurs primarily during the late fall, winter, and early spring in those geographic areas where soils are not frozen throughout the winter (Kladivko et al., 1991; Drury et al., 1993; Goss et al., 1993, 1998; Fenelon and Moore, 1998; Kladivko et al., 1999). Between 78 and 100% of annual flow occurred after harvest during the dormant season of October to May in Ontario (Patni et al., 1996). An 8-yr study in the UK showed 84% of the nitrate lost in subsurface drainage occurred between fall seeding of annual cereals and the first application of N in the spring (Goss et al., 1988). In Iowa, 45 to 85% of the

Table 2
Amount of subsurface drainage water as affected by annual precipitation and cropping system. (adapted from Randall et al., 1997).

Crop System	Year					
	1988	1989	1990	1991	1992	1993
	mm					
Continuous Corn	0	0	20	178	132	442
Corn-Soybean	0	0	18	274	122	488
Soybean-Corn	0	0	28	218	175	478
Alfalfa	0	0	0	41	56	320
CRP[1/]	0	0	0	43	86	510
Average of Row Crop Systems	0	0	22	223	143	469
Average of Perennial Crop Systems	0	0	0	42	71	415
% of Normal Annual Precipitation[2/]	69	76	95	125	117	160

[1/] CRP = Conservation Reserve Program (mixture of grass and alfalfa).
[2/] 1961-90 Normal = 647 mm.

annual nitrate loss through subsurface drainage occurred in the spring and fall when crops were not actively growing (Bjorneberg et al., 1996). In Minnesota where soils usually remain frozen from early December through late March, subsurface drainage occurs primarily between mid-April and early July (Gast et al., 1978). The 3-month period of April, May, and June accounted for 65% of the annual drainage volume (Fig. 1) and 70% of the annual nitrate loss from corn rotated with soybean in a 13-yr (1987-99) Minnesota study (Randall, 2000). The high proportion of annual subsurface drainage flow occurring from late fall through the spring without an actively growing crop has profound implications on N management, especially fall-applied N and residual nitrate in the soil profile after harvest.

Nitrate-N concentrations in subsurface drainage water do not appear to vary consistently with daily drain flow volumes but do show seasonal and yearly variability (Harris et al., 1984; Kladivko et al., 1991). Factors affecting this variability include crop uptake of N, residual nitrate in the soil from the previous year, and amount and distribution of rainfall. Goolsby et al. (1997) noted that the concentration and flux of nitrate in rivers of the Mississippi River Basin tend to be highest in the spring when stream flow is highest. Increased flows and elevated concentrations in agricultural tile drains were speculated as contributing to this relationship. Annual average nitrate-N concentrations in the Des Moines River from 1980 through 1990 ranged from 2.0 mg L^{-1} in 1989 to 9.1 mg L^{-1} in 1982 with an 11-yr average of 5.6 mg L^{-1} (Keeney and DeLuca, 1993). Maximum daily nitrate-N concentration ranged from 5.9 mg L^{-1} in 1989 to 14.5 mg L^{-1} in 1982.

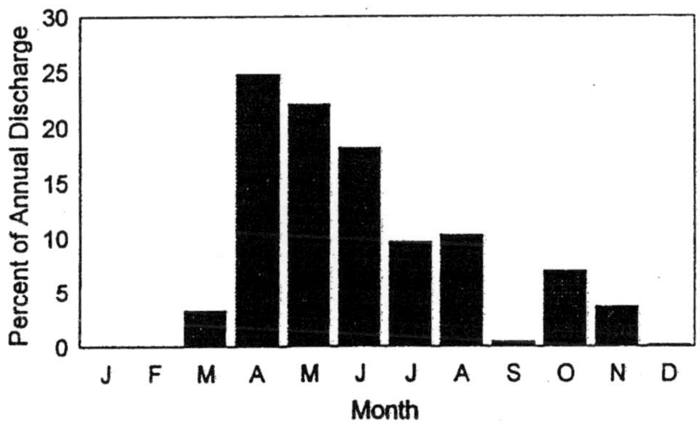

Fig. 1. **Monthly distribution of subsurface tile drain discharge averaged across a 13-yr (1987-99) period for corn.**

Nitrate-N losses in subsurface tile drainage water vary greatly among years due primarily to differences in annual and growing season precipitation. Fourteen studies conducted for 3 to 11 years in North America and Europe show annual nitrate-N losses ranging from 0 to 139 kg ha^{-1} yr^{-1} (Table 3). Moreover, a few days of high-flow events can lead to most of the annual nitrate loss in some years (David et al., 1997). Because of the influence of precipitation on nitrate losses, long-term drainage research, which integrates the effect of climatic variability, is vital to our understanding of nitrate losses from agricultural production systems to subsurface drainage.

Nitrate-N concentrations and losses are also greatly affected by dry and wet climatic cycles (Randall, 1998; Logan et al., 1994). These effects are seen at the river scale as well as at the edge-of-field scale. In Iowa, excessive nitrate-N concentrations in 1990 in the Raccoon River followed 2 yr of less-than-normal precipitation in 1988 and 1989 (Lucey and Goolsby, 1993). High concentrations in streamflow persisted as streamflow decreased to baseflow conditions, implying that substantial quantities of nitrate-N were being leached from the soil and transported by subsurface flow in 1990. Drainage plots in Minnesota with continuous corn grown from 1986 through 1992 clearly showed the edge-of-field effects of dry and wet climatic cycles (Randall, 1998). Annual flow-weighted nitrate-N concentration in 1986 was 14 mg L^{-1} while drainage totaled 402 mm (Table 1). Dry conditions during 1987-1989, when April-October rainfall was 25% below normal, resulted in <50 mm drainage yr^{-1} and annual average nitrate-N concentrations ranging from 9 to 15 mg L^{-1}. Residual soil N (RSN) totaled 225 kg ha^{-1} in the 0 to 1.5 m profile in October 1989. In 1990 and 1991, April-October rainfall averaged 36% above normal and generated annual drainage volumes >480 mm yr^{-1}. In addition, nitrate-N concentrations in the drainage water doubled from the previous three dry years to 24 mg L^{-1} in these two wet years. At the end of 1991, RSN was 50% lower than

Table 3
Range of annual nitrate-N losses in subsurface tile drainage water in European and North American research studies.

Source	Location	Years Studied	Crop(s)	Nitrate-N Loss
				kg ha^{-1} yr^{-1}
Gast et al., 1978	Minnesota, USA	3	CC	6 to 25
Bergstrom, 1987	Sweden	4	Barley	<1 to 27
Goss et al., 1988	England	8	Wheat	6 to 73
Kladivko et al., 1991[1/]	Indiana, USA	3	CC	28 to 31
Drury et al., 1993	Ontario, CAN	3	CC	7 to 29
Logan et al., 1994	Ohio, USA	4	C-Sb	<1 to 86
Randall and Iragavarapu, 1995	Minnesota, USA	11	CC	1 to 139
Bjorneberg et al., 1996[2/]	Iowa, USA	3	CC	19 to 100
		3	C – Sb	17 to 53
		3	Sb – C	8 to 52
Patni et al., 1996	Ontario, CAN	4	CC	10 to 39
Randall et al., 1997	Minnesota, USA	6	CC	0 to 91
			C – Sb	0 to 88
			Sb – C	0 to 81
Bjorneberg et al., 1998[2/]	Iowa, USA	3	C – Sb	3 to 33
Catt et al., 1998	England	4	Mixed [3/]	1 to 48
			Cereals [4/]	4 to 22
			Grass	1 to 10
Jaynes et al., 1999	Iowa, USA	4	C – Sb	5 to 51
Kladivko et al., 1999[1/]	Indiana, USA	3	CC	22 to 50
Watson et al., 2000	Ireland	9	Ryegrass	16 to 52
Randall et al., 2000	Minnesota, USA	4	CC	12 to 20

1/ 20 m tile spacing only.
2/ Chisel plow tillage only.
3/ Mixed crops of barley, beans, and wheat including winter cover crops.
4/ Winter barley, winter oats, and winter wheat.

at the end of the dry years. In the third consecutive wet year (1992), more than 400 mm of water drained from the plots, nitrate-N concentrations in the drainage water returned to 14 mg L^{-1}, and RSN totaled only 50 kg ha^{-1}. Nitrate-N loading in the subsurface drainage water each year was greatly affected by both nitrate-N concentration and drainage volume. These data clearly indicate a buildup of RSN in the soil profile during dry years when drainage was limited. Much of the RSN buildup could likely be attributed to mineralization of soil organic matter, annual additions of fertilizer N, and limited uptake of N by the poor yielding corn. In the subsequent wet years, substantial losses of nitrate-N occurred in subsurface drainage due to high concentrations of nitrate-N and high drainage volumes.

The general effects of precipitation on nitrate-N losses can also be illustrated using basin-wide water quality monitoring data collected in the Minnesota River basin, a 4 million hectare agricultural basin draining to the Upper Mississippi River basin (Mulla, 1997). Mean annual precipitation increased from about 560 mm on the western side to 810 mm on the eastern side across this distance, which produces a corresponding and dramatic increase in the discharge from subsurface tile drains into ditches and streams that eventually flow into the Minnesota River. Water quality monitoring data from 1977 – 1994 show that nitrate-N concentrations range from 0.36 mg L^{-1} in the headwaters on the western side to 4.6 mg L^{-1} at the mouth of the river on the eastern end where it enters the Mississippi River. Fewer than 1% of the water quality samples collected since 1977 from the western portion of the basin have a nitrate-N concentration that exceeds 10 mg L^{-1}. About 10% of the water quality samples collected since 1977 exceed 10 mg L^{-1} on the eastern side of the basin. Differences in nitrate-N contributions across the basin in response to a gradient in precipitation are even larger when nitrate-N loads are compared. Four watersheds located in the wetter eastern portion of the basin account for 75% of the total nitrate-N load in the entire basin, yet they drain only 31% of the total basin area. Six watersheds on the drier western side of the basin collectively generate only 7% of the nitrate-N load.

3.2 Soil mineralization

Mineralization of soil organic matter can contribute a substantial amount of nitrate that is susceptible to loss in subsurface, tile drainage. Drain gauges built in 1870 at the Rothamsted Experimental Station in the UK were isolated laterally and vertically for the collection of drainage water from undisturbed soils. Drainage water collected from 1877 to 1915 on these soil blocks that had not been cropped, fertilized, or cultivated averaged from 39 to 45 kg nitrate-N ha^{-1} yr^{-1} during the first seven years. Nitrate-N leakage decreased to 30 to 35 kg ha^{-1} yr^{-1} in the last 28 years of the study. Mineralization of soil organic matter coupled with atmospheric deposition of N and non-symbiotic N_2 fixation were the primary sources of nitrate in this long-term study. (Addiscott, 1988). Four field drainage plots in Minnesota were fallowed (no crop grown and no N applied) with periodic tillage each year from 1987 until 2000. Nitrate-N concentration in the tile drainage water averaged 57 mg L^{-1} in 1990 following three dry years with no drainage. Flow-weighted annual concentrations decreased to 38, 25, and 23 mg L^{-1} in 1991, 1992, and 1993, respectively, and have plateaued at about 20 mg L^{-1} since (Randall, 2000).

Numerous studies have shown substantial losses of nitrate during the winter and early spring prior to N fertilizer application for fall-sown cereals (see section 3.1). Much of the nitrate lost during this period has been attributed to mineralization, but residual soil nitrate after the previous crop also may have contributed significantly. Loss of fertilizer N by leaching after spring topdressing is usually less than loss in autumn and winter of N released by the mineralization of organic matter, provided that good agronomic practices are adopted (Addiscott et al., 1991; Goss et al., 1993). The magnitude of the leaching loss of N released by mineralization in autumn and winter depended on the method of tillage, nutrient uptake by the current crop, volume of drainage, and previous crop.

Tile drainage from continuous corn plots that received only 20 kg N ha^{-1} yr^{-1} in Minnesota contained annual flow-weighted nitrate-N concentrations of 13, 19, and 19 mg L^{-1} in 1973, 1974, and 1975, respectively (Gast et al., 1978). No drainage occurred in 1976, an extremely dry year. In 1977, with slightly above-normal rainfall, nitrate-N concentrations averaged 28 mg L^{-1} from these plots. Drainage studies in Iowa indicated that nitrate-N losses to subsurface drainage water occur primarily as a result of asynchronous production and uptake of nitrate in the soil and the presence of large quantities of potentially mineralizable N in soil organic matter (Cambardella et al., 1999). Keeney and DeLucca (1993) examined nitrate-N concentrations in the Des Moines River in 1945, 1955, 1976, and from 1980 through 1990 and found the average nitrate-N concentration to have changed little in the last 45 years (5.0 mg L^{-1} in 1945 to 5.6 mg L^{-1} in 1980-90). They concluded that intensive agricultural practices that enhance mineralization of soil N coupled with subsurface, tile drainage are the major contributors of nitrate-N rather than solely fertilizer N. Somewhat similar conclusions were drawn by David et al. (1997) who surmised that agricultural disturbance leading to high mineralization rates and N fertilization combined with subsurface, tile drainage contributed significantly to nitrate export in the Embarras River in Illinois. In their 6-yr study, an average of 49% (range from 25 to 85%) of the large pool of nitrate remaining after harvest was leached through drain tiles and exported by the River.

3.3 Cropping systems

Nitrate-N concentrations in and losses to subsurface drainage water have frequently been shown to be affected by cropping systems. In one of the earliest Ontario studies, nitrate losses in tile drainage were highest with continuous corn, intermediate with a corn-oat (*Avena sativa* L.)-alfalfa (*Medicago sativa)*-alfalfa rotation, and lowest with continuous bluegrass *(Poa pratensis* L.) (Bolton et al., 1970). Logan et al. (1980) reported nitrate losses were highest with N-fertilized corn, intermediate for soybean or systems where other crops were in rotation, and lowest with alfalfa. Tile drainage from alfalfa fields in California also contained very low concentrations and losses of nitrate-N (Letey et al., 1977). Subsurface drainage from row-crop systems (continuous corn and a corn-soybean rotation) fertilized with N based on a spring soil test was compared to that from perennial crops (alfalfa and a CRP grass-alfalfa mix) in a Minnesota study (Randall et al., 1997). Four-yr flow-weighted nitrate-N concentrations averaged 28 mg L^{-1} for continuous corn, 22 to 23 mg L^{-1} for the corn-soybean rotation and <2 mg L^{-1} for the alfalfa and CRP systems. Due to greater flow volumes (Table 2) and nitrate concentrations from the row crops, nitrate losses from the row crops ranged from 30 to 50X greater than from the perennial crops (Table 4). Similar results were reported by Drury et al., (1993) who found nitrate concentrations and losses of 12 to 17 mg L^{-1} and 38 kg ha^{-1}, respectively, for continuous corn compared to 1.2 to 2.6 mg L^{-1} and 4 kg ha^{-1} for bluegrass during a 2-yr period in Ontario. Nitrate-N concentrations in drainage from alfalfa were also shown to be much lower compared to corn and soybeans in Iowa (Baker and Melvin, 1994).

Table 4
Effect of crop system on 4-yr flow-weighted nitrate-N concentrations and total nitrate-N loss (adapted from Randall et al., 1997).

Crop System	4-Yr Flow-weighted Nitrate-N Concentration	4-Yr Total Nitrate-N Loss
	mg L^{-1}	kg ha^{-1}
Continuous corn	28	217
Corn-soybean	23	204
Soybean-corn	22	202
Alfalfa	1.6	7
CRP	0.7	4

Nitrate losses to drainage water from corn-soybean rotations have been compared to continuous corn in Iowa and Ohio. A 3-yr study in Iowa with recommended N rates of 202 kg ha^{-1} for continuous corn and 168 kg ha^{-1} for corn after soybean showed markedly higher nitrate-N concentrations and losses for continuous corn (Table 5) (Kanwar et al., 1997). These authors further concluded that cropping system had a much greater effect on nitrate losses than did tillage. A 4-yr study in Ohio showed nitrate-N concentrations in subsurface drainage water from soybean to be as high or higher than with corn in a corn-soybean rotation, especially in the spring (Logan et al., 1994). They concluded that a significant portion of the nitrate in tile drainage is due to N carried over from the previous corn crop. Potato fields in New Brunswick, Canada were also found to be quite leaky with nitrate-N losses ranging from 5 to 33 kg ha^{-1} for April through December in 10 site-years (Milburn et al., 1990).

Table 5
Three-yr average nitrate-N concentrations and losses as affected by crop rotation in Iowa. (adapted from Kanwar et al., 1997).

Crop rotation	Nitrate-N[1/]	
	Concentration	Loss
	mg L^{-1}	kg ha^{-1}
Continuous corn	30	58
Corn-soybean	18	28
Soybean-corn	18	29

[1/] Averaged across four tillage systems.

Perennial crops have been shown to reduce nitrate losses to tile drainage compared to cereal and bean crops. Robbins and Carter (1980) showed lower nitrate-N concentrations with alfalfa compared to dry bean *(Phaseolus vulgaris* L.) in Idaho. In Swedish studies, Bergstrom (1987) found much lower drainage volume and nitrate-N concentrations and flux with two perennial ley crops, fescue *(Festuca pratensis* L.) and alfalfa compared to barley *(Hordeum distichum* L.). The fescue and alfalfa leys acted as optimal catch crops, mostly due to their extended growing season. Because grass absorbs N whenever mineralization is occurring, grassland is potentially a less leaky system than arable farming (Addiscott et al., 1991), especially when the grass is unfertilized (Ryden et al., 1984). However, plowing a grassland releases mineral N from the organic matter and much nitrate can potentially leach to drainage water, particularly from older stands.

The effect of three cropping systems [continuous winter cereals (barley, oats, and wheat), mixed crops (winter and spring cereals, spring beans, winter cover crops, and winter fallow), and an unfertilized, ungrazed grass ley] on nitrate-N losses to subsurface, tile drainage was measured on hydrologically isolated field plots in England (Catt et al., 1998). Four-year average nitrate-N concentrations and total nitrate-N losses were substantially lower for the grass plots than for the plots planted to continuous winter cereals (Table 6). Nitrate-N concentrations and losses in the drainflow from the grass plots were about 90% lower than the mean for the annual crops in the second, third, and fourth years. Losses of nitrate-N during the 4-yr period were 160% greater for the mixed cropping systems receiving an annual average fertilizer N application rate of 93 kg ha^{-1} compared to the continuous winter cereals that received an annual fertilizer N rate of 138 kg ha^{-1}. This dramatic effect of cropping systems occurred consistently both in the winter when losses were large and in the spring after N fertilization when losses were small. These researchers concluded that on heavy clay soils in the UK a cropping regime of continuous winter cereals offers the best compromise between profitable crop production and minerals nitrate loss to surface waters.

Table 6
Influence of three European cropping systems on the 4-yr average nitrate-N concentration and total nitrate-N loss (adapted from Catt et al., 1998).

Crop System	Avg. Nitrate-N Concentration	Total Nitrate-N Loss Winter[1/]	Spring[2/]	Annual
	mg L^{-1}	kg ha^{-1}		
Mixed crops	33	102	11	113
Winter cereals	30	39	4	43
Grass ley	5	13	1	14

1/ Between harvest and the first spring application of fertilizer N.
2/ Between the first spring application of fertilizer N and harvest.

Cover crops planted in the fall and killed prior to spring planting have been shown to effectively reduce downward movement of nitrate. Winter rye *(Secale cereale)* was planted in the fall as a cover crop following corn harvest in a Minnesota drainage experiment (Strock et al., 2000). The rye was then chemically killed prior to planting soybeans the next spring. Water and nitrate loss through subsurface tile drains was reduced 23% and 50%, respectively, compared with the corn-soybean rotation without the rye cover crop in a year of adequate rainfall. Martinez and Guiraud (1990) concluded from a lysimeter study that keeping the soil covered with a growing crop during the autumn and winter is considered to be an efficient means to reduce nitrate leaching from arable land in contrast to conducting fall tillage in wheat production systems. Although less nitrate-N was leached from winter cover crops compared to winter fallowing in a mixed winter and spring cropping system in England, greater amounts of nitrate were leached from the cover crop system compared to continuous winter cereals (Catt et al., 1998). Poor synchronization between mineralization of the cover crop residues and periods of N uptake by the subsequent spring and winter-planted cereals was hypothesized as the reason for increased nitrate loss. These results indicate that the entire cropping system needs to be considered when deciding whether to establish cover to minimize nitrate leaching.

In summary, these studies show substantially higher nitrate-N concentrations in row crops, especially continuous corn, compared to perennial crops that have an extended period of greater root activity (water and nutrient uptake) and where cycling of N is optimized.

3.4 Rate of N application

Applying the proper rate of N for a crop is a major management decision facing crop producers. Using too little N for a highly responsive crop such as corn or wheat results in lower yields, poorer grain quality, and reduced profits. When too much N is applied, crop yields and quality are not impacted, but profit can be reduced somewhat and negative environmental consequences likely will occur.

Long-term research provides the guidance necessary to make N rate decisions. Nitrogen rate recommendations also include credits for N from sources such as manure and N fixed by legumes. These N credits are then subtracted from the total amount of N required by the crop to provide a fertilizer N rate recommendation. Even though the examples used in the following discussion focus on fertilizer N, it should be remembered that these principles also relate to N supplied by manure and legume fixation.

Many have shown that the mass of nitrate-N leached and/or lost in subsurface tile drainage increases as N fertilization rates increase (Angle et al., 1993; Baker and Johnson, 1981; Gast et al., 1978). The relationship between the annual fertilizer N rate for continuous corn and the annual flow-weighted nitrate-N concentration in subsurface tile water is shown in Fig. 2 for a Minnesota study (Randall, 2000). Annual N rates were begun in 1975 but no drainage occurred in 1975 and 1976 due to very dry weather. Thus, at the beginning of 1977 increasingly high amounts of residual soil nitrate (RSN) remained in the soil profile with each additional amount of N applied. Consequently, very high concentrations of nitrate-N were found in the 12 cm of drainage water in 1977. Nitrate-N concentrations in the drainage water were lower in 1978 and were reduced further in 1979 as drainage volume increased and yields improved. Annual flow-weighted N concentrations from the 0-kg N plots ranged from 13 to 16 mg L^{-1}, again indicating the role that soil mineralization played during this dry to wet climatic cycle in this high organic matter soil. Averaged across the 3 yr when tile flow occurred, nitrate-N concentrations in the drainage water were increased by 16 mg L^{-1} when the N rate was increased from 112 to 224 kg ha^{-1} and by 20 mg L^{-1} when N rate was increased from 224 to 336 kg ha^{-1}. If 190 kg N ha^{-1} was the recommended N rate for a yield goal of 10 Mg ha^{-1}, but the grower decided to apply an additional 45 kg N ha^{-1} for "insurance" purposes, based on these data, nitrate-N concentrations in the drainage water would be projected to increase about 7 mg L^{-1}. If an annual N credit of 100 kg ha^{-1} from manure were ignored and a total of 290 kg N ha^{-1} was applied annually, nitrate-N concentrations could be expected to increase by about 17 mg L^{-1}. On the other hand, if the N rate was reduced 10% to 170 kg ha^{-1}, nitrate-N concentrations could be expected to decrease by about 3 mg L^{-1} with a yield reduction of about 0.3 to 0.4 Mg ha^{-1}.

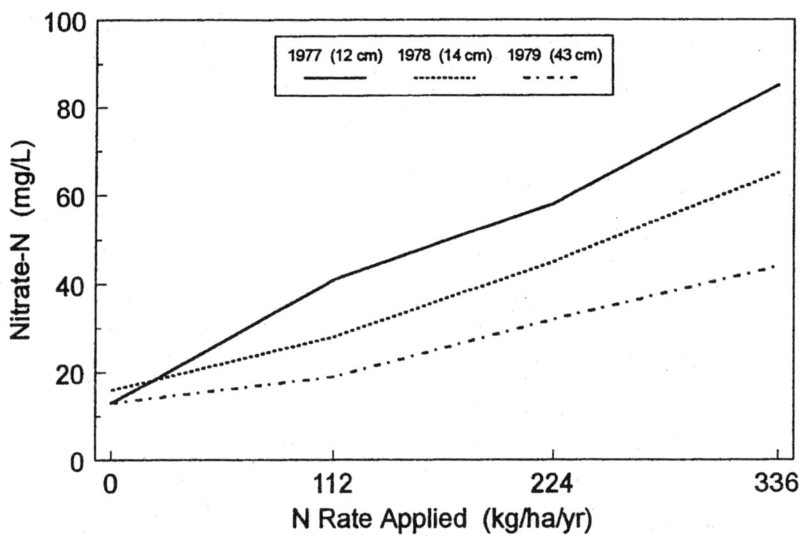

Fig. 2. **Nitrate-N concentration in tile drainage water as affected by N rate for continuous corn in Minnesota.**

Although abnormally dry conditions prevailed for the first two years of the above study, the results clearly show the effect of increasing N rate on the concentration of nitrate-N in tile drainage water. Residual soil nitrate that accumulates in the soil profile during dry periods is the major source of the nitrate lost in tile drainage. Accounting for RSN following dry years by using spring soil N tests (Blackmer et al., 1989; Bundy et al., 1993; Magdoff, 1984;

Schmitt and Randall, 1994) could be quite helpful to growers. Unless the nitrate has been leached below the sampling zone, these tests should be able to provide information that would lead to reductions in the rate of fertilizer N recommended, resulting in lower nitrate-N losses in tile drainage water.

Improved manure management, including uniform application of known nutrient amounts and immediate incorporation, is critical if the optimum N rate is to be achieved in livestock production systems. Altogether too often manure is applied with a disposal objective in mind rather than with a utilization objective. When this occurs, rates of N as manure tend to be high and are not distributed evenly across the field. Consequently, credit is not given for N in the manure and the total rate of N (fertilizer plus manure) becomes excessive. When the nutrient content of manure is known and best management practices are used in land application, manure does not lead to greater nitrate losses to subsurface, tile drainage than does N from commercial fertilizer (Randall et al., 2000). If manure is applied at greater than agronomic rates, elevated concentrations of nitrate will occur in the drainage water.

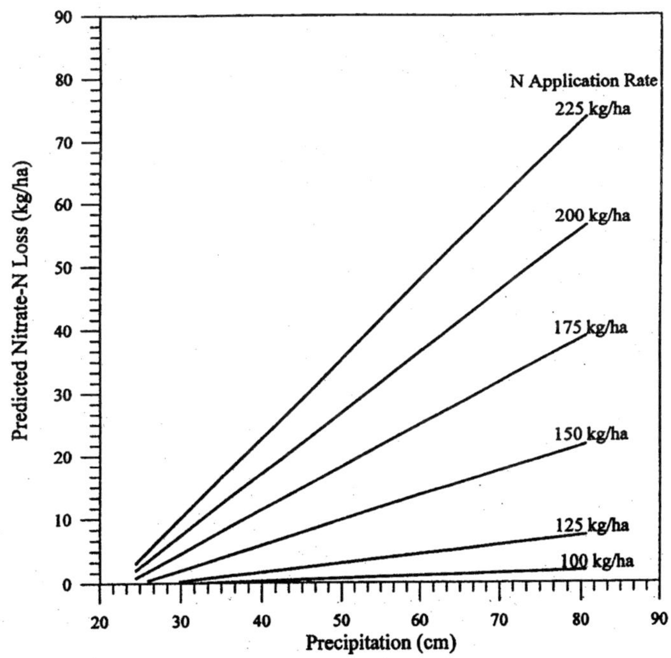

Fig. 3. **Predicted nitrate-N loss in subsurface tile drainage as affected by N rate and April-August precipitation in Minnesota.**

The relationship between growing season precipitation (April – August) and predicted nitrate-N losses in tile drainage in Minnesota for various N rates is shown in Fig. 3. These relationships were obtained by running the ADAPT model for 82 years of precipitation data and plotting predicted drainage losses of nitrate-N versus precipitation (Davis et al., 2000). As expected, the predicted nitrate-N losses in wet years were much greater than in dry years for a given rate of applied N, and the magnitude of nitrate-N losses increased as N application rate increased. In dry years, nitrate-N losses through tile drainage were quite low for all N application rates because of a lack of precipitation to drive nitrate leaching. During normal years (50 cm of precipitation), nitrate-N losses were reduced from about 35 kg ha^{-1} to about 10 kg ha^{-1} when N fertilizer application rates were reduced from 225 kg ha^{-1} to 150 kg ha^{-1}.

3.5 Time of N application

Agronomically and environmentally speaking, spring applications are frequently superior to fall application because N loss between application and N uptake by the crop is less. However, many U.S. corn growers, especially in the northern part of the Corn Belt, desire to apply N in the fall because they usually have more time and field conditions are better. Early planting of corn as soon as the soils are fit in the spring is desirable for highest yields and profit. Consequently, the window of opportunity for spring N application becomes very narrow (Randall and Schmitt, 1998). Soil compaction can also be a deterrent to spring application of N. In Europe, N applied in autumn, either as mineral fertilizer (Goss et al., 1993) or as animal manure (Thompson et al., 1987) is very vulnerable to leaching in the winter.

Nitrogen was applied as 15-N depleted ammonium sulfate in the fall and spring for continuous corn to determine the effect of N application time and rate on nitrate losses to subsurface drainage and corn yields in Minnesota (Randall and Mulla, 2001). Corn yields from the late fall application (early November) of 134 and 202 kg N ha^{-1} averaged 8% lower than with spring (late April) application (Table 7). In addition, annual losses of nitrate-N in the tile drainage water averaged 36% higher (9 kg ha^{-1} yr^{-1}) with fall application compared to spring application. Averaged across time of application, yields and nitrate-N losses in the drainage water were 17 and 30% higher for the 202-kg rate compared to the 134-kg rate. At the end of the study, 65% of the N being lost in the drainage from the 268-kg fall treatment was derived from the fertilizer, whereas only 15% of the N in the drainage water lost from the 134-kg spring treatment was derived from the fertilizer (Buzicky et al., 1983).

Split application of N should theoretically result in increased N efficiency and reduced nitrate losses because of greater synchronization between time of application and crop uptake. Evidence in the literature to support this concept is mixed, however. Baker and Melvin (1994) reported losses of nitrate-N to be higher for split application compared to a preplant application for continuous corn. Losses with split application for the corn-soybean rotation were lower in the year of application but tended to be higher in the following year when soybeans followed corn. In another Iowa study, Bjorneberg et al (1998) concluded that combining a split N fertilizer management strategy based on the pre-sidedress nitrate soil test

Table 7
Effect of N rate and time of application on nitrate-N losses to subsurface drainage and corn yield in Minnesota.

N[1/]		Annual Loss of	
Rate	Time	Nitrate-N in Drainage	5-Yr Yield Avg.
kg ha^{-1}		kg ha^{-1} yr^{-1}	Mg ha^{-1}
0	0	8	4.1
134	Fall	30	8.2
134	Spring	21	9.4
202	Fall	38	10.0
202	Spring	29	10.5

[1/] Ammonium sulfate applied about 1 Nov or 1 May.

(PSNT) with no-tillage practices can have positive environmental benefits without reducing corn yields in a corn-soybean rotation.

3.6 Nitrification inhibitors

Nitrification inhibitors are sometimes added to ammonium fertilizers [anhydrous ammonia (AA) and urea] to retard or slow the conversion of ammonium to nitrate after fertilizer application. Nitrapyrin (N-Serve) has been the most commonly used nitrification inhibitor in the U.S. and has been a component in many N research studies. Many of these studies have focused on the agronomic effects of nitrapyrin coupled with leaching conditions. Anhydrous ammonia was applied in four treatments [(late fall, late fall + nitrapyrin, spring preplant, and split (40% preplant + 60% sidedress)] to drainage plots in Minnesota from 1987 through 1993 (Randall and Mulla, 2001). Subsurface tile drainage did not occur in 1987 through 1989 due to very dry conditions. Flow weighted nitrate-N concentrations across the 4-yr flow period (1990-93) averaged 20, 17, 16, and 16 mg L^{-1} for the four treatments, respectively (Table 8). Yields were increased significantly in the very wet years by the addition of nitrapyrin to the fall application. These data obtained from poorly drained, fine-textured soils during wetter than normal years suggest that application of AA in the spring or in late fall along with a nitrification inhibitor (N-Serve) would: (1) reduce nitrate-N concentration by about 3 to 4 mg L^{-1}, (2) reduce nitrate-N flux by about 18 kg ha^{-1} yr^{-1} and (3) increase corn yields about 0.6 Mg ha^{-1} compared to a late fall application of AA without a nitrification inhibitor. Earlier fall applications of AA when soil temperatures are warmer and conversion to nitrate (nitrification) is faster would be expected to produce even greater losses of nitrate to drainage water and poorer yields.

Response to nitrapyrin may be particularly dependent on time of N application. Quesada et al. (2000) reported the agronomic and economic effects of nitrapyrin applied with ammonia in the spring during a 10-yr period in Iowa. Grain yield responses occurred with

Table 8
Effect of time of N application and nitrapyrin on nitrate-N losses and corn yield in a corn-soybean rotation. (adapted from Randall and Mulla, 2001).

N Treatment[1/]	Avg. Annual F.W. nitrate-N Conc	Total Nitrate-N Lost	Avg. Yield
	mg L^{-1}	kg ha^{-1}	Mg ha^{-1}
Fall	20	264	8.0
Fall + nitrapyrin	17	208	8.6
Spring	16	177	8.6
Split	16	190	9.0

[1/] Anhydrous ammonia applied 25 Oct or 1 May.

nitrapyrin in one year for continuous corn but did not occur for corn in rotation with soybean. These data suggest that nitrapyrin would not likely be beneficial in reducing nitrate losses to tile drainage when applied with ammonia in the spring.

3.7 Tillage

Numerous studies have been conducted in Canada, Europe, and North America to determine the effect of tillage on nitrate losses to subsurface drainage water. The effect of tillage depends largely on the cropping system and the time of year when most of the drainage occurs. Corn and soybean systems in the Corn Belt and southern Canada generally show greater amounts of drain flow for no tillage and other very reduced tillage systems, whereas nitrate-N concentrations tend to be greater with the more conventional tillage systems, i.e., moldboard plow and chisel plow (Bjorneberg et al., 1996; Patini et al., 1996; Randall and Iragavarapu, 1995). Multiple tillage systems were compared in a continuous corn and a corn-soybean rotation for a 3-yr period in Iowa (Table 9) while moldboard plow (MP) and no-till (NT) systems were compared for continuous corn in Minnesota (11 years) and Ontario (40 months) (Table 10). Nitrate-N losses among the tillage treatments in all three studies were not statistically significant, leading to the general conclusion that tillage does not affect total nitrate flux in subsurface drainage in these corn and soybean systems. However, a 3-yr study with continuous corn in Ontario (Drury et al., 1993) did show slightly decreased concentrations and losses of nitrate-N for ridge tillage and no tillage compared to moldboard plow tillage.

Cereal production systems in Europe, where drainage during the late fall and winter predominates, have shown significant affects of tillage on nitrate losses in subsurface drainage. In the UK, annual losses through mole and tile drainage were smaller from direct drilled (no-till) wheat than from plowed plots on a clay soil (Goss et al., 1988, 1993). Nitrate losses during the 6-yr period were reduced by 18% with direct drilling compared to plowing

Table 9
Drain flow, nitrate-N concentrations, and nitrate-N losses in subsurface tile drainage as influenced by tillage in three Iowa cropping systems. (adapted from Bjorneberg et al., 1996).

Tillage	3-Yr Average		
		Nitrate-N	
	Drain flow	Concentration	Loss
	mm	mg L^{-1}	kg ha^{-1}
		Continuous corn	
Moldboard plow	129	38	47
Chisel plow	194	32	65
Ridge tillage	207	25	54
No tillage	263	23	63
		Corn after soybean	
Moldboard plow	131	20	28
Chisel plow	178	20	35
Ridge tillage	145	17	24
No tillage	132	16	24
		Soybean after corn	
Moldboard plow	152	21	32
Chisel plow	182	20	35
Ridge tillage	168	16	25
No tillage	168	14	25

Table 10
Effect of tillage for continuous corn on tile flow and nitrate-N losses in subsurface drainage in Minnesota and Ontario. (adapted from Randall and Iragavarapu, 1995 and Patini et al., 1996).

Parameter	Minnesota		Ontario	
	MP	NT	MP	NT
Annual drainage (mm)	279	315	119	174
Mean nitrate-N conc. (mg L^{-1})	15	13	25	21
Annual nitrate-N loss (kg ha^{-1})	43	41	32	36
N lost as a % of N applied	21	20	21	24

(Table 11). Tillage prior to planting enhanced mineralization providing a larger pool of nitrate in the fall, which along with the 20 kg ha^{-1} application of fertilizer N leached into subsurface drainage during the winter. Nitrate losses after the application of 157 kg N ha^{-1} in the spring totaled only 15 to 25% of the annual loss and were not different between the two tillage systems. Reducing the depth of autumn tillage from 200 mm to 100 mm or incorporating harvest residues compared to burning did not significantly decrease nitrate leaching in 3 or 4 years in a continuous winter cereal cropping system (Goss et al., 1998). Nitrate leaching in spring sown small grain crops (barley, oats, and wheat) was 24 kg N ha^{-1} greater with early fall (September) moldboard plow tillage compared to spring tillage during a 3-yr period in Sweden (Stenberg et al., 1999). Based on the results from these European studies, it seems likely that fall tillage in the southern and eastern region of the Corn belt may also stimulate over-winter losses of nitrate, especially following soybeans.

Table 11
Nitrate-N loss in subsurface drainflow as affected by tillage for fall sown cereals in England from 1981-86. (adapted from Goss et al., 1988).

Tillage system	6-Yr Avg. Nitrate-N Loss		
	Winter	Spring	Annual
	---------------- kg ha^{-1} ----------------		
Plow	33	6	39
Direct drill (No till)	24	8	32

3.8 Drain tile spacing and depth

The optimum spacing between subsurface, tile drains is generally a function of the hydraulic properties of the soil, annual precipitation, rate or speed of drainage desired, depth of placement, and installation cost. As farms and farm equipment get larger, the importance of timeliness and suitable soil moisture conditions for conducting field operations becomes more critical for crop producers. Soil compaction is likely reduced and trafficability enhanced with closer drain spacings that remove excess water more quickly. Lower yields are often thought to be due to inadequate drainage in these poorly drained soil landscapes. Consequently, many farmers install additional drain tile to narrow the spacing between tiles with the expectation that crop yields will be improved because of enhanced drainage.

A 6-yr study on a poorly drained Clermont silt loam soil in Indiana showed drain spacing to markedly affect nitrate losses in subsurface drainage water (Kladivko et al., 1991; Kladivko et al., 1999). Annual nitrate-N losses averaged across the 6-yr were 32, 41, and 60 kg ha^{-1} for the 20-m, 10-m, and 5-m drain spacings, respectively. These differences were largely affected by drain flow which averaged 130, 166, and 242 mm yr^{-1} for the three spacings, respectively. Nitrate-N concentrations were affected very little by drain spacing. Yield measurements taken during a 4-yr period (1984-87) at this site showed lower corn yields on the 5-m spacings than on the 10-m or greater spacings (Larney et al., 1989). Averaged across a 10-yr period, corn yields were not different among the three drain spacings. Although few data exist, inferences drawn from this field work suggest that narrowing tile drainage to spacing <20 m could result in greater losses compared to wider spacings. Recent modeling studies, however, suggest nitrate losses in subsurface drainage are not reduced as much with changes in tile spacing as compared to reductions in rate of N applied. Davis et al. (2000) reached this conclusion after conducting long-term (80 yr) simulations using the ADAPT model. Similar conclusions were drawn from the DRAINMOD-N model using climatic variation during a 24-yr period (Zhao et al., 2000). Additional research is needed to more clearly define the agronomic and environmental influence of tile spacing in subsurface drainage systems.

Controlled drainage is a management practice that adjusts or controls water height within a subsurface, tile drainage system. Controlling the drainage by some mechanical means allows timely drainage but also maximizes storage of water within the field for utilization by the crop. Field studies in North Carolina and Ontario have shown substantial reductions in losses of nitrate from subsurface drainage where controlled drainage has been employed.

Nitrate reductions in controlled drainage systems result from two factors: (i) volume of drainage water leaving a field is reduced and (ii) denitrification within the soil profile is promoted because a higher water table level is maintained. Water table depth in a controlled drainage system should be no lower than 90 cm below the soil surface to reduce nitrate losses (Gilliam et al., 1997). In North Carolina, reductions in nitrate losses were due to less water leaving the field via tile flow in moderately well-drained soils (Gilliam et al., 1979). In Ontario, nitrate loss reductions resulted from both less flow and lower nitrate concentrations because of enhanced denitrification. During a 3-yr period, Drury et al. (1996) found controlled drainage to reduce flow volume by 24%, nitrate concentration by 25%, and nitrate loss by 43%. In a 2-yr study, Mejia and Madramootoo (1998) found nitrate in effluent from controlled drainage at 50 and 75-cm depths to be reduced by 84 and 77%, respectively, in 1995 and 61 and 52%, respectively, in 1996. Nitrate loads to the receiving lake were reduced 95 and 30%, respectively, in 1996 compared to free drainage. Results from these studies demonstrated that controlled drainage is an effective method of reducing nitrate losses from subsurface drainage in relatively flat landscapes favorable to their installation.

4. SUMMARY

Field studies have been conducted in Europe since the mid-1800s and in North America for the last 40 years to assess the role of subsurface tile drainage on promoting nitrate losses from agricultural production areas to surface waters. Generally speaking, factors influencing nitrate losses in subsurface drainage can be divided into two categories - - uncontrollable and controllable. Uncontrollable factors consist of precipitation and soil mineralization which can be manipulated to some degree. Controllable factors are those management practices used by crop producers to best fit the needs of their enterprise.

The following summarizes the effects of many individual factors on nitrate losses in sub- surface drainage from agricultural production systems.

- Precipitation has a huge effect on drainage flow volume and nitrate losses. Annual nitrate-N losses can range from 0 in dry years to >100 kg ha^{-1} in years with much above-normal precipitation. Losses are also affected by dry and wet climatic cycles with greatest losses occurring in wet years following dry years. The temporal distribution of precipitation (fall vs. winter vs. spring vs. summer) can affect some N and crop management decisions, i e., time of N application, tillage, etc. A substantial proportion of the annual nitrate loss may occur within only a few days when soils are saturated and very large precipitation events happen.

- Soil mineralization contributes nitrate that is susceptible to loss in subsurface drainage. Nitrate losses can be very high, especially following dry years, if crop uptake is not synchronized with availability of the mineralized nitrate.

- Nitrate losses from the landscape are highly related to cropping system. Row crops, i.e., corn and soybean, yield much greater drainage volumes and nitrate-N concentrations in the drainage water than do perennial crops, i.e., alfalfa and CRP grass/legume mix. Nitrate-N losses can be 30 to 50X higher from these row crops compared to these perennial crops. Cereal production systems in Europe also yield greater nitrate-N concentrations and losses in drainage water than do unfertilized grass by crops. This is especially true for spring-planted cereals when a winter crop has not been established.

- Rate of N application affects nitrate losses in drainage more than any other nutrient management decision. Nitrate losses increase as the rate of N application increases. At application rates greater than needed by the crop, the excess N will likely remain in the soil profile where it is highly susceptible to loss, especially in wetter-than-normal years. Nitrate losses are not different among N sources as long as the rate of application is similar.

- Nitrate losses to drainage as affected by time of N application are greatly influenced by temporal distribution of precipitation, evapotranspiration demand, leaching, and source of N. Fall application of N is particularly risky if leaching is significant between application and crop uptake. If over-winter leaching is absent due to frozen soils and spring conditions are dry, losses from fall-applied N will not be different from spring-applied N. Anhydrous ammonia and manure are likely to be less affected by time of application due to slower nitrification than those fertilizers that nitrify more quickly. Sidedress and late split applications of N to corn tend to give greater nitrate concentrations in drainage in the succeeding year compared to preplant applications.

- Nitrification inhibitors added to fall-applied anhydrous ammonia can reduce nitrate losses and improve corn yields in those years when leaching is synchronous with persistence of the inhibitor. If leaching occurs after the inhibitor has degraded, protection against nitrate losses will no longer be provided. Spring application of nitrification inhibitors has generally not been beneficial on poorly drained soils.

- Tillage systems for row crops in the Northern US generally do not affect the amount of nitrate lost in drainage. Drainage flow volume is often greatest for the very reduced systems while nitrate concentrations are greatest for the conventional, plow-tillage systems. In cereal production systems in Europe, where over-winter drainage dominates, fall tillage increases nitrate losses compared to no tillage. This again shows the need to consider climate and time of leaching when making crop and nutrient management decisions to minimize nitrate losses.

- Limited field studies show greater nitrate losses as spacing between tile lines is reduced. However, modeling studies suggest minimal effect of drain spacing and depth compared to rate of N applied. Controlled drainage systems that manage water table depth throughout the season can significantly lower nitrate losses in drainage water.

- Long-term, subsurface drainage research, which integrates the effects of climatic variability, soil properties, and various cropping systems, is vital to our understanding of nitrate losses to subsurface drainage. Educators and policy makers must consider this research as they deal with the occurrence of nitrates from agricultural production systems in surface waters.

REFERENCES

Addiscott, T.M. 1988. Long-term leakage of nitrate from bare unmanured soil. Soil Use and Management. 4:91-95.

Addiscott, T. M., A. P. Whitmore, and D. S. Powlson. 1991. Farming, Fertilizers and the Nitrate Problem. CAB International, Wallingford, UK.

Angle, J. S., C. M. Gross, R. L. Hill, and M. S. McIntosh. 1993. Soil nitrate concentrations under corn as affected by tillage, manure, and fertilizer applications. J. Environ. Qual. 22:141-147.

Antweiler, R.C., D.A. Goolsby, and H.E. Taylor. 1995. Nutrients in the Mississippi River. pp. 73-86. *In* R.H. Meade (ed.) Contaminants in the Mississippi River, 1987-92. U.S. Geological Survey, Circ. 1133.

Baker, J. L., K. L. Campbell, H. P. Johnson, and J. J. Hanway. 1975. Nitrate, phosphorus, and sulfate in subsurface drainage water. J. Environ. Qual. 4:406-412.

Baker, J. L., and H. P. Johnson. 1981. Nitrate-nitrogen in tile drainage as affected by fertilization. J. Environ. Qual. 10:519-522.

Baker, J. L., and S. W. Melvin. 1994. Chemical management, status and findings. pp. 27-60. *In* Agricultural Drainage Well Research and Demonstration Project – Annual Report and Project Summary. Iowa Dept. of Agric. and Land Stewardship and Iowa State Univ.

Bergstrom, L. 1987. Nitrate leaching and drainage from annual and perennial crops in tile-drained plots and lysimeters. J. Environ. Qual. 16:11-18.

Bjorneberg, D. L., R. S. Kanwar, and S. W. Melvin. 1996. Seasonal changes in flow and nitrate N loss from subsurface drains. Trans. ASAE 39:961-976.

Bjorneberg, D. L., D. L. Karlen, R. S. Kanwar, and C. A. Cambardella. 1998. Alternative N fertilizer management strategies effect on subsurface drain effluent and N uptake. Appl. Eng. Agric. 14(5):469-473.

Blackmer, A. M., D. Pottker, M. E. Cerrato, and J. Webb. 1989. Correlations between soil nitrate concentrations in late spring and corn yields in Iowa. J. Prod. Agric. 2:103-109.

Bolton, E. F., J. W. Aylesworth, and F. R. Hore. 1970. Nutrient losses through tile drains under three cropping systems and two fertility levels on a Brookston clay. Can. J. Soil Sci. 50:275-279.

Bundy, L. G., M. A. Schmitt, and G. W. Randall. 1992. Predicting N fertilizer needs for corn in humid regions: Advances in the Upper Midwest. *In* B. R. Bock and K. R. Kelley (eds.). Predicting N Fertilizer Needs for Corn in Humid Regions. Bull. Y-226. Nat. Fert. and Environ. Ctr., TVA, Muscle Shoals, AL 35660.

Burkhart, M. R., and D. E. James. 1999. Agricultural-nitrogen contributions to hypoxia in the Gulf of Mexico. J. Environ. Qual. 28:850-859.

Buzicky, G. C., G. W. Randall, R. D. Hauck, and A. C. Caldwell. 1983. Fertilizer N losses from a tile-drained mollisol as influenced by rate and time of 15-N depleted fertilizer application. p. 213. *In* Agronomy abstracts. ASA, Madison, WI.

Cambardella, C. A., T. B. Moorman, D. B. Jaynes, J. L. Hatfield, T. B. Parkin, W. W. Simpkins, and D. L. Karlen. 1999. Water quality in Walnut Creek Watershed:Nitrate-nitrogen in soils, subsurface drainage water, and shallow groundwater. J. Environ. Qual. 28:25-34.

Catt, J. A., K. R. Howse, D. G. Christian, P. W. Lane, G. L. Harris, and M. J. Goss. 1998. Strategies to decrease nitrate leaching in the Brimstone Farm Experiment, Oxfordshire, UK, 1988-93: the effects of winter cover crops and unfertilized grass leys. Plant and Soil 203:57-69.

David, M. B., and L. E. Gentry. 2000. Anthropogenic inputs of nitrogen and phosphorus and riverine export for Illinois, USA. J. Environ. Qual. 29:494-508.

David, M. B., L. E. Gentry, D. A. Kovacic, and K. M. Smith. 1997. Nitrogen balance in and export from an agricultural watershed. J. Environ. Qual. 26:1038-1048.

Davis, D. M., P. H. Gowda, D. J. Mulla, and G. W. Randall. 2000. Modeling nitrate-nitrogen leaching in response to nitrogen fertilizer rate and tile drain depth or spacing for southern Minnesota, USA. J. Environ. Qual. 29:1568-1581.

Diaz, R. J., and R. Rosenberg. 1995. Marine benthic hypoxia: A review of its ecological effects and the behavioral responses of benthic macrofauna. Oceanogr. Marine Biol. 33:245-303.

Drury, C. F., D. J. McKenney, W. I. Findlay, and J. D. Gaynor. 1993. Influence of tillage on nitrate loss in surface runoff and tile drainage. Soil Sci. Soc. Am. J. 57:797-802.

Drury, C. F., C. S. Tan, J. D. Gaynor, T. O. Oloya, and T. W. Welacky. 1996. Influence of controlled drainage-subirrigation on surface and tile drainage nitrate loss. J. Environ. Qual. 25:317-324.

Fausey, N. R., L. C. Brown, H. W. Belcher, and R. S. Kanwar. 1995. Drainage and water quality in Great Lakes and Cornbelt states. J. Irrig. and Drain. Engrg. 121(4):283-288.

Fenelon, J. M., and R. C. Moore. 1998. Transport of agrichemicals to ground and surface water in a small central Indiana watershed. J. Environ. Qual. 27:884-894.

Gast, R. G., W. W. Nelson, and G. W. Randall. 1978. Nitrate accumulation in soils and loss in tile drainage following nitrogen applications to continuous corn. J. Environ. Qual. 18:258-261.

Gilliam, J. W., D. L. Osmond, and R. O. Evans. 1997. Selected Agricultural Best Management Practices to Control Nitrogen in the Neuse River Basin, North Carolina Agricultural Research Service Technical Bulletin 311, North Carolina State University, Raleigh, NC.

Gilliam, J. W., R. W. Skaggs, and S. B. Weed. 1979. Drainage control to diminish nitrate loss from agricultural fields. J. Environ. Qual. 8:137-142.

Goolsby, D. A., W. A. Battaglin, and R. P. Hooper. 1997. Sources and transport of nitrogen in the Mississippi River Basin. *In* Proc. of Conf. "From the Corn Belt to the Gulf.... Agriculture and Hypoxia in the Mississippi River Watershed". St. Louis, Mo. 14-15 July.

Goss, M. J., P. Colburn, G. L. Harris, and K. R. Howse. 1988. Leaching of nitrogen under autumn sown crops and the effects of tillage. pp. 269-282. *In* D. S. Jenkinson and K. A. Smith (ed.) Nitrogen efficiency in agricultural soils. Elsevier Applied Science. England.

Goss, M. J., K. R. Howse, D. G. Christian, J. A. Catt, and T. J. Pepper. 1998. Nitrate leaching: modifying the loss from mineralized organic matter. J. of Soil Science 49:649-659.

Goss, M. J., K. R. Howse, P. W. Lane, D. G. Christian, and G. L. Harris. 1993. Losses of nitrate-nitrogen in water draining from under autumn-sown crops established by direct drilling or mouldboard ploughing. J. of Soil Sci. 44:35-48.

Hallberg, G. R., J. L. Baker, and G. W. Randall. 1986. Utility of tile-line effluent studies to evaluate the impact of agricultural practices on groundwater. pp. 298-326. *In* Proc. Conf. Agricultural Impacts on Groundwater, Omaha, NE. 11-13 Aug. 1986. Nat'l Well Water Assoc., Dublin, OH.

Harris, G. L., M. J. Goss, R. J. Dowdell, K. R. Howse, and P. Morgan. 1984. A study of mole drainage with simplified cultivation for autumn-sown crops on a clay soil. 2. Soil water regimes, water balances, and nutrient loss in drain water, 1978-80. J. Agric. Sci., Camb. 102:561-581.

Jaynes, D. B., J. L. Hatfield, and D. W. Meek. 1999. Water quality in Walnut Creek Watershed: Herbicides and nitrate in surface waters. J. Environ. Qual. 28:45-59.

Jury, W. A. 1975a. Solute travel-time estimates for tile-drained fields: I. Theory. Soil Sci. Soc. Am. Proc. 39:1020-1024.

Jury, W. A. 1975b. Solute travel-time estimates for tile drained fields: II. Application to experimental studies. Soil Sci. Soc. Am. Proc. 39:1024-1028.

Kanwar, R. S., J. L. Baker, and D. G. Baker. 1988. Tillage and split N-fertilization effects on subsurface drainage water quality and crop yields. Trans. ASAE 31:453-460.

Kanwar, R. S., T. S. Colvin, and D. L. Karlen. 1997. Ridge, moldboard, chisel, and no-till effects on tile water quality beneath two cropping systems. J. Prod. Agric. 10:227-234.

Keeney, D. R., and T. H. DeLuca. 1993. Des Moines River nitrate in relation to watershed agricultural practices: 1945 versus 1980s. J. Environ. Qual. 22:267-272.

Kladivko, E. J., J. Grochulska, R. F. Turco, G. E. Van Scoyoc, and J. E. Eigel. 1999. Pesticide and nitrate transport into subsurface tile drains of different spacings. J. Environ. Qual. 28:997-1004.

Kladivko, E. J., G. E. Van Scoyoc, E. J. Monke, K. M. Oates, and W. Pask. 1991. Pesticide and nutrient movement into subsurface tile drains on a silt loam soil in Indiana. J. Environ. Qual. 20:264-270.

Kroening, S. E. 1996. Nitrogen and phosphorus in streams in part of the Upper Mississippi River Basin, Minnesota and Wisconsin, 1984-93. *In* Proc. Rivertech 96 1st Conf. On New/Emerging concepts for Rivers. W. H. C. Maxwell et al (eds.). Chicago, IL.

Larney, F. J., E. J. Kladivko, E. J. Monke, and J. V. Mannering. 1989. Corn growth and yield behavior with distance from subsurface drains in drier than normal growing seasons. Trans. ASAE 32:579-587.

Lawes, J. B., J. H. Gilbert, and R. Warington. 1882. On the amount and composition of rain and drainage water at Rothamsted. Part III. J. of the Royal Agric. Soc. of England. 18:1-71.

Letey, J., J. M. Blair, D. Devitt, L. J. Lund, and P. Nash. 1977. Nitrate-nitrogen in effluent from agricultural tile drains in California. Hilgardia 45:289-319.

Logan, T. J., D. J. Eckert, and D. G. Beak. 1994. Tillage, crop and climatic effects on runoff and tile drainage losses of nitrate and four herbicides. Soil & Tillage Res. 30:75-103.

Logan, T. J., G. W. Randall, and D. R. Timmons. 1980. Nutrient content of tile drainage from cropland in the North Central Region. North Central Reg. Res. Publ. 268. OARDC Res. Bull. 1119. OARDC, Wooster, OH.

Lucey, K. J., and D. A. Goolsby. 1993. Effects of climatic variations over 11 years on nitrate-nitrogen concentrations in the Raccoon River, Iowa. J. Environ. Qual. 22:38-46.

Magdoff, F. R., D. Ross, and J. Amadon. 1984. A soil test for nitrogen availability to corn. Soil Sci. Soc. Am. J. 48:1301-1304.

Martin, J. D., C. G. Crawford, J. W. Frey, and G. A. Hodgkins. 1996. Water-quality assessment of the White River Basin, Indiana – Analysis of selected information on nutrients, 1980-92. U. S. Geol. Surv. Water Resour. Invest. Rep. 96-4192. USGS, Denver, Co.

Martinez, J., and G. Guiraud. 1990. A lysimeter study of the effects of a ryegrass catch crop during winter wheat/maize rotation on nitrate leaching and on the following crop. J. Soil Sci. 41:5-16.

Mejia, M. N., and C. A. Madramootoo. 1998. Improved water quality through water table management in Eastern Canada. J. Irrig. Drain. Engrg. 121:283-288.

Milburn, P., J. E. Richards, C. Gartley, T. Pollock, H. O'Neill, and H. Bailey. 1990. Nitrate leaching from systematically tiled potato fields in New Brunswick, Canada. J. Environ. Qual. 19:448-454.

Mulla, D. J., (ed.). 1997. Minnesota river basin water quality overview. 8 pp. Univ. of Minnesota Extension Service. FO-7079E.

Omerinik, J. M. 1977. Nonpoint source-stream nutrient level relationships – A nationwide study. EPA-600/3-P77-105, 151 p. U. S. Environmental Protection Agency. Corvallis, OR.

Patni, N. K., L. Masse, and P. Y. Jui. 1996. Tile effluent quality and chemical losses under conventional and no-tillage - Part 1: Flow and nitrate. Trans. ASAE 39:1665-1672.

Quesada, J. P., R. Killorn, and A. M. Dierdickx. 2000. Response of corn grown in two crop rotations to different N rates and nitrapyrin. p. 274. *In* Agronomy abstracts. ASA. Madison, WI.

Rabalais, N. N., R. E. Turner, D. Justic, Q. Dortch, W. J. Wiseman, Jr., and B. K. Sen Gupta. 1996. Nutrient changes in the Mississippi River and system responses on the adjacent continental shelf. Estuaries 19:386-407.

Rabalais, N. N., W. J. Wiseman, Jr., and R. E. Turner. 1994. Comparison of continuous records of near-bottom dissolved oxygen from the hypoxic zone along the Louisiana coast. Estuaries 14 (4):850-861.

Randall, G. W. 2000. Nitrogen management and its influence on N losses to surface water through subsurface tile lines. pp. 14-32. *In* Proc. 30th North Central Extension – Industry Soil Fertility Conf., St. Louis, MO. 15 Nov. 2000.

Randall, G. W. 1998. Implications of dry and wet cycles on nitrate loss to subsurface tile drainage. pp. 53-60. *In* Proc. 7th Annual Drainage Symposium. Drainage in the 21st Century. 8-10 March, 1998. Orlando, FL

Randall, G. W., and T. K. Iragavarapu. 1995. Impact of long-term tillage systems for continuous corn on nitrate leaching to tile drainage. J. Environ. Qual. 24:360-366.

Randall, G. W., T. K. Iragavarapu, and M. A. Schmitt. 2000. Nutrient losses in subsurface drainage water from dairy manure and urea applied for corn. J. Environ. Qual. 29:1244-1252.

Randall, G. W., D. R. Huggins, M. P. Russelle, D. J. Fuchs, W. W. Nelson, and J. L. Anderson. 1997. Nitrate losses through subsurface tile drainage in CRP, alfalfa, and row crop systems. J. Environ. Qual. 26:1240-1247.

Randall, G. W., and D. J. Mulla. 2001. Nitrate-N in surface waters as influenced by climatic conditions and agricultural practices. J. Environ. Qual. 30:(Accepted).

Randall, G. W., and M. A. Schmitt. 1998. Advisability of fall-applying nitrogen. pp. 90-96. *In* Proc. of the 1998 Wisconsin Fertilizer, Aglime, and Pest Management Conf. Middleton, WI. 20 Jan. 1998.

Richard, T. L., and T. S. Steenhuis. 1988. Tile drain sampling of preferential flow on a field scale. J. Contam. Hydrol. 3:307-325.

Robbins, C. W., and D. L. Carter. 1980. Nitrate-nitrogen leached below the root zone during and following alfalfa. J. Environ. Qual. 9:447-450.

Ryden, J. C., P. R. Ball, and E. A. Garwood. 1984. Nitrate leaching from grassland. Nature 311:50-53.

Schmitt, M. A., and G. W. Randall. 1994. Developing a soil nitrogen test for improved recommendations for corn. J. Prod. Agric. 7:328-334.

Scribner, E. A., D. A. Goolsby, E. M. Thurman, M.T. Meyer, and W. A. Battaglin. 1996. Concentrations of selected herbicides, herbicide metabolites, and nutrients in outflow from selected midwestern reservoirs, April 1992 through September 1993. U. S. Geol. Surv. Open-File Rep. 96-393. USGS, Denver, CO.

Stenberg, M., H. Aronson, B. Linden, T. Rydberg, and A. Gustafson. 1999. Soil mineral nitrogen and nitrate leaching losses in soil tillage systems combined with a catch crop. Soil Tillage
Res. 50:115-125.

Strock, J. S., P. M. Porter, and M. P. Russelle. 2000. Reduced nitrate leaching from tile drainage using a rye cover crop. p. 323. *In* Agronomy abstracts. ASA. Madison, WI.

Thompson, R. B., J. C. Ryden, and D. R. Lockyer. 1987. Fate of nitrogen in cattle slurry following surface application or injection to grassland. J. of Soil Sci. 38:689-700.

Turner, R. E., and N. N. Rabalais. 1994. Coastal eutrophication near the Mississippi River delta. Nature 368:619-621.

Watson, C. J., C. Jordan, S. D. Lennox, R. V. Smith, and R. W. J. Steen. 2000. Inorganic nitrogen in drainage water from grazed grassland in Northern Ireland. J. Environ. Qual. 29:225-232.

Zhao, S. L., S. C. Gupta, D. R. Huggins, and J. F. Moncrief. 2000. Predicting subsurface drainage, corn yield, and nitrate nitrogen losses with DRAINMOD-N. J. Environ. Qual. 29:817-825.

Zucker, L. A., and L. C. Brown (Eds.). 1998. Agricultural drainage: Water quality impacts and subsurface drainage studies in the Midwest. Ohio State Univ. Ext. Bul. 871.

Chapter 6. Nitrogen in Groundwater Associated with Agricultural Systems

Michael R. Burkart[a] and Jeffrey D. Stoner[b]

[a] U.S. Department of Agriculture, Agricultural Research Service, National Soil Tilth Laboratory, Ames, Iowa

[b] U.S. Geological Survey, National Water Quality Assessment Program, Nutrient Synthesis Office, Denver, Colorado

1. INTRODUCTION

Nitrogen, particularly in the form of nitrate, is the most common contaminant in aquifer systems (Freeze and Cherry, 1979). Hallberg (1989) points to agriculture as the most substantial anthropogenic source of nitrate, and Keeney (1986) suggests this is caused by the intensive and extensive land-use activities associated with crops and animal production. The discussion of the occurrence of nitrogen in groundwater beneath agricultural systems is presented by examining the factors influencing aquifer vulnerability to nitrogen contamination, and by characterizing the geographic distribution of groundwater contamination by nitrogen. Factors that influence aquifer vulnerability are presented in the context of exposure to nitrogen sources from general agricultural systems and hydrologic conditions that facilitate transfer of those sources to groundwater. This analysis focuses on the occurrence of nitrate in the United States because data are readily available on many variables needed for such an analysis. Data from the U.S. Geological Survey National Water Quality Assessment Program (NAWQA, Gilliom et al., 1995); the Census of Agriculture; the National Resources Inventory; and the State Soils Geographic Database [STATSGO (U.S. Department of Agriculture, 1994)] provide an unique opportunity to directly relate nitrogen in groundwater to agricultural systems at a national or continental scale. Results of international research and monitoring are introduced to compare the occurrence of nitrogen in similar agricultural and hydrologic systems supported by literature and data available from the United Nations Food and Agriculture Organization.

1.1 Groundwater and Well Water

Selection of groundwater chemistry information is critical to understanding whether the aquifer is contaminated or whether wells used for drinking water have intercepted some contaminated ground or surface water adjacent to the well. Some excellent studies have provided information about nitrate concentrations in wells used for private and community drinking water (Richards et al., 1996, U.S. Environmental Protection Agency, 1990; Kross, et al. 1990; Monsanto Company, 1990, LeMasters and Doyle, 1989, and Exner and Spalding, 1985). These studies provide valuable human-health information, but less information about the general condition of aquifers that form the groundwater resource. Other regional, national, and statewide studies of the quality of groundwater resources have included assessments of ambient conditions in aquifers (Nolan and Stoner, 2000, Mueller and Helsel, 1996, Kolpin et al., 1996, and Burkart and Kolpin, 1993).

Aquifers are subsurface materials that store and transmit groundwater from recharge areas to discharge areas. Recharge areas often cover large parts of the landscape, whereas discharge areas generally are relatively small, such as surface water bodies and withdrawal wells. Aquifers and individual wells can be contaminated by substantially different processes. Aquifers can be contaminated by agricultural-chemical use over large parts of recharge areas. Properly constructed wells down gradient from recharge areas can withdraw water with dissolved contaminants derived from those areas. Agricultural chemicals can contaminate improperly constructed wells without appreciably affecting the aquifer. This contamination can occur when chemicals present near a well move from the surface down the outside of the well casing or laterally into the well through hydrologic units that are not isolated during well construction. The following discussion will only include processes by which aquifers can be impacted by nitrogen derived from agricultural systems and leached to aquifers in recharge areas.

1.2 Forms of Nitrogen in Groundwater

The forms of nitrogen generally measured in groundwater include nitrate (NO^-_3), nitrite (NO^-_2), and ammonia (NH^+_3) ions. Most analyses combine NO^-_3 and NO^-_2 and investigators report this as NO^-_3 because NO^-_2 occurs in substantially smaller concentrations in groundwater than NO^-_3. Nitrite also is an intermediate product of both nitrification and denitrification that is relatively unstable (Keeney, 1986), which helps explain its limited occurrence in groundwater. Nitrification is an oxidizing process and denitrification a reducing process with respect to NO^-_3, but both are biologically mediated. Organic nitrogen is rarely measured and not well known in groundwater (Korom, 1992). A generally accepted hypothesis is that measurable NH^+_3 and organic nitrogen rarely occur in groundwater because the required biological activity to produce them is minimal in groundwater systems. Nolan and Stoner (2000) reported that nitrate was detected more than 13 times as often as NH^+_3 and organic nitrogen in shallow groundwater of major aquifers of the United States, based on a detection threshold of 0.2 mg/L as nitrogen. In fact, concentrations of ammonia in groundwater rarely exceeded 0.1 mg/L, indicating chemical instability. This chapter will deal dominantly with nitrate (reported as nitrate + nitrite) with reference to NH^+_3 occurrence where limited information is available.

1.3 Nitrate Contamination Levels

Contamination is the occurrence of NO^-_3 that exceeds a generally accepted concentration attributable to natural conditions. The authors are not aware of studies to examine minimum natural thresholds of NH^+_3, perhaps because its occurrence is too infrequent and concentrations are comparatively small. Nitrate is the most common contaminant in aquifers (Freeze and Cherry, 1979) and has been the most frequently mentioned groundwater contaminant associated with human activities throughout the world for several decades. The large number of NO^-_3

measurements may be due to the establishment of international standards for drinking water for this ion and its wide distribution in the environment (Feth, 1966). The concentrations of nitrate in waters unaffected by human activities were shown to be less than 10 mg/L of NO^-_3 by Feth (1966). A wide range of natural or background concentrations in groundwater has been reported for specific geographic locations from as small as 0.2 mg/L NO^-_3 in Ohio (Baker et al., 1989) to as much as 100 mg/L NO^-_3 in the Sahel of Africa (Edmunds and Gaye, 1997). Nitrogenous minerals have been reported in geologic materials that could provide natural sources of nitrate to groundwater in the northern Great Plains of the United States (Boyce et al., 1976 and Ferguson et al., 1972) and in the San Joaquin Valley, California (Strathouse et al., 1980), for example. Extensive analysis of historical data from the U.S. for the *National Water Summary* by Madison and Brunett (1985) concluded that concentrations in excess of 3 mg/L may be indicative of human inputs. A more recent analysis of U.S. Geological Survey national data from shallow groundwater (<30 m) beneath forest and rangeland concluded that 2.0 mg/L is a probable threshold for background concentration of NO^-_3 (Mueller and Helsel, 1996).

1.4 Temporal Factors in Nitrate Contamination

Little monitoring data exists to interpret temporal trends of nitrate in ground water because few monitoring programs have been designed to look at the quality of groundwater over time. Some examples have documented increased nitrate concentrations that relate to increased use of fertilizer and irrigation in the Snake River Plain of Idaho and the San Joaquin Valley of California (Fuhrer et al., 1999). Studies in the San Joaquin Valley showed that from the 1950s through 1980, the use of nitrogen fertilizer increased from 51,756 to 338,230 metric tons per year. This was accompanied by an increase of nitrate concentrations in groundwater from less than 2 to about 5 mg/L for the same period of time.

The complexities in the distribution of nitrate even in relatively simple hydrogeologic settings can confound interpretations of how groundwater nitrate relates to agricultural practices at the land surface. Recently, accurate methods of determining absolute groundwater dates for recharge as long ago as the 1940s have improved our understanding of groundwater contamination relative to the history of agricultural practices. Small atmospheric concentrations of man-made chlorofluorocarbons (CFCs) have been increasing steadily for more than 50 years in the United States, and have been used to estimate the age of the groundwater within 2 years under ideal conditions (Plummer et al., 1993). They can be used to resolve recharge dates less than 10 years old, a fact needed to assess water-quality conditions in relatively shallow aquifers (Busenberg and Plummer, 1992; Dunkle et al, 1993; Reilly et al, 1994; Cook et al., 1995; Boehlke and Denver, 1996; Oster et al., 1996; and Tesoriero et al., 2000). Application of age-dating technology to aquifers under irrigated cropland showed larger nitrate concentrations, many greater than 10 mg/L, with younger groundwater that was consistent with the history of increased fertilizer and irrigation applications starting about 30 years ago (Stoner et al, 1997). Groundwater older than 36 years was sampled from deeper parts of this unconfined sand and gravel aquifer. This deeper water had significantly lower nitrate concentrations. Other studies have linked nitrate contamination to agricultural practices using tritium dating methods having less accurate resolution of age dating (Burow et al, 1998, Hinkle, 1997, and Savoca et al., 2000). Many of these studies incorporated analysis of the groundwater flow system, and possible effects of denitrification support interpretations based on tritium measures of residence time.

2. GROUNDWATER VULNERABILITY TO NITROGEN

The principles upon which groundwater vulnerability may be estimated include both specific vulnerability to sources of nitrogen from agricultural systems and intrinsic features of hydrologic susceptibility (Water Science and Technology Board, 1993; Zaporozec, 1994). Specific vulnerability to agricultural systems is a function of contaminant factors such as the quantity, rate, timing, and methods of nitrogen application, irrigation, and other agricultural management characteristics. Intrinsic susceptibility is a function of hydrogeologic factors such as proximity of an aquifer to the land surface, hydrologic properties of soil, and the amount, timing, and location of aquifer recharge. Understanding the juxtaposition of both specific vulnerability and intrinsic susceptibility is necessary to adequately define groundwater vulnerability.

2.1 Specific Vulnerability Factors and Processes Associated with Agricultural Systems

Manure and inorganic fertilizer are the principal sources of agricultural nitrogen that are easiest to document and compare globally. Mobile nitrogen, generally in the form of NO^-_3, can also be generated *in situ* by mineralization of soil organic matter, crop residues, legume fixation, and redeposition of ammonia from nearby sources such as manure and crop loss during senescence (Schepers and Mosier, 1991). However, defining the distribution of these *in situ* sources is beyond the scope of this chapter. A substantial factor that has allowed the growth of the world production of food and fiber has been the expanded use of inorganic nitrogen fertilizer for crop production. Rates of nitrogen fertilizer use and changes by world regions (FAO, 2000) are shown in Figure 1. The most outstanding

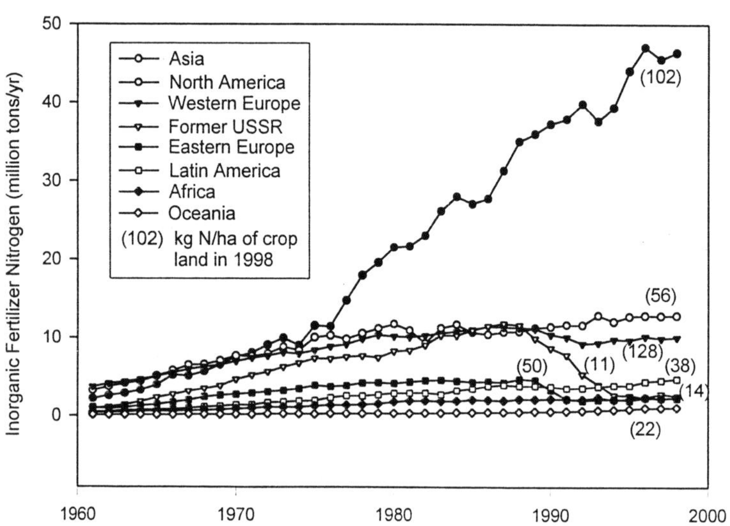

Figure 1. Use of inorganic nitrogen fertilizer by region since 1960.

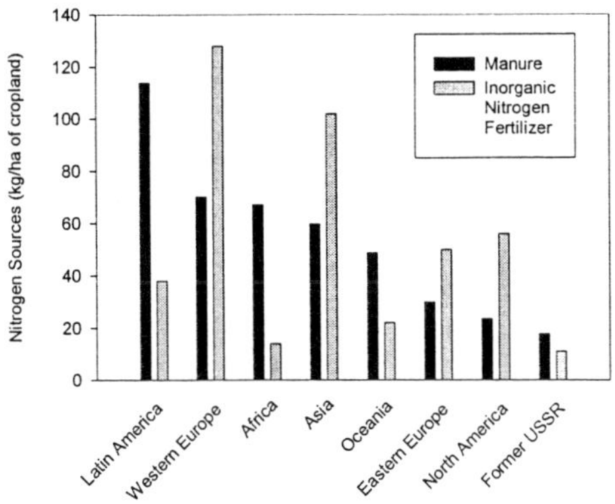

Figure 2. Nitrogen available from animal manure during 1999 by world regions.

feature in these data is the dramatic increase of fertilizer in Asia since the 1970s, although Western Europe currently uses the largest unit-area amount on cropland. Both American continents continue to increase their use of inorganic nitrogen fertilizer, although at rates less than those prior to the 1980s. Also interesting are declines seen in Europe (both Western and Eastern), and the former USSR since 1989. These trends may be useful to project long-term changes in nitrogen contamination of groundwater throughout the world. Estimates of nitrogen available from livestock manure (Fig. 2) show a different global distribution from that of inorganic fertilizer. The estimates were based on FAO statistics (2000) on the number of animals in several categories and the estimated amount excreted by each animal (Lander et al., 1998). The ratio of source-load of manure-N can be classified in two quite different systems (Fig. 2). In North America, Asia, Western and Eastern Europe, approximately twice as much nitrogen comes from inorganic fertilizer compared to manure. In the remaining regions, the ratio is inverted with more than twice the nitrogen from manure except in the former USSR (only 1.7). The trend of increasing concentration of livestock production in the United States (U.S. Department of Commerce, 1997) is also concentrating the manure generated by livestock. Concentration of manure production is accompanied by a proportionate concentration of nitrogen sources available for leaching to groundwater. The concentration and storage of manure also increases nitrogen losses to the atmosphere (Lander et al., 1998). This loss to the atmosphere of nitrogen will not likely reduce the non-point source load of nitrogen because the deficit will be made up with inorganic fertilizer applications. In addition, up to 75% of locally derived atmospheric NH^{+}_{3} and NH^{+}_{4} may be redeposited within 400 km (Ferm, 1998). If the trend of increasing size of livestock operations is global, there may be an accompanying trend of increasing nitrogen contamination of groundwater from both point sources of manure storage systems and non-point sources of manure disposal on fields near large livestock facilities. The processes in agricultural systems that generate nitrate support both plant growth and water contamination. These processes act on both imported sources and nitrate generated *in situ*. Fertilizer and manure are the primary imported sources (Power and Schepers, 1989) and organic-matter mineralization and fixation are the principal processes generating nitrate within the soil (Schepers and Mosier, 1991). Crop uptake and microbial assimilation are the dominant processes that immobilize nitrate in the unsaturated zone. Immobilization by soil microorganisms may be offset by the opposing process of mineralization, both of which generally occur continuously (Keeney, 1986). During periods when neither crops nor soil microorganisms are active, available nitrate will leach through highly permeable soils to the water table when water from precipitation or irrigation exceeds evaporation. In many systems, imported nitrogen sources are added to the nitrogen pool at precisely these vulnerable periods in spring and fall. The barren-ground periods before crop canopies develop and the time after harvest are also periods of substantial rainfall in many temperate climates. This rainfall provides the recharge water to leach nitrate.

Denitrification is the dominant process that can reduce nitrate in saturated materials beneath agricultural systems. This microbially mediated process occurs most readily above 10° C (Firestone, 1982) and generally requires both reduced oxygen levels (Meisinger and Randall, 1991) and readily available carbon (Parkin, 1987) or other electron donors. Denitrification is an active process in saturated soils with organic carbon and microbial activity that consume dissolved oxygen (Meisinger and Randall, 1991). Rates vary widely in aquifers because many good aquifers have large hydraulic conductivities which often preclude the presence of carbon sources for the depletion of oxygen or support of denitrification. Examples in unconsolidated sand aquifers have shown that denitrification is more likely to reduce nitrate concentrations with increased depth (Komor and Anderson, 1993) and remove as much as 50% of the nitrate leached below the root zone (Puckett et al. , 1999). However, the latter study showed that the aquifer received about three times as much nitrogen as would be expected under background conditions.

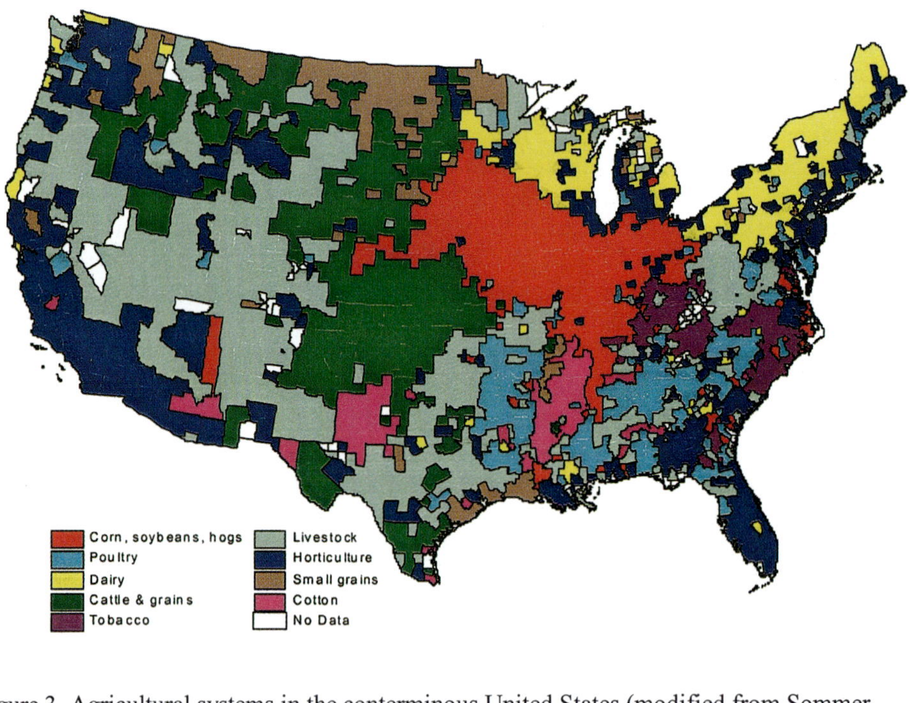

Figure 3. Agricultural systems in the conterminous United States (modified from Sommer and Hines, 1991).

An analysis of nitrate behavior in shallow groundwater of southeastern United States (Nolan, 1999) reported inverse relations between nitrate concentrations and dissolved oxygen on one hand

and dissolved organic carbon, iron, manganese, and ammonia in groundwater on the other. In contrast, denitrification does not occur throughout southeastern United States aquifers as evident by low nitrate concentrations with higher concentrations of dissolved oxygen, some of which were in karst areas. Yusop et al., (1984) showed that denitrification was not a prominent process affecting water quality beneath sandy materials in Belgium. Substantial differences in subsurface denitrification rates were related to slope position in aeolian deposits (Geyer et al., 1992).

2.2 Specific vulnerability in the United States

A convenient way of defining specific vulnerability to agricultural nitrogen sources and management is to use clusters of relatively homogeneous agricultural systems (Fig. 3). The diverse agricultural systems in the U.S. were classified using cluster analysis (Sommer and Hines, 1991). The analysis included 19 farm enterprise variables, five resource variables, and three farm-nonfarm interaction variables. The analysis measured differences among counties across all 27 variables, grouped counties with the greatest similarities, and produced 12 clusters of U.S. agricultural systems that have analogs on other continents. A further generalization of these clusters to a total of nine agricultural systems was made by combining clusters of "part-time cattle" and "sheep, cattle, and other livestock" into a general "livestock" category. "Fruit, other crops, and vegetables" and "nursery products" were placed in a horticulture category. The resulting pattern of systems for the U.S. is illustrated in Figure 3. These clusters also make a convenient framework in which groundwater measurements can be summarized and related to relatively homogeneous agricultural systems.

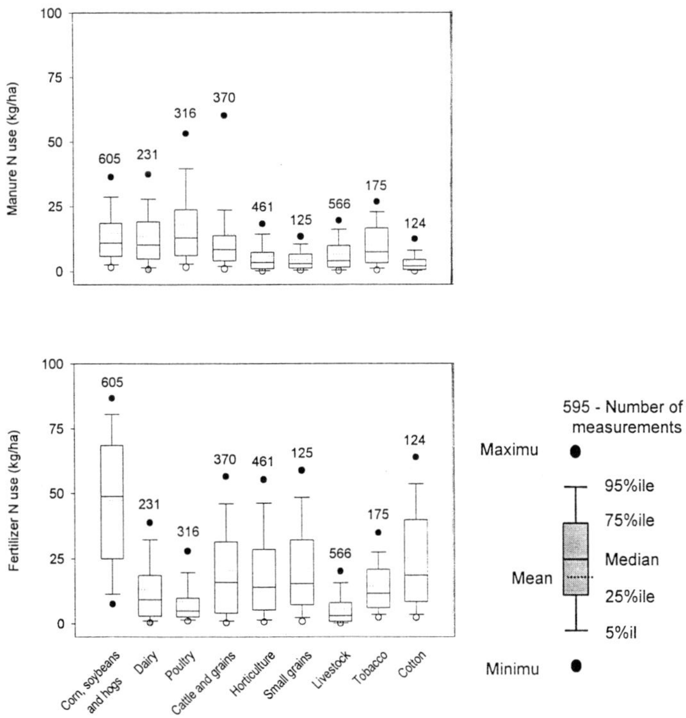

Figure 4. Sources of nitrogen in agricultural systems of the conterminous United States during 1997.

The geographic distribution of imported nitrogen sources for 1997, the latest Census of Agriculture year, was summarized by agricultural systems as shown in Figure 4. Inorganic fertilizer estimates were provided by David Lorenz, U.S. Geological Survey (written communication) and are estimated sales of all forms of fertilizer by county. Manure data were estimated using data on livestock numbers from the Census of Agriculture (U.S. Department of Commerce, 1997) and general estimates of the nitrogen content of manure (Lander et al., 1998) from each class of animals. Both inorganic fertilizer and manure estimates were normalized by county area and counties were aggregated by agricultural systems mapped in Figure 3. A recent analysis of the groundwater risk of nitrate contamination (Nolan et al., 1997) used fertilizer, manure, and wet atmospheric deposition of nitrogen to define risk groups. This analysis of shallow (<30 m) wells showed that counties with well drained soils and sources of nitrogen larger than 21 kg/ha had a significantly larger concentration of nitrate and frequency of concentrations exceeding 10 mg/L than counties with less than 21 kg/ha nitrogen sources. When manure and fertilizer nitrogen are aggregated into agricultural systems (Fig. 4), only two systems had median values that exceed the high risk criteria used by Nolan et al., (1997); corn, soybeans, and hogs (59.7 kg/ha); and cattle and grains (24.41 kg/ha). However, cotton (20.6 kg/ha) and dairy (19.5 kg/ha) had median values close to this threshold.

2.3 Agricultural management factors contributing to specific vulnerability

The presence of cropland may be a good indicator of groundwater vulnerability to nitrate contamination. Cropland management incorporates imported nitrogen sources and the agricultural practices that mobilize nitrogen in soil-organic matter during critical periods with reduced plant cover. Row-crop agricultural systems constitute the largest and most extensively managed land-use in the U.S. More than 177 million ha in the 48 contiguous states are used for crops (U.S. Department of Commerce, 1997) such as corn, cotton, soybeans, and wheat. Similarly, large fractions of other continents are used to produce major row crops. Keeney (1986) states that these systems provide vast areas of nonpoint sources of nitrogen. In addition to the external nitrogen inputs needed to sustain row-crop production, tillage and other management activities promote the mineralization of soil organic matter and crop residue into nitrate providing an *in situ* source (Schepers and Mosier, 1991). These crops are generally managed by various types of soil tillage and weed control that leave the land bare of vegetation for extended periods during the year. In many climates, this bare period coincides with substantial rainfall or snow melt that can enhance leaching of nitrate to groundwater. The bare-ground periods immediately before plant emergence and after harvest coincide with periods of no crop uptake. However, active microbial communities continue converting organic matter to nitrate during warms parts of these periods. Where climate and soil conditions allow multiple crops, leaching potential is not likely reduced if the imported nitrogen exceeds the demands of these additional crops.

Irrigation can contribute substantially to groundwater contamination because it increases the potential for recharge and nitrate leaching. U.S. counties (Fig. 5) where more than 50 % of the cropland is irrigated are concentrated in several areas that are vulnerable to nitrate contamination. Larger concentrations of nitrate and greater frequency of excess nitrate occur in groundwater in

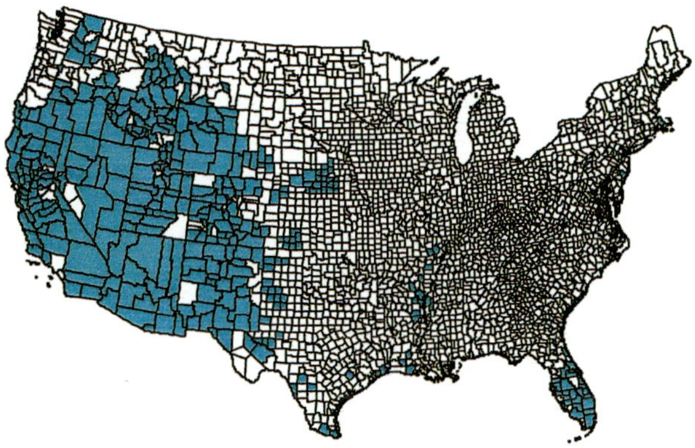

Figure 5. Counties in the conterminous United States in which at least 50% of the cropland is irrigated.

these areas than in areas without irrigation (Waddell et al, 2000; Bhatt, 1997; Spalding and Exner, 1993; Hamilton and Helsel, 1995; and Eckhoff and Bergman, 1995). Irrigation using groundwater is most practical in areas where aquifers are very near the land surface. The additional recharge afforded by irrigation in excess of crop needs facilitates leaching of nitrate to groundwater. In some irrigation systems, leaching is intentionally encouraged to remove soluble salts imported with irrigation waters (Power and Schepers, 1989). Other irrigated systems are located where permeable soils require frequent application of nutrients because of the high rates of leaching. Consequently, irrigation in many areas represents multiple contributions to vulnerability by providing both the water and additional nitrogen sources to increase leaching to groundwater. Irrigation is frequently used on crops with large N-fertilizer demand such as corn, potatoes, and vegetables, further compounding the vulnerability of groundwater under irrigation.

Global examples of irrigation impacts on groundwater are sufficient to indicate that irrigation is a universal contributor to nitrate contamination. The distribution of irrigated cropland in regions of the world (Fig. 6) may indicate that Asia will be a region to experience its greatest impact.

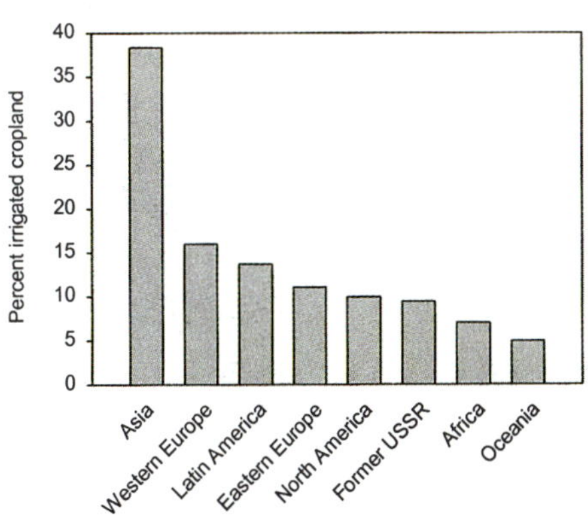

Figure 6. Percent of irrigated cropland in principal regions of the world.

Unfortunately, few studies from Asia have been able to distinguish the impact of irrigation from those resulting from multiple cropping or large nitrogen sources. Several investigations have examined nitrate leaching under various agricultural systems in the Great Plains of North America. Hamilton and Helsel (1995) found median concentrations of nitrate in Nebraska under irrigated corn to be slightly less than 10 mg/L. Irrigation water that contains more than 20 mg/L nitrate in this same region results in the addition of 60 kg/ha nitrogen under a common irrigation schedule (Power and Schepers, 1989). In South Dakota, 38 % of groundwater samples exceeded 10 mg/L nitrate under irrigated wheat and corn (Bhatt, 1997). In Montana, Eckhoff and Bergman (1995) found nitrate in excess of 5 mg/L to be common under irrigated safflower or sugar beets and more than 10 mg/L under irrigated small grains. Substantial nitrate leaching was also documented under irrigation where feedlot manure was the source of nitrogen in Canada (Chang and Entz, 1996). Groundwater beneath an irrigated horticultural crop system in Spain (Guimera, 1998) was found to contain as much as 160 mg/L NO^{-}_{3}-N in a setting where irrigation withdrawals cause recirculation of nitrate-loaded water. Irrigated horticultural systems in Chile (Schalscha et al., 1979) were reported to produce concentrations of 20-35 mg/L nitrate in water below the root zone and 9 to 14 mg/L in shallow wells. Irrigated systems for a variety of cropping systems in India (Khurshid and Khan, 1982) commonly produced more than 10 mg/L nitrate in groundwater, and several areas commonly had in excess of 20 to as much as 500 mg/L nitrate.

2.4 Intrinsic Susceptibility of Groundwater

Three classes of shallow aquifers in the U.S. were mapped to show the extent of aquifers most susceptible to agricultural-nitrogen contamination. Some shallow aquifer classes that may have similar susceptibility to agricultural nitrogen such as non-carbonate fractured rocks could not be as consistently mapped with the confidence of these classes. Shallow aquifers have been identified as particularly susceptible because large-scale surface activities, such as agriculture, often occur immediately above recharge areas. The proximity of these shallow aquifers to the surface facilitates direct transport of contaminants from surface activities to groundwater. In many agricultural systems, these activities are carried out in soils that are the unsaturated materials immediately above the water table or the top of the groundwater flow system. Such close proximity is commonly associated with shallow carbonate, unconsolidated sand and gravel, and alluvial aquifers.

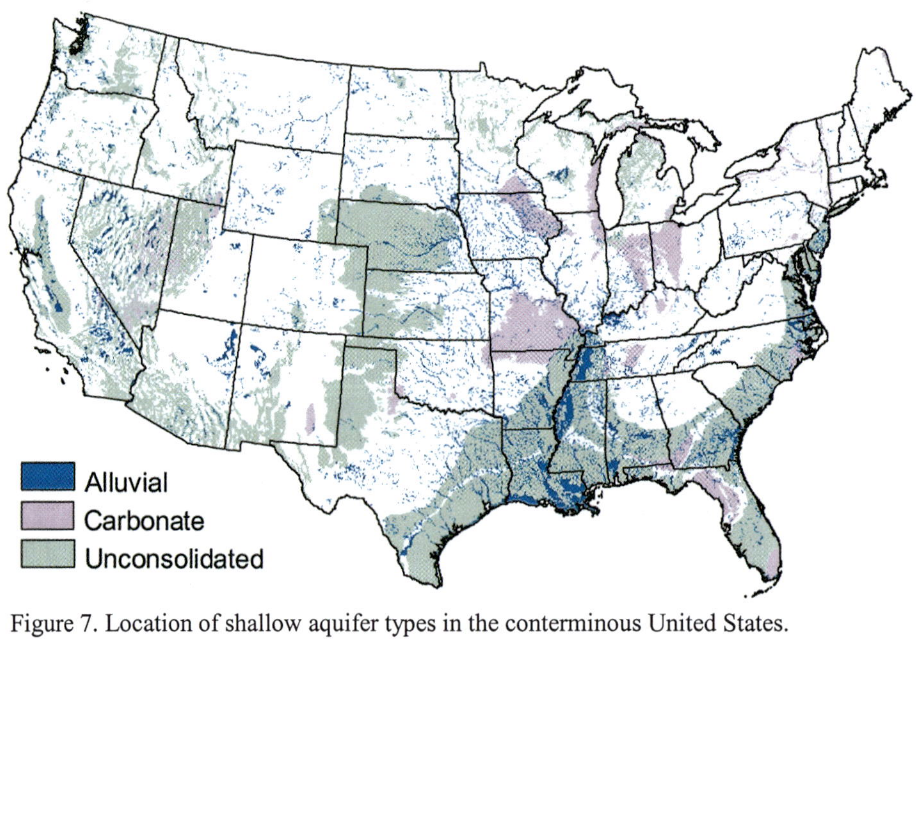

Figure 7. Location of shallow aquifer types in the conterminous United States.

Carbonate Aquifers are bedrock aquifers most commonly formed in limestone, dolomite, and chalk. Karst features, such as solution-enlarged fractures, sink holes, and caves, form in these rocks at land surface and in the subsurface. Boundaries of this class of aquifers (Fig. 7) were adapted from carbonate-rock aquifers shown on the Principal Aquifers map of the United States (U.S. Geological Survey, 2000). Water levels in these aquifers may be deep, even though they are commonly unconfined in the outcrop and subcrop areas shown. Where carbonate aquifers are near the land surface they are particularly susceptible to nitrate contamination because of the direct and effective recharge flow-paths from thin soil cover to and through the aquifers via solution features. Geographically diverse examples exist of nitrate contamination associated with a variety of agricultural systems operating over these aquifers. Foster et al. (1982) report some of the most severe nitrate contamination associated with arable land in an eastern England karst region. Nitrate-sensitive areas are also related to arable land over the Great Oolite aquifer of the United Kingdom (Evans et al., 1993). About 18% of the grazing and pasture in the Appalachian region of the U.S. is underlain by extensive carbonate aquifers where Boyer and Pasquarell (1996) found a strong relationship between nitrate concentration and agricultural land.

Unconsolidated Sand and Gravel Aquifers are found in a variety of depositional environments such as glacial outwash, coastal plain sediments, and aeolian sands. The map of these aquifers (Fig. 7) includes the semi-consolidated and unconsolidated aquifers from the Principal Aquifers map of the U.S. National Atlas (U.S. Geological Survey, 2000). In the areas of the U.S. with continental glacial deposits the map was generated by calculating sand content from seive variables in STATSGO (U.S. Department of Agriculture, 1994). STATSGO map units in which the dominant soil contained more than 50 % sand were interpreted to overlie shallow unconsolidated aquifers. Frequent and high nitrate concentrations have been related to a variety of agricultural systems located over outwash aquifers. These systems include livestock and horticulture (Zebarth et al., 1998) and potatoes (Hill, 1982) in Canada; potatoes and corn in North-central United States (Landon et al., 1995 and Prunty and Greenland, 1997); and seed corn and horticulture in southern Michigan (Kehew et al., 1996). Nitrate contamination of coastal unconsolidated aquifers has been well documented along the entire eastern U.S. coastal plain (Weil et al., 1990, Craig and Weil, 1993, Lichtenberg and Shapiro, 1997, Reay et al., 1992, Tyson et al., 1995) as well as in similar aquifers in Spain (Guimera et al. 1995). Aeolian sands like the Nebraska Sand Hills, Quaternary sands of northern India (Kakar, 1981) and dune deposits in areas bordering the North Sea and northern Atlantic (Andersen and Kristiansen, 1984) are also classified as unconsolidated aquifers. Nitrate contamination appears to be less well documented in aeolian sands, perhaps because these sands form landscapes that are not conducive to substantial agricultural development.

Alluvial Aquifers are generally unconfined and consist of unconsolidated sand and gravel deposits interbedded with finer-grained deposits. They are distinguished from other unconsolidated aquifers for this discussion because of their direct hydraulic connection to streams. This class of aquifers was mapped (Fig. 7) using flood frequency variables found in the STATSGO soils database similar to those mapped by Burkart et al. (1999). STATSGO map units with dominant soils that were occasionally or frequently flooded were compiled to represent the location of alluvial aquifers. These aquifers are commonly found adjacent to and underlying rivers throughout the world. They often are limited to the flood plains of major rivers and may range from several hundred meters to several kilometers wide along a river. Because these aquifers are at or very near the land surface, they can provide a convenient and generally plentiful quantity for water supplies. However, their proximity to the land surface, which is commonly flat in alluvial valleys, also exposes them to the potential for direct contamination resulting from overlying land use including agriculture. Alluvial aquifers and other shallow unconsolidated aquifers have been shown to be among the most vulnerable to agrichemical contamination in the U.S. (Burkart and Kolpin, 1993). Other studies have shown corn production to be directly related to excess nitrate (Schepers et al., 1991) in alluvial and terrace aquifers of the Great Plains of the U.S. Agricultural-nitrate contamination of similar aquifers has been reported on other continents including Africa (Adentunji, 1994), Europe (Pekny et al., 1989) and Asia (Kakar, 1981).

2.5 An Example Linking Specific Vulnerability and Intrinsic Susceptibility Factors

A number of vulnerability or risk classification systems based on overlays of land use and susceptibility characteristics have been developed for the United States. Kellogg et al. (1994) used agrichemical sources and soil characteristics to predict leaching potential. Nolan et al. (1997) used a combination of nitrogen loading, population density, soil drainage, and land use

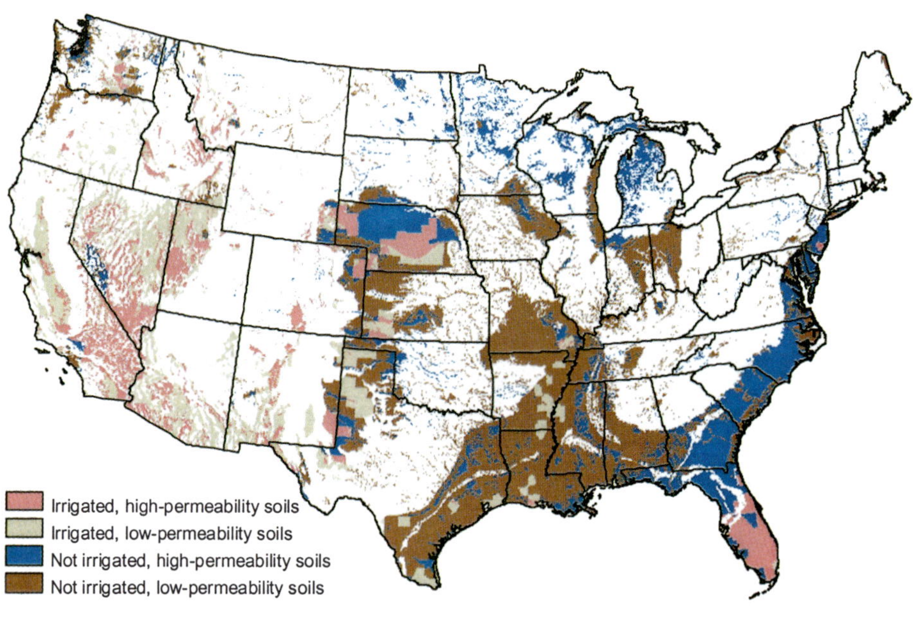

Figure 8. Distribution of irrigation and soil permeability in areas of the conterminous United States underlain by shallow aquifers.

to classify and map the risk of nitrate contamination of groundwater. The study by Nolan et al. (1997) included water quality data to verify that the areas with highest and lowest risks coincided with areas where nitrate concentrations and frequency of nitrate exceeding 10 mg/L were also highest and lowest. Burkart et al. (1999) proposed an overlay method to assess vulnerability as one of a variety of methods for characterizing groundwater vulnerability to agrichemical contamination.

A geographical information system overlay was used as an example here to map areas with relative vulnerability to nitrate contamination of groundwater and define one context in which water-quality data can be aggregated. The vulnerability classification (Fig. 8) was generated by overlying maps of all three shallow aquifers (Fig. 7), areas dominated by soils with permeability greater than 64 mm/hr (high-permeability soils, Fig. 8), and counties with more than 50 % irrigated cropland (Fig. 5). The result is four vulnerability classes that utilize one specific vulnerability factor, irrigation, and two intrinsic susceptility factors, shallow aquifers and permeable soils.

3. DISTRIBUTION OF GROUNDWATER NITROGEN IN SHALLOW AQUIFERS OF THE UNITED STATES

Groundwater-quality analyses were aggregated to characterize the distribution of nitrate in agricultural systems and in four classifications of aquifer vulnerability (Fig. 8). These data were assembled for a ground-water nutrient retrospective analysis as part of the U.S. Geological Survey NAWQA Program. The database contains historical analyses from more than 10,000 wells representing the groundwater quality in 20 NAWQA study areas and selected regional studies in the 48 conterminous states. These data resulted from a nationally consistent set of selection criteria that produced high quality data on both nitrate and ammonia concentrations.

The analysis presented here included only wells with depths of 30 m or less. It is hypothesized that deeper wells generally yield older water that would be less likely to contain nitrate related to recent land use. This selection of shallow wells resulted in a data set of 3125 wells with nitrate analyses in 493 counties. An analysis of a similar subset of these data (Bernard T. Nolan, personal communication) determined that nitrate concentrations from wells deeper than 30 m were significantly smaller than that from shallow wells. These groundwater data were analyzed to determine if there were differences in the nitrate concentrations among the nine agricultural systems shown in Figure 3. The values for both nitrate and ammonia were each aggregated by agricultural system to define any significant differences in the central tendency of nitrate concentrations among systems. Almost 14% of the total samples exceeded 10 mg/l nitrate-N, the maximum contaminant level for public drinking water supplies in the United States. Almost 24% of the wells located within the agricultural system of corn, soybeans, and hogs exceeded this standard (Fig. 9). Other systems in which this standard was exceeded by more than 10% include cattle and grains, poultry, small grains, dairy, and horticulture. Few ammonia concentrations were greater than 0.1 mg/L and differences among agricultural systems could not be readily distinguished. Consequently, only nitrate analyses are presented here.

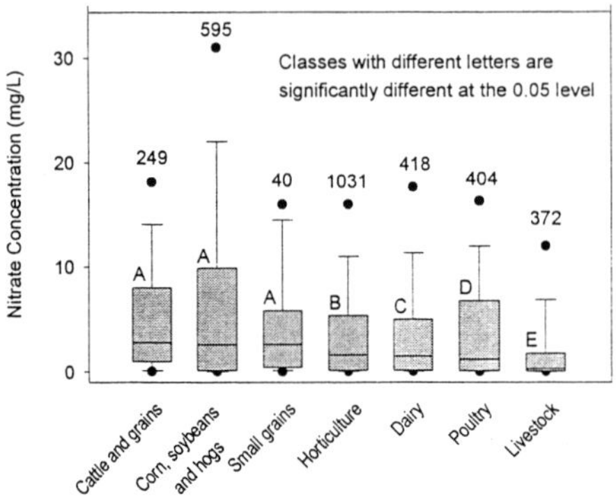

Figure 9. Nitrate concentrations under agricultural systems in the United States

Non-parametric statistics were used because nitrate concentrations were not assumed to be normally distributed. Results of a Kruskal-Wallis test indicated there were significant differences among the nitrate concentrations associated with agricultural systems at the 0.05 level. Figure 9 shows the distribution of nitrate concentrations among agricultural systems. Results of Tukeys multiple variable comparison test performed on the ranks of nitrate concentrations show that groundwater concentrations in three systems (cattle and grains; corn, soybean and hogs; and small grains) were significantly larger than all other systems at the 0.05 level. Unfortunately, too few nitrate samples were available to evaluate either tobacco or cotton systems. However, nitrate concentrations among the cattle and grains system; corn, soybean and hogs system; and small grains system were not significantly different.

Nitrate concentrations were significantly larger (at the 0.05 level) in counties with greater than 50% irrigated cropland than in non-irrigated counties when analyzed for all samples under all systems combined (Total on Fig. 10). This difference was defined using Tukeys multiple comparison test conducted on the ranks of nitrate concentration. This test also confirmed the significance of differences in nitrate concentrations between irrigated and non-irrigated corn, soybeans, and hogs system. The apparent differences between irrigated and non-irrigated small grain systems (Fig. 10) were not statistically significant due to the very small number of samples from irrigated areas. No samples of irrigation associated with dairy or tobacco systems were available and too few from poultry or cotton to compare. Two other regional studies that found significant differences between irrigated and non-irrigated agriculture in the United States (Kolpin, 1997 and Power and Schepers, 1989) were coincidentally concentrated in the corn, soybean, and hogs system.

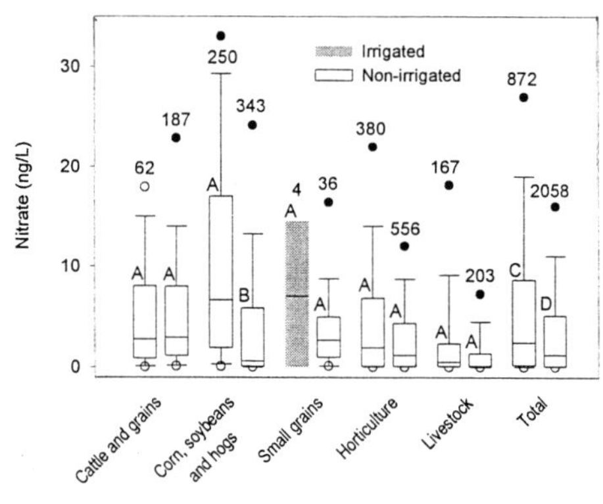

Figure 10. Nitrate concentrations in irrigated and non-irrigated agricultural systems in the United States. Classes with different letters (A,B) are significantly different at the 0.05 level.

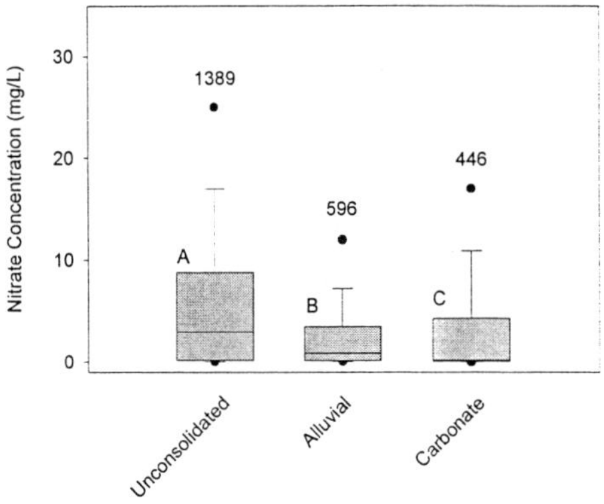

Figure 11. Nitrate concentrations grouped by shallow aquifer types.

Nitrate concentrations were analyzed to show variations among samples drawn from shallow aquifer types; unconsolidated, alluvial, and carbonate (Fig. 11). There were significant differences among nitrate concentrations from the three aquifer types at the 0.05 level using Tukey's multiple variable comparison performed on nitrate ranks. Unconsolidated aquifers were found to have the largest nitrate concentrations followed by alluvial aquifers and carbonate aquifers (Fig. 11). Carbonate aquifers, when close to the land surface, can be directly connected to the surface through karst features such as enlarged fracture systems and sink holes that provide direct recharge paths not available in the other types of aquifers. However, although generally thin, the soils, colluvium, and glacial deposits that overlie these aquifers may provide a sufficient barrier to nitrate leaching to protect many carbonate aquifers. Unconsolidated and alluvial aquifers are both composed of sand and gravel, but may differ in the nature and thickness of overlying materials. The significantly larger nitrate concentrations found in unconsolidated aquifers may result from the overlying soils being developed directly in the sand and/or gravel. Alluvial aquifers, on the other hand, can be buried under varying thicknesses of fine-grained flood-plain deposits that are less permeable than sand, contain substantial organic matter fractions, and have low dissolved oxygen, typical of groundwater discharge areas. These differences in overlying materials or terrain and related flow systems are sufficient to reflect significant differences in the nitrate concentrations.

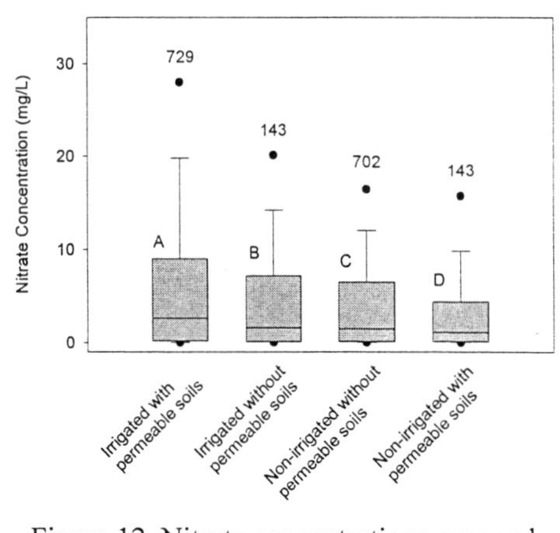

Figure 12. Nitrate concentrations grouped by aquifer vulnerability classes.

Analysis of the four vulnerability classes (Fig. 8) shows the cumulative effects of soil permeability and irrigation on the distribution of nitrate concentrations in shallow aquifers (Fig. 12). It was hypothesized that vulnerability to nitrate contamination increased when shallow aquifers were overlain by soils with permeability exceeding 64 mm/hr. It was further hypothesized that irrigation provided an increase in the potential for nitrate contamination. Four vulnerability classes were defined using permutations of irrigation intensity and presence or absence of high-permeability soils. There were significant differences among samples from all four vulnerability classes at 0.05 level using Tukeys multiple variable comparison applied to ranks of nitrate concentration. Shallow aquifers in highly irrigated areas yielded significantly larger nitrate concentrations than those in non-irrigated areas regardless of overlying soil permeable (> 64 mm/hr). Since nitrate concentrations in non-irrigated areas with low-permeability soils were significantly larger than in areas with permeable soils is not intuitive. This apparent conflict with the hypothesis that permeable soils are more susceptible to contamination may result from influence of other factors, particularly the combination of agricultural systems associated with irrigation and less permeable soils.

4. CONCLUSIONS

This chapter focused on processes by which aquifers can be affected by nitrogen derived from agricultural systems. The primary form of nitrogen of concern for drinking water, nitrate, is costly to remove in water treatment. Many major aquifers used for urban drinking water are buried deep beneath large population centers. These aquifers are geographically removed from recharge areas near agricultural systems. However, shallow aquifers in urban areas may be contaminated by atmospheric and turf fertilizer sources of nitrogen. Groundwater sources for large municipal water supplies can be and often are blended with several sources allowing dilution of any nitrate contamination. Most rural drinking-water supplies, however, are served through individual or a limited number of wells that are usually completed in shallow aquifers. These shallow aquifers are commonly recharged beneath agricultural activities. Limited research results have shown that once a shallow aquifer has been contaminated by nitrate, it may take decades for the groundwater quality to improve even after pollution controls have been implemented. Few programs exist that routinely monitor private groundwater systems for contamination from agricultural nitrogen. This makes it difficult for well owners to know trends in nitrate contamination of their aquifer over time. Consequently, local and regional understanding about vulnerable aquifers beneath certain agricultural systems becomes critical information for preventive and effective protection of water supplies, particularly in rural settings.

Two lines of evidence support several factors that contribute to groundwater vulnerability to nitrate contamination in agricultural settings. Research from several regions of the world provides a collection of spatially anecdotal information to hypothesize globally applicable hydrologic and agricultural factors contributing to groundwater vulnerability. Preliminary analysis of a United States data set compiled by the U.S. Geological Survey NAWQA Program from a variety of sources confirms these hypotheses for most agricultural systems.

Shallow unconfined aquifers have been most susceptible to nitrate contamination associated with agricultural systems. Unconsolidated aquifers and alluvial aquifers are more vulnerable, although shallow carbonate aquifers provide a smaller but substantial contamination risk. In areas dominated by irrigation, shallow aquifers are more vulnerable to nitrate contamination than areas without irrigation. The presence of permeable soils over shallow aquifers compounds the risk of contamination in irrigated areas.

Three agricultural systems (cattle and grains; corn, soybean and hogs; and small grains) produced significantly larger concentrations of groundwater nitrate than other agricultural systems. However, significant differences of nitrate concentrations among these three systems could not be confirmed. Irrigation, particularly in corn, soybean, and hogs systems was found to have consistently larger groundwater-nitrate concentrations in the United States data as well as in studies from outside this country.

Varying time lags exist in shallow groundwater responses to changes in agricultural inputs at the surface. If trends in increased fertilizer use and groundwater nitrate in the United States are repeated in other regions of the world, Asia may experience increasing problems because of recent and substantial increases in fertilizer use in that region. Both American continents also continue to increase their use of inorganic nitrogen fertilizer, albeit at rates less than those seen prior to the 1980s and those presently seen in Asia. Scientists and policymakers should be interested in learning if there will be a reduction in the trend of increasing concentrations of nitrate in groundwater where fertilizer inputs have been reduced. The most rapid responses may be seen in areas with extensive macropore flow where land-use changes may produce the earliest changes in groundwater quality. It will be particularly interesting to monitor changes in groundwater nitrate in both Western and Eastern Europe as well as in the former USSR where fertilizer use overall has dropped since the early 1990s. Groundwater nitrate measurements in these regions may provide tests of hypotheses that reduced nitrate contamination will follow reduced inorganic fertilizer inputs. Fertilizer-use trends may be useful to estimate long-term changes in nitrogen contamination of groundwater throughout the world. Use of these trends to strategically locate long-term monitoring will help answer questions about whether and when proportional changes in concentrations of nitrate will follow these changes in fertilizer.

If the trend in concentrated livestock production seen in the United States is global, there may be an accompanying trend of increasing nitrogen contamination locally in groundwater. Concentrated livestock operations provide both point sources of nitrogen in the immediate area of the confinement as well as larger areas of intense non-point sources as fields close to facilities become used for manure disposal.

A major contributor to groundwater vulnerability is the distribution of irrigated cropland. Regions were this practice expands, such as in Asia, may experience its greatest impact. More data and research will be needed in Asia to determine if patterns of water quality degradation in irrigated areas is repeated in this region.

ACKNOWLEDGMENTS

The authors thank David James and Robert Jaquis of the USDA, National Soil Tilth Laboratory for their statistical support and for generating the maps and graphs that illustrate this chapter and David Lorenz, U.S. Geological Survey, who prepared estimates of fertilizer use in the United States We also appreciate the thoughtful and thorough manuscript reviews by U.S. Geological

Survey scientists Bernard T. Nolan and Dana W. Kolpin. Their comments and suggestions were very helpful to improve the early drafts of this chapter.

REFERENCES

Adetunji, M.T. 1994. Nitrogen application and underground water contamination in some agricultural soils of South Western Nigeria. Fertilizer Research 37(2):159-163.

Andersen, L.J. and H. Kristiansen. 1984. Nitrate in groundwater and surface water related to land use in the Karup Basin, Denmark. Environmental Geology 5(4):207-212.

Baker, D.B., L.K. Wallrabenstein, R.P. Richards and N.L. Creamer. 1989. Nitrate and pesticides in private wells of Ohio: A state atlas. The Water Quality Laboratory, Heidelberg College, Tiffin, Ohio.

Bhatt, K. 1997. Occurrence and distribution of nitrate and pesticides in Bowdle aquifer, South Dakota. Environmental Monitoring and Assessment 47:223-237.

Boehlke, J.K., and J.M. Denver. 1996. Combined use of groundwater dating, chemical, and isotopic analysis to resolve the history and fate of nitrate contamination in two agricultural watersheds, Atlantic Coastal Plain, Maryland: Water Resources Research, 31:9:2319-2339.

Boyce, J.S., J. Muir, A.P. Edwards, E.C. Seim, and R.A. Olson. 1976. Geologic nitrogen in Pleistocene loess of Nebraska. Jour. Environmental Quality. 5:93.

Boyer, D.G. and G.C. Pasquarell. 1996. Agricultural land use effects on nitrate concentrations in a mature karst aquifer. Water Resources Bulletin 32(3):565-573.

Burkart. M.R. and D.W. Kolpin. 1993. Hydrologic and land-use factors associated with herbicides and nitrate in near-surface aquifers. Journal of Environmental Quality 22:646-656.

Burkart, M.R., D.W. Kolpin, and D.E. James. 1999. Assessing groundwater vulnerability to agrichemical contamination in the Midwest U.S. Water Science and Technology. 39:3:103-112.

Burow, K.R., S.V. Stork, and N.M. Dubrovsky. 1998. Nitrate and pesticides in ground water in the Eastern San Joaquin Valley, California--occurrence and trends: U.S. Geological Survey Water-Resources Investigations Report 98-4040, 33 p.

Busenberg, E., and L.N. Plummer. 1992. Use of chlorofluoromethanes (CCL 3 F and CCL 2 F 2) as hydrologic tracers and age-dating tools--The alluvium and terrace system of central Oklahoma: Water Resources Research, 28:9:2257-2283.

Chang, C. and T. Entz. 1996. Nitrate leaching losses under repeated cattle feedlot manure applications in southern Alberta. J. Environ. Qual. 25:145-153.

Cook, P.G., D.K. Solomon, L.N. Plummer, E. Busenberg, and S.L. Schiff. 1995. Chlorofluorocarbons as tracers of groundwater transport processes in a shallow, silty sand aquifer: Water Resources Research, 31:3:425B434.

Craig, J.P. and R.R. Weil. 1993. Nitrate leaching to a shallow mid-Atlantic Coastal Plain aquifer as influenced by conventional no-till and low-input sustainable grain production systems. Water Science and Technology 28(3-5):691-700.

Dunkle, S.A., L.N. Plummer, E. Busenberg, P.J. Phillips, J.M. Denver, P.A. Hamilton, R.L.

Michel, and T.B. Coplen. 1993. Chlorofluorocarbons (CCl 3 F and CCl 2 F 2) as dating tools and hydro-logic tracers in shallow groundwater of the Delmarva Peninsula, Atlantic Coastal Plain, United States: Water Resources Research, 29:12:3837B3860.

Eckhoff, J.L.A. and J.W. Bergman. 1995. Nitrate in soil and groundwater under irrigated crops. Montana AgResearch, Fall 1995, p. 9-12.

Edmunds, W.M. and C.B. Gaye. 1997. Naturally high nitrate concentrations in groundwaters from the Sahel. Journal of Environmental Quality 26(5):1231-1239.

Evans, D., I.R. Moxon, and J.H.C. Thomas. 1993. Groundwater nitrate concentrations from the Old Chalford nitrate-sensitive area, West Oxfordshire. Journal of the Institution of Water and Environmental Management 7(5):506-512.

Exner, M.E. and R.F. Spalding. 1985. Ground-water contamination and well construction in southeast Nebraska. Ground Water. 23:26-34.

FAO. 2000. FAO Statistical Databases. FaoStat. Food and Agriculture Organization of the United Nations, Rome, Italy, http://apps.fao.org.

Ferguson, H, P.L. Brown, and M.R. Miller. 1972. Saline seeps on nonirrigated lands of the northern plains, in Control of Agriculture-related Pollution in the Great Plains. Great Plains Agricultural Council Publ. 60. University of Nebraska, Lincoln. 169p.

Ferm, Martin. 1998. Atmospheric ammonia and ammonium transport in Europe and critical loads: a review. Nutrient Cycling in Agroecosystems. 51:5-17.

Feth, J.H. 1966. Nitrogen compounds in natural water B A review. Water resources Research, 2:1:41-58.

Firestone, M.K. 1982. Biological denitrification. p. 289-326. in F.J. Stevenson (ed.) Nitrogen in agricultural soils. Agron. Monogr. 22. ASA, CSSA, and SSSA. Madison, WI.

Foster, S.S.D., A.C. Cripps, and A. Smith-Carington. 1982. Nitrate leaching to groundwater. Philosophical Transactions of the Royal Society of London 296:477-489.

Freeze, R. Allen and John A. Cherry. 1979. Groundwater. Prentice-Hall, Inc. Englewrood Cliffs, NJ. 604 p.

Fuhrer, G.J., R.J. Gilliom, P.A. Hamilton, J.L. Morace, L.H. Nowell, J.F. Rinella, J.D. Stoner, and D.A. Wentz. 1999. The quality of our Nation's water: nutrients and pesticides: U.S. Geological Survey Circular 1225. 82 p.

Geyer, D.J., C.K. Keller, J.L. Smith, and D.L. Johnstone. 1992. Subsurface fate of nitrate as a function of depth and landscape position in Missouri Flat Creek watershed, U.S.A. Journal of Contaminant Hydrology 11:127-147.

Gilliom, R.J., W.M. Alley, and M.E. Gurtz. 1995. Design of the National Water-Quality Assessment Program: Occurrence and Distribution of Water-Quality Conditions: U.S. Geological Survey Circular 1112. 33 p.

Guimera, J., O. Marfa, L. Candela, and L. Serrano. 1995. Nitrate leaching and strawberry production under drip irrigation management. Agriculture, Ecosystems and Environment 56:121-135.

Guimera, J. 1998. Anomalously high nitrate concentrations in ground water. Ground Water 36(2):275-282.

Hallberg, G.R. 1989. Nitrate in groundwater in the United States. p. 35-74. In R.F. Follett (ed.) Nitrogen management and groundwater protection. Elsevier, Amsterdam.

Hamilton, P.A. and D.R. Helsel. 1995. Effects of agriculture on ground-water quality in five regions of the United States. Ground Water 33(2):217-226.

Hill, A.R. 1982. Nitrate distribution in the ground water of the Alliston Region of Ontario, Canada. Ground Water 20(6):696-702.

Hinkle, S.R., 1997, Quality of shallow ground water in alluvial aquifers of the Willamette Basin, Oregon, 1993-95: U.S. Geological Survey Water Resources Investigations Report 98-4082-B, p. 24.

Kakar, Y.P. 1981. Nitrate pollution of groundwater in southern and south-western Haryana, India. Studies in Environmental Science 17:125-129.

Keeney, D.R. 1986. Sources of nitrate to ground water. CRC Critical Reviews in Environmental Control 16(3):257-304.

Kehew, A.E., W.T. Straw, W.K. Steinmann, P,G, Barrese, G. Passarella, and W-S. Peng. 1996. Ground-water quality and flow in a shallow glaciofluvial aquifer impacted by agricultural contamination. Ground Water 34(3):491-500.

Kellogg, R. L., M. Maizel, and D. W. Goss. 1994. "The Potential for Leaching of Agrichemicals Used in Crop Production: A National Perspective." Journal of Soil and Water Conservation 49 (3):294-298.

Kolpin, D.W. 1997. Agricultural chemicals in groundwater of the Midwestern United States: relations to land use. J. Environ. Qual. 26:1025-1037.

Kolpin, D.W., K.E. Zichelle, and E.M. Thurman. 1996. Water-quality data for nutrients, pesticides, and volatile organic compounds in near-surface aquifers of the Midcontental United States, 1992-1994. U.S. Geological Survey Open--File Report 96-435. 47 p.

Komor, S.C. and Anderson, Jr., H.W. 1993. Nitrogen isotopes as indicators of nitrate sources in Minnesota sand-plain aquifers. Ground Water 31(2):260-270.

Korom, S.F. 1992. Natural denitrification in the saturated zone B A review. Water Resources Research. 28:6:1657-1668.

Khurshid, S. and Khan, A.H. 1982. Study of the groundwater nitrate pollution in some rural areas of Delhi and Gurgaon. Indian Journal of Zoology 10(1):37-40.

Kross, B.C., G.R. Hallberg, D.R. Bruner, R.D. Libra, K.D. Rex, L.M.B. Weih, M.E. Vermace, L.F. Burmeister, N.H. Hall, K.L. Cherryholmes, J.K. Johnson, M.I. Selim, B.K. Nations, L.S. Seigley, D.J. Quade, A.G. Dudler, K.D. Sesker, M.A. Culp, C.F. Lynch, H.F. Nicholson, and J.P. Hughes. 1990. The Iowa State-wide Rural Well-water Survey Water-quality Data: Initial Analysis. Iowa Department of Natural Resources. Technical Information Series 19.

Lander, C.H., D. Moffitt, and K. Alt. 1998. Nutrients available from livestock manure relative to crop growth requirements. U. S. Department of Agriculture, Natural Resources Conservation Service, Resource Assessment and Strategic Planning Working Paper 98-1. http://www.nhq.nrcs.usda.gov/land/pubs/nlweb.html.

Landon, M.K., G.N. Delin, L. Guo, C.P. Regan, J.A. Lamb, J.L. Anderson, and R.H. Dowdy. 1995. Occurrence of agricultural chemicals in ground water at the Princeton, Minnesota Management Systems Evaluation Area. p. 434-438.

LeMasters, G. and D.J. Doyle. 1989. Grade A dairy farm well water quality survey: Madison, Wisconsin. Wisconsin Department of Agriculture, Trade and Consumer Protection, and Wisconsin Agricultural Statistics Service. 36 p.

Lichtenberg, E. and L.K. Shapiro. 1997. Agriculture and nitrate concentrations in Maryland community water system wells. J. Environ. Qual. 26:145-153.

Madison, R.J. and J.O. Brunett. 1985. Overview of the occurrence of nitrate in ground water of the United States in National Water Summary 1984--Hydrologic events, selected water quality trends, and ground-water resources. U.S. Geological Survey Water-Supply Paper 2275. p. 93-105.

Meisinger, J.J. and G.W. Randall. 1991. Estimating nitrogen budgets for soil-crop systems. in Managing Nitrogen for Groundwater Quality and Farm Profitability. R.F. Follett, D.R. Keeney, and R.M. Cruse (eds.) Soil Science Society of America, Madison, Wisconsin. 357 p.

Monsanto Company. 1990. The national alachlor well water survey (NAWWS): Data summary. Monsanto Tech. Bull. July, 1990. Monsanto Co. St. Louis, Missouri.

Mueller, D.K. and D.R. Helsel. 1996. Nutrients in the Nation=s waters -- too much of a good thing? U.S. Geological Survey Circular 1136. 24p.

Nolan, B.T. 1999. Nitrate behavior in ground waters of the Southeastern USA: Journal of Environ. Quality. 28:(5):1518-1527.

Nolan, B.T., B.C. Ruddy, K.J. Hitt, and D.R. Helsel. 1997. Risk of nitrate in ground waters of the United States--A national perspective: Environmental Science and Technology. 31:(8):2229-2236.

Nolan, B.T. and J.D. Stoner. 2000. Nutrients in groundwater of the conterminous United States, 1992-1995: Environmental Science and Technology. 34:(7):1156-1165.

Oster, H., C. Sonntag, and K.O. Munnich. 1996. Groundwater age dating with chlorofluorocarbons: Water Resources Research, 1996, 32:10:2989-3001.

Parkin, T.B. 1987. Soil microsites as a source of denitrification variability. Soil Science Society of America Journal. 51:L1194-1199

Pekny, V., J. Skorepa, and J. Vrba. 1989. Impact of nitrogen fertilizers on groundwater quality - some examples from Czechoslovakia. Journal of Contaminant Hydrology 4:51-67.

Plummer, L.N., R.L. Michel, E.M. Thurman, and P.D. Glynn,. 1993. Environmental tracers for age dating young ground water, in Alley, W.M., ed., Regional ground-water quality: New York, Van Nostrand Reinhold, p. 255B294.

Power, J.F. and J.S. Schepers. 1989. Nitrate contamination of groundwater in North America. Agriculture, Ecosystems and Environment 26:165-187.

Prunty, L. and R. Greenland. 1997. Nitrate leaching using two potato-corn N-fertilizer plans on sandy soil. Agriculture, Ecosystems and Environment 65:1-13.

Puckett, L.J., T.K. Cowdery, D.L. Lorenz, and J.D. Stoner. 1999. Estimation of nitrate contamination of an agro-ecosystem outwash aquifer using a nitrogen mass-balance budget. Journal of Environmental Quality 28(6):2015-2025.

Reay, W.G., D.L. Gallagher, and G.M. Simmons, Jr. 1992. Groundwater discharge and its impact on surface water quality in a Chesapeake Bay inlet. Water Resources Bulletin

28(6):1121-1134.

Richards, R.P.,D.B. Baker, N.L. Creamer, J.W. Kramer, D.E. Ewing, B.J. Merryfield, and L.K. Wallrabenstein. 1996. Well water quality: well vulnerability, and agricultural contamination in the Midwestern United States. Journal of Environmental Quality. 25:389-402.

Reilly, T.E., L.N. Plummer, P.J. Phillips, and E. Busenberg. 1994. The use of simulation and multiple environmental tracers to quantify groundwater flow in a shallow aquifer: Water Resources Research, 30:2: 421B433.

Savoca, M.E., E.M. Sadorf, S.M. Linhart, and K.K.B. Akers. 2000. Effects of land use and hydrogeology on the water quality of alluvial aquifers in eastern Iowa and southern Minnesota, 1997: U.S. Geological Survey Water--Resources Investigations Report 99-4246, 38 p.

Spalding, R.F. and M.E. Exner. 1993. Occurrence of nitrate in groundwater -- A review. J. Environ. Qual. 22:392-402.

Schalscha, E.B., Vergara, I., Schirado, T., and Morales, M. 1979. Nitrate movement in a Chilean agricultural area irrigated with untreated sewage water. J. Environ. Qual. 8(1):27-30.

Schepers, J.S., M.G. Moravek, E.E. Alberts, and K.D. Frank. 1991. Maize production impacts on groundwater quality. Journal of Environmental Quality 20(1):12-16.

Schepers, J.S. and A.R. Mosier. 1991. Accounting for nitrogen in nonequilibrium soil-crop systems. pp. 125-128. in Managing Nitrogen for Groundwater Quality and Farm Profitability. R.F. Follett, D.R. Keeney, and R.M. Cruse (eds.). Soil Science Society of America, Madison, Wisconsin. 357 p.

Sommer, J.E. and F.K. Hines. 1991. Diversity in U.S. agriculture: A new delineation by farming characteristics. U.S. Department of Agriculture Economic Research Service, Agricultural Economic Report 646. 19 p.

Stoner, J.D., T.K. Cowdery, and L.J. Puckett. 1997. Ground-water age dating and other tools used to assess land-use effects on water quality: U.S. Geological Survey--Water Resources Investigations Report 97-4150, 6 p.

Strathouse, S.M., G. Sposito, P.J. Sullivan, and L.J. Lund. 1980. Geologic nitrogen: a potential geochemical hazard in the San Joaquin Valley, California. Jour. Environmental Quality. 9:1:54-60.

Tesoriero, A.J., H. Liebscher, and S.E. Cox. 2000. Mechanism and rate of denitrification in an agricultural watershed: Electron and mass balance along groundwater flow paths. Water resources Research. 36:6:1545-1559

Tyson, A.W., P. Bush, R. Perkins, and W. Segars. 1995. Nitrate occurrence in Georgia=s domestic wells. in Proceedings, Clean Water -- Clean Environment -- 21st Century. Volume II: Nutrients, Kansas City, Missouri, March 5-8, 1995. Working Group on Water Quality -- U.S. Department of Agriculture, Washington, DC. p. 231-234.

U.S. Department of Agriculture. 1994. State Soil Geographic (STATSGO) Data Base: Data Use Information. Nat. Cartography and GIS Center, U.S. Dept. of Agriculture, Natural Resource Conservation Service, Ft. Worth, Texas. 105 p.

U.S. Department of Commerce. 1997. 1997 Census of Agriculture. Geographic Area Series,

Volume 1, 1A, 1B, 1C. CD-ROM Set. U.S. Department of Agriculture, National Agricultural Statistics Service.

U.S. Environmental Protection Agency. 1990. National Survey of Pesticides in Drinking Water Wells -- Phase I Report. EPA 570/9-90-015. Office of Water. Washington DC.

U.S. Geological Survey. 2000. National Atlas of the United States. http://www.nationalatlas.gov/

Yusop, M.K., O. Van Cleemput, and L. Baert. 1984. Nitrogen fertilisation and nitrate pollution of groundwater in sandy soils. Environmental Pollution (Series B: Chemical and physical) 7(1):43-48.

Waddell, J.T., S.C. Gupta, J.F. Moncrief, C.J. Rosen, and D.D. Steele. 2000. Irrigation- and nitrogen-management impacts on nitrate leaching under potato. Journal of Environmental Quality 29:251-261.

Water Science and Technology Board. 1993. Ground Water Assessment: Contamination Potential Under Conditions of Uncertainty. National Research Council. National Academy Press. Washington, D.C. 204 p.

Weil, R.R., R.A. Weismiller, and R.S. Turner. 1990. Nitrate contamination of groundwater under irrigated coastal plain soils. Journal of Environmental Quality 19(3):441-448.

Zaporozec, Alexander. 1994. Concept of groundwater vulnerability. p. 3-8. in Guidebook on Mapping Groundwater Vulnerability. J. Vrba and A. Zaporozec, eds. International Contributions to Hydrogeology. Volume 16. International Association of Hydrogeologists, London, UK.

Zebarth, B.J., B. Hii, H. Liebscher, K. Chipperfield, J.W. Paul, G. Grove, and S.Y. Szeto. 1998. Agricultural land use practices and nitrate contamination in the Abbotsford aquifer, British Columbia, Canada. Agriculture, Ecosystems & Environment 69(2):99-112.

Nitrogen in the Environment: Sources, Problems, and Management
R.F. Follett and J.L. Hatfield (Eds)
Published by Elsevier Science B.V.

Chapter 7. The Importance and Role of Watersheds in the Transport of Nitrogen

T. J. Sauer[a], R. B. Alexander[b], J. V. Brahana[c], and R. A. Smith[b]

[a]USDA-ARS, National Soil Tilth Laboratory
2150 Pammel Drive, Ames, IA 50011-4420

[b]U.S. Geological Survey, 413 National Center
12201 Sunrise Valley Drive, Reston, VA 20192

[c]University of Arkansas, Department of Geosciences
114 Ozark Hall, Fayetteville, AR 72701

A recent report to Congress concerning water quality in the U.S. indicated that 35, 45, and 44% of the assessed rivers and streams, lakes, and estuaries, respectively, were impaired by one or more pollutants (U.S. Environmental Protection Agency, 1999). Nutrients, primarily nitrogen (N) and phosphorus (P), contributed to the impairment of 30% or 135,000 km of the nation's impaired rivers and streams, 44% of the impaired lakes, and 23% of the impaired estuaries. Excessive nutrient loads are implicated in the eutrophication of lakes and reservoirs in the United States and coastal ecosystems where N is most limiting to primary productivity (Vitousek et al., 1997; Carpenter et al., 1998). Efforts are currently underway to establish Total Maximum Daily Load (TMDL) values for pollutants, including nutrients, of impaired water bodies as described under Section 303(d) of the Clean Water Act of 1972.

The movement of N in the terrestrial environment is intimately related to the movement of water. Water in the form of precipitation, flowing across the soil surface as runoff, and percolating through soil layers to ground water can all be significant carriers of organic and inorganic N constituents. The relative importance of these transport mechanisms is a complex function of N sources and transformations, hydrologic processes, climate patterns, and land use. While some elements of the N cycle can be studied in the laboratory under controlled experimental conditions, many can be studied in a meaningful way only in the natural and culturally affected environments of watersheds. By considering N transport over a range of spatial and temporal scales, it is possible to improve our understanding of the factors affecting the fate of N in watersheds, including the effects of land use and N sources (point, nonpoint, agricultural, urban, organic, and inorganic), N transformations (mineralization, nitrification, denitrification, and immobilization), and transport mechanisms (runoff, percolation to ground water, and ground water transport). Knowledge of the variability in N transport in relation to these factors is critical to developing and implementing effective strategies for mitigating unacceptably high N inputs to receiving waters.

The strengths of watershed-scale evaluations include easily definable hydrologic boundaries, identification of N sources with respect to water flow patterns, and a convenient, integral measure of water-quality response at a single point (the basin outlet). The objective of this chapter is to synthesize current understanding of the major processes and controls affecting N transport in watersheds. The scope is limited to technical issues of flow and chemistry of fixed (i.e., biologically reactive) organic and inorganic N forms in watersheds. We begin with a background discussion of watershed hydrology (Section 1) and the effects of N on ecosystems and human health (Section 2). We then describe the major sources of reactive N to watersheds (Section 3), and summarize the principal terrestrial and aquatic processes affecting N transport (Section 4). Section 5 illustrates the effects of various natural and cultural properties of watersheds on the yield of N in surface waters (N yield is defined as the N mass observed at the outlet of a watershed expressed per unit of drainage area). We conclude with a discussion of the results of empirical modeling methods that have been used to separate the effects of N supply and loss processes and estimate the fate of N sources in watersheds.

1. WATERSHED HYDROLOGY

A watershed (catchment or drainage basin) is an area of land where all of the precipitation that falls, less the water lost to evaporation and deep aquifer recharge, eventually flows to a single outlet. A watershed encompasses both surface and subsurface components of water drainage that contribute to stream discharge. On a global perspective, watersheds vary dramatically in physical features including area, shape, drainage pattern, aspect, orientation, and elevation (Schumm, 1977; Black, 1996). Geomorphic features of watersheds reflect the geologic formations and soils present and the erosive forces that have reshaped these materials. In regions with low relief and homogenous surficial materials such as areas of the Midwestern U.S., watersheds are often pear-shaped with a dendritic drainage pattern, as there is little difference in resistance to erosion to influence the headward cutting of stream channels (Black, 1996). Other, less random drainage patterns are the direct result of the varying erodibility of soil and rock. Structural differences in underlying formations can create well-defined, regular patterns as the channels develop following the path of least resistance to erosion. Slope aspect and watershed orientation (general direction of main stream channel) become important at higher altitudes and latitudes and especially with regard to snow hydrology. In the Northern Hemisphere, snowmelt will occur later on slopes with a north aspect in steep, east-west oriented watersheds as compared to slopes with a south aspect. Streamflow during snowmelt in watersheds that have a northern orientation may also be impeded by unmelted ice downstream.

Several numerical parameters have been used to describe the physical characteristics of watersheds. These include stream order, drainage density, and area and shape relations (Schumm, 1977; Linsley et al., 1982; Moseley and McKerchar, 1993). A classification of stream order was first proposed by Horton (1945) to describe the amount of branching within a basin. A first-order stream is small with no tributaries, a second-order stream has only first-order tributaries, and a third-order stream has tributaries of first- and second-order, etc. The order of a watershed is determined by the order of its principal stream. Drainage density refers to the total length of streams divided by the drainage area. A highly dissected basin will have a high drainage density and stream discharge that responds more quickly to precipitation events than less-dissected basins. A low drainage density may indicate erosion-resistant or highly permeable soils and low relief. Several area and shape relationships have been developed to create scales with which to compare watershed shapes with each other and with known shapes such as a circle or ellipse. Generally, ratios of basin parameters such as channel length, basin area, perimeter, and relief are calculated to provide indices, which are often dimensionless numbers, to allow relative comparison between watersheds. The numerical parameters used to describe the physical features of watersheds are in turn correlated with storm runoff, as measured by stormflow hydrographs. Functional relationships have been developed between runoff characteristics (e.g. time to initiation of runoff, time to peak flow, discharge at peak flow, total runoff volume, and time to recession) and storm characteristics and physical watershed features (Moseley and McKerchar, 1993).

Streamflow or discharge is a composite of surface (overland flow) and subsurface (baseflow) contributions. On the surface, flow follows the topography, from high elevations to lower elevations along interconnected pathways that provide the steepest gradient down. In the subsurface, discharge to the surface may be concentrated at permeability contrasts such as soil and rock interfaces and through preferential flow paths such as macropores, worm and root channels, bedding planes, fractures, and caves. In the subsurface as on the surface, ground water flow is driven by hydraulic gradients and the favored pathways are those that provide the least resistance to flow. Along all pathways, chemical interaction may occur between the water and solid, liquid, and gas components present in the water. The nature and extent of such interactions depend on the specific biogeochemical environment and residence time. Hydrologic and geochemical processes are rarely uniform over the area of a watershed. This is evidenced by numerous investigations of such features as variable source areas of runoff (Anderson and Burt, 1978; Bernier, 1985), riparian-zone processes (Hill, 1996; Cirmo and McDonnell, 1997; Devito et al., 2000), and karst hydrogeology (LeGrand and Stringfield, 1973; White, 1988). Riparian and karst settings are also notable in that they include frequent and significant interaction between surface and subsurface water, often with important implications for N transport.

2. NITROGEN IMPACTS ON WATER QUALITY

Three well-documented water-quality concerns are related to loadings of N to surface and ground waters. The presence of high levels of nitrate (NO_3) in drinking water has been linked to two different human-health concerns. The risk of methemoglobinemia in infants due to ingestion of high NO_3 drinking water is well understood and recognized. An increased incidence of stomach cancer and non-Hodgkin's lymphoma due to NO_3 intake is less certain (U.S. Environmental Protection Agency, 1976; Heathwaite et al., 1993). Nonetheless, a drinking water standard of 10 mg L^{-1} for NO_3-N, as established by the U.S. Environmental Protection Agency, is now widely accepted.

A second area of concern is the toxic effect of ammonia (NH_3) on freshwater aquatic life (U.S. Environmental Protection Agency, 1976). It has been known since the early 1900's that NH_3 is toxic to fish and that this effect varies with water pH and temperature. A concentration of 0.02 mg L^{-1} as un-ionized NH_3 is the current standard for NH_3 in freshwater for the U.S..

The third and perhaps most significant water quality concern with respect to N is the overenrichment or eutrophication of surface waters. Eutrophication and its attendant problems of algal blooms, subsequent low dissolved-oxygen concentrations, and fish kills has been described in an extensive body of literature. Overabundance of P is the most common cause of eutrophication in fresh waters, although exceptions are known (Hecky and Kilham, 1988; Correll, 1998). In coastal marine waters, either N or P and possibly other nutrients, such as silicon, may be limiting, whereas in the open ocean, N is generally considered the key nutrient controlling primary production (Correll, 1998; Burkart and James, 1999; Council for Agricultural Science and Technology, 1999). Overenrichment of N has been implicated in the development of anoxic and hypoxic zones in shallow coastal waters in Europe, North America, and Asia. Excessive phytoplankton production in these areas leads to oxygen depletion when the organic residues decompose, often with devastating effects on local fisheries.

3. NITROGEN SOURCES TO WATERSHEDS

The inputs of biologically available forms of N to terrestrial and aquatic freshwater ecosystems have increased globally by more than a factor of two over the past two centuries as cultural activities that fix N have rapidly expanded. Nitrogen fixation refers to the conversion of dinitrogen gas (N_2) to NH_3 either naturally via N-fixing plants (legumes), or through cultural processes such as the manufacture of N fertilizer and combustion of fossil fuels. Fertilizer application, cultivation of leguminous crops, and fossil-fuel combustion, represent 57, 29, and 14% of the culturally derived N, respectively (Galloway et al., 1995; Vitousek et al., 1997). Cultural inputs are unevenly distributed around the world, with the highest concentrations in areas of intensive agriculture and industrial processing (Matthews, 1994). The largest increases have occurred in the latter half of the 20th century as the industrial production of N for use as fertilizer increased many fold. More than 50% of all the industrially fixed N applied as fertilizer through 1990 was used during the decade of 1980-90 (Vitousek et al., 1997). In the United States, fertilizer use has increased by a factor of about 12 since the 1950s, with much of the increase occurring prior to 1980 (Goolsby et al., 1999). Natural sources of N, principally biological fixation by non-cultivated leguminous plants and lightning fixation, represent 64 to 93% (90 to 140 teragrams N) of the total culturally fixed N at the global scale (Galloway et al., 1995; Vitousek et al., 1997), but vary geographically with vegetation and land use. Natural sources of N are typically small (< 10%) in relation to cultural sources in many developed regions of the world, such as in the United States (Jordan and Weller, 1996).

The agricultural food chain is the principal pathway for culturally derived N to enter the terrestrial and aquatic ecosystems of developed watersheds. More than 90% of the culturally derived N in the United States enters croplands and pastures through fertilizer application, crop fixation, and atmospheric deposition on agricultural lands (Jordan and Weller, 1996). Nearly 50% of the N applied in fertilizer is recycled in food and feed products (Keeney, 1982; Howarth et al., 1996) that are consumed by livestock and humans. Livestock consume the vast majority of the N in harvested crops and forages, most of which is excreted in feces and urine. From 10-40% of the N in animal manures is volatilized (Terman, 1979), and much of that subsequently enters nearby watersheds in NH_3 deposition from the atmosphere (Howarth et al., 1996). Manure that is applied to cultivated or pasture lands enters watersheds in organic-N, NO_3-N, or NH_3-N, (Haynes and Williams, 1993; Jordan and Weller, 1996; Kellogg et al., 2000). Less than 15% of the N consumed by livestock is subsequently ingested by humans in meat, eggs, and milk (Jordan and Weller, 1996). Much of the N in human wastes is recycled into the hydrosphere through on-site septic systems or is discharged to streams and rivers in the effluent of wastewater treatment plants.

In addition to animal manures and human wastes, which largely involve the terrestrial recycling of culturally derived N, mineralized organic N (i.e. N that is biologically converted from organic to inorganic forms) in soil is potentially an important recycled N source to watersheds and aquatic ecosystems (Burkart and James, 1999; Goolsby et al., 1999). Organic N deposits in soils reflect the recent and long-term accumulation of N from fertilizers and biologically fixed N, immobilized by soil microbes and plant residues. Although N mineralization occurs naturally, cultivation may initially expose the soil to much higher rates of mineralization that are equivalent to or even greater than annual N-fertilizer application rates (Burkart and James, 1999). On lands that have been cultivated over extended periods, the mineralization of N in soil organic matter may approach equilibrium with agricultural inputs (Paul et al., 1997).

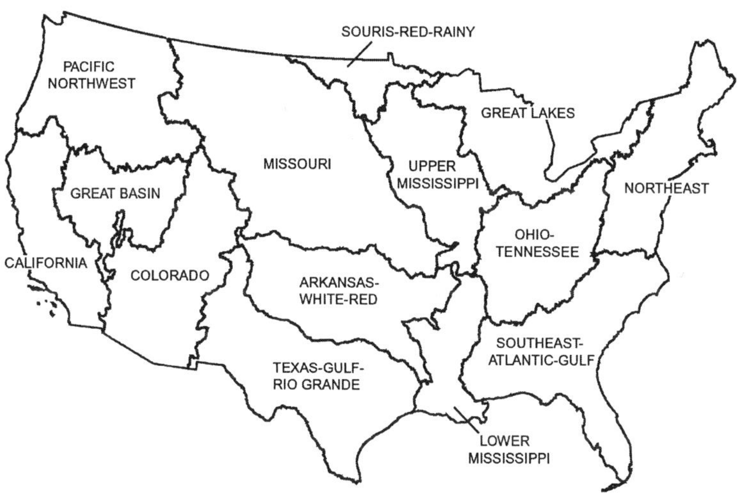

Figure 1. Water-resources regions of the conterminous United States. [Modified from Seaber et al. (1987).]

Despite the extensive terrestrial cycling of N in soils, vegetation, livestock, and humans, estimates of N transfers and the net releases of N to watersheds by major cultural activities have been the focus of intensive research and are now known for many areas of the United States (Howarth et al., 1996; Jordan and Weller, 1996; Burkart and James, 1999; Goolsby et al., 1999; U.S. Environmental Protection Agency, 1999; Kellogg et al., 2000). Estimates of N transfers often require assuming average values for N concentrations in organic materials and rate constants for N transformations, the use of state or county level census data, and extrapolation from field-scale measurements. Recent estimates of cultural inputs of fixed N to major regional watersheds of the United States (Jordan and Weller, 1996; Figure 1) are presented in Table 1. Table 1 separately reports "newly" fixed N inputs, reflecting *in situ* fixation by crops and the initial terrestrial application of N (fertilizer and NO_3 deposition in precipitation), and the releases of previously fixed (terrestrially recycled) N in livestock manure and human wastes. Also reported are estimates of net food and feed transfers by region, which are included in the releases of recycled N in livestock manure and human wastes. Fertilizer typically contributes about 50% of the "newly" fixed N inputs in the watersheds (sum of fertilizer, crop fixation, and deposition) with the highest contributions in the highly agricultural California region and lowest in the highly populated Northeast region. Crop fixation accounts for a third or more of the total inputs of newly fixed N, with some of the highest contributions occurring in the Northeast, Upper Mississippi, and Missouri.

Table 1. Cultural inputs of newly fixed and recycled N and net imports of food and feed N in major water-resource regions of the conterminous United States. [Modified from Jordan and Weller (1996). Municipal and industrial inputs of N from Giannesi and Peskin, (1984); livestock manure N from Puckett et al. (1998).]

Region	Total Area (10^6 ha)	Nitrogen (kg ha^{-1} yr^{-1})								
		Newly Fixed Nitrogen						Recycled Nitrogen Releases to Land and Water		
		Fixation by Agr. Biota	Fertilizer	Nitrate Deposition	Total Fixed N	Food / Feed Imports	Total Net Inputs	Point Sources Indust.	Point Sources Munic.	Livestock Manure
Northeast	42	8.3	6.0	6.8	21	10	31	1.3	7.7	12.0
Southeast Atlantic-Gulf	68	5.8	13	5.0	24	5.2	29	0.2	1.2	7.4
Great Lakes	30	15	18	6.8	40	-7.2	33	0.7	3.0	5.7
Ohio-TN	52	15	20	5.8	41	-11	30	1.9	2.3	19.6
Upper Miss.	48	27	40	5.4	72	-37	35	0.2	1.3	18.0
Lower Miss.	25	13	19	5.4	37	-13	24	1.1	1.2	3.1
Souris-Red Rainy	15	11	23	1.8	36	-19	17	<0.1	0.2	2.6
Missouri	130	12	13	2.8	28	-10	18	<0.1	0.3	8.6
Ark-Red	64	7.4	12	2.8	22	-0.7	21	0.1	0.3	10.6
Texas-Gulf-Rio Grande	80	3.3	7.4	2.8	14	1.1	15	0.3	1.3	9.3
Colorado	65	1.6	1.6	1.4	4.6	0.5	5.1	<0.1	0.2	2.8
Great Basin	36	2.4	0.9	0.9	4.2	0.5	4.7	<0.1	0.1	1.3
Pacific NW	70	3.8	6.4	1.4	12	-1.6	10	0.1	0.4	3.1
California	41	3.4	12	1.4	17	3.5	21	0.2	2.2	6.0

Atmospheric deposition of NO_3 is much lower than agricultural inputs, typically contributing from 10 to about 20% of the total inputs in most regions. The highest atmospheric contributions (32%) are found in the Northeast region, where deposition rates are high and fertilizer inputs are among the lowest. The inclusion of additional oxidized N compounds (NOy, including wet and dry deposition) could be expected to increase the estimates of deposition inputs in Table 1 by as much as a factor of 2 (Howarth et al., 1996). Approximately 20% of the total inputs of culturally derived N are transported in agricultural products nationwide in food and feed imports (Table 1). In most regions, exports of N in agricultural products nearly balance the imports of N in these products. The major exceptions include the Northeast, where imports represent nearly 50% of the newly fixed N inputs, and the Upper Mississippi region, where 51% of the newly fixed N inputs are transported to other regions of the country. The large imports of food and feed in the Northeast can account for the unusually large N releases in livestock manures and municipal/industrial wastes in this region. Nitrogen inputs from municipal and industrial wastes are also relatively high in the Great Lakes, Ohio-Tennessee, and California regions. The Ohio-Tennessee, Arkansas-Red-White, Texas, and Colorado regions show the largest releases of N in livestock manures in comparison to the newly fixed N input to these regions.

4. NITROGEN CYCLING AND LOSSES IN TERRESTRIAL AND AQUATIC ECOSYSTEMS

Biologically available forms of N are highly mobile in the environment, and are subject to extensive biogeochemical cycling in terrestrial and aquatic ecosystems (Vitousek et al., 1997). Nitrogen cycling in terrestrial and aquatic ecosystems involves an intricate array of biogeochemical processes that can vary spatially and temporally in the environment in both rate and direction. Individual processes and the entire N cycle for selected systems have been the subject of numerous studies, many of which have been summarized in comprehensive review articles and monographs including Keeney (1973 and 1983), Stevenson (1982 and 1994), Floate (1987), Russelle (1992), Powlson (1993), and Vitousek et al. (1997). Discussion here will be limited to a brief description of principal N transformations affecting N transport from watersheds.

Immobilization is the assimilation of inorganic N by plants and microorganisms to form organic N compounds whereas mineralization is the decomposition of organic N to ammonium (NH_4). Nitrification is the microbial oxidation of NH_4 to nitrite (NO_2) and NO_3 whereas, conversely, denitrification is the reduction of NO_3 to NO_2, nitrous oxide (N_2O), and dinitrogen gas (N_2). Nitrification is important from an N transport perspective in that it involves the transformation of a relatively immobile species (NH_4) to a highly mobile one (NO_3). Lastly, N fixation is the conversion of N_2 to NH_3, either naturally via N-fixing plants (legumes), or through cultural processes via the manufacture of N fertilizer.

Nitrogen cycling dynamics and pathways differ within and between terrestrial, freshwater, and marine ecosystems. Nonetheless, some similarities persist and often dominate N dynamics in the environment. Since most agriculturally productive soil environments have extended periods of aerobic conditions, mineralization of organic N to form NH_4 is generally followed by nitrification. Thus, in many terrestrial settings with significant N present, N as NO_3 is commonly found at relatively high concentrations even though it is also the form of N preferred for uptake by many plants. Since NO_3 is also highly mobile in the hydrosphere, it is often the dominant form of N in freshwater systems. Denitrification of NO_3 occurs under anaerobic conditions such as are found in flooded soils, riparian areas, and in the sediment of streams, lakes, and reservoirs. From a watershed perspective, the dominant processes of the N cycle vary not only by location, but also seasonally at the same location.

Nitrogen from natural and cultural sources is removed from runoff and subsurface flows in the terrestrial and aquatic ecosystems of watersheds by many biogeochemical processes. Denitrification permanently removes N from watersheds by converting N to less reactive gaseous forms (NO, N_2O, or N_2) that escape to the atmosphere. Other means of N removal in watersheds, including the uptake of N by vegetation, burial of organic matter on the landscape, and storage of N on floodplains and in reservoirs and ground water, represent temporary storage sites for N over time scales ranging from fractions of a day to decades. Over long periods, these storage sites are likely to gradually release un-denitrified N to streams and rivers. Variability in the reported quantities of N removed in watersheds may in part reflect variations in the temporal and spatial scales over which these loss processes operate in both terrestrial and aquatic ecosystems (Seitzinger, 1988; Correll et al., 1992; Hill, 1996; Harvey and Wagner, 2000). However, most multi-year studies report the loss of large fractions of the N inputs to watersheds for a range of spatial scales, based on comparisons of inputs with the N yields from watersheds in streams and rivers (Galloway et al., 1995; Puckett, 1995; Howarth et al., 1996; Jordan and Weller, 1996; Vitousek et al., 1997; Goolsby et al., 1999). In large North American and European watersheds (basin sizes from 340,000 to 3.2 million km^2), comparisons of total inputs of N with stream yield indicate that 65 to 90% of the inputs (mean = 75%) are removed by terrestrial and aquatic processes (Howarth et al., 1996). Similar losses of N have also been observed in small watersheds of mixed land use (Jaworski et al., 1992; Jaworski et al., 1997) and in small, forested and agricultural catchments (Howarth et al., 1996). Because forest ecosystems are N limited, forested watersheds are capable of storing considerable quantities of N in biomass and soils. However, large variations have been observed in the percentage of loss, ranging from a few percent to more than 100 percent of N inputs (Johnson, 1992). This wide range may be explained by variations in the biological demand for N, which can fluctuate in response to such factors as N depositional history, forest successional stage, and species composition (Johnson, 1992; Stoddard, 1994; Howarth et al., 1996; Williams et al., 1996) as well as the effects of temperature on nitrification and other N transformations (Murdoch et al., 1998).

Many natural and cultural properties of watersheds may explain spatial and temporal variations in the rates of denitrification, nitrification, mineralization, and N storage and their effects on N transport in streams. These include factors such as land use, climate (precipitation and evaporation), the oxygen and carbon content of soils and stream sediments, and stream morphology (channel density, channel size, and water travel time). Watershed properties that affect the quantity, velocity, and direction of water movement along surface and subsurface flow paths (climate and geology) may have an especially important influence on N transport. Certain flow paths are more likely to remove N from the flow stream than others, such as in stream riparian and hyporheic zones where biochemical conditions may enhance denitrification. The effects of these various watershed characteristics on stream N yield are discussed in the following section.

5. EFFECT OF WATERSHED CHARACTERISTICS ON N TRANSPORT IN STREAMS

The reported N yield from watersheds (in units of kg km^{-2} yr^{-1}) throughout the world is highly variable, spanning more than four orders of magnitude (Beaulac and Reckhow, 1982; Meybeck, 1982; Smith et al., 1997; Caraco and Cole, 1999), and may be explained by a variety of watershed characteristics affecting the supply and removal of N in terrestrial and aquatic systems. In this section, we discuss the effects of many of the principal watershed properties on spatial variations in N yield, including stream discharge, climate, geology, soil properties, land surface topography, stream morphology, natural and cultural sources, and land use.

5.1. Stream discharge

The relation between stream N yield and discharge (the net quantity of water made available to streams via precipitation minus evaporation) illustrates the aggregate effects of surface and subsurface characteristics of watersheds. Streams in watersheds in more humid areas generally transport larger amounts of N and water per unit of drainage area than those in more arid regions. The mean annual yield of N in streams is nearly proportional to the mean annual stream discharge for watersheds around the globe (Caraco and Cole, 1999). For developed watersheds in the United States with a range of cultural N sources (Figure 2), stream discharge and total N yield span nearly four orders of magnitude and display a strong positive

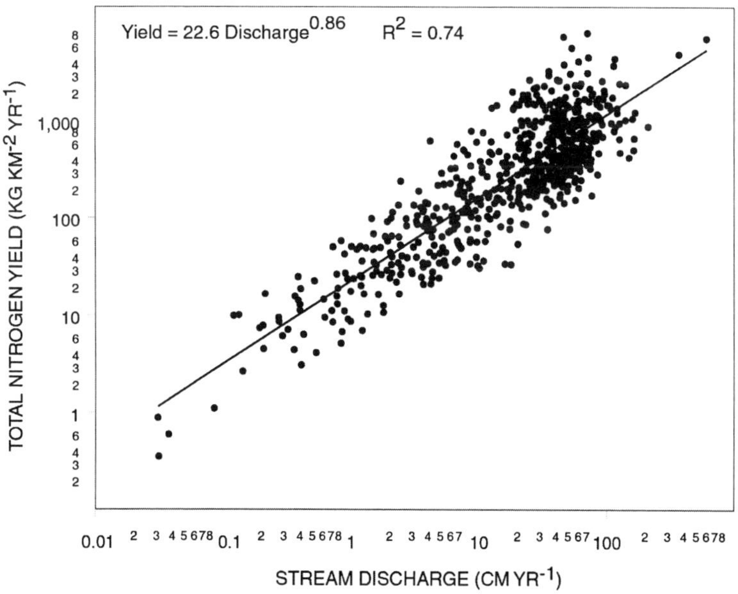

Figure 2. Relation of stream yield of total N to stream discharge for developed watersheds of the United States.

relation ($R^2 = 0.74$)—the exponent on discharge is slightly less than one (0.86). Similar rates of N yield per unit discharge have been observed in developed and undeveloped watersheds of the world (Caraco and Cole, 1999) and for relatively undeveloped watersheds in the United States (R.A. Smith, written communication). The slope of many of the observed relations generally spans a rather narrow range, with exponents from 0.80 to 0.87. Undeveloped tropical watersheds in South America with lower atmospheric N inputs (Lewis et al., 1999) also show similar to somewhat lower exponents for NO_3 (0.80) and total N (0.63). The intercept of the yield-discharge relations differs depending upon the magnitude of cultural inputs of N to the watersheds and units of discharge in the log linear model.

At individual stream locations, N yield also varies considerably in response to storms as well as seasonal and annual fluctuations in precipitation and streamflow. These responses have been extensively documented in the literature (Beaulac and Reckhow, 1982; Mueller et al., 1995; Alexander et al., 1996; Goolsby et al., 1999; U.S.Geological Survey, 1999). Although larger spatial than temporal variability in N yield is generally observed (Beaulac and Reckhow, 1982), temporal changes in N yield at individual sites have important implications for the management of sources. A recent study isolated the effects of year-to-year variations in streamflow on NO_3 yield at 104 monitoring locations along the East and Gulf coasts of the United States (Alexander et al., 1996). Although the mean annual NO_3 yield at all sites varied by as much as two orders of magnitude in response to year-to-year fluctuations in flow, variations at most sites ranged from 20 to 40% of the mean NO_3 yield. The change in NO_3 yield in response to annual streamflow variations was nearly linear and proportional in most watersheds (i.e., a one percent change in flow corresponded to nearly equivalent percentage change in NO_3 yield), although many streams displayed nonlinear responses. The variance in NO_3 yield at the sites (expressed as a percentage of the mean yield) was negatively correlated with the mean annual streamflow and nonurban land use of the watersheds; the largest variability in yield was observed in watersheds with arid conditions and large diffuse sources of N.

5.2. Climate

Climate explains much of the variability in stream N yield-discharge relations for watersheds. Climate influences the distribution and composition of vegetation and soils, which affect the supply of organic and inorganic N forms to watersheds (Beaulac and Reckhow, 1982; Downing et al., 1999). The productivity of natural and cultivated vegetation tends to be higher in wetter, more temperate climates, and fertilizer-intensive crops are also generally grown in these areas. The rates of water movement over the land surface, through the subsurface, and in stream channels also govern N residence times and loss in watersheds. Water transport may affect the rates of biogeochemical processing of N by controlling the contact and exchange of N-enriched water with sites suitable for denitrification, such as anoxic soils, benthic stream sediments, channel hyporheic and riparian zones, wetlands, and aquifers (Harvey et al., 1996; Hill, 1996). Water travel time, which is strongly correlated with discharge, has been found to be an important predictor of N loss in streams and reservoirs (Kelly et al., 1987; Howarth et al., 1996; Alexander et al., 2000a). Nitrogen losses in streams are also correlated with stream discharge (expressed per unit of drainage area), based on observations in large watersheds in Germany (Behrendt, 1996).

Changes in global climate that may occur in response to recent and anticipated rises in atmospheric levels of CO_2 and other greenhouse gases will potentially affect stream N yield through changes in precipitation and ambient temperatures and their corresponding effects on such factors as stream discharge, biological activity, and land use (Murdoch et al., 2000). Although most general circulation climate models are generally in agreement that temperatures and precipitation will rise over global scales, regional variations are expected to be large (Gleick and Adams, 2000). For example, recent predictions of precipitation through 2030 from two climate models of North America (Gleick and Adams, 2000) indicate large regional differences in the magnitude and even the direction of changes in precipitation, emphasizing the large uncertainty in current predictive models. Nevertheless, the predicted climate-related changes in precipitation or temperature are far reaching and could be expected to have notable effects on nutrient cycling in the terrestrial and aquatic ecosystems of watersheds, the nature of which are discussed in many recent reviews and analyses (Moore et al., 1997; Mulholland et al., 1997; Schindler, 1997; Gleick and Adams, 2000; Murdoch et al., 2000). Stream discharge is one of the major watershed properties likely to be affected by global warming, and is generally more sensitive to changes in precipitation than to temperature-induced changes in evapotranspiration (Wolock and McCabe, 1999). Changes in discharge would affect the quantity and rates of water movement along surface and subsurface flow paths that control the rates of N removal. Both spatial and temporal N yield-discharge relations (e.g., Fig. 2; Alexander et al., 1996) suggest that the long-term changes in N yield could be expected to be nearly proportional to the changes in stream discharge, although climate-related changes in land use and the rates of biochemical processing of N may cause more non-linear, short-term responses in yield. Changes in temperature may also be expected to affect terrestrial and aquatic rates of productivity and N uptake (Mulholland et al., 1997; Murdoch et al., 1998; Murdoch et al., 2000), and could change the density of microbial communities in soils and stream sediments, which govern the rates of nitrification and denitrification (Murdoch et al., 2000). Moreover, shifts in land use in response to changing precipitation and temperature, such as changes in the location of row-crop agriculture and reservoir construction, are additional factors that could affect N yield from watersheds (Murdoch et al., 2000).

5.3. Physiography and Subsurface Hydrology

Variability in N yield (both explained and unexplained by stream discharge) is related to various physiographic features of watersheds that govern the residence times of water and N, including soil properties, geology, and landscape topography (Beaulac and Reckhow, 1982). Many of these features control streamflow in watersheds according to the concept of variable-source areas (Beven and Kirkby, 1979; Wolock, 1993). Such features have been cited as important factors affecting ground- and surface-water interactions and N yield in streams at local and regional spatial scales (Bohlke and Denver, 1995; Winter et al., 1998). Variable-source-area models such as TOPMODEL (Beven and Kirkby, 1979; Wolock, 1993) stress the importance of slope, relief, soil permeability, soil moisture content, and depth to the water table, in defining water infiltration and overland flow. According to these models, overland flow typically occurs where the subsurface movement of water is impeded, such as in low-lying areas and soils of low permeability.

The effects of soil permeability on water and N flow in unsaturated soils have been clearly demonstrated by lysimeter studies in small agricultural watersheds (Howarth et al., 1996). Rates of N leaching in sandy soils have been reported to be 2 or more times those in loam or clay soils (Cameron, 1983; Sogbedji et al., 2000). In large watersheds with high cultural N inputs, studies have found that permeable soils and rocks result in low NO_3 yield in streams (U.S. Geological Survey, 1999). For example, a relatively low NO_3 yield was observed in the Lost River in Indiana, where the shallow permeable karst bedrock rapidly diverted N into the subsurface (U.S. Geological Survey, 1999). Low N yields in the Prairie and Shell Creeks in Nebraska were explained by a relatively flat terrain and sandy/silty soils that rapidly transport N into the shallow ground water system (U.S. Geological Survey, 1999).

These results are consistent with those from empirical models of stream monitoring data over regional scales (Mueller et al., 1997; Smith et al., 1997). These studies show an inverse relation between mean annual N yields in streams and soil permeability. Tile drainage systems, which have been used extensively on poorly drained croplands in the mid-continent region of the United States (Mueller et al., 1997; Goolsby et al., 1999; Skaggs and van Schilfgaarde, 1999), generally reduce the travel times of N to streams and rivers (Kladivko et al., 1991). Artificial drainage of otherwise poorly drained crop or grazing land by surface channels or subsurface drains can exacerbate N transport from the soil root zone and expedite delivery to surface-water bodies and/or shallow aquifers (Durieux et al., 1995; de Vos et al., 2000). Land drainage networks effectively bypass the natural filtering effects of wetlands and riparian areas and provide direct conduits of surface runoff to streams and lakes. Conversely, any N that is diverted to the subsurface in response to the hydrogeologic properties of watersheds has the potential to be denitrified (Hill, 1996). Subsurface N may also re-appear later in the baseflow of streams. For example, ground waters contribute nearly 50% of the N flux to streams in the Chesapeake Bay watershed; these streams include waters with residence times of 10 to 20 years (Michel, 1992; Bohlke and Denver, 1995; Focazio et al., 1998). In a study of 27 watersheds in the Chesapeake Bay Watershed, Jordan et al. (1997) found that NO_3 concentration increased with increasing proportion of baseflow to streamflow, suggesting that NO_3 transport was promoted by ground water flow in these areas.

One of the most dynamic responses of watersheds to precipitation and runoff occurs in stream riparian areas and especially in wetlands (Lowrance et al., 1984), where soils rapidly saturate to become the initial sites for overland flow (this is reflected by variable-source-area models of flow generation). The storage and gradual release of water in riparian areas also control baseflow during the recession of peak flows and over more extended periods (Lowrance et al., 1985). Riparian areas have been shown to significantly reduce the quantities of N (more than 80%) transported from upland areas to streams in overland flow and ground waters (Peterjohn and Correll, 1984; Correll et al., 1992; Lowrance, 1992); however, the quantities of N removed are highly variable (Hill, 1996). The age of forests, the types of vegetation and soils, and the geology in riparian areas contribute to this variability. Riparian areas that most effectively remove N have permeable surface soils and shallow impermeable layers that produce shallow subsurface ground water flows with long residence times and extensive contact with roots and soils (Hill, 1996). The removal of N in ground water via denitrification is also controlled by biogeochemical properties of aquifers (e.g. flow paths, organic carbon and oxygen supply, and density of denitrifying bacteria) that are independent of riparian locations, soils, or other land surface characteristics (Korom, 1992; Postma et al., 1991; Bohlke and Denver, 1995; Hill, 1996). Thus, N may be removed in the subsurface by processes that are not readily predicted from land use or other mapped surface features. Moreover, the effect of riparian areas on N transport is uncertain because most studies that report decreases in NO_3 concentration do not report water discharge.

Another type of ground water flow path involves the disrupted drainage patterns characteristic of karst terrain. Karst terrain includes distinctive features such as sinkholes, caves, and springs that develop when soluble rock, often carbonates, occur near the surface. Approximately 15% of the continental U.S. has karst features, including parts of the Appalachian Mountains, interior lowlands and plateaus in Kentucky, Indiana, and Tennessee, the coastal plain of Florida and Georgia, the Edwards Plateau of Texas, and the Ozark Highlands (Davies and LeGrand, 1972). Karst features allow for rapid conveyance of water from the surface to the aquifer and often within the fractured aquifer itself (LeGrand and Stringfield, 1973; White, 1988). The potentially short water residence time in karst aquifers may limit the opportunity for biogeochemical transformations of N constituents. Owing to the potential for capture of runoff in karst terrain, land-use practices affecting N distribution and runoff on the soil surface may directly affect N transport to shallow aquifers.

Brahana et al. (1999a) and Sauer et al. (1999a) describe the development of a research watershed (Savoy Experimental Watershed) in karst terrain within the Ozark Highland region of northwestern Arkansas. One sub-basin of this watershed has physiographic features (mantled karst, ridge and valley topography) and land-use (hardwood forest and pasture) typical of the Ozark Highlands. Two continuously flowing springs (Copperhead and Langle) discharge on opposing sides of the watershed outlet, which drains directly into the Illinois River. Dye-tracing studies have demonstrated that both springs capture runoff via conduits in limestone beneath porous gravel in the valley floor during storm events and rapidly transmit the intercepted water to the springs and, from there, overland to the Illinois River (Sauer et al., 1998; Brahana et al., 1999b). Figure 3 presents discharge, total Kjeldahl N (TKN)

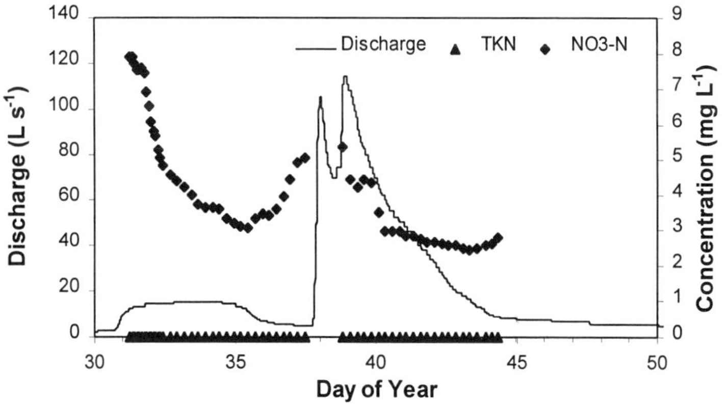

Figure 3. Discharge and total Kjeldahl nitrogen (TKN) and nitrate-nitrogen (NO_3-N) concentrations for Copperhead Spring for a 20-day interval in 1999.

concentration, and NO_3-N concentration for Copperhead Spring during two events over a 20-d interval in 1999. Nitrate was the dominant N species in the spring flow throughout the measurement interval, as concentrations of TKN and NH_3-N (data not shown) were less than 0.1 mg L^{-1}. Temporal variations in NO_3-N concentration are typical of runoff events (higher concentrations early with gradual decrease) indicating again that spring flow during storm events is dominated by captured runoff.

Table 2. Nitrate-N concentrations in samples taken during base flow conditions over two years at four locations in the Savoy Experimental Watershed.

Date	Shallow seep	Copperhead Spring	Langle Spring	Illinois River
	mg L^{-1}			
02-01-98	2.9	3.4	1.2	2.3
05-22-98	2.0	3.2	0.8	1.1
05-28-98	1.5	0.8	0.4	0.6
06-04-98	4.7	7.6	2.2	2.8
06-11-98	4.7	8.4	2.1	3.0
06-25-98	4.9	9.2	1.2	2.5
09-09-98	6.0	12.4	8.8	3.1
12-08-98	4.8	6.2	2.1	3.3
01-14-99	5.8	6.0	3.3	4.3
04-29-99	2.1	1.6	0.4	1.8
07-27-99	7.0	10.3	3.5	4.1
09-24-99	3.9	7.5	3.2	2.8
Mean	4.2	6.1	2.6	2.7
Maximum	7.0	12.4	8.8	4.3
Minimum	1.5	0.8	0.4	0.6

Table 2 lists NO_3-N concentrations in samples taken during base flow conditions for a two year period from Copperhead and Langle Springs, a shallow seep discharging in the valley slope above the soil/rock interface, and the Illinois River. Water from the seep represents interflow at the bottom of the root zone, which under non-storm flow conditions is diluted by ground water already in the shallow karst aquifer. The higher NO_3–N concentrations for Copperhead Spring reflect the more-intensely managed grazing lands within its recharge area. Discharge from Langle Spring had lower NO_3-N concentrations than the Illinois River for all sample dates except two whereas, conversely, Copperhead Spring's discharge has higher NO_3-N concentrations for all sample dates except one. These data illustrate the potential interaction between surface and subsurface water in karst settings and the subsequent implications for N transport. Runoff from upland areas flows into the valley but a portion is captured by the springs, mixed with ground water, and discharged from the springs to the Illinois River.

Water-quality research in karst settings in other locations in the U.S. has found correlations with land-use and has documented interactions between flow dynamics and N losses. Nitrate concentrations measured in several springs of a karst region in West Virginia were found to have a strong linear correlation (R^2 = 0.96) with percent agricultural land use in the spring basins (Boyer and Pasquarell, 1995). Kalkhoff (1995) found subbasins of Roberts Creek in northeastern Iowa with karst hydrology generally lost water and had lower NO_3 concentrations in streamflow as compared to those subbasins underlain with till and shale materials. Seepage from the stream to ground water in the karst subbasins of Roberts Creek reduced discharge and flow velocity in the stream thereby causing increased residence time of the water.

5.4. Stream Channels and Reservoirs

The effects of stream channels and their riparian areas on N yield from moderate- to large-sized watersheds (> 200 km^2 in size) have been observed in empirical models relating mean annual N yield to point and diffuse sources and various descriptors of stream hydrography (Omernik et al., 1981; Osborne and Wiley, 1988; Smith et al., 1997; Tufford et al., 1998). Several studies (Omernik et al., 1981; Osborne and Wiley, 1988; Tufford et al., 1998) accounted for the effects of channels and riparian areas on N yield by developing measures of the proximity of N sources to stream channels. The researchers reported higher accuracy for models with greater weights assigned to sources in the riparian areas of streams than to sources located outside of these areas.

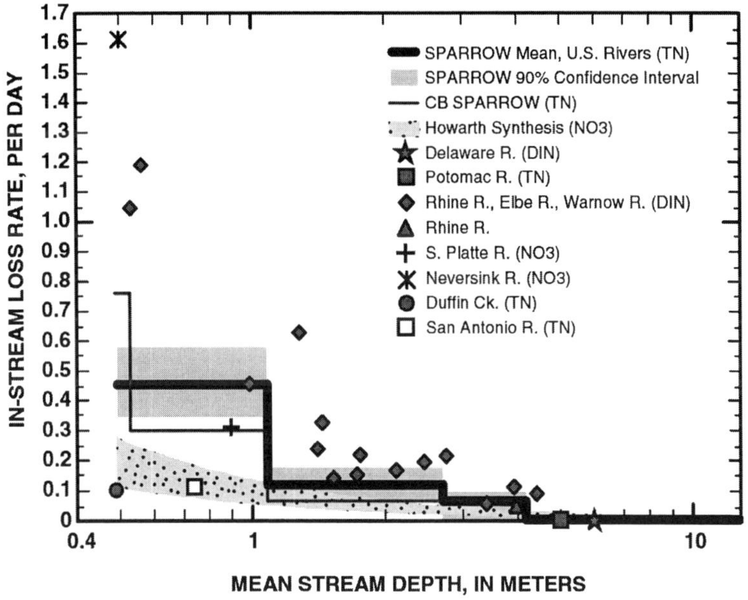

Figure 4. Nitrogen-loss rate in streams (per day of water travel time) in relation to stream channel depth [From Alexander et al. (2000a).]

A study of N transport in rivers of the United States used a mechanistic model structure (Smith et al., 1997) to empirically estimate the attenuation of N sources from upstream watersheds as a function of the physical properties of the watersheds (soils, temperature, and drainage density) and stream channels (water time of travel and channel size). Estimates of in-stream N loss were inversely related to stream channel size and ranged from 0.45 per day of water travel time in small streams to 0.005 per day in large rivers (Figure 4). When stream channel depth was used as an explanatory factor, these estimates were found to generally agree with those from mass balance and experimental studies that are available for selected North American and European streams (Alexander et al., 2000a). The inverse relation between N loss and channel depth may be explained by the effect of channel size (depth and water volume) on particulate N settling times and denitrification (Kelly et al., 1987; Rutherford et al., 1987; Harvey et al., 1996; Howarth et al., 1996; Alexander et al., 2000a). The natural rates of N loss via denitrification and settling are generally expected to be smaller in deeper channels where stream waters have less contact with the benthic sediment. Larger variability is observed in the rates of N loss in shallow streams, which likely indicates variability in the stream conditions responsible for N removal. These conditions include the hyporheic exchange of waters, organic and oxygen content of sediment, density of denitrifying populations, and water-column NO_3 concentrations. These results suggest that the proximity of N sources to large streams and rivers, as measured by water travel times in small tributaries, has a major effect on the downstream transport of N. Sources entering large streams and their nearby tributaries may be transported over very long distances in watersheds (Alexander et al., 2000a).

The physical and hydraulic properties of lakes (e.g., water residence time and depth) are also related to the observed rates of N loss in North America and Europe (Kelly et al., 1987; Howarth et al., 1996; Windolf et al., 1996; Seitzinger et al., unpublished data) and in New Zealand (McBride et al., 2000). Rates of N loss varied over a wide range from less than 10% to about 90% in these studies, and declined with increases in measures of the rates of water transport through reservoirs [i.e., the ratio of mean depth to water residence time—water displacement (Howarth et al., 1996; Kelly et al., 1987) and the ratio of reservoir discharge to surface area—areal hydraulic load (McBride et al., 2000)]. These lake properties affect the contact and exchange of water with the benthic sediment, which influences the rates of particulate N settling and denitrification (Kelly et al., 1987; Windolf et al., 1996). This implies that the mechanisms for the net removal of N are generally consistent with those in streams (Howarth et al., 1996; Seitzinger et al., unpublished data).

5.5. Cultural Sources and Land use

Human sources of N (fossil fuel combustion, fertilizer, human wastes, and livestock manures) and land use are known to have a major effect on N yield in surface waters (Beaulac and Reckhow, 1982; Peierls et al., 1991; Howarth et al., 1996; Vitousek et al., 1997; Carpenter et al., 1998; Seitzinger and Kroeze, 1998; Caraco and Cole, 1999; McFarland and Hauck, 1999; Arnheimer and Lidén, 2000; Castillo et al., 2000). Variability in N yield may be caused by spatial variations in the intensity and timing of N inputs to watersheds as well as differences in land management activities. Nitrogen concentration in streams and rivers of the United States have risen two- to ten-fold since the early part of the 20th century because of increased cultural inputs of N, and similar increases have been noted in European rivers and lakes (Howarth et al., 1996; Vitousek et al., 1997). The N yield of streams in relatively undisturbed watersheds of the North Atlantic region (Howarth et al., 1996) has been recently estimated to range from 0.8 to 2.3 kg ha^{-1} yr^{-1}. A study of background concentrations and yield from 66 relatively undeveloped, forest, grass and range land watersheds in the conterminous United States (sizes range from 6 to 2,700 km^2) over the period 1976 to 1997 (Clarke et al., 2000) indicates a range in the yield of total N that is similar to that reported by Howarth et al. (1996) (Table 3). Yield typically ranged from about 0.5 to 2.1 kg ha^{-1} yr^{-1} (based on interquartile range of mean annual yields). Yields larger than 2.1 kg ha^{-1} yr^{-1} and as high as 8.4 kg ha^{-1} yr^{-1} were observed in the eastern United States, where the rates of atmospheric deposition are highest (up to 4 kg ha^{-1} yr^{-1} wet NO_3). The yield of total N, NO_3, and NH_3 all increase with stream discharge (ranging from <1 to about 160 cm yr^{-1}) and atmospheric deposition.

Comparisons of N yields from relatively undeveloped watersheds with those from developed watersheds in North America reveal significant differences that can be traced to human activities. For example, N yield is frequently reported to be more than a factor of two higher in agricultural and urban watersheds in comparison to less-developed watersheds, including those predominantly in forest and rangeland (Beaulac and Reckhow, 1982; Mueller et al., 1995; U.S. Geological Survey, 1999). Historical data from two U.S. Geological Survey (USGS) water-quality monitoring networks illustrate these effects. These data provide a geographically representative description of N conditions in streams and rivers of the conterminous United States (Smith et al., 1993). The networks include 506 sites in the National Stream Quality Accounting Network (NASQAN) for the period from 1975 to 1992 (Alexander et al., 1998) and 185 sites in the National Water-Quality Assessment Program (NAWQA) network for the period 1993 to 1995. The yield of total N was consistently 3 to 4 times higher in developed watersheds than in undeveloped watersheds (Tables 3 and 4). The median N yield for the developed watersheds (3.3 kg ha^{-1} yr^{-1}) was 3.8 times higher than the median yield for undeveloped watersheds (0.86 kg ha^{-1} yr^{-1}). Some of the highest yields in both developed and undeveloped watersheds occur in the eastern United States, where atmospheric deposition is high. Smaller differences are observed between the stream N conditions in developed and undeveloped watersheds of the western United States than in other regions because of the relatively small inputs of cultural sources of N and more arid conditions in these western regions (Table 1; Clarke et al., 2000).

Table 3. Stream yields of N (kg ha^{-1} yr^{-1}) in 66 undeveloped watersheds of the United States. [Nitrate-nitrite and ammonia are dissolved. Data from Clarke et al., 2000.]

Metric	Percentiles					Percentiles - Fraction of Total Nitrogen		
	Min.	25th	50th	75th	Max.	25th	50th	75th
TN	<0.01	0.49	0.86	2.07	8.38			
Nitrate-nitrite	<0.01	0.11	0.24	0.52	5.83	0.14	0.27	0.55
Ammonia	<0.01	0.04	0.08	0.12	0.33	0.05	0.08	0.11
Organic	<0.01	0.16	0.33	1.07	5.07	0.32	0.60	0.75
Runoff (cm/yr)	0.1	22.0	34.1	58.4	163.1			

Table 4. Stream yields of N (kg ha^{-1} yr^{-1}) in 691 developed watersheds of the United States. [Nitrate-nitrite and ammonia are dissolved. The stations are located in developed watersheds representing a wide range of land cover types: 191 are predominantly agricultural, 34 are primarily urban, and 455 are classified as containing a mixture of land-cover types. Watersheds in the National Stream Quality Accounting Network range in drainage basin size from about 15 to 2.9 million km^2 with a median size of 11,000 km^2 (interquartile range from 3,100 to 37,000 km^2). Watersheds in the National Water Quality Assessment Program are typically smaller in size, ranging from about 15 to 220,000 km^2 with a median size of 1,300 km^2; interquartile range from 150 to 6,400 km^2.]

Metric	Percentiles					Percentiles - Fraction of Total Nitrogen		
	Min.	25th	50th	75th	Max.	25th	50th	75th
TN	<0.01	1.05	3.28	7.36	81.08			
Nitrate-nitrite	<0.01	0.22	1.06	3.33	79.02	0.25	0.40	0.60
Ammonia	<0.01	0.05	0.16	0.38	7.84	0.04	0.06	0.08
Organic	<0.01	0.50	1.51	2.88	58.02	0.33	0.52	0.70
Drainage Area (km^2)	13	1,585	7,268	28,381	2,953,895			
Runoff (cm yr^{-1})	0.03	6.8	27.7	49.2	598.3			

On the basis of USGS data, the median N yield in predominantly agricultural basins (5.9 kg ha^{-1} yr^{-1}; n=191) and urban watersheds (6.0 kg ha^{-1} yr^{-1}; n=34) was more than twice as large as the median N yield in watersheds of mixed land use (2.7 kg ha^{-1} yr^{-1}; n=455).

Moreover, the median yield from agricultural and urban watersheds was more than six times the median N yield in relatively undeveloped watersheds (0.9 kg ha^{-1} yr^{-1}; Table 3). Figure 5 illustrates the relation between agricultural land area and N yield. An increase in agricultural land area from a few percent to nearly 100 percent corresponded to more than a five-fold increase in stream yields. For watersheds with similar percentages of agricultural land Agricultural management practices also can have a major effect on N transport. The addition of fertilizers and organic matter (manure and biosolids) to grassland ecosystems, which are naturally N limited, improves their utility for the grazing of livestock, but contributes to large watershed yields of N. Timmons and Holt (1977) showed that annual stream N yield from ungrazed native little bluestem prairie (*Andropogon scoparius* Michx.) was only 0.8 kg ha^{-1}. By contrast, N yield from 2 grazed rangeland watersheds in Central Oklahoma ranged from 1.7 to 5.2 kg ha^{-1} y^{-1} (Olness et al., 1980). Higher N yield from grazed watersheds is often due to grazing animal urine, which is known to increase both runoff losses of N and NO_3 leaching (Schepers and Francis, 1982; Stout et al., 1997; Sauer et al., 1999b). Annual N yield of 2 to 9 kg ha^{-1} y^{-1} has been observed where fertilizer or manure N additions were made to improve forage production on grazing lands (Kilmer et al., 1974; McLeod and Hegg, 1984; Nelson et al., 1996). Nitrate is typically the dominant form of N transported from grazing lands, often with significant concentrations in both runoff and ground water flow (Sharpley and Syers, 1981; Owens et al., 1983; Cuttle et al., 1992), however, there is a considerable range in N yields because of the effects of other N sources as well as differences in the rates of N processing in watersheds related to many of the factors discussed previously (Beaulac and Reckhow, 1982).

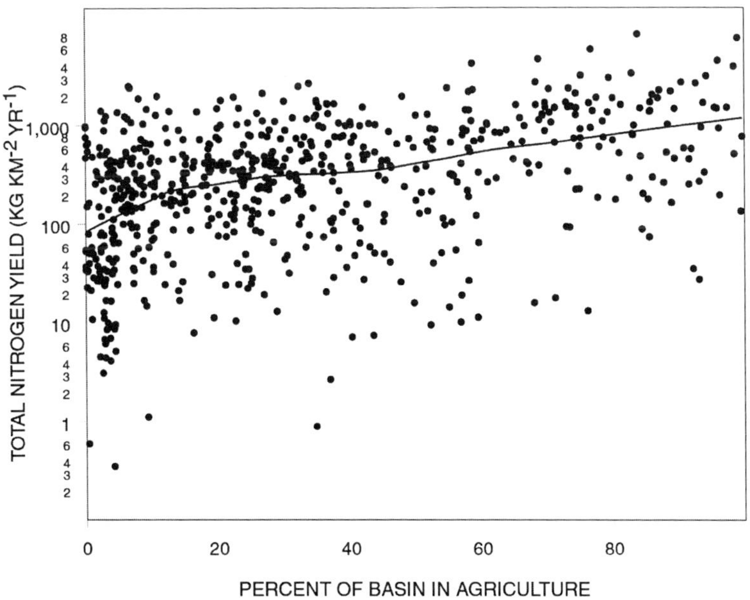

Figure 5. Relation of stream yield of total N to the percentage of basin area in agriculture for developed watersheds of the United States. The fitted line is obtained from a LOWESS smoothing technique (Cleveland, 1979). The LOWESS line displays the central tendency of the data, and provides an approximate description of the univariate relation. A more complex multivariate relation would be required to accurately predict stream N yield as a function of agricultural intensity.

Although N yield from forested watersheds can be low, watersheds disturbed by activities such as logging or development can be a significant source of NO_3 (Hallberg and Keeney, 1993). The high demand for NO_3-N by vegetation can result in a greater proportion of N yield in the organic form. Timmons et al. (1977) measured nutrient transport from an aspen-birch (*Populus tremuloides* Michx., and *Betula papyrifera* Marsh.) forest and found 80% of the total N load in runoff (1.25 of 1.56 kg ha^{-1} y^{-1}) was organic N. An average of 67% of the N yield in runoff from upland pasture and forest sites in a grazed watershed in the Ozark Highlands was in the organic form (Sauer et al., 2000). Organic N transported to surface-water bodies is subject to further transformations (mineralization, nitrification, and denitrification) in aquatic or benthic environments.

The amount and timing of N loads in streams also has been correlated with row crop acreage and N management practices. Schilling and Libra (2000) monitored NO_3-N concentrations in 15 Iowa watersheds with row crops covering 24 to 87% of the watersheds area. Average annual NO_3-N concentrations were directly related ($p < 0.0003$) to row crop area. Linear regression showed that an estimate of average annual NO_3-N concentration in surface water could be obtained by multiplying a watershed's row crop percentage by 0.1. Nitrate-N concentrations in streams in 10 states of the upper-midwestern U.S. were positively correlated with streamflow, upstream areas of corn (*Zea mays* L.), and N-fertilizer application rates (Mueller et al., 1997). Others (Becher et al., 2000, David and Gentry, 2000) also have found correlations between N fertilizer use and N yield in agricultural watersheds of the Midwest U.S. Figure 6 shows seasonal changes in average NO_3-N concentrations in stream

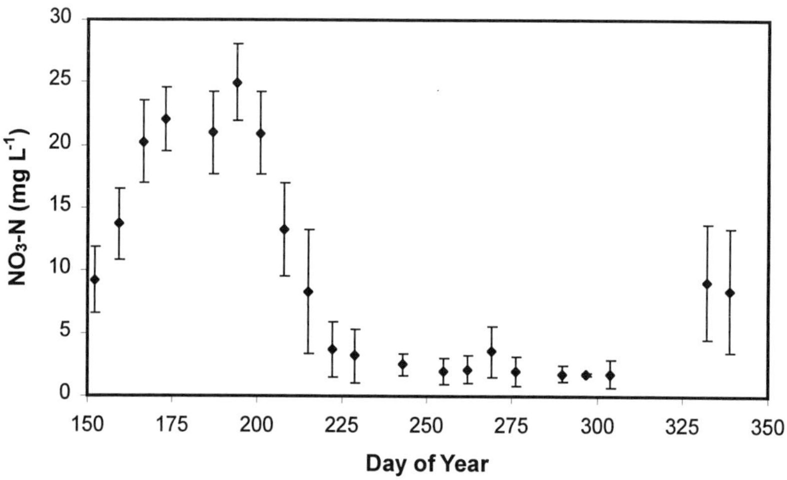

Figure 6. Mean nitrate-nitrogen (NO_3-N) concentration from 13 sampling sites along Tipton Creek in central Iowa during 2000. Error bars represent 1 standard deviation from the mean.

water for a 202 km^2 agricultural watershed in central Iowa (T.J. Sauer, unpublished data). Nitrate-N values in Figure 6 are daily means of stream-water samples collected from 13 locations on each date. Ammonia-N concentrations in samples collected on these dates were insignificant (< 1 mg L^{-1}). All samples except those on days 152, 166, and 194 were collected during baseflow conditions with stream discharge less than 150 L s^{-1}. Samples on days 152, 166, and 194 were collected as stream discharge was decreasing following runoff events.

This watershed (Tipton Creek) typifies the intense row-crop management of the upper-Midwest U.S., with 84% of the area being in corn or soybean (*Glycine max* Merr.) production. The increase in NO_3-N concentration in streamflow during late spring/early summer in cropped watersheds like Tipton Creek has been attributed to nitrification of N in fertilizers and animal manures (Becher et al., 2000; Castillo et al., 2000). In this instance, fertilizer and/or manure would typically be applied to fields sometime between days 100 and 140 to provide nutrients for corn during the growing season. Another process that may contribute to the trends observed in Figure 6 is mineralization of organic N after tillage and as the soil warms in spring.

5.6. Watershed Size

Much of the research on the fate of N in watersheds has focused on small catchments (Sharpley and Syers, 1981; Johnson, 1992; Hill, 1996; Pionke et al., 1996), where the natural and cultural influences on stream N yield are more spatially uniform, and N sources, transformations, and hydrologic flow paths are more readily discerned. Considerable variability has been observed in N yields from these catchments because of the wide range of sampled watershed properties (Beaulac and Reckow, 1982; Johnson, 1992; Hill, 1996). Relatively little information, however, has emerged about how N yields vary with watershed size. At progressively larger spatial scales, stream yields reflect the effects of an increasingly complex range of N sources and biogeochemical processes. This makes it difficult to quantify how the effect of any individual factor changes with watershed size.

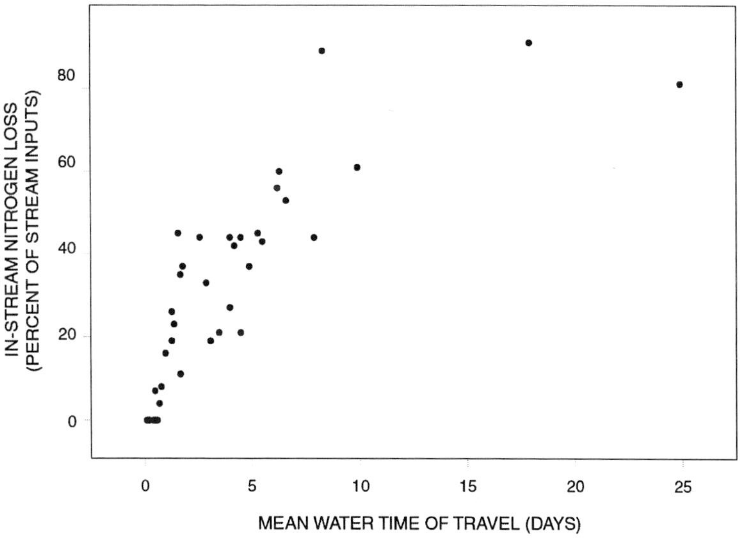

Figure 7. Relation of in-stream total N loss to the water time of travel in coastal watersheds of the United States. [Model predictions from Alexander et al., 2000b.]

A few studies (Alexander et al., 2000a,b; Seitzinger et al., unpublished data) have used empirical data from a range of watershed sizes to quantify the effects of in-stream N removal processes (denitrification and N storage) on the transport of N through drainage networks. As water is carried downstream, N is continually removed from the water column through contact with the benthic sediment. Although the rate of N removal per unit of water travel time declines significantly with increases in channel size (Alexander et al., 2000a; see Section 6.4), the fraction of N inputs to streams that is removed generally increases with cumulative water travel time in streams, which is positively correlated with drainage basin size (Alexander et al., 2000b). Figure 7 illustrates this concept on the basis of a study of 40 coastal watersheds in the United States in which the SPARROW model was used (Smith et al., 1997; Alexander et al., 2000b). Nitrogen loss, expressed as a percentage of the N delivered to streams, ranged from negligible quantities to 90% or more, and monotonically increased with the mean travel time of water in streams of the watersheds. Travel times can be as much as 24 days in several large, arid watersheds in Texas. Nitrogen losses of less than 10% were estimated for the smaller watersheds with less than about 2 to 3 days of mean water travel time. More than 50% of the N delivered to streams was removed in watersheds having mean water travel times greater than about 7 days. The estimates of N loss in Figure 7 reflect the cumulative removal of N over the range of stream sizes in these watersheds. Much lower N losses are expected for similar water travel times in large rivers, such as the Mississippi and its major tributaries, for which low first-order N loss rates have been estimated (Alexander et al. 2000a; Fig. 4).

Changes in the intensity of land use with watershed size also have discernable effects on stream N yields. Nitrogen yields are typically higher in small, upland watersheds that are intensively managed than in larger, heterogeneous watersheds. In two large U.S. river basins (Fig. 8), stream N yields are as much as 2 to 10 or more times higher in smaller tributary watersheds, many with predominantly agricultural and urban land use, than observed at downstream locations on the mainstem of the two rivers. Two of the mainstem sites located in the upper reaches of the South Platte River show the effects of urban sources. These land-use patterns reinforce the effects of in-stream N removal processes on stream N yields. In some watersheds, increases in the intensity of cultural N sources in lower reaches can cause stream N yield to increase in a downstream direction. For example, Castillo et al. (2000) found that NO_3 concentrations in the River Raisin in Michigan increased from the headwaters to the river mouth and were strongly correlated with the ratio of agricultural to forest land upstream. In such cases, the intensity of land use has a predominant effect on stream N yield, and overcomes the effects of in-stream loss processes.

5.7. Nitrogen Forms in Streams

The quantities of stream N in the forms of NO_3, NH_3, and organic N differ with the magnitude of cultural inputs of N and other watershed characteristics. Based on estimates of N yield from relatively undeveloped watersheds in the United States (Clarke et al., 2000; Table 3) organic N typically accounts for more than 60% of the N. Other studies have also noted the predominance of organic N in the streams draining relatively undisturbed forests (Vitousek et al., 1997). However, the organic-N content of streams in minimally developed watersheds display considerable spatial variability (Table 3), and large organic fractions are not uncommon in more developed watersheds (Table 4). In undeveloped watersheds, the highest organic-N fractions (>70%) were observed in the southeastern and Texas coastal plains and the southern central portion of the United States, whereas the lowest organic-N fractions (<50%) were observed in forested and rangeland watersheds of the Appalachians and arid areas of the northern central portion of the United States (Clarke et al., 2000). Nitrate represents a majority of the remaining N in undeveloped watersheds, typically representing at least a quarter of the total N (Table 3). Ammonia is typically less than 8% of the total.

world (Caraco and Cole, 1999), the proportions of organic N and NO_3-N were found to be roughly equivalent. More complex multivariate relations would be required to accurately predict N forms in streams.

6. SOURCE CONTRIBUTIONS TO STREAM YIELD

A longstanding problem in quantifying the relative importance of specific natural and cultural sources of N to the stream yield from watersheds has centered on understanding the effects of land use, climate, and the biogeochemical processing of N in terrestrial and aquatic ecosystems over a range of spatial scales. At larger spatial scales, source inputs have commonly been used to characterize source contributions to streams (Jaworski et al., 1992; Puckett, 1995), but these methods do not account for the appreciable differences that exist in the rates of N processing and transport in watersheds as reflected in measurements of stream yield (Beaulac and Reckhow, 1982). A variety of watershed models have been used to resolve the interactions between N supply and loss processes. At large watershed scales, where the applicability and reliability of fine-scale deterministic models is more uncertain (Rastetter et al., 1992), empirical models that are calibrated to stream measurements of N have frequently been used to quantify N sources and losses in watersheds. Examples include spatial regression models of stream N yield on population density (Peierls et al., 1991), net anthropogenic sources (Howarth et al., 1996), atmospheric deposition (Howarth et al., 1996; Jaworski et al., 1997), and models containing a range of explanatory variables describing both N sources and watershed characteristics (Lystrom et al., 1978; Omernik et al., 1981; Osborne and Wiley, 1988; Mueller et al., 1997; Smith et al., 1997; Tufford et al., 1998; Goolsby et al., 1999).

Estimates of the sources of N in streams of the major water-resources regions of the United States (Smith and Alexander, 2000), based on the application of the SPARROW model (Smith et al., 1997; Alexander et al., 2000a,b), are illustrated in Table 5. This model provides separate quantification of a range of major N sources and accounts for the terrestrial and aquatic losses of N as a function of watershed properties. Details of the model structure and calibration to N measurements from 400 stream monitoring sites are given in Smith et al. (1997), and discussions of the model verification are given in Alexander et al. (2000a,b), National Research Council (2000), and Stacy et al. (2000). The major N sources in streams as defined by the model include agricultural diffuse sources (fertilizer and livestock manures), atmospheric deposition, municipal and industrial point sources, and other sources associated with nonagricultural lands. In addition to applied fertilizers, the fertilizer source may also include inputs of fixed N in leguminous crop residues and other mineralized soil N from cultivated lands (Alexander et al., 2000b). Atmospheric sources include wet deposition of inorganic NO_3-N as well as additional N contributions from wet organic and dry inorganic N (Alexander et al., 2000a,b). Nonagricultural runoff includes the remaining sources of N (i.e., not quantified by point sources and other diffuse model terms), delivered to streams in the overland flow and ground water from urban, forested, wetlands, and barren lands. The runoff from forested lands may include N supplied from natural fixation.

The sources of N to stream yield vary greatly among the regions (Table 5), and show a general correspondence to the inputs of newly fixed and recycled N inputs described in Table 1. Estimates of N loss in watersheds, based on the median stream yield (Table 5) and the total

Table 5. Point- and nonpoint-source contributions to total N yield from watersheds in major water-resource regions of the conterminous United States. Total yield is the median stream yield from hydrologic cataloging units in each region. The median and quartile values for the source contributions within each region are expressed as a percentage of the total yield. [Modified from Smith and Alexander (2000).]

Region	Total Yield	Percentage of Total Yield									
		Point Sources		Fertilizer		Animal Agriculture		Atmosphere		Nonagricultural Runoff	
	(kg ha^{-1} yr^{-1})	Median	Quartiles	Median	Quartiles	Median	Quartiles	Median	Quartiles	Median	Quartiles
Northeast	8.0	4.2	1.5 – 19	13	6.2 – 18	10	4.2 – 19	31	22 – 40	26	17 – 36
Southeast Atlantic-Gulf	5.9	2.7	1.1 – 8.2	26	17 – 38	14	8.8 – 21	21	15 – 28	26	19 – 34
Great Lakes	8.0	4.3	1.3 – 12	22	7.8 – 41	10	4.6 – 17	25	16. 34	17	6.5 – 40
Ohio-TN	11	2.1	0.7 – 7.2	26	11 – 48	15	10 – 21	25	18 – 39	16	7.6 – 25
Upper Miss.	13	0.8	0.5 – 1.6	55	40 – 66	21	15 – 27	13	11 – 17	3.6	2.1 – 10
Lower Miss.	7.6	2.3	1.0 – 11	40	14 – 64	6.3	3.2 – 10	22	14 – 28	18	8.0 – 28
Red Rainy	3.5	0.3	0.1 – 0.6	75	57 – 81	5.2	2.8 – 9.0	9.3	7.4 – 14	7.2	3.4 – 20
Missouri	2.1	<0.1	<0.1 - <0.1	30	8.8 – 51	20	15 – 25	16	12 – 20	29	9.5 – 55
Ark-Red	3.9	0.8	0.2 – 1.9	29	20 – 46	23	17 – 29	18	14 – 23	20	12 – 28
Texas-Gulf-Rio Grande	2.1	0.3	<0.1 – 2.6	20	4 – 37	15	10 – 21	16	12 – 20	37	18 – 66
Colorado	1.1	0.1	<0.1 – 0.4	2.0	0.8 – 8.8	7.7	3.6 – 12	10	7.7 – 16	74	64 – 80
Great Basin	0.9	<0.1	<0.1 - <0.1	3.6	0.9 – 9.2	9.3	5.6 – 15	6.4	5.4 – 8.1	78	61 – 86
Pacific NW	4.2	<0.1	<0.1 - <0.1	12	5.5 – 30	11	7.3 – 14	13	8.0 – 16	57	34 – 69
California	4.8	1.2	0.3 – 6.7	21	8.9 – 52	12	7.6 – 17	8.7	5.5 – 13	35	16 – 62
United States	4.7	0.8	0.5 – 3.4	22	7.5 – 45	14	8.2 – 21	16	11 – 23	28	13 – 56

net N input to the regions (Table 1), range from 62 to 89% of the net inputs of N (median=76%). Agricultural sources (fertilizer and livestock manures) are the largest contributors to stream yield in most of the regions, representing more than 40% in the Ohio-Tennessee, Southeast-Gulf, Upper and Lower Mississippi, Souris-Red-Rainy, Missouri, Arkansas-Red, and Texas regions. Livestock manures contribute large quantities of N in watersheds in the Ohio-Tennessee, Upper Mississippi, Missouri, Arkansas-Red, and Texas regions (Table 5); these contributions are consistent with the estimated large inputs of N in livestock manures in these regions in relation to the total net inputs of N (newly fixed N plus net food/feed imports) reported in Table 1. Atmospheric N contributes more than a quarter of the stream yield in most of the watersheds in the Northeast region, and is a dominant source in watersheds of the Great Lakes and Ohio-Tennessee region. Nonagricultural diffuse sources contribute a majority of the N to the stream yields in the Colorado, Great Basin, and Pacific Northwest regions, where cultural inputs of N are generally low. Nonagricultural diffuse contributions are also important in the Northeast and Southeast-Gulf, where watersheds generally receive large natural sources of organic N from forest vegetation. Point sources, generally among the smallest contributors in most watersheds, are the highest in the densely populated Northeast, Ohio-Tennessee, and Great Lakes regions; this is generally consistent with estimates of the inputs to watersheds in these regions from municipal and industrial wastewater treatment plants (Table 1). These results are also consistent with other studies of moderate to large watersheds, which find municipal and industrial point sources to be a relatively small source of N to streams (Puckett 1995; Howarth et al., 1996; Goolsby et al., 1999). However, in small, highly urbanized watersheds, municipal and industrial wastewaters frequently account for significantly larger shares of the N in streams (U.S. Geological Survey, 1999; Alexander et al. 2000b).

7. SUMMARY

Stream N yields have been assessed in watersheds through detailed process-oriented studies at the local scale and over larger, regional scales using statistical techniques. These approaches have been applied to natural and culturally affected environments in watersheds to elucidate the hydrologic and biogeochemical factors that affect N transport. The biogeochemical processing of N has been studied over a range of spatial and temporal scales in watersheds to enable the interpretation of data trends and development of conceptual and numerical models of N yield. Surface and subsurface hydrology, climate, physiography, and basin size all affect the partitioning of precipitation between infiltration and runoff and subsequent water flow paths. Natural and cultural sources of N and their subsequent transformations influence the amount and mobility of N constituents in soil, plant materials, and water. Watersheds represent a physical coupling of the hydrologic and source components in a continuous, dynamic system. Management of land resources based on principles derived from watershed-scale studies is a key component of ongoing efforts to improve the efficiency of N use and limit adverse water quality impacts from excessive N loadings to surface, subsurface, and marine waters.

REFERENCES

Alexander, R.B., R.A. Smith, and G.E. Schwarz. 2000a. Effect of stream channel size on the delivery of nitrogen to the Gulf of Mexico. Nature, 403:758-761.

Alexander, R.B., R.A. Smith, G.E. Schwarz, S.D. Preston, J.W. Brakebill, R. Srinivasan, and P.A. Pacheco. 2000b. Atmospheric nitrogen flux from the watersheds of major estuaries of the United States: an application of the SPARROW watershed model. *In* R. Valigura, et al. (eds.) Nitrogen loading in coastal water bodies: an atmospheric perspective. American Geophysical Union Water Monograph 57, Washington, DC.

Alexander, R.B., J.R. Slack, A.S. Ludtke, K.K. Fitzgerald, and T.L. Schertz. 1998. Data from selected U.S. Geological Survey national stream water quality monitoring networks, Water Resour. Res.34:2401-2405.

Alexander, R.B., P.S. Murdoch, and R.A. Smith. 1996. Streamflow-induced variations in nitrate flux in tributaries to the Atlantic Coastal zone. Biogeochem. 33:149-177.

Anderson, M.G., and T.P. Burt. 1978. Toward more detailed field monitoring of variable source areas. Water Resour. Res. 14:1123-1131.

Arnheimer, B., and R. Lidén. 2000. Nitrogen and phosphorus concentrations from agricultural catchments – influence of spatial and temporal variables. J. Hydrol. 227:140-159.

Beaulac, M.N., and K.H. Reckhow. 1982. An examination of land use-nutrient export relationships. Water Res. Bull. 18:1013-1024

Becher, K.D., D.J. Schnoebelen, and K.K.B. Akers. 2000. Nutrients discharged to the Mississippi River from eastern Iowa watersheds, 1996-1997. J. Am. Water Resour. Assoc. 36:161-173.

Behrendt, H. 1996. Inventories of point and diffuse sources and estimated nutrient loads—a comparison for different river basins in central Europe. Water Sci. Tech 33:99-107.

Bernier, P.Y. 1985. Variable source areas and storm-flow generation: an update of the concept and a simulation effort. J. Hydrol. 79:195-213.

Bevin, K.J., and M.J. Kirkby. 1979. A physically based, variable contributing area model of basin hydrology. Hydro. Sci. Bull. 24: 43-69.

Black, P. E. 1996. Watershed hydrology. 2nd ed. Ann Arbor Press, Chelsea, MI.

Bohlke, J.K., and J.M. Denver. 1995. Combined use of ground water dating, chemical, and isotopic analyses to resolve the history and fate of nitrate contamination in two agricultural watersheds, Atlantic coastal plan, Maryland. Water Resour. Res. 31:2319-2339.

Boyer, D.G., and G.C. Pasquarell. 1995. Nitrate concentrations in karst springs in an extensively grazed area. Water Resour. Bull. 31:729-736.

Brahana, J.V., P.D. Hays, T.M. Kresse, T.J. Sauer, and G.P. Stanton. 1999a. The Savoy Experimental Watershed – early lessons for hydrogeologic modeling from a well-characterized karst research site. p. 247-254 *In* A.N. Palmer et al. (ed.) Karst modeling. Karst Waters Institute Special Publication 5.

Brahana, J.V., T.J. Sauer, P.D. Hays, T.M. Kresse, P. Little, and J. Funkhouser. 1999b. Hydrologic monitoring to characterize dominant controls of ground-water flow and transport in an area of confined animal production on a mantled karst terrane, Northwestern, Arkansas. p. 30 *In* F.D. Wilde et al. (ed.) Effects of animal feeding operations on water resources and the environment. U.S. Geological Survey Open-File

Report 00-204.

Burkart M.R., and D.E. James. 1999. Agricultural-nitrogen contributions to hypoxia in the Gulf of Mexico. J. Environ. Qual. 28: 850-859.

Cameron, K.C. 1983. Nitrate leaching: some fundamentals. p. 15-21 *In* Proc. Agronomy Society of New Zealand.

Caraco N.F., and J.J. Cole. 1999. Human impact on nitrate export: an analysis using major world rivers. Ambio. 28:167-170

Carpenter, S., N.F. Caraco, D.L. Correll, R.W. Howarth, A.N. Sharpley, and V.H. Smith. 1998. Nonpoint pollution of surface waters with phosphorus and nitrogen. Issues in Ecology, No. 3. 12 pp.

Castillo, M.M., J.D. Allan, and S. Brunzell. 2000. Nutrient concentrations and discharges in a Midwestern agricultural catchment. J. Environ. Qual. 29:1142-1151.

Cirmo, C.P., and J.J. McDonnell. 1997. Linking the hydrologic and biogeochemical controls of nitrogen transport in near-stream zones of temperate-forested catchments: a review. J. Hydrol. 199:88-120.

Clarke, G.M., D.K. Mueller, and M.A. Mast. 2000. Nutrient concentrations and yields in undeveloped stream basins of the United States. J. Amer. Water Res. Assoc. 36:849-860.

Cleveland, W.S. 1979. Robust locally weighted regression and smoothing scatterplots. J. Am. Stat. Assoc. 74:829-836.

Correll, D.L. 1998. The role of phosphorus in the eutrophication of receiving waters: a review. J. Environ. Qual. 27:261-266.

Correll, D.L., T.E. Jordan, and D.E. Weller. 1992. Nutrient flux in a landscape: Effects of coastal land use and terrestrial community mosaic on nutrient transport to coastal waters. Estuaries 15: 431-442.

Council for Agricultural Science and Technology. 1999. Gulf of Mexico hypoxia: land and sea interactions. Council for Agricultural Science and Technology, Ames, IA.

Cuttle, S.P., M Hallard, G. Daniel, and R.V. Scurlock. 1992. Nitrate leaching from sheep-grazed grass/clover and fertilized grass pastures. J. Agric. Sci. (Cambridge) 119:335-343.

David, M.B., and L.E. Gentry. 2000. Anthropogenic inputs of nitrogen and phosphorus and riverine export for Illinois, USA. J. Environ. Qual. 29:494-508.

Davies, W.E., and H.E. LeGrand. 1972. Karst of the United States. p. 467-505 *In* M. Herak and V.T. Stringfield (ed.) Karst – Important karst regions of the northern hemisphere. Elsevier, New York.

Devito, K.J., D. Fitzgerald, A.R. Hill, and R. Aravena. 2000. Nitrate dynamics in relation to lithology and hydrologic flow path in a river riparian zone. J. Environ. Qual. 29:1075-1084.

de Vos, J.A., D. Hesterberg, and P.A.C. Raats. 2000. Nitrate leaching in a tile-drained silt loam. Soil Sci. Soc. Am. J. 64:517-527.

Downing, J.A., M. McClain, R. Twilley, J.M. Melack, J. Elser, N.N. Rabalais, W.M. Lewis, Jr., R.E. Turner, J. Corredor, D. Soto, A. Yanez-Arancibia, J.A. Kopaska, and R.W. Howarth. 1999. The impact of accelerating land-use change on the N-cycle of tropical aquatic ecosystems: current conditions and projected changes. Biogeochem. 46:109-148.

Durieux, R.P., H.J. Brown, E.J. Stewart, J.Q. Zhao, W.E. Jokela, and F.R. Magdoff. 1995. Implications of nitrogen management strategies for nitrate leaching potential: roles of nitrogen source and fertilizer recommendation system. Agron. J. 87:884-887.

Floate, M.J.S. 1987. Nitrogen cycling in managed grasslands. p. 163-172 *In* R.W. Snaydon (ed.) Managed grasslands – analytical studies. Elsevier, New York.

Focazio, M.J., L.N. Plummer, J.K. Bohlke, E. Busenberg, L.J. Backman, and D.S. Powars. 1998. Preliminary estimates of residence times and apparent ages of ground water in the Chesapeake Bay watershed, and water-quality data from a survey of springs. U.S. Geological Survey Water-Resources Investigations Report 97-4225.

Galloway, J.N., W.H. Schlesinger, H. Levy, A. Michaels, and J.L. Schnoor. 1995. Nitrogen fixation: anthropogenic enhancement-environmental response. Global Geochem. Cycles 9: 235-252.

Gianessi, L.P., and H.M. Peskin. 1984. An overview of the RFF Environmental Data Inventory: methods, sources, and preliminary results. Resources for the Future, Washington DC.

Gleick, P.H., and D.B. Adams. 2000. Water: The potential consequences of climate variability and change for water resources of the United States. The report of the water sector assessment team of the national assessment of the potential consequences of climate variability and change. U.S. Global Change Research Program. Pacific Institute for Studies in Development, Environment, and Security. Oakland, CA.

Goolsby, D.A., W.A. Battaglin, G.B. Lawrence, R.S. Artz, B.T. Aulenbach, R.P. Hooper, D.R. Keeney, and G.J. Stensland. 1999. Flux and sources of nutrients in the Mississippi-Atchafalaya River basin. Report of Task Group 3 to the White House Committee on Environment and Natural Resources, Hypoxia Work Group. Federal Register 64: 23834-23835.

Hallberg, G.R., and D.R. Keeney. 1993. Nitrate. p. 297-322 *In* W.M. Alley (ed.) Regional ground-water quality. Van Nostrand Reinhold, New York.

Harvey, J.W., B.J. Wagner, and K.E. Bencala. 1996. Evaluating the reliability of the stream tracer approach to characterize stream-subsurface water exchange. Water Resour. Res. 32:2441-2451.

Harvey, J.W., and B.J. Wagner. 2000. Quantifying hydrologic interactions between streams and their subsurface hyporheic zones. p. 3-44 *In* J.B. Jones and P.J. Mulholland (ed.) Streams and ground waters, Academic Press, San Diego, CA.

Haynes, R.J., and P.H. Williams. 1993. Nutrient cycling and soil fertility in the grazed pasture ecosystem. Adv. Agron. 49:119-199.

Heathwaite, A.L., T.P. Burt, and S.T. Trudgill. 1993. Overview – the nitrate issue. p. 3-21 *In* A.L. Heathwaite et al. (ed.) Nitrate: processes, patterns and management. John Wiley and Sons, New York.

Hecky, R.E., and P. Kilham. 1988. Nutrient limitation of phytoplankton in freshwater and marine environments: a review of recent evidence on the effects of enrichment. Limnol. Oceanogr. 33:796-822.

Hill, A.R. 1996. Nitrate removal in stream riparian zones. J. Environ. Qual. 25: 743-755.

Horton, R.E. 1945. Erosional development of streams. Geol. Soc. Am. Bull. 56:281-283.

Howarth, R.W., G. Billen, D. Swaney, A. Townsend, N. Jaworski, K. Lajtha, J.A. Downing, R. Elmgren, N. Caraco, T. Jordan, F. Berendse, J. Freney, V. Kudeyarov, P. Murdoch, and Z. Zhu. 1996. Regional nitrogen budgets and riverine N and P fluxes for the drainages to the North Atlantic Ocean: natural and human influences. Biogeochem. 35:75-139.

Jaworski, N.A., P.M. Groffman, A.A. Keller, and J.C. Prager. 1992. A watershed nitrogen and phosphorus balance: the Upper Potomac River basin. Estuaries 15:83-95.

Jaworski, N.A., R.W. Howarth, and L.J. Hetling. 1997. Atmospheric deposition of nitrogen oxides onto the landscape contributes to coastal eutrophication in the northeast United States. Environ. Sci. Techno. 31:1995-2004.

Johnson, D.W. 1992. Nitrogen retention in forest soils. J. Environ. Qual. 21:1-12

Jordan, T.E., and D.E. Weller. 1996. Human contributions to terrestrial nitrogen flux: assessing the sources and fates of anthropogenic fixed nitrogen. BioScience 46:655-664.

Jordan, T.E., D.L. Correll, and D.E. Weller. 1997. Relating nutrient discharges from watersheds to land use and streamflow variability. Water Resour. Res. 33:2579-2590.

Kalkhoff, S.J. 1995. Relation between stream-water quality and geohydrology during base-flow conditions, Roberts Creek watershed, Clayton County, Iowa. Water Resour. Bull. 31:593-604.

Keeney, D.R. 1973. The nitrogen cycle in sediment-water systems. J. Environ. Qual. 2:15-29.

Keeney, D.R. 1982. Nitrogen management for maximum efficiency and minimum pollution. p. 605-649 *In* F.J. Stevenson (ed.) Nitrogen in agricultural soils. Agronomy Monograph 22, American Society of Agronomy, Madison, WI.

Keeney, D.R. 1983. Transformations and transport of nitrogen. p. 48-64 *In* F.W. Schaller and G.W. Bailey (ed.) Agricultural management and water quality. Iowa State University Press, Ames, IA.

Kellogg, R.L., C.H. Lander, D.C. Moffitt, N. Gollehon. 2000. Manure nutrients relative to the capacity of cropland and pastureland to assimilate nutrients: spatial and temporal trends for the United States. USDA-NRCS and ERS. Washington, DC.

Kelly, C.A., J.W.M. Rudd, R.H. Hesslein, D.W. Schindler, P.J. Dillon, C.T. Driscoll, S.A. Gherini, and R.E. Hecky. 1987. Prediction of biological acid neutralization in acid-sensitive lakes. Biogeochem. 3:129-141.

Kilmer, V.J., J.W. Gilliam, J.F. Lutz, R.T. Joyce, and C.D. Eklund. 1974. Nutrient losses from fertilized grassed watersheds in Western North Carolina. J. Environ. Qual. 3:214-219.

Kladivko, E.J., G.E. Van Scoyoc, E.J. Monke, K.M. Oates, and W. Pask. 1991. Pesticide and nutrient movement into subsurface tile drains on a silt loam soil in Indiana. J. Environ. Qual. 20:264-270.

Korom, S.F. 1992. Natural denitrification in the saturated zone: a review: Water Resour. Res. 28:1657-1668.

LeGrand, H.E., and V.T. Stringfield. 1973. Karst hydrology – a review. J. Hydrol. 20:97-120.

Lewis, W.M., J.M. Melack, W.H. McDowell, M. McClain, and J.E. Richey. 1999. Nitrogen yields from undisturbed watersheds in the Americas. Biogeochem 46:149-162.

Linsley, R.K., Jr., M.A. Kohler, and J.L.H. Paulhus. 1982. Hydrology for engineers. 3rd ed. McGraw-Hill, New York.

Lowrance, R.R. J. Todd, J. Fail, Jr., O. Hendrickson, Jr., R. Leonard, and L. Asmussen. 1984. Riparian forests as nutrient filters in agricultural watersheds. BioScience 34:374-377.

Lowrance, R., R. Leonard, and J. Sheridan. 1985. Managing riparian ecosystems to control nonpoint pollution. J. Soil Water Conserv. 40:87-97.

Lowrance, R.R. 1992. Groundwater nitrate and denitrification in a coastal riparian forest. J. Environ. Qual. 21:401-405.

Lystrom, D.J., F.A. Rinella, D.A. Rickert, and L. Zimmermann. 1978. Regional analysis of the effects of land use on stream-water quality, methodology and application in the Susquehanna River Basin, Pennsylvania and New York. U.S. Geological Survey Water-Resources Investigations 78-12.

Matthews, E. 1994. Nitrogenous fertilizers: global distribution of consumption and associated emissions of nitrous oxide and ammonia. Global Biogeochem. Cycles 8:411-439.

McBride, G., R.B. Alexander, S. Elliot, and U. Shankar. 2000. Regional scale modeling of water quality, *Water and Atmosphere*, NIWA, Auckland, New Zealand.

McFarland, A.M.S., and L.M. Hauck. 1999. Relating agricultural land uses to in-stream stormwater quality. J. Environ. Qual. 28:836-844.

McLeod, R.V., and R.O. Hegg. 1984. Pasture runoff water quality from application of inorganic and organic nitrogen sources. J. Environ. Qual. 13:122-126.

Meybeck, M. 1982. Carbon, nitrogen and phosphorus transport by world rivers. Amer. J. of Sci. 282: 401-450.

Michel, R.L. 1992. Residence times in rivers basins as determined by analysis of long-term tritium records. J. Hydrology 130:367-378.

Moore, M.V., M.L. Pace, J.R. Mather, P.S. Murdoch, R.W. Howarth, C.L. Folt, C.Y. Chen, H.F. Hemond, P.A. Flebbe, and C.T. Driscoll 1997. Potential effects of climate change on freshwater ecosystems of the New England/Mid-Atlantic Region. Hydro. Processes. 11: 925-948.

Moseley, M.P., and A.I. McKerchar. 1993. Streamflow. p. 8.1-8.39 *In* D. R. Maidment (ed.) Handbook of hydrology. McGraw-Hill, New York.

Mueller, D.K., P.A. Hamilton, D.R. Helsel, K.J. Hitt, and B.C. Ruddy. 1995. Nutrients in ground water and surface water of the United States-an analysis of data through 1992. U.S. Geological Survey Water-Resources Investigations Report 95-4031.

Mueller, D.K., B.C. Ruddy, and W.A. Battaglin. 1997. Logistic model of nitrate in streams of the upper-midwestern United States. J. Environ. Qual. 26:1223-1230.

Mulholland, P.J., G.R. Best, C.C. Coutant, G.M. Hornberger, J.L. Meyer, P.J. Robinson, J.R. Stenberg, R.E. Turner, F. Vera-Herrera, and R.G. Wetzel. 1997. Effects of climate change on freshwater ecosystems of the South-eastern United States and the Gulf Coast of Mexico. Hydro. Processes. 11: 949-970.

Murdoch, P.S., D.A. Burns, and G.B. Lawrence. 1998. Relation of climate change to the acidification of surface waters by nitrogen deposition. Environ. Sci Techn. 32:1642-1647.

Murdoch, P.S., J.S. Baron, and T.L. Miller. 2000. Potential effects of climate change on surface water quality in North America. J. Am. Water Resour. Assoc. 36: 347-366.

National Research Council. 2000. Clean coastal waters: understanding and reducing the effects of nutrient pollution. Ocean Studies Board and Water Science and Technology Board, National Academy of Sciences, Washington, DC.

Nelson, P.N., E. Cotsaris, and J.M. Oades. 1996. Nitrogen, phosphorus, and organic carbon in streams draining two grazed catchments. J. Environ. Qual. 25:1221-1229.

Olness, A., E.D. Rhoades, S.J. Smith, and R.G. Menzel. 1980. Fertilizer nutrient losses from rangeland watersheds in Central Oklahoma. J. Environ. Qual. 9:81-86.

Omernik, J.M., A.R. Abernathy, and L.M. Male. 1981. Stream nutrient levels and proximity of agricultural and forest land to streams: some relationships. J. Soil Water Conserv. 36:227-231.

Osborne, L.L., and M.J. Wiley. 1988. Empirical relationships between land use/cover and stream water quality in an agricultural watershed. J. Environ. Manage. 26:9-27.

Owens, L.B., R.W. Van Keuren, and W.M. Edwards. 1983. Nitrogen loss from a high-fertility, rotational pasture program. J. Environ. Qual. 12:346-350.

Paul, E.A., K.A. Paustian, E.T. Elliott, and C.V. Cole (eds.) 1997. Soil organic matter in temperate agroecosystems: Long-term experiments in North America, CRC Press, New York.

Peierls, B.L., N.F. Caraco, M.L. Pace, and J.J. Cole. 1991. Human influence on river nitrogen. Nature 350:386-387.

Peterjohn, W.T., and D.L. Correll. 1984. Nutrient dynamics in an agricultural watershed: observations on the role of a riparian forest. Ecology 65:1466-1475.

Pionke, H.B., W.J. Gburek, A.N. Sharpley, and R.R. Schnabel. 1996. Flow and nutrient export patterns for an agricultural hill-land watershed. Water Resour. Res. 32:1795-1804.

Postma, D., C. Boesen, H. Kristiansen, and F. Larsen. 1991. Nitrate reduction in an unconfined sandy aquifer: water chemistry, reduction processes, and geochemical modeling. Water Resour. Res. 27:2027-2045.

Powlson, D.S. 1993. Understanding the soil nitrogen cycle. Soil Use and Manage. 9:86-94.

Puckett, L.J., R.B. Alexander, and K.J. Hitt. 1998. County-based estimates of nitrogen and phosphorus content of animal manure in the United States for 1982, 1987, and 1992, version 1.0.0, data available on the World Wide Web <http://water.usgs.gov/lookup/getspatial?manure>.

Puckett, L.J. 1995. Identifying the major sources of nutrient water pollution. Envir. Sci Techn. 29:408-414.

Rastetter, E.B., A.W. King, B.J. Cosby, G.M. Hornberger, R.V. O'Neill, and J.E. Hobbie. 1992. Aggregating fine-scale ecological knowledge to model coarser-scale attributes of ecosystems. Ecological Apps. 2:55-70.

Russelle, M.P. 1992. Nitrogen cycling in pasture and range. J. Prod. Agric. 5:13-22.

Rutherford, J.C., R.B. Williamson, and A.B. Cooper. 1987. Nitrogen, phosphorus, and oxygen dynamics in rivers. p. 139-165 *In* A.B. Viner (ed.) Inland waters of New Zealand. Science Information Publishing Centre. Department of Scientific and Industrial Research. Wellington, New Zealand.

Sauer, T.J., J.V. Brahana, T.M. Kresse, S.R. Compston, D.K. Winters, J.F. Murdoch, and P.F. Vendrell. 1999a. The Savoy Experimental Watershed: water quality assessment of a basin dedicated to interdisciplinary animal waste management research. *In* Proc. Arkansas Water Resources Center Annual Research Conference. Fayetteville, AR.

Sauer, T.J., T.C. Daniel, P.A. Moore, Jr., K.P. Coffey, D.J. Nichols, and C.P. West. 1999b. Poultry litter and grazing animal waste effects on runoff water quality. J. Environ. Qual. 28:860-865.

Sauer, T.J., T.C. Daniel, D.J. Nichols, C.P. West, P.A. Moore, Jr., and G.L. Wheeler. 2000.

Runoff water quality from poultry litter-treated pasture and forest sites. J. Environ. Qual. 29:515-521.

Sauer, T.J., P.A. Moore, Jr., K.P. Coffey, and E.M. Rutledge. 1998. Characterizing the surface properties of soils at varying landscape positions in the Ozark Highlands. Soil Sci. 163:907-915.

Schepers, J.S., and D.D. Francis. 1982. Chemical water quality of runoff from grazing land in Nebraska: I. Influence of grazing livestock. J. Environ. Qual. 3:351-354.

Schilling, K.E., and R.D. Libra. 2000. The relationship of nitrate concentrations in streams to row crop land use in Iowa. J. Environ. Qual. 29:1846-1851.

Schindler, D.W. 1997. Widespread effects of climatic warming on freshwater ecosystems in North America. Hydro. Processes. 11: 1043-1067.

Schumm, S. A. 1977. Drainage basin morphology. Dowden, Hutchinson and Ross, Stroudsburg, PA.

Seaber, P.R., F.P. Kapinos, and G.L. Knapp. 1987. Hydrologic units maps, U.S. Geological Survey Water Supply Paper 2294.

Seitzinger, S.P. 1988. Denitrification in freshwater and coastal marine ecosystems: ecological and geochemical significance. Limnol. Oceanogr. 33:702-724.

Seitzinger, S.P., and C. Kroeze. 1998. Global distribution of nitrous oxide production and N inputs in freshwater and coastal marine ecosystems. Global Biogeochem. Cycles 12:93-113.

Sharpley, A.N., and J.K. Syers. 1981. Amounts and relative significance of runoff types in the transport of nitrogen into a stream draining an agricultural watershed. Water, Air and Soil Pollut. 15:299-308.

Skaggs, R.W., and J. van Schilfgaarde, eds. 1999. Agricultural drainage. Agronomy Monograph 38. American Society of Agronomy, Madison, WI.

Smith, R.A., and Alexander, R.B. 2000. Sources of nutrients in the Nation's watersheds. Proc. of the National Resource, Agriculture, and Engineering Service, A conference for nutrient management consultants, extension educators, and producer advisors, March 28-30, 2000. Camp Hill, PA.

Smith, R.A., G.E. Schwarz, and R.B. Alexander. 1997. Regional interpretation of water-quality data. Water Resour. Res. 33:2781-2798.

Smith, R.A., R.B. Alexander, and K.J. Lanfear. 1993. Stream water quality in the conterminous United States--Status and trends of selected indicators during the 1980's. *In* R.W. Paulson et al. (ed.) U.S. Geological Survey Water Supply Paper 2400.

Sogbedji, J.M., H.M. van Es, C.L. Yang, L.D. Geohring, and F.R. Magdoff. 2000. Nitrate leaching and nitrogen budget as affected by maize nitrogen rate and soil type. J. Environ. Qual. 29:1813-1820.

Stacy, P.E., H.S. Greening, J.N. Kremer, D. Peterson, and D.A. Tomasko. 2000. Contributions of atmospheric nitrogen deposition to U.S. estuaries: summary and conclusions. *In* R. Valigura et al. (eds.) Nitrogen loading in coastal water bodies: an atmospheric perspective. American Geophysical Union Water Monograph 57. Washington, DC.

Stevenson, F.J. (ed.) 1982. Nitrogen in agricultural soils. American Society of Agronomy, Madison, WI.

Stevenson, F.J. 1994. Organic forms of nitrogen. p. 59-95 *In* Humus chemistry. John Wiley, New York.

Stoddard, J.L. 1994. Environmental chemistry of lakes and reservoirs. p. 223-284. *In* L.A. Baker (ed.) Advances in chemistry series 237. American Chemical Society, Washington DC.

Stout, W.L., S.A. Fales, L.D. Muller, R.R. Schnabel, W.E. Priddy, and G.F. Elwinger. 1997. Nitrate leaching from cattle urine and feces in Northeast USA. Soil Sci. Soc. Am. J. 61:1787-1794.

Terman, G.L. 1979. Volatilization losses of nitrogen as ammonia from surface-applied fertlizers, organic amendments, and crop residues. Adv. Agron. 31:189-223.

Timmons, D.R., and R.F. Holt. 1977. Nutrient losses in surface runoff from a native prairies. J. Environ. Qual. 6:369-373.

Timmons, D.R., E.S. Verry, R.E. Burwell, and R.F. Holt. 1977. Nutrient transport in surface runoff and interflow from an aspen-birch forest. J. Environ. Qual. 6:188-192.

Tufford, D.L., H.N. McKellar, Jr., and J.R. Hussey, Jr. 1998. In-stream nonpoint source nutrient prediction with land-use proximity and seasonality. J. Environ. Qual. 27:100-111.

U.S. Environmental Protection Agency. 1976. Quality criteria for water. U.S. Government Printing Office, Washington, DC.

U.S. Environmental Protection Agency. 1999. National water quality inventory: 1998 report to congress. EPA 841-R-00-001. Washington, DC.

U.S. Geological Survey. 1999. The quality of our Nation's waters - nutrients and pesticides. USGS Circular 1225.

Vitousek, P.M., J.D. Aber, R.W. Howarth, G.E. Likens, P.A. Matson, D.W. Schindler, W.H. Schlesinger, and D.G. Tilman. 1997. Human alteration of the global nitrogen cycle: sources and consequences. Ecological Applications 7:737-750.

White, W.B. 1988. Geomorphology and hydrology of karst terrains. Oxford University Press, New York.

Williams, M.W., J.S. Baron, N. Caine, R. Sommerfield, and R. Sanford. 1996. Nitrogen saturation in the Rocky Mountains. Envir. Sci Techn. 30:640-646.

Windolf, J., E. Jeppesen, J.P. Jensen, and P. Kristensen. 1996. Modelling of seasonal variation in nitrogen retention and in-lake concentration: a four-year mass balance study in 16 shallow Danish lakes. Biogeochem. 33:25-44.

Winter, T.C., J.W. Harvey, O.L. Franke, and W.M. Alley. 1998. Ground water and surface water: a single resource. U.S. Geological Survey Circular 1139.

Wolock, D.M. 1993. Simulating the variable-source-area concept of streamflow generation with the watershed model Topmodel. U.S. Geological Survey Water-Resources Investigations Report 93-4124.

Wolock, D.M., and G.J. McCabe. 1999. Estimates of runoff using water-balance and atmospheric general circulation models. J. Amer. Water Resour. Assoc. 35: 1341-1350.

Nitrogen in the Environment: Sources, Problems, and Management
R.F. Follett and J.L. Hatfield (Eds)
Published by Elsevier Science B.V.

Chapter 8. Nitrogen transport and fate in European streams, rivers, lakes and wetlands

B. Kronvang[a], J.P. Jensen[b], C.C. Hoffmann[a] and P. Boers[c]
[a]) National Environmental Research Institute, Department of Streams and Riparian Areas
Vejlsoevej 25, DK-8600, Silkeborg, Denmark

[b]) National Environmental Research Institute, Department of Lake and Estuarine Ecology
Vejlsoevej 25, DK-8600, Silkeborg, Denmark

[c]) Institute for Inland Water Management and Wastewater Treatment, P.O. Box 17, 8200 AA, Lelystad, The Netherlands

This chapter provides an overview of the present state of nitrogen pollution in European streams, rivers and lakes. The main focus of the chapter is on diffuse sources of nitrogen. Diffuse sources are today the main concern in many European catchments, and measures need to be developed to protect drinking water supplies and maintain the environmental quality of rivers, lakes and coastal waters. In 1991, the European Union adopted the Nitrate Directive aimed to reduce or prevent nitrate pollution of water due to application and storage of inorganic fertilizers and manure on farmland. The EU countries have identified zones vulnerable to nitrate pollution, and most countries have adopted action plans to reduce nitrogen pollution. Thus, throughout Europe catchment managers are combating nitrogen pollution from both point and diffuse sources, a task requiring thorough knowledge of the mechanisms governing nitrogen loss arising from different uses of land, as well as of the fate of nitrogen in groundwater and surface waters. Based on examples from different European catchments and comprehensive data sets gathered from ongoing nutrient monitoring programmes in Denmark and the Netherlands, this chapter illustrates the most important aspects to be considered.

1. INTRODUCTION

Elevated nitrogen concentrations in European surface waters have mostly been related to modern agricultural practices, in particular the use of nitrogen fertilizers (Neill, 1989; Edwards et al., 1990; Wright et al., 1991; Stibe and Fleicher, 1991; Kronvang et al., 1995). However, in some countries and catchments, nitrogen discharge from point sources such as sewage treatment plants and industry still contributes significantly to riverine nitrogen loading (Kristensen and Hansen, 1994; Iversen et al., 1997).

The elevated riverine nitrogen loading has been associated with increased primary production and nuisance algal growth in coastal zones and semi-enclosed and enclosed areas of European seas (Kronvang et al., 1993; OSPARCOM, 1992; Mee, 1991). Examples of the consequences of eutrophication are increased frequency of algal blooms (sometimes toxic), increased water turbidity, oxygen depletion in deeper waters and mass kills of fish and benthic fauna (Kronvang et al., 1993; European Environment Agency, 1995).

The estimated European riverine gross flux of nitrogen amounts to 2.5 - 6.5 million tonnes per year (European Environment Agency, 1995). Part of the riverine nitrogen loading is, however, removed during its passage from source to the open sea (Kronvang *et al.*, 1999). Knowledge of the fate of the riverine nitrogen transport (mostly as nitrate) is important for allowing accurate estimations of nitrogen emissions and the resulting net escape to the open sea. Hence, quantification of the transformation of nitrate under anoxic conditions into N_2O and N_2 gases in rivers, lakes, wetlands and estuaries is an important issue.

2. PROBLEM IDENTIFICATION

2.1 Nitrogen sources in European catchments

Around 1990, the input from agriculture to the total emission of nitrogen to the aquatic environment in 10 European countries ranged between 24% and 81%, the highest values being recorded in Denmark, Latvia and The Netherlands, and the lowest in Finland and Sweden (European Environment Agency, 1995). The remaining nitrogen sources are point source emissions, nitrogen emissions from undisturbed land (forest, mountains, tundra) and atmospheric deposition on surface waters (lakes, rivers). The relative importance of the different nitrogen emissions in different European regions is illustrated in Table 1. In densely populated regions, point source discharge of nitrogen contributes significantly to the riverine nitrogen loading (e.g. River Po, Italy). In intensively agri-cultivated regions such as Denmark and The Netherlands, diffuse nitrogen loss from arable land is the dominant nitrogen source, whereas diffuse nitrogen loss from undisturbed land dominates in sparsely populated and cultivated regions such as the Gulf of Botnia, Sweden.

Table 1. Sources of total nitrogen in surface waters in selected catchments. (Redrawn from Iversen et al., 1997, and Bøgestrand, 1999).

Region	Period	Catchment area (km^2)	Annual loading (kg N ha^{-1})	Point sources	Agriculture	Undis-turbed land	Atmos-pheric depositi on
Gulf of Botnia, Sweden	1982-89	116,103	2.1	3%	1%	88%	7%
Göta älv, Sweden	1982-87	50,181	4.3	19%	22%	33%	26%
Austrian part of Danube	1994	80,731	9.9	36%	46%	18%	-
Germany	1989-91	356,950	29	39%	53%	8%	-
River PO, Italy	1989	69,400	35	43%	54%	3%	-
Rhine and Meuse, The Netherlands	-	37,000	37	25%	75%	-	-
Denmark	1989-98	43,022	18.9	7%	82%	10%	1%

2.2 Nitrogen consumption in European agriculture

Today, more than 40% of the European land area is used for agricultural production (Table 2). The land use varies, however, much from country to country, and may even vary from region to region within each individual country. For example, arable land constitutes about 59% of the total land area in Denmark, 40% in Spain and Italy, 34% in Germany, 24% in The Netherlands, 6% in Sweden and 3% in Norway. Similarly, farming intensity and crop types vary from region to region. The amount of arable land per inhabitant in Europe (0.38 ha) is only half that found in the USA (0.76 ha) and only slightly higher than in Africa (0.27 ha).

Table 2. Distribution of land use in Europe including the European part of the Russian Federation (Veldkamp et al., 1995).

Type of land use	Area (km^2)	%
Arable land	3,343,838	24.3
Grassland	1,513,557	15.7
Permanent crops	144,508	1.5
Coniferous + mixed forest	2,488,327	25.8
Deciduous forest	664,587	6.9
Urban areas	117,887	1.2
Inland waters	203,196	2.1
Other (mountain, tundra, etc.)	2,175,516	22.5

Average consumption of chemical nitrogen fertilizers per hectare agricultural land increased in Europe with 75% between 1970 and 1989, followed by a decrease during the early 1990s (Fig. 1). However, large variations in the consumption of nitrogen fertilizers are found, the highest values being recorded in the north-western part of Europe and the lowest in

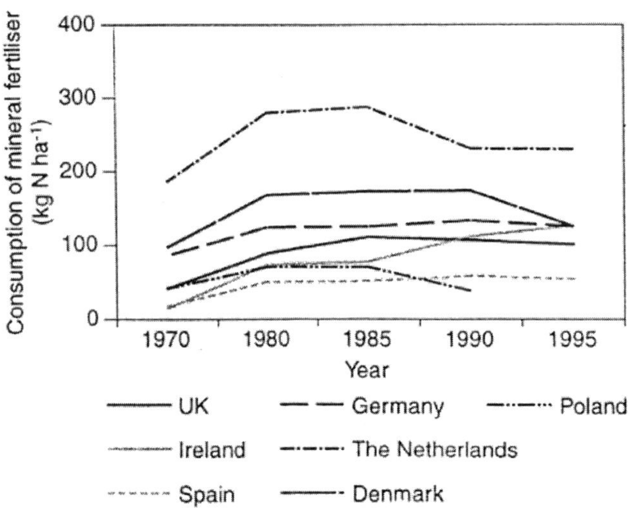

Figure 1. Average consumption of chemical nitrogen fertilizers in Europe during 1970-1995.

the southern and eastern part of Europe. Similarly, the application of manure to agricultural land varies widely between European countries, the highest values occurring in The Netherlands (>200 kg N ha^{-1}) and the lowest - and less than 40 kg N ha^{-1} - in parts of southern and eastern Europe. Hence, the average nitrogen surplus in European agriculture varies considerably from country to country (Table 3).

Table 3. Average nitrogen surplus in different European countries calculated as balances of inputs (mineral fertilizers, manure, biological fixation and atmospheric deposition) and outputs (harvested crops). (Source: EEA, 2000).

	Nitrogen surplus (kg N ha^{-1})		
Country	1990	1993	1995
Austria	-	-	16
Belgium	106	109	103
Denmark	93	92	72
Finland	-	-	51
France	47	54	57
Germany	105	101	102
Greece	84	61	58
Ireland	47	60	62
Italy	62	83	76
Netherlands	229	212	213
Portugal	27	23	22
Spain	40	37	37
Sweden	-	-	38
UK	40	39	40
EU 12/15	60	60	60

2.3 Nitrogen concentration and trends in European rivers

Descriptive statistics for the concentration of total nitrogen, nitrate nitrogen and ammonium nitrogen in European rivers during 1988-1991 is shown in Table 4. Dissolved inorganic nitrogen (nitrate and ammonium) constitutes 88% of the total nitrogen concentration in river water. The median total nitrogen concentration in 43 near pristine European rivers was much lower (0.33 mg N l^{-1}) than the median total nitrogen concentration reported for all 329 European rivers (2.12 mg N l^{-1}), indicating an anthropogenic influence from, for instance, agriculture and sewage effluents.

The concentration of nitrate has increased in many European rivers, especially during the period 1950-1990 (Fig. 2). The increase in riverine nitrate concentrations is mainly attributable to a corresponding increase in the use of nitrogen in chemical fertilizers in most European countries (Fig. 1). Also, during the same period, enhanced emission of nitrogen from the burning of fossil fuels and agricultural activity has, however, increased the atmospheric dry and wet deposition of nitrogen. This has mainly resulted in an increase in riverine nitrogen concentrations in remote and sparsely populated areas of Europe.

Table 4. Descriptive statistics of the annual mean nitrogen concentration in European rivers. (Redrawn from Kristensen and Hansen, 1994).

	Number of river stations	Percentage of river stations with concentrations not exceeding (mg N l^{-1})					
		Mean	10%	25%	50%	75%	90%
All rivers							
Total N	329	3.07	0.30	0.80	2.12	4.50	7.07
Nitrate N	654	2.63	0.25	0.70	1.80	3.90	5.72
Ammonium N	580	0.67	0.03	0.07	0.18	0.45	1.42
Near pristine rivers							
Total N	43	0.40	-	0.19	0.33	0.39	-
Nitrate N	39	0.30	-	0.05	0.10	0.22	-

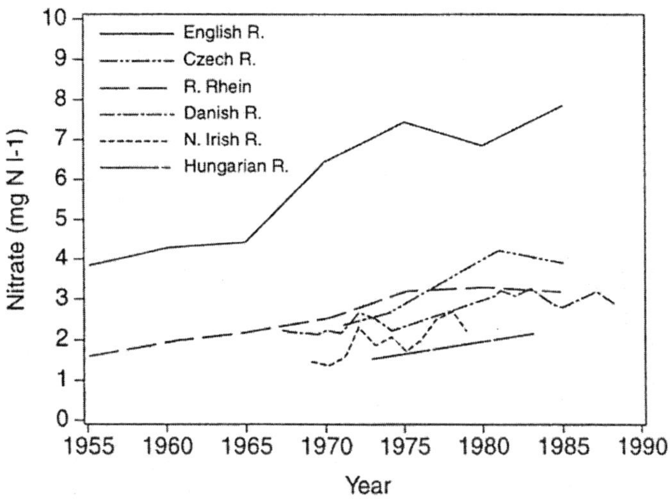

Figure 2. Trends in the nitrate concentration of selected European rivers. (Redrawn from Kristensen and Hansen, 1994).

On average, 70% of European rivers experienced an increase in nitrate concentrations between 1978-1988 and 1988-90 (Kristensen and Hansen, 1994). The increase was most pronounced in eastern and southern Europe, possibly because the use of nitrogen fertilizers peaked later here than in the north-western European countries (European Environment Agency, 1995). Moreover, a general downward trend in riverine nitrogen concentrations has been recorded in northern European countries (e.g. Denmark) due to regulations on the use of nitrogen in chemical fertilizer and manure (Andersen et al., 1999). Thus, during the period 1989-1996 a general downward trend in nitrogen concentrations in Danish streams draining agricultural dominated catchments was reported by Larsen et al. (1999). Since 1990, the economic change in many Eastern European countries has also led to a strong decline in the use of nitrogen fertilizers. A concomitant decrease in riverine nitrate concentrations cannot yet be clearly detected in the different parts of Europe (European Environment Agency, 1999). However, the most significant downward trend in riverine nitrate concentrations has been recorded in Eastern European rivers (European Environment Agency, 1999)

3. SOURCES AND RETENTION OF NITROGEN IN A LARGE RIVER - THE RHINE

Table 5. Land use in the Rhine catchment. CH: Switzerland; D: Germany; L: Luxembourg; F: France: NL: The Netherlands.

Country	Total area*	Urban area		Forest area		Agricultural area			
						arable land		grass land	
	x1000 km^2	x1000 km^2	%	x1000 km^2	%	x1000 km^2	%	x1000 km^2	%
CH	34	3	9	11	33	5	15	14	41
D	102	22	22	40	39	27	26	13	13
L	2.6	0.3	12	0.9	35	0.5	21	0.7	27
F	23	1.3	6	8.8	40	5.9	27	5.0	23
NL	24	7	31	2	9	4	16	9	36
TOTAL	188	34	18	63	34	42	23	42	22

*: note that total area does not equal the sum of urban, forest and agricultural areas. This is due to both uncertainties in area estimates and the fact that other, less important, forms of land use is not included in the table.

The catchment area (186,000 km^2) of the Rhine (length 1,220 km) covers parts of nine countries with a total of 50 million inhabitants. The average discharge at the Dutch-German border is 2300 $m^3 s^{-1}$ (Fig. 3). Table 5 summarises the land use structure of the catchment.

Concentrations of N-compounds are measured at several stations along the Rhine (Fig. 4). Especially for Lobith, long time series are available. The average concentrations vary along the length of the river. In 1993, the measured concentrations ranged from 1.7 - 5.1 mg total N l^{-1}. The highest concentrations were measured downstream in the Netherlands.

The concentrations also vary with time and are partly correlated with discharge (Van Dijk et al., 1996). Annual average concentrations are calculated as the discharge weighted means. Total N concentrations measured at Lobith increased to about 7.5 mg N l^{-1} (1980-1985) followed by a decrease to 4.9 mg N l^{-1} (1993) (Fig. 4).

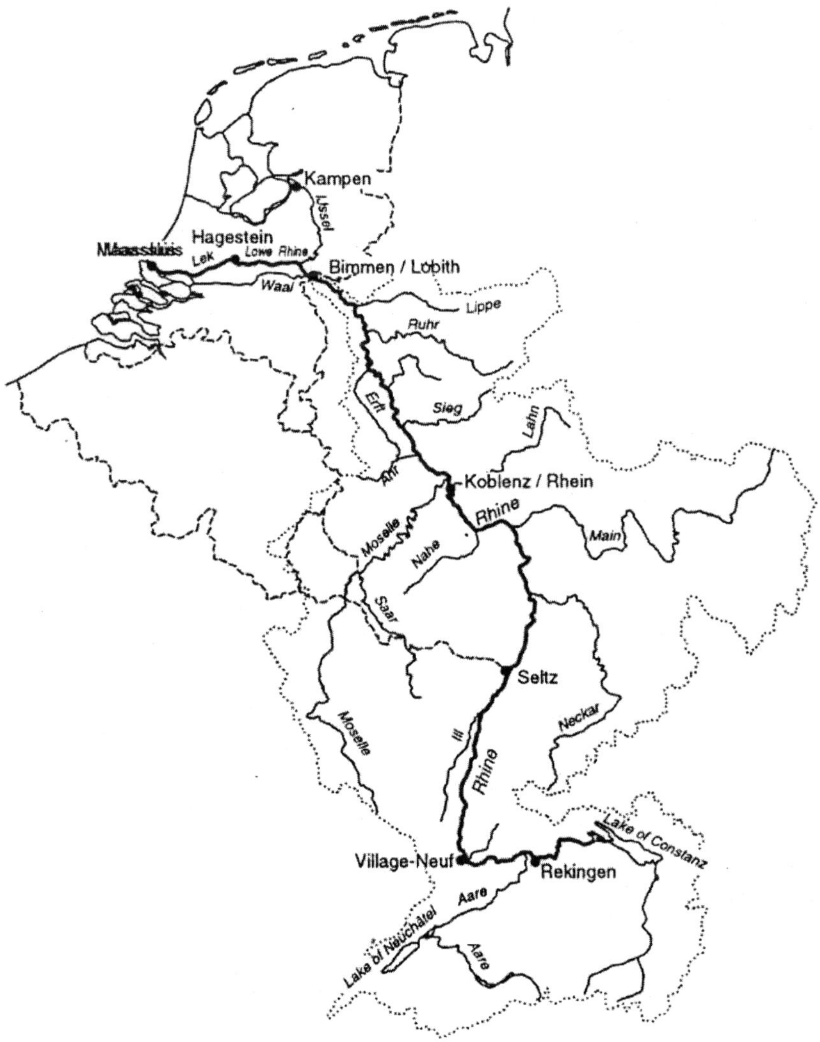

Figure 3. Catchment area of the Rhine showing tributaries and sampling stations.

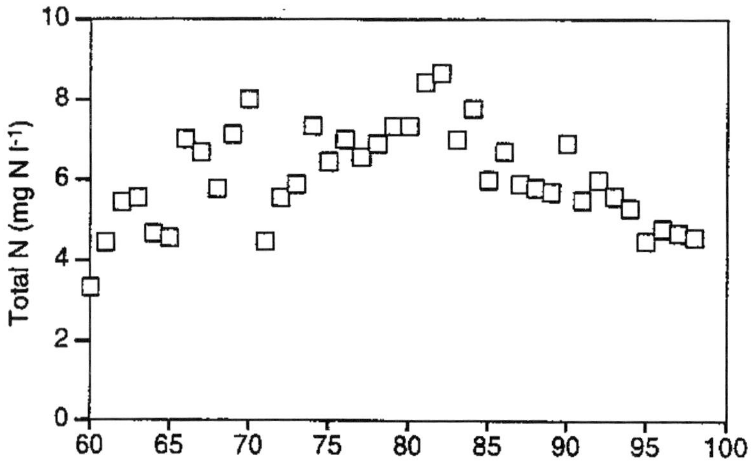

Figure 4. Annual average concentrations of total N in the Rhine at Lobith from 1960 - 1995.

The elevated nitrogen concentrations in the Rhine originate from emission. Estimates of the most important sources of nitrogen for the year 1992 are summarized in Table 6. Unfortunately, no complete set of data is available. Most of the point sources are monitored yearly, but the data on diffuse sources, mainly agriculture, are updated less frequently. Moreover, the data from the different countries were obtained using different methods (IRC, 1992; Borgvang et al., 1998).

France uses the balancing method (IRC, 1992): the total loading of a river at a certain point is measured and the emission from all point sources subtracted, the remainder being attributed to diffuse sources. Since all errors in both loading and point sources are accumulated, the value obtained for diffuse emissions is inaccurate as well as underestimated, retention in the watercourses not being taken into account.

Table 6
Estimates of diffuse and point sources of nitrogen for the year 1992.

Country	nitrogen loading (kton N/y)				
	WWTP's	industry	agriculture	other diffuse sources	totals
CH	17.5	1	10.8[1]	10.5	39.8
D	120	31	124.5	52	327
F	29	4.6	26.6[1]	-	60.2
NL	10	2.0	22.7	8.2	42.9
total	176.5	38.6	184.6	70.7	470.4

1: data for 1985.

Germany, Switzerland and The Netherlands all use nutrient leaching models (BUWAL, 1993; Kroes *et* al., 1996; Werner et al., 1991). These models differ greatly in approach and complexity. The performance of these models has never been compared (Borgvang et al., 1998). Generally speaking, such models can be calibrated directly for small fields only. For larger areas, retention processes in the surface waters have to be taken into account, which introduces new uncertainties. Application of nutrient leaching models on the regional and national scale requires data on fertilizer application rates, land use, soil properties, hydrology, etc. Uncertainties in these data constitute an additional source of error in the calculated emission data on nutrient leaching.

Germany and Switzerland use relatively simple models based upon the principle of emission factors. These factors relate nutrient emission to fertilizer application rates, land use, soil properties and slope, and are mostly derived from observations at field scale. The Netherlands, due to the complex hydrology of the mostly flat soils, uses a much more complex deterministic model comprising a model for soil hydrology (Kroes et al., 1996; Boers et al., 1997).

Most of the data are from 1992 and some from 1985 (IRC, 1992). Agriculture and domestic wastewater are the most important sources of nitrogen. The total emission in 1992 was about 470 ktons N y^{-1}, 427 ktons were emitted upstream of Lobith. The same year, the riverine loading at Lobith amounted to 303 kton N. The most important reasons for the difference are probably retention in the catchment and errors in emission estimations.

4. NITROGEN CYCLING IN SMALL CATCHMENTS

The nitrogen cycling in 21 small and predominantly arable catchments was investigated during the period 1989-1996 (Fig. 5). The monitored catchments are representative of the typical Danish farming systems ranging from plant production on the predominantly loamy soils on the islands to animal production on the sandy soils in Jutland. The catchments also represent the typical soil types of Denmark and the gradients in climate and hydrology (Table 7). Agricultural practices at field level were investigated by a questionnaire survey conducted in 1993/94. The data collected were used for calculating nitrogen leaching from the root zone applying an empirical leaching model (Andersen *et* al., 1999).

Nitrogen leaching from the root zone on agricultural areas was calculated for all 21 catchments both for 1993/94 and the seven-year period 1989/90 to 1995/96. The latter was calculated by assuming same agricultural practice as in 1993/94 and letting climate vary from year to year in each of the catchments. Measurements of total nitrogen concentrations and discharge in the stream draining each catchment were conducted during the whole study period, enabling us to calculate the nitrogen loss from the catchment. Moreover, the annual nitrogen loss from agricultural areas within the catchments was calculated by applying measured annual nitrogen losses from non-agricultural areas within the catchments (Kronvang et al., 1995 and 1996).

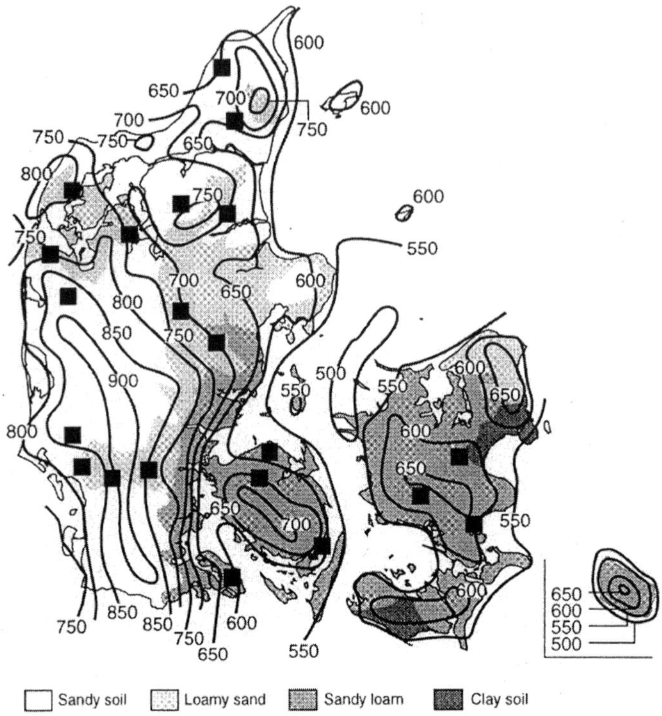

Figure 5. The 21 small catchments incorporated as parts of the Danish Aquatic Monitoring Programme are situated in different parts of Denmark and cover the dominant soil types and gradients in precipitation.

Table 7. Description of catchment area, soil type, runoff and hydrological regime of the two classes of catchments situated in the sandy western and loamy eastern region of Denmark.

	Sandy western region of Denmark	Loamy eastern region of Denmark
Number of catchments	10	11
Average catchment area	12.6 km^2	13.6 km^2
Average proportion of sandy soils	86%	25%
Average proportion of organic soils	7.6%	1.1%
Average annual runoff	289 mm	242 mm
Baseflow index	0.74	0.55
Average annual interflow[1]	30%	55%
Average annual groundwater flow[1]	70%	45%

[1]) Calculated by setting up a precipitation-runoff model for each catchment.

4.1 Nitrogen budgets

The monitoring data enable us to establish an overall nitrogen budget for the catchments situated in the sandy western region and the loamy eastern region for 1993/94 (Fig. 6). The nitrogen balance for 1993/94 reveals that the application of nitrogen is highest on agricultural areas within the sandy western catchments, mainly due to higher production and use of nitrogen in animal manure (Fig. 6). Although the nitrogen surplus was considerably greater for the sandy catchments than the loamy catchments, the modelled average annual nitrogen leaching from the root zone was only slightly higher for the sandy (71 kg N ha^{-1}) than for the loamy catchments (66 kg N ha^{-1}). Nitrogen retention in the root zone seems therefore to be of greater significance in the sandy catchments than in the loamy catchments possibly because of higher application of animal manure in the former.

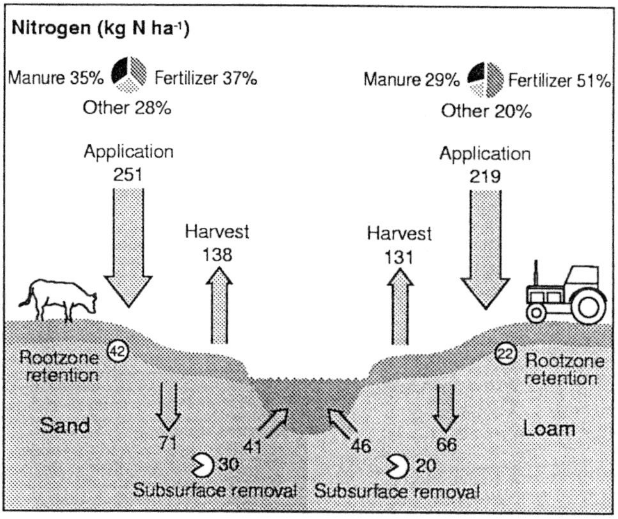

Figure 6. The average nitrogen input balance and nitrogen flows for 10 sandy catchments and 11 loamy catchments in Denmark from autumn 1993 to autumn 1994 (agricultural data) and the calendar year 1994 (nitrogen export data).

The average nitrogen export from the agricultural areas within the catchments measured at stream stations during 1994 was slightly higher for the loamy catchments (46 kg N ha^{-1}) than from the sandy catchments (41 kg N ha^{-1}). For both the sandy and loamy catchments, average nitrogen export from the agricultural areas within the catchments was lower than the modelled average nitrogen leaching (Fig. 6). The difference obtained can be ascribed to both subsurface nitrogen removal processes (denitrification) or, during this short-term period, maybe to hydrological inertia (time lag) within the catchments. The latter presumably being most pronounced in sandy catchments where the proportion of stream water derived from groundwater is highest (Table 7).

4.2 Subsurface nitrogen removal

The year of 1994 was a relatively wet year (mean annual precipitation 880 mm) compared to the average (694 mm) for the seven-year period 1989-1996. Therefore, both the modelled nitrogen leaching from the root zone on agricultural land and the measured nitrogen export from the agricultural areas within the catchments were considerably higher than the average for the period 1989-96. To obtain a better description of the missing link between nitrogen leaching and nitrogen export, we investigated the differences for the period 1989-96. Average subsurface nitrogen retention (measured as the difference between nitrogen leaching and nitrogen export divided by the former) was significantly ($p<0.001$) correlated with the proportion of sandy soils (S) and average runoff in the 21 catchments (Fig. 7).

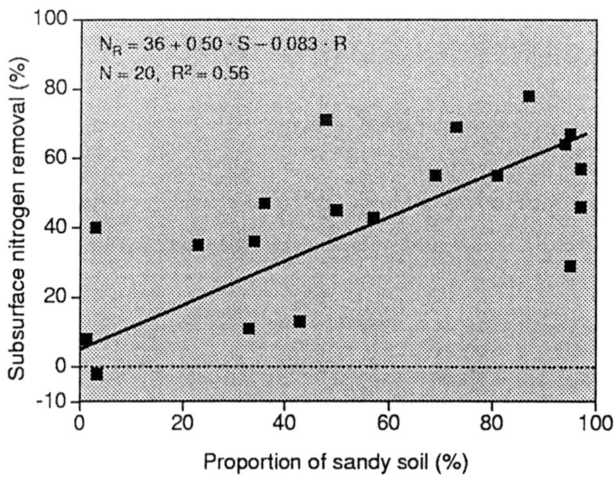

Figure 7. Relationship between average annual subsurface nitrogen removal and the proportion of sandy soils within 21 small Danish headwater catchments.

We believe that the nitrogen retention in subsurface soils is due to subsurface removal of nitrate nitrogen in groundwater rather than a hydrological time lag (groundwater residence time) in the small headwater catchments investigated. However, in larger catchments groundwater residence time may be of great significance in the comparison of comparing nitrogen leaching with nitrogen export. Subsurface nitrogen retention is in these cases of vital importance for the linkage between changes in agricultural practices and trends in riverine nitrogen concentrations or loading.

5. IMPORTANCE OF AGRICULTURAL LAND FOR NITROGEN EXPORT

The average annual export of total-N, nitrate-N and ammonium-N from 1989-98 in relation to the proportion of agricultural land in 70 Danish catchments is shown in Table 8. The nitrogen export from the monitored catchments reveals an increase concurrently with an increase in the proportion of agricultural land (Table 8). Thus, the average annual export of total-N, nitrate-N and ammonium-N increases by a factor 9, 14 and 5, respectively, when the proportion of agricultural land increases from 0-20% to >70%. Although the average catchment area and average runoff also experience an increase, there is no doubt that a substantial proportion of the agricultural nitrogen surplus is lost to surface waters.

Table 8. Average annual export of total nitrogen, nitrate nitrogen and ammonium nitrogen from catchments with different proportions of agricultural land and where nitrogen emission from point source is less than 0.5 kg N ha^{-1} during the ten-year period 1989-98. Also shown are number of catchments, mean catchment area and mean annual runoff from each of the land use classes.

Proportion of agricultural land	Number of catchments	Catchment area (km^2)	Runoff (mm)	Total-N export (kg N ha^{-1})	Nitrate-N export (kg N ha^{-1})	Ammonium-N export (kg N ha^{-1})
< 20%	5	7.4	183	2.4 ± 1.9	1.5 ± 1.6	0.09 ± 0.06
20-40%	5	3.3	185	6.4 ± 3.4	4.7 ± 2.6	0.26 ± 0.47
40-60%	6	8.7	284	14.5 ± 6.2	12.2 ± 5.2	0.22 ± 0.08
60-70%	8	37.6	200	14.9 ± 6.0	14.2 ± 5.1	0.25 ± 0.17
70-80%	20	30.2	286	21.9 ± 6.7	20.5 ± 9.8	0.41 ± 0.28
> 80%	26	12.5	234	21.1 ± 10.8	19.0 ± 9.2	0.41 ± 0.28

If we correct the measured nitrogen export for differences in average annual runoff (discharge weighted concentration) we still see a significant increase in the nitrogen input to surface waters with increasing proportions of agricultural land (Table 9). The observed increase in average annual discharge-weighted concentration of nitrogen is again most pronounced for nitrate-N (factor 11), followed by total-N (factor 7) and ammonium-N (factor 3.5) (Table 9).

Table 9. Average annual discharge-weighted concentration of total nitrogen, nitrate nitrogen and ammonium nitrogen from catchments with different proportions of agricultural land and where the point source nitrogen emission is less than 0.5 kg N ha^{-1} during the ten-year period 1989-98.

Proportion of agricultural land	Number of catchments	Total-N (mg N l^{-1})	Nitrate-N (mg N l^{-1})	Ammonium-N (mg N l^{-1})
< 20%	5	1.4 ± 0.8	0.8 ± 0.7	0.05 ± 0.04
20-40%	5	3.5 ± 1.7	2.7 ± 1.5	0.11 ± 0.18
40-60%	6	5.4 ± 1.8	4.4 ± 1.5	0.09 ± 0.04
60-70%	8	7.8 ± 2.2	6.9 ± 2.2	0.12 ± 0.05
70-80%	20	8.4 ± 2.8	7.5 ± 2.9	0.15 ± 0.10
> 80%	26	9.8 ± 3.9	8.6 ± 3.6	0.18 ± 0.10

6. NITROGEN REMOVAL IN LAKES

Lakes play an important role as sinks in the transport of land-based nitrogen to downstream coastal and marine areas. The reported rates of nitrogen removal have, however, varied significantly (8-81%, Seitzinger, 1988), although quite constant and high loss rates have been found in 69 Danish shallow eutrophic lakes (Jensen *et al.*, 1990). The present section aims to bring an overview of the magnitude of nitrogen removal in shallow lakes and the important controlling factors. Quantification of the nitrogen removal is based on mass balances from 22 Danish lakes with 10 years data and 4 Dutch lakes with 11-13 years data. For more details on methods and calculations, we refer to Jensen *et al.* (1990,1992) and Jeppesen *et al.* (1998).

Table 10. Average annual nitrogen loading and retention in 23 Danish shallow lakes and 4 lakes in the Netherlands. Included is also a description of lake morphology and average hydraulic retention time.

Name of lake	Mean depth (m)	Lake area (km^2)	Retention time (years)	Nitrogen loading (mg N m^{-2} d^{-1})	Nitrogen retention (mg N m^{-2} d^{-1})	Nitrogen retention (% of loading)
			Denmark			
Lake Arreskov	1.9	3.2	1.46	39.4	23.2	58.3
Lake Arresø	3.1	39.9	4.14	32.1	17.9	58.6
Lake Borup	1.1	0.1	0.07	409.1	48.1	15.5
Lake Bryrup Langsø	4.6	0.4	0.23	523.2	241.6	48.1
Lake Dons Nørresø	1.0	0.4	0.05	450.8	113.8	27.6
Lake Engelsholm	2.6	0.4	0.23	195.2	125.9	64.9
Lake Fuglesø	2.0	0.1	0.17	513.8	226.2	49.4
Lake Fårup	5.6	1.0	0.46	123.1	71.7	57.7
Lake Gundsømagle	1.2	0.3	0.09	552.6	151.6	28.7
Lake Hejrede	0.9	0.5	0.15	246.0	53.4	23.7
Lake Hinge	1.2	0.9	0.05	400.2	55.7	14.6
Lake Jels Oversø	1.2	0.1	0.02	1592.6	147.2	9.0
Lake Kilen	2.9	3.3	0.77	98.0	69.8	71.4
Lake Langesø	3.1	0.2	0.58	229.5	108.1	49.7
Lake Lemvig	2.0	0.2	0.09	590.1	172.9	30.5
Lake Ravn	15.0	1.8	2.33	190.8	100.4	60.1
Lake St. Søgård	2.7	0.6	0.23	445.7	109.7	25.2
Lake Søgård	1.6	0.3	0.07	1005.4	226.1	20.5
Lake Søholm	6.5	0.3	1.78	90.9	50.2	57.7
Lake Tissø	8.2	12.3	1.06	224.6	145.5	68.1
Lake Tystrup	9.9	6.6	0.55	650.3	243.1	40.5
Lake Vesterborg	1.4	0.2	0.08	743.6	121.2	18.5
Lake Ørn	4.0	0.4	0.05	357.6	37.1	10.3
Mean	3.6	3.2	0.64	421.9	115.7	39.5
			The Netherlands			
Lake Veluwemeer	1.3	32.4	0.13	126.7	65.5	51.4
Lake Woldewijd	1.5	18.0	0.32	49.5	18.1	37.1
Lake Nuldernauw	1.5	9.5	0.12	163.5	69.5	42.6
Lake Drontermeer	1.1	5.4	0.02	611.8	122.0	21.0
Mean	1.4	16.3	0.15	237.9	68.8	38.0
Grand Mean	2.5	9.8	0.4	329.9	92.3	38.8

Most of the nitrogen removal in lakes is due to denitrification, and only a minor part to permanent burial in the sediments (Dudel & Kohl, 1992; Van Luijn *et al.*, 1996). In Danish Lake Søbygård, Jensen *et al.* (1992) estimated that around 90% of the nitrogen removal could be attributed to denitrification. In the remaining part of this section, the term N-removal will be used for the sum of denitrification and burial of nitrogen in lakes.

Average N-removal for the 26 lakes is estimated to 92.3 mg N m^{-2} d^{-1} (Table 10). A substantial amount of the nitrogen entering the lakes is removed, the relative N-removal being as high as 38.8%. This value is quite corresponds well with the value reported for 69 shallow Danish lakes (mean: 43%) (Jensen *et al.*, 1990).

The N-removal rate ranges from 17.9 to 243.1 mg N m^{-2} d^{-1} as the nitrogen loading ranges from 32.1 to 1592.6 mg N m^{-2} d^{-1}, while the relative nitrogen removal ranges from 9.0 to 71.4 % removal of the nitrogen loading. No significant differences are found between the Dutch and the Danish lakes in either absolute or relative N-removal rates. The intra-lake variation seems to be as high as the inter-lake variation in the two countries (Table 10).

6.1 Abiotic factors controlling nitrogen removal

Nitrogen removal depends on the nitrogen loading to the lake and the rate of removal is generally higher with higher loading ($P<0.0001$, Fig. 8). The removal rate can, however, vary markedly, even at similar nitrogen levels of loading. The main reason for this variation is chiefly explained by differences in hydraulic retention time, a factor upon which relative nitrogen removal is highly dependent (Fig. 8).

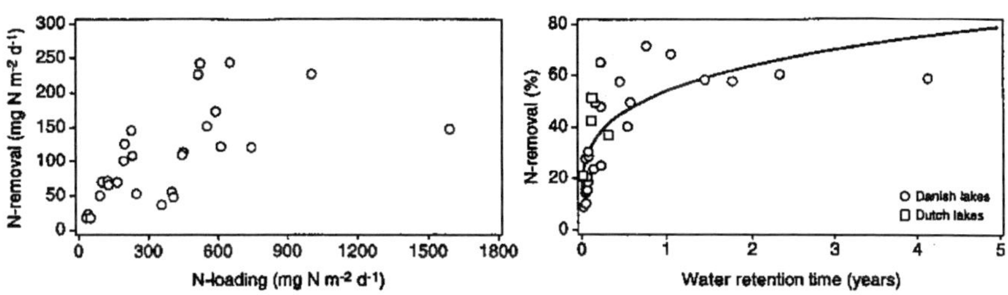

Figure 8. Nitrogen removal vs. nitrogen loading in Danish and Dutch shallow lakes and the relative nitrogen removal compared to hydraulic retention time.

A simple empirically derived relationship, depending solely on the hydraulic retention time in the lake (τ_W), may account for 60% of the observed variation in the relative nitrogen removal ($N_{ret}(\%)$):

$$N_{ret}(\%) = 53.9 * \tau_W^{0.235},\ r^2=0.60,\ P<0.0001.$$

This relationship is comparable with the model proposed by Jensen *et al.* (1990) and Windolf *et al.* (1996).

6.2 Biotic factors controlling nitrogen removal

Besides the abiotic factors controlling N-removal, also the biological structure of a lake may markedly influence N-removal. In two of the Danish lakes included, the biological structure changed dramatically during the investigation period (Jeppesen *et al.*, 1998). In Lake Arreskov, fish kill in winter 1991-92 caused a shift from a turbid plankton-dominated stage to a clearwater and hence macrophyte-dominated stage. The relative N-removal increased from 26-38% before to 48-62% afterwards the fish kill. Similarly in Lake Engelsholm, N-removal increased from 49-53% to 59-66% following a partial removal of the planktivorous fish stock in 1992-94.

Various factors resulted in the increased N-removal in the two lakes (Jeppesen *et al.*, 1998): i) a decrease in organic nitrogen in the lakes and outlets due to the decrease in the nitrogen incorporated in the phytoplankton; ii) reduced resuspension due to a decrease in the number of fish foraging in the sediment and an increase in benthic algal growth; iii) higher denitrification in the sediment, reflecting less competition between denitrifiers and phytoplankton for nitrate, enhanced N retention by phyto- and zoobenthos and enhanced sediment nitrification due to higher oxygen concentrations. These two cases clearly demonstrate the very complicated interactions between N-removal and lake biological structure. Differences in biological structure may thus be part of the explanation of differences in the reported N-removal rates, especially in shallow lakes.

7. NITROGEN REMOVAL IN FRESHWATER WETLANDS

In Denmark, studies of nitrogen removal in freshwater wetlands have been undertaken since the mid-1980s. Both natural fens and meadows with different hydrological regimes as well as restored and constructed wetlands have been investigated.

7.1 Natural wetlands

In natural freshwater wetlands receiving groundwater recharge (i.e. minerotrophic wetlands), nitrogen removal varies from 57 kg N to >2100 kg N ha^{-1} yr^{-1}. The relative removal varies only from 56 to 97%, without any clear correlation between loading and efficiency (Table 11 and 12). Thus, there is a hugh capacity of Danish freshwater wetland soils for nitrogen removal through denitrification. The nitrogen loading to the wetlands reflects upland characteristics such as land use, precipitation surplus, drainage conditions, soil type, the groundwater flow pattern, etc. Some of the overall factors characterizing wetlands recharged by groundwater and examples on their ability to remove nitrogen will be given in this section.

At the River Stevns, the concentration of nitrate in recharging groundwater to the meadow varied from 15–30 mg NO_3^--N l^{-1}. Nearly 74% of all upland groundwater was discharged directly into River Stevns through drainage pipes. Only 57 kg N were retained in the meadow, of which 31 kg N ha^{-1} y^{-1} was denitrified, while on average 89 kg N were removed from the meadow through haymaking. Thus, nitrogen removal from the meadow was higher than the input the meadow (total input: 60 kg N - total output: 120 kg N).

The wet meadow at the Rabis brook is recharged by nitrate-rich groundwater, which at the valley slope breaks through to the soil surface and irrigates the meadow naturally (Table 11). This is possibly the reason why the relative nitrogen removal is so low (56%), since the nitrate has to move by advective flow or by diffusion to the active denitrification sites close to the soil surface.

In the Gjern River catchment area, studies have been made for several years of wetlands with different hydraulic regimes (Table 11). Especially in a 73 metre wide water-covered fen (area B, Table 11), nitrogen turnover has been studied intensively and particularly high rates of denitrification have been found in the area around the river valley slope. Over a distance of only 13–17 m, the nitrate concentration falls from approx. 25 to 0.01 mg NO_3^--N l^{-1}, corresponding to a denitrification rate of 1-5 g N m^{-2} d^{-1}, depending on where in the zone of enhanced denitrification sampling is undertaken (Blicher-Mathiesen, 1998; Hoffmann, 1998a; Hoffmann *et al.*, 2000b). Only a few other studies have hitherto reported denitrification rates of this magnitude, e.g. Cooper (1990)(8.1 g N m^{-2} d^{-1}), Haycock and Burt (1993) (0.74 g N m^{-2} d^{-1}), Haycock and Pinay (1993) (up to 10 g N m^{-2} d^{-1}) and Jørgensen *et al.* (1988) (2.1 g N m^{-2} d^{-1}).

Table 11. Nitrate removal in different riparian wetlands with groundwater recharge (flow through). %: percentage of incoming nitrate loading removed (From Hoffmann, 1998).

Locality	Nitrate removal rates (kg NO_3^-- N ha^{-1} $year^{-1}$)	Reduction (%)
River Stevns, meadow	57	95
Rabis brook, meadow	398	56
River Gjern:		
A, meadow	140	67
B, fen (1993)	2100	97
B, fen (5 years)	1079	97
C, meadow (5 years)	541	96
D, meadow (5 years)	398	97

7.2 Rehabilitation of fens, wetlands and wet meadows in floodplains

In most European countries watercourses have been modified by man to improve certain features, for example flood control, drainage of surrounding land, navigation, etc. In countries like Denmark with an intensive agricultural production, more than 90 per cent of the total river network has been regulated to some extent (Iversen *et al.*, 1993). Straightening and channelization of watercourses was conducted to ensure sufficient drainage of the floodplains.

Table 12. Examples of nitrogen removal rates measured in two Danish river valleys with groundwater recharge following re-meandering of the river channel. Both studies were conducted in the first year following the river restoration (1995/96). (From Hoffmann et al. 1998, 2000a).

Locality	Kg NO_3^--N ha^{-1} $year^{-1}$	%
River Brede, large-scale, meadow (63 ha)	92	71
Headwaters of River Gudenaa, large-scale, meadow (57 ha)	8.4	57

Today we are restoring many of our rivers by reinstating their former meandering course (Kronvang *et al.*, 1998). Hence, former fens, wetlands and wet meadows are reinstated in our river valleys by re-meandering the river channel, elevating the river bed and disconnecting drains and ditches which also lead to increased nitrogen removal (Table 12). The nitrogen removal rates obtained at two river restoration sites in Denmark are shown in Table 12. Considerable variation in the groundwater flow pattern both along the river and from riverside to riverside was found, implying that nitrogen transport and removal vary significantly (Hoffman *et al.*, 1998; Hoffmann *et al.*, 2000a).

7.3 Irrigation of meadows with drainage or river water

Experiments involving the irrigation of meadows and reed forests with drainage water or river water have also yielded promising results with respect to nitrate reduction. The extent of nitrate removal primarily depends on the amount of water infiltrating the soil. Therefore, the size of the area to be irrigated needs to be adjusted to the amount of water it is expected to receive. In an irrigation experiment alongside the River Stevns, where all the water input to the area infiltrated the soil, 99% of the nitrate was denitrified in the uppermost 2 cm of the soil profile (Table 13) (Hoffmann *et al.*, 1993). In the same study, nitrate reduction was also measured on a plot of the meadow that had been fed with tile drainage water for approx. 100 years via a drainage conduit terminating on the river valley slope. Over a distance of 45 metres, the nitrate concentration fell from 11.3 mg NO_3^--N l^{-1} at the conduit outflow to 0.1 mg NO_3^--N l^{-1} midway into the meadow, i.e. a reduction of 99% (Hoffmann *et al.*, 1993).

Table 13. Nitrate removal by irrigation with time drainage water or stream water. The removal rates obtained during irrigation with tile drainage water takes into account the periodicity of tile drainage runoff (in Denmark, highest runoff in winter and spring: October to May).

Locality	Kg NO_3^--N ha^{-1} $year^{-1}$	%
Glumsø, reedswamp§	520	65
Glumsø, reedswamp§	975	62
Glumsø, reedswamp§	2725	54
Glumsø, large-scale, reedswamp§	569	94
River Stevns, meadow*	350	99
River Stevns, meadow with drainage pipes!	(conc.) 11.3	99
Syv brook, meadow	300	72
River Storå, restored meadow	530	48
River Gjern, meadow*§ (min)	34	88
River Gjern, meadow*§ (max)	200	98

* Short term experiment.
§ Different hydraulic loading and different nitrate loading.
! Concentration given in mg NO_3^- -N l^{-1}.

When part of the irrigation water is discharged as surface runoff, the relative nitrate removal decreases to between 48 and 72% according to the studies hitherto undertaken at Syv Bæk brook (Hoffmann, 1991), Lake Glumsø (Hoffmann, 1986; Jørgensen *et al.*, 1988) and the River Storå (Fuglsang, 1993; 1994) (Table 13), this being attributable to the fact that denitrification does not occur in the surface water due to the prevailing oxic conditions. Any nitrogen removal occurring in the surface water must therefore take place through algal uptake, uptake into the microbial pool or sedimentation of particulate nitrogen.

Irrigation of riparian areas with tile drainage water can only be undertaken in periods during which water is flowing in the drains. In consequence, the operational period highly depends on local conditions such as the amount of precipitation, soil type etc. The annual rates given for the meadows at the River Stevns and Syv brook shown in Table 13 are thus calculated for the period during which they were actually irrigated by drainage water, i.e. 120 and 200 days, respectively.

A short-term irrigation and flooding study at the lower reaches of the River Gjern shows that nitrate reduction occurs in the uppermost part (0-2 cm) of the soil in areas subjected to regular flooding or irrigation events (Hoffmann, 1996; Hoffmann, 1998b). Although the infiltration capacity is low at the Gjern study site, the results show reduced nitrate values even during short-term flooding/irrigation events. It means that apart from being important for sedimentation, naturally meandering rivers and their riparian areas serve as a functional unit and a stabilising ecological factor for the aquatic environment.

8. MEASURES TO REDUCE EMISSIONS OF NITROGEN FROM POINT AND DIFFUSE SOURCES IN EUROPE

Measures to reduce point source emissions of nitrogen focus on urban wastewater treatment plants (UWWT) as well as various key industries. The Urban WasteWater Treatment Directive (91/271/EEC) adopted by the European Commission in 1991 is a key directive for water management in the European Union. The Directive sets minimum standards for the collection, treatment and disposal of wastewater dependent on the size of discharge, and the type and sensitivity of receiving waters. In the case of nitrogen, a maximum annual average threshold of 15 mg N l^{-1} is set for smaller UWWT's (10,000-100,000 PE), whereas the threshold is lower (10 mg N l^{-1}) for larger UWWT's (> 100,000 PE).

Measures to reduce nitrogen emissions from agriculture are more difficult to implement, both in a political and practical perspective. A variety of policies has been applied in the EU and the different countries: (i) legislation; (ii) economic instruments; (iii) information. In 1991, the European Commission adopted the Nitrate Directive aimed to reduce or prevent the nitrate pollution of water caused by the application and storage of inorganic fertilizers and manure on farmland. The Nitrate Directive requires that member states: (i) identify vulnerable areas to nitrate pollution; (ii) establish Action Plans governing the time and rate of fertilizer and manure application, and conditions of manure storage in vulnerable zones; (iii) implement monitoring programmes to assess the effectiveness of action programmes; (iiii) establish Codes of Good Agricultural Practice to be implemented by farmers on a voluntary basis in other areas.

The impact of the Nitrate Directive will of course depend on the interpretation of requirements by the EU Member States, especially that of the area extent of vulnerable zones. As an example, Austria, Denmark, Germany and The Netherlands have designated the whole territory, France 46% vulnerable zones of the whole territory, while Portugal has designated 5 and the UK 69 vulnerable zones.

Table 14. Measures applied across Europe to combat nitrogen pollution.

Countries approach	A	B	D	DK	E	FIN	F	GR	IRL	I	NL	N	P	PL	S	UK
Point sources																
UWWT tertiary treatment		,	,	,	,	,	,		,		,	,		,	,	,
Industry – BAT	,		,	,												,
Agriculture																
Manure restrictions	,	,	,	,	,		,	,	,	,	,	,			,	,
Manure storage	,	,	,	,			,	,	,		,	,			,	,
Silage storage			,	,				,	,		,	,			,	,
Fertilizer restrictions	,	,	,	,			,	,	,		,	,			,	,
Atmospheric input																
Atmospheric sources	,	,	,	,	,						,	,	,	,	,	,
Industry – BAT	,		,	,										,	,	,

A: Austria; B: Belgium; D: Germany; DK: Denmark; E: Spain; FIN: Finland; F: France; GR: Greece; IRL: Ireland; I: Italy; NL: The Netherlands; N: Norway; P: Portugal; PL: Poland; S: Sweden; UK: United Kingdom.

Table 15. Action Plans and their major elements adopted in Denmark to combat nitrogen pollution from agriculture.

Action Plans	Year of adoption	Major implemented measures to combat diffuse nitrogen pollution
NPO Action Plan	1985	Elimination of direct discharges from farms
		Livestock harmony on the farm level
Action Plan on the Aquatic Environment I	1987	9 month storage facility for slurry
		65% winter green fields
		Crop and fertilizer plans
Plan for Sustainable agricultural development	1991	Standard N fertilisation values for crops
		Standard values for N in animal manure
		Required utilisation of N in animal manure (30-45%)
		Fertilizer accounts at farm level
Action Plan on the Aquatic Environment II	1998	Demands for an overall 10% reduction of N application to crops
		Demands for catch crops
		Demands for transforming 16,000 hectares farmland to wetlands
		Reiterated demands for utilisation of N in animal manure (40-55%)

The measures applied in selected European countries for combating nitrogen pollution from point and diffuse sources are shown in Table 14. Most countries have initiated tertiary wastewater treatment and imposed regulations on manure restrictions and storage.

In Denmark, the government has adopted four major Action Plans since the early 1980s to reduce by 50% nitrogen pollution of the aquatic environment (Table 15). The Action Plan on the Aquatic Environment I from 1987 demanded nitrogen removal at all major sewage treatment plants (> 15,000 PE). The regulations imposed to ensure a reduction in nitrogen pollution from agriculture have been reiterated in all subsequent Action Plans (Table 15). Besides the multiple measures implemented, Denmark has also adopted three other plans: (i) a Plan for Afforestation (1987) aimed to double the forest area during the next 60-80 years (5,000 hectares per year); (ii) the Strategy on Marginal Land (1987) aimed to restore 20,000 ha of former wetlands over a 10-20 year period; (iii) EU Set-Aside schemes. These plans will of course also contribute to the reduction of the nitrogen pollution from farmland.

9. CONCLUSIONS AND PERSPECTIVES

Concern about elevated nitrogen concentrations and loading to groundwater and surface water in Europe has prompted the introduction of many reduction strategies at international, national and local levels. Measures to reduce point source discharges of nitrogen focus on tertiary treatment of urban wastewater as well as various key industries have been implemented in most countries within the European Union (EU) (see Tables 14 and 15).

Today, a high proportion of the total anthropogenic nitrogen loading to the aquatic environment consists of nitrogen loading from agriculture (cf. Table 1). However, reducing the nitrogen input from agriculture is difficult for both technical and political reasons. Measures may be introduced/implemented at the source to reduce the large nitrogen surplus (as seen in many European countries, see Table 3). Another approach may be to introduce transport measures to reduce the level of nitrogen before it reaches surface and thus increase the nitrogen removal potential along the routes followed by nitrogen from soil to water.

The latter may be allowed by reinstating formerly drained lakes and wetlands in which the potential for nitrate removal through the denitrification process normally is high (see Tables 10-14). So far results obtained in Denmark and The Netherlands have shown that the potential for nitrogen removal in lakes and freshwater wetlands is high. In Denmark the Second Action Plan on the Aquatic Environment (see Table 15) included the following initiative: to remove a total of 5600 tonnes nitrogen y^{-1} by rehabilitating 16,000 hectares of drained and otherwise reclaimed wetlands within five years.

REFERENCES

Andersen, H.E., Kronvang, B. and Larsen, S.E. (1999) Agricultural practices and diffuse nitrogen pollution in Denmark: Empirical leaching and catchment models. *Water Science and Technology* **39**(12): 257-264.

Blicher-Mathiesen, G. (1998) Nitrogen removal in riparian areas. PhD thesis, Aarhus University, Biological Institute, Department of Microbial Ecology and National Environmental Research Institute, Department of Streams and Riparian Areas, 94 pp.

Borgvang SA, Selvik JR, Skiple A, Kroglund M and Bratli JL (1998) Harmonised reporting system for nutrients (HARP), Workshop report. NIVA Report O-970075, Oslo.

Boers PCM, Boogaard HL,Hoogeveen J, Kroes JG, Noij IGAM, Roest CWJ, Ruijgh EFW and Vermulst JAPH (1997) Huidige en toekomstige belasting van het oppervlaktewater met stikstof en fosfaat vanuit de landbouw. (Present and future emission of nutrients from agriculture (in Dutch with an English summary). RIZA report 97.013, Lelystad.

BUWAL (1993) Der Stickstoffhaushalt in der Schweiz (The nitrogen balance of Switzerland). Bundersamt für Umwelt, Wald und Landschaft (BUWAL) report 209, Bern.

Cooper, A.B. (1990) Nitrate depletion in the riparian zone and stream channel of a small headwater catchment. *Hydrobiologia* **202**: 13 - 26.

Dudel, G. & Kohl, J.G. (1992) The nitrogen budget of a shallow lake (Grosser Müggelsee, Berlin). *Int. revue ges. Hydrobiol.* **77**: 43-72.

Edwards, A.C., Pugh, K., Wright, G., Sinclair, A.H. and Reaves, G.A. (1990) Nitrate status of two major rivers in N.E. Scotland with respect to land use and fertiliser addition. *Chemistry and Ecology* **4:** 97-107.

European Environment Agency (1995) *Europe's Environment – The Dobris Assessment.* Stanners. D. and Bourdeau, P. (Eds.). European Environment Agency, Copenhagen, 676 p.

European Environment Agency (1999) Nutrients in European ecosystems. Thyssen, N. (Ed.) Environmental Assessment Report No. 4, European Environment Agency, Copenhagen, 155 p.

Fuglsang, A. (1994) De våde enge kan tilbageholde kvælstof (in Danish). Vand og Jord, No. 1(2), pp. 52 - 54.

Fuglsang, A. (1993) Resultater for marginaljordsprojekt ved Storå i Fyns Amt, 1990 - 92 (in Danish). Memorandum, Funen County.

Haycock, N.E. and Burt, T.P. (1993) Role of floodplain sediments in reducing the nitrate concentration of subsurface run-off: A case study in the Cotswolds, UK. *Hydrological Processes* **7**: 287 - 295.

Haycock, N.E. and Pinay, G. (1993) Groundwater nitrate dynamics in grass and poplar vegetated riparian buffer strips during winter. *J. Environ. Qual.* **22**: 273 - 278.

Hoffmann, C.C. (1985) Fosfor og kvælstof dynamik under kontrollerede hydrauliske betingelser i en rørsump ved Glumsø sø (in Danish). Dissertation. Copenhagen University, Freshwater Biology Laboratory.

Hoffmann, C.C. (1986) Nitrate reduction in a reedswamp receiving water from an agricultural watershed. Proc. 13th Nordic Symp. on Sediments, Anaboda, Sweden, 1985.

Hoffmann, C.C. (1991) Water and nutrient balances for a flooded riparian wetland. In: Nitrogen and Phosphorus in Fresh and Marine Waters - Project Abstracts of the Danish NPo Research Programme, C13b, pp. 203-220. NPo-forskning fra Miljøstyrelsen, C - Abstracts.

Hoffmann, C.C., Dahl, M., Kamp-Nielsen, L. and Stryhn, H. (1993) Vand- og stofbalance i en natureng (in Danish). Environmental Project No. 231. Danish Environmental Protection Agency, Copenhagen, 152pp.

Hoffmann, C.C. (1996) Fate of phosphate, nitrate, and other elements during short-term flooding of a riparian meadow. In: Sediment and phosphorus, Erosion and delivery, transport and fate of sediments and sediment-associated nutrients in watersheds (Eds. Kronvang, B. Svendsen, L.M and Sibbesen, E.), pp. 135 - 142. Proceedings from an international workshop held in Silkeborg, Denmark, October 9-12, 1995. Ministry of Environment and Energy, 150 pp. - NERI technical Report No 178.

Hoffmann, C.C. (1998a) Nutrient retention in wet meadows and fens. PhD thesis, University of Copenhagen, Freshwater Biological Institute, and National Environmental Research Institute, Department of streams and Riparian Areas, 134 pp.

Hoffmann, C.C. (1998b) Nitrate removal in a regularly flooded riparian meadow. *Verh. Internat. Verein. Limnol.* **26:** 1352-1358.

Hoffmann, C.C., Pedersen, M.L., Kronvang, B.K. and Øvig, L. (1998) Restoration of the Rivers Brede, Cole and Skerne: A joint Danish and British EU-LIFE demonstration project, IV - Implications for nitrate and Iron transformation. *Aquatic Conservation: Marine and Freshwater Ecosystems* **8**(1): 223-240

Hoffmann, C.C., Pedersen, M.L. and Laubel, A.L. (2000a) Headwater restoration of the River Gudenå – 2. Implications for nutrients in riparian areas. Verh. Internat. Verein. Limnol. 27 Vol. 1, pp.602-609.

Hoffmann, C.C., Rysgaard, S. and Berg, P. (2000b) Denitrification rates in riparian fens predicted by $^{15}NO_3^-$ microcosm studies, *in situ* measurements, and modeling. *J. Environ. Qual.* (in press).

IRC (1992) Tätigkeitsbericht 1992. Report Internationale Kommission zum Schutze des Rheins gegen Verunreinigung, Koblenz.

Iversen, T.M., Kronvang, B., P. Markmann, B.L. Madsen and M.B. Nielsen, 1993: Re-establishment of Danish streams: Restoration and maintenance measures. *Aquatic Conservation: Marine and Freshwater Ecosystems* **3:** 73-92.

Iversen, T.M., Kjeldsen, K., Kristensen, P., Haan, B., Oirschot, M., Parr, W. and Lack, T. (1997) *Integrated Environmental Assessment on Eutrophication – A Pilot Study.* National Environmental Research Institute, Denmark, Technical Report No. 207, 101 p.

Jensen, J.P., P. Kristensen and E. Jeppesen (1990) Relationships between nitrogen loading and in-lake concentrations in shallow Danish lakes. *Verh. Internat. Verein. Limnol.* **24**: 201-204′.

Jensen, J.P., E. Jeppesen, P. Kristensen, P.B. Christensen and M. Søndergaard (1992) Nitrogen loss and denitrification as studied in relation to reductions in nitrogen loading in a shallow, hypertrophic lake (Lake Søbygård, Denmark). *Int. Revue gesamt. Hydrobiologia* **77**: 29-42.

Jeppesen, E., Jensen, J.P., Søndergaard, M., Lauridsen, T.L., Møller, P. Hald & Sandby, K. (1998) Changes in Nitrogen Retention in Shallow Eutrophic Lakes Following a Decline in Density of Cyprinids. *Archiv für Hydrobiologie* **142**: 129-151.

Jørgensen, S.E., Hoffmann, C.C. and Mitsch, W.J. (1988) Modelling nutrient retention by a reedswamp and wet meadow in Denmark. In: M.J. Mitsch, S.E. Jørgensen, and M. Straskraba (Eds) Wetland modelling, pp. 133-151. Elsevier.

Kroes, JG, Boers PCM, Boogaard HL, Ruijgh E, Vermulst JAPH (1996) Inpact of manure policies on leaching of nitrogen and phosphorus from rural areas to Dutch ground and surface waters: methodology and application on a national scale. Proceedings worksshop on nitrate directive, 20 - 22 June 1996, The Hague, the Netherlands.

Kristensen, P. and Hansen, O.H. (1994) *European Rivers and Lakes – Assessment of their Environmental State*, European Environment Agency, EEA Environmental Monographs 1, 122 p.

Kronvang, B., Ærtebjerg, G., Grant, R., Kristensen, P. Hovmand, M & Kirkegaard, J. (1993) Nationwide monitoring of nutrients and their ecological effects: state of the Danish aquatic environment. *AMBIO* **22:** 176-187.

Kronvang, B., Grant, R., Larsen, S.E., Svendsen, L.M. and Kristensen, P (1995) Non-point source nutrient losses to the aquatic environment in Denmark. Impact of agriculture. *Marine and Freshwater Research* **46**: 167-177.

Kronvang, B., Græsbøll, P., Larsen, S.E., Svendsen, L.M. and Andersen, H.E. (1996) Diffuse nutrient losses in Denmark. *Water Science and Technology* **33**(4-5): 81-88.

Kronvang, B.,Svendsen, L.M., Brookes, A., Fisher, K., Møller, B., Ottosen, O., Newson, M. and Sear, D. 1998. Restoration of the rivers Brede, Cole and Skerne: A joint Danish and British EU-LIFE project, III - Channel morphology, hydrodynamics and transport of sediment and nutrients. *Aquatic Conservation: Marine and Freshwater Ecosystems* **8**: 209-222.

Kronvang, B., Hoffmann, C.C., Svendsen, L.M., Windolf, J., Jensen, J.P. and Dørge, J. (1999) Retention of nutrients in river basins. *Aquatic Ecology* **33**: 29-40.

Larsen, S.E:, Kronvang, B., Windolf, J. and Svendsen, L.M. (1999) Trends in diffuse nutrient concentrations and loading in Denmark: Statistical trend analysis of stream monitoring data. *Water Science and Technology* **39**(12): 197-205.

Mee, L.D. (1992) The Black Sea in crisis: a need for concerted international action. *AMBIO* **21**: 278-285.

Neill, M. (1989) Nitrate concentrations in river waters in the south-east of Ireland and their relationship with agricultural practices. *Wat. Res.* **23**, 1339-1355.

OSPARCOM (1992) Nutrients in the Convention Area. Oslo and Paris Commission. ISBN 0946956 23 5.

Seitzinger, S.P. (1988) Denitrification in freshwater and coastal marine ecosystems: Ecological and geochemical significance. *Limnol. Oceanogr.* **33**: 702-724.

Stibe, L. and Fleischer, S. (1991) Agricultural production methods – impact on drainage water nitrogen. *Verh. Int. Ver. Limnol.* **24**, 1749-1752.

Van Dijk GM, Stalnacke P, Grimvall A, Tonderski A, Sundblad K and Schäfer A (1996) Long-term trends in nitrogen and phosphorus concentrations in the Lower River Rhine. *Arch. Hydrobiol. Suppl.* **113**: 99-109.

Van Luijn, L., Boers, P.C.M. and Lijklema, L. (1996) Comparison of denitrification rates in lake sediments obtained by the N_2 flux method, the ^{15}N isotope pairing technique and the mass balance approach. *Water Research* **30**: 893-900.

Werner W, Olfs HW, Auerswald K and Iserman K (1991) Stickstoff- und Phosphoreintrag in Oberflächengewässer über diffusse Quellen (Nitrogen and phosphurus emissions into surface waters from diffuse sources; in German). In: Studie über Wirkungen und Qualitätsziele von Nährstoffen in Flieszgewässern. Academia Verlag St. Augustin.

Windolf, J., E. Jeppesen, J.P. Jensen and P. Kristensen (1996) Modelling of seasonal variation in nitrogen retention: a four-year mass balance study in 16 shallow lakes. *Biogeochemistry* **33**: 25-44.

Wright, G.G., Edwards, A.C., Morrice, J.G. and Pugh, K. (1991) North east Scotland river catchment nitrate loading in relation to agricultural intensity. Chemistry and Ecology 5, 263-281.

Nitrogen in the Environment: Sources, Problems, and Management
R.F. Follett and J.L. Hatfield (Eds)
Published by Elsevier Science B.V.

Chapter 9. Nitrogen Effects on Coastal Marine Ecosystems

John R. Kelly

U.S. Environmental Protection Agency (US EPA),
National Health and Environmental Effects Research Laboratory,
Mid-Continent Ecology Division, 6201 Congdon Blvd., Duluth MN 55804

1. INTRODUCTION: NITROGEN AND COASTAL MARINE SYSTEMS*

In the late 1960s the problem of nutrient inputs to freshwater systems, and scientific debate about it, was reaching a peak. Although a critical volume on aquatic eutrophication edited by Gene Likens (1972) did not exclude papers on marine and estuarine systems, the principal focus was clear (Likens, p.vi):

"I hope that the information provided in this volume may be useful to decisionmakers and thus contribute to a slowing of the eutrophication rate in our *lakes and streams*." [italics added]

Even before the freshwater decision-makers became fully focused on setting limits on phosphorus (P) loading by a seminal experiment of David Schindler (1974), Ryther and Dunstan (1971) suggested that nitrogen (N) played the more critical role in coastal marine systems. We now know that there are low-salinity portions of some estuaries, as well as some special individual marine cases, which are P-sensitive (e.g. Howarth 1988; Krom et al. 1991; Doering et al. 1995; Malone et al. 1996, Boynton 2000). Similarly, there are various freshwater systems that are N-sensitive (e.g. Elser et al. 1990). Moreover, it is clear that there are possible ecological consequences of relative availability of silicate (Si) and N in coastal marine systems (e.g. Officer and Ryther 1980, Ryther and Officer 1981, Doering et al. 1989, Conley 2000, Rabalais et al. 2000). But the overwhelming evidence, including observational evidence and controlled mesocosm experiments, confirms a central role of N in establishing biological responses to nutrient loading in coastal waters (Howarth 1988; Oviatt et al. 1995; Doering et al. 1995; NRC 2000).

Issues of shifted element ratios and their moderation of N effects are fascinating and substantial (cf. Nixon et al. 1980, Nixon 1981, Doering et al. 1989, Seitzinger 2000, Rabalais et al. 2000). But it is challenging enough to review effects relating to different levels of N, which is the principal focus of this chapter.

N circulation through the global environment has been growing dramatically (Vitousek et al. 1997). Delivery via rivers to receiving systems has increased greatly in the last half of the 20th century (e.g. Howarth et al. 1996, CENR 2000, other chapters in this volume). Unlike P, there is a significant atmospheric pathway for N. Local- to longer-range atmospheric transport and deposition to many regions, including coastal waters, has risen as a consequence of fossil fuel burning, agricultural practices, and other human activities. Riverine and atmospheric enrichments, combined

* The term "coastal marine" or "coastal systems" is used in this chapter as shorthand for estuaries, shallow embayments, and lagoons, as well as more open nearshore and shelfwater ecosystems along oceanic-terrestrial margins, as distinct from similar coastal systems on large inland seas like the Great Lakes.

with burgeoning coastal zone populations of humans and their associated land use changes and wastewater releases, make coastal systems a frontline receiving system for much of the enormous global N enrichment experiment now in progress. What is happening in coastal ecosystems and what do we expect to happen if N loads continue to rise?

In the same volume cited above (Likens 1972), Clifford Mortimer gave some opening comments (p.viii), which provide a fabric for this chapter:

"Although the analogy can not be pressed too far, it will be noted that the research aimed toward cure of cancer–like eutrophication, a problem of unwanted proliferation of cells–devotes considerably more effort and resources to the study of cell processes than to description of symptoms. Until now, the reverse has been the case in eutrophication research."

It is not that recognition of coastal marine eutrophication had its genesis in 1971; indeed it began at least at the latter part of the 19th century (cf. historical perspectives of Nixon et al. 1986, Nixon 1992, 1998). But the paper of Ryther and Dunstan, with work started much earlier by Ryther (1954), was an epiphany which did spur an emphasis on N research that has continued and increased for the last three decades (Nixon 1995). At the beginning of the 21st century, we now have very little problem identifying Mortimer's "symptoms" of N overenrichment, due to innumerable related conferences, symposia and associated volumes, articles, books, and virtually uncountable numbers of greyer-literature reports (Lauff 1967, NAS 1969, Nielson and Cronin 1981, Boynton et al. 1982, Carpenter and Capone 1983, Chesapeake Bay Program 1983, Kennedy 1984, Valiela 1984, Rosenberg 1985, Nixon et al. 1986, Kullenberg 1986a, Nixon 1988, Howarth 1988, GESAMP 1990, Elliot and Ducrotoy 1991, Vollenweider et al. 1992, NRC 1993, Nixon 1995, Bricker and Stevenson 1996, Nixon et al. 1996, Anderson and Garrison 1997, NRC 2000, Hobbie 2000). At core, symptoms are similar to those noted early on by John Ryther in the oyster beds and waters adjacent to the N-polluting duck farms of Long Island.

This chapter emphasizes that we now know quite a bit about coastal eutrophication* and Mortimer's "proliferation of [autotrophic] cells." We have struggled to provide good loading estimates and nutrient budgets. We have observed, probably in tens of thousands of places and in more than a million samples the concentrations of nitrogenous nutrients in coastal waters. There has been strong progress made in connecting inputs, concentrations, and effects *in concert* in selected places. But we cannot yet predict the stimulation of some of the most undesired cells (e.g. toxic dinoflagellates) or the precise point at which adverse secondary consequences of cell proliferation will occur in any given system. And we do not have a *generalized and quantitative* description of adverse effects of N loading, in part because there are a wide variety of coastal systems.

Also complicating the picture of N as a pollutant is that to some level and to some beholders, the effects of N loading are desirable. This is the agricultural paradigm so effectively written about by

* Scott Nixon (1995) offered a definition of coastal eutrophication as "an increase in the rate of supply of organic matter to an ecosystem." It was offered, in part, because the term has had considerable ambiguity in usage and to emphasize that it is a process, not a state. In context, "eutrophication" doesn't necessarily equate to "undesired effects." In fact, Nixon suggested the definition to be "value neutral." Accepting it, one should talk of the "consequences" of eutrophication as part of the possible set of responses to, or effects of, N enrichment.

Scott Nixon—the notion that fertilization enhances productivity and leads to higher yields of desired species (see Ketchum 1969, Sutcliffe 1972, 1973; Sutcliffe et al. 1977, 1983; Nixon et al. 1986, Nixon 1988, 1992, 1995). Nature repeatedly has shown that it can produce other than desired results, but we do not precisely know the *positive* limits of fertilization. Notwithstanding, the basic ingredients of the recipe for "enriching the sea to death" (Nixon 1998) are known. Observations over the past few decades indicate that many individual system's limits have been passed to realize an oxygen-depleted, mortality-inducing recipe. Examination of such cases, among other lines of evidence, should help resolve a fundamental question: when and where does that unfortunate death recipe result? More subtle effects than fish kills occur, so there are related fundamental questions: when and where does N stimulate undesired species changes or wholesale food web shifts?

The state of knowledge is such that it does not yet allow us to answer the above questions to satisfaction for many systems. This is unfortunate, for such answers are critical to an ability to set N limits that would be protective. There is, however, a huge, and growing, world literature; this review draws heavily from it, even as it reflects my own experience and an admittedly U.S./north temperate bias. I hope my small perspective on the trends in N loading and responses of coastal systems stimulates continued synthesis, as we try to provide the science to achieve Likens' simple goal, for coastal marine systems.

2. SYMPTOMS OF N ENRICHMENT

A recent summary (NRC 2000) lists commonly known ecological responses to N enrichment. These include:

- Increased plant biomass and primary productivity
- Increased oxygen demand and hypoxia or anoxia
- Shifts in benthic community structure caused by anoxia and hypoxia
- Changes in plankton community structure caused directly by nutrient enrichment
- Stimulation of harmful algal blooms (HABs)
- Degradation of seagrass and algal beds, formation of macroalgal mats
- Coral reef destruction

Also listed as a concern, not with N *per se*, but with one vector for it, human sewage, is a potential increase in disease and pathogen species. With the exception of coral reefs, and the substitution of a term like "nuisance" for "harmful" and "macrophytes" for "seagrass," these effects are the classic symptoms of lake eutrophication (Wetzel 1983).

I have grouped effects for review and discussion in this chapter into five prime categories of response to N loading. These have served generally as focal points and endpoints for research:

- Chlorophyll
- Phytoplankton primary production
- Dissolved oxygen
- Benthic producers (Submerged Aquatic Vegetation [SAV], macroalgae) in

shallow waters

- HABs, as part of change in phytoplankton species composition

After a short discussion on N loading (Section 3), I examine each of these five fundamental effects of concern, using examples of where they have been noted and/or have been increasing (Section 4). The probable role of N is suggested, and I try to capture the different kinds of evidence that can link it to the problem. Evidence includes what I refer to as "epidemiological" associations, a spatio-temporal co-occurrence, either local or regionalized. Other evidence includes: time trends of N and effects observed at individual or multiple sites; empirical patterns that emerge from comparing conditions across sites, which begins to assess the generality of the coupling between input and response; and finally, experimentally-observed linkages (primarily in microcosm or mesocosm* experiments), which help confirm and in some cases quantify the nature of the relationship. The strength and kind of evidence linking N and each problem varies, but to the degree possible I indicate some situations where a quantitative linkage has been established. In a summary (Section 5), I speculate on the quantitative sequence of events with increasing loading.

The five effect categories are compiled in a very simple conceptual model to frame how symptoms relate to N loading (Figure 1). I have not attempted to include all the ecological components, flows, confounding factors, etc. in a spaghetti-like picture of interactions that captures more of the true complexity of "sophisticated" constructs or model formulations. Briefly, water column chlorophyll, phytoplankton primary production, and other algal increases are viewed as a direct, nutrient uptake response. Algal increases, representing increased levels of organic matter, secondarily promote low DO through increased decomposition and respiration. Increased algae shade SAV in shallow water to produce a secondary effect of seagrass decline through light reduction. Competitions among the algal community may ultimately promote toxic or nuisance blooms of harmful algae. A concert of secondary effects acts further on food webs/fisheries, but even the direct and first-level indirect effects of N loading (Figure 1) have been difficult to quantify broadly. In Section 5, I touch on such further ramifications of these effects, but they have consequence to aesthetics, human health, valued estuarine and marine populations, food webs, diversity, and ecosystem sustainability. Although we ultimately care most about these, they are some of the most difficult to assess (see CENR 2000).

Any consideration of coastal systems and their potential responses (Figure 1) must also recognize some special, complicating aspects. These systems are generally very open to flow of water and materials, including organisms, from both "upstream" and "downstream" sources (due to tides and circulation changes, as well as biological transport or active migration). Most coastal systems have many sub-areas and pockets of different habitats, so spatial and temporal variability is a confounding problem in their fundamental ecological characterization and in definition of their response to inputs. Coastal systems also represent a set of fairly bewildering diversity in size, shape, and other physical,

---

* Mesocosms are considered to be contained systems (tanks, ponds) larger than bottle or laboratory-size (i.e. "micro"-cosms), which capture some or many of the environmental features and realism of a natural system (usually outside exposed to natural lighting). Systems are usually replicated and manipulated for controlled experiments. Example systems, cited in this chapter in relation to nutrient enrichment experiments, include the MERL (Marine Ecosystem Research Laboratory) systems (2.63 m^2 area, 5 m deep, with coupled pelagic and soft-bottom communities; cf. Nixon et al. 1984, 1986) and several shallow pond/lagoon/tank systems used for macrophyte or seagrass studies (e.g., Twilley et al. 1985, Short 1987; Short et al. 1995; Taylor et al. 1995a,b).

Figure 1. Conceptual model of N loading and effects discussed in this chapter.

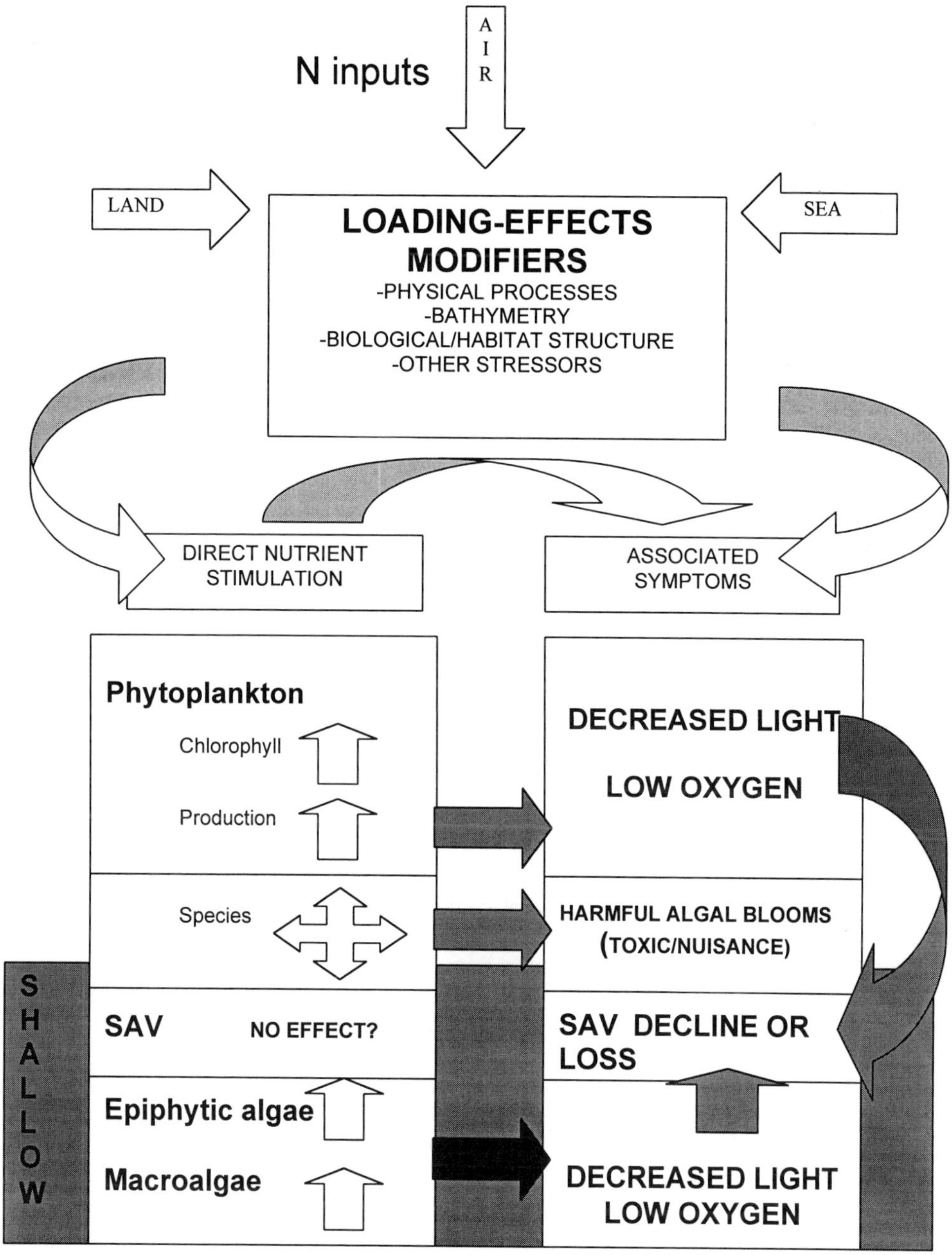

chemical, and biological characteristics. Monbet (1992) suggests that responses "vary from estuary to estuary, from segment to segment within a given estuary, and from time to time within any segment of an estuary." Perception of estuaries each as unique is echoed through the literature. The notion of "yes, but that doesn't hold for my system," is a common one and is bolstered by recognition that "the extreme variation in response to any level of loading clearly demonstrates the importance of other factors that determine differences between estuaries" (NRC 2000). Continued intensive field studies and site-specific modeling of select systems will undoubtedly reveal much more we should know in terms of N and ecological responses. There also has been a strong recognition that we cannot study each of the many thousands of systems intensively. We need to be able to group similar systems in terms of their vulnerability to N enrichment. The notion of ecological classification is thus in the vanguard of the attempt to aggregate across the diversity of systems and to develop more generally a quantitative relationship of coastal responses to enrichment (cf. Jay et al. 2000, NRC 2000).

3. NUTRIENT LOADING TO COASTAL SYSTEMS

3.1. Multiple sources, uncertainties, and high nutrient loading

To quantify effects and develop N loading–biological response relationships, one needs to start with nutrient inputs. Quantifying inputs to coastal systems has not been a small task, for several reasons. The possible sources of N are many. Obvious point sources were initially and easily tracked, but usable methods and models to assess non-point N surface flows, as well as atmospheric deposition, have taken considerable effort to develop and apply. Offshore exchanges and groundwater inputs are still not easily or routinely assessed. Moreover, all sources have changed markedly in a brief span of history. In addition, there are many forms to analyze and which contribute to "total" N (TN = ammonia, nitrate, nitrite, dissolved organics, particulate matter, organic, and inorganic). Now standard analytical methods were not all standard until after 1970 and raging debates have ensued as to whether only dissolved inorganic N (DIN) loading, or also organic N forms, are stimulatory nutrient sources. So, estimating TN (like total P for lakes) has not always been the goal of those assessing loading or responses; many examples cited here use DIN. In addition to these factors, many coastal systems are large, so spatial and temporal assessment of sources is not a small matter.

Even with incomplete N budgets and sources not as well characterized as for some other systems, we have known for some time that coastal systems receive high nutrient loading. Figure 2 shows estuarine systems for which land-derived inputs were summarized in the mid-1980s. Coastal systems often integrate flows and inputs from large watersheds, so from their position in the landscape, we could expect many of them to receive relatively high nutrient inputs compared to other systems. There is a significant range in loading for coastal systems, but it is not as wide as observed in lakes. Eutrophic/hypereutrophic lakes reach the same high levels of loading as many estuaries, but oligotrophic lakes are far less enriched (up to over two orders of magnitude lower). Less-enriched systems, such as some lakes and forests, tend to receive relatively high N loads (and thus have a higher N/P input ratio), because a majority of their input is from atmospheric sources (Kelly and Levin 1986). Many estuaries receive inputs from terrestrial sources at rates well above those applied to intensively fertilized agricultural fields.

In spite of the difficulties of source assessment, we now believe we have good input budgets for total N (and DIN) for a few coastal systems. The most complete loading estimates have a smattering of measurement, modeling, averaging across years, and some measure of best professional judgment (e.g. Nixon et al. 1995, 1996). Compared to values in the summary of Figure 2, total inputs are probably higher for most coastal systems, in part due to inclusion of several sources that were not well known or quantified in the mid-1980s. For example, atmospheric inputs are substantial to some

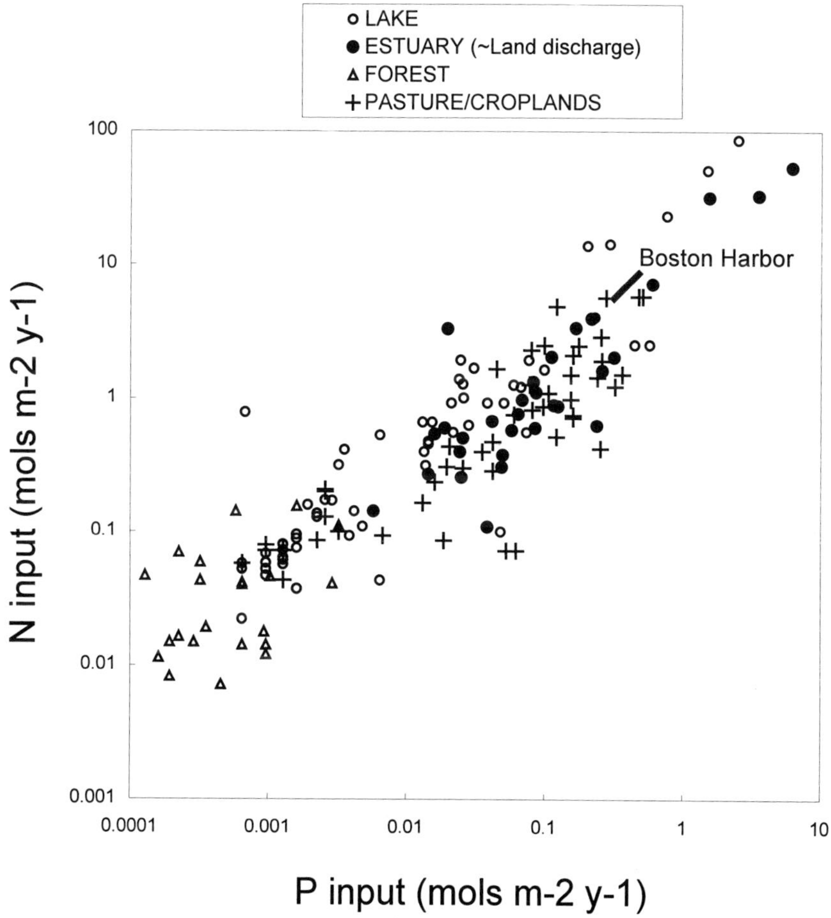

Figure 2. Nitrogen and phosphorus input to a variety of terrestrial and aquatic ecosystems. Modified from Kelly and Levin (1986). Estuarine systems show DIN or TN input from land and in some cases, atmosphere. Boston Harbor is from land-derived sources only (Kelly 1997), showing an example range for a system considering only DIN (lower end of bar) or TN (upper end of bar).

systems (principally larger, more open water ones), and also have been increasing (e.g. Paerl and Whitall 1999). Groundwater inputs have also been quantified and are significant in certain systems (e.g. Valiela et al. 1997a).

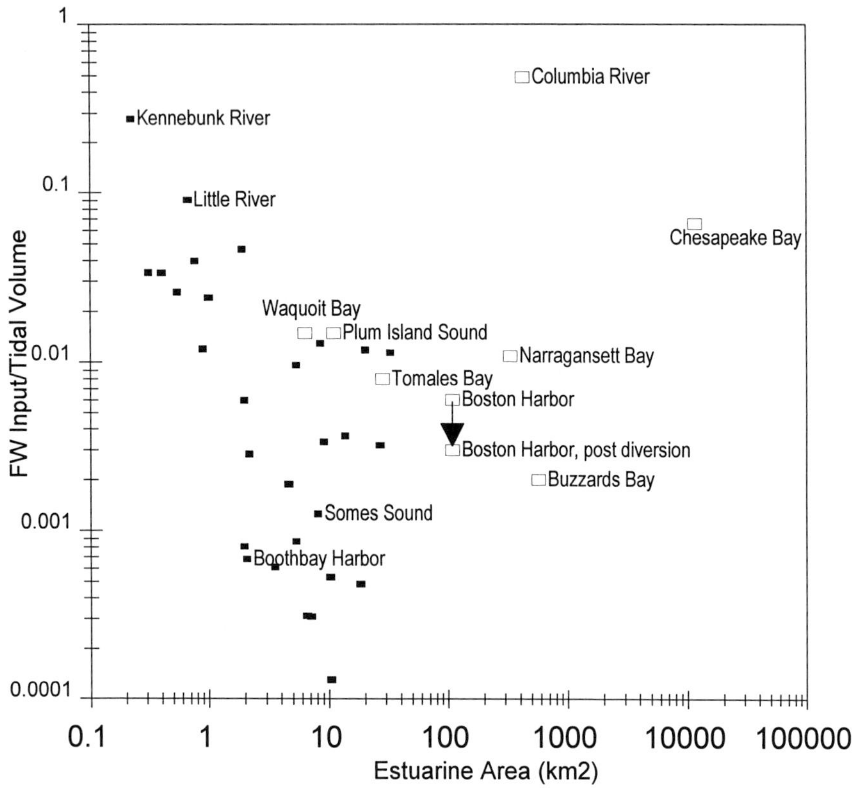

Figure 3a. Freshwater volume : tidal volume ratio in small and large coastal ecosystems. Data are from a summary of Maine (closed squares) and other larger northeast systems (Narragansett Bay, Boston Harbor, Buzzards Bay) by Kelly (1997b), along with intensive coastal LMER sites around the US described by Jay et al. 1997. Systems of different size range from river-dominated to ocean-dominated.

Most recently, in the course of developing complete coastal nutrient budgets, it has become broadly recognized that loading from the seaward, as well as the landward, edge can be very substantial (Garside et al. 1976, Nixon 1997, Kelly 1998). Boston Harbor, Narragansett Bay, and other northeastern US systems are an appropriate region to focus on ocean inputs because of large tidal ranges and, in comparison, relatively low freshwater inputs (Figure 3a). For example, Boston Harbor has a freshwater (Fw) to tidal volume ratio <0.01. At this ratio, the concentration of N in freshwater must be ~100 times that in the tidal floodwater to provide equivalent loading; even with Boston's large effluent discharge to the Harbor (now being diverted offshore) this turns out not to be the case. Inclusion of ocean loading to the budget based only on land and atmosphere sources raises

Boston Harbor's N input estimate by ~100 to 200% (DIN and TN, respectively; Kelly 1998). Many systems do not have the direct wastewater load of Boston and many have Fw/tidal ratios far less, indicating greater potential for ocean-domination of loading. Not all of the tidal volume input actually mixes with the water within an embayment, and this must be accounted for in assessing ocean loading. The ratio nonetheless is a first-order illustrator of relative source strengths. Figure 3a suggests why several systems have "River" as part of their name and that Chesapeake Bay is much more freshwater-driven than many northeastern sites.

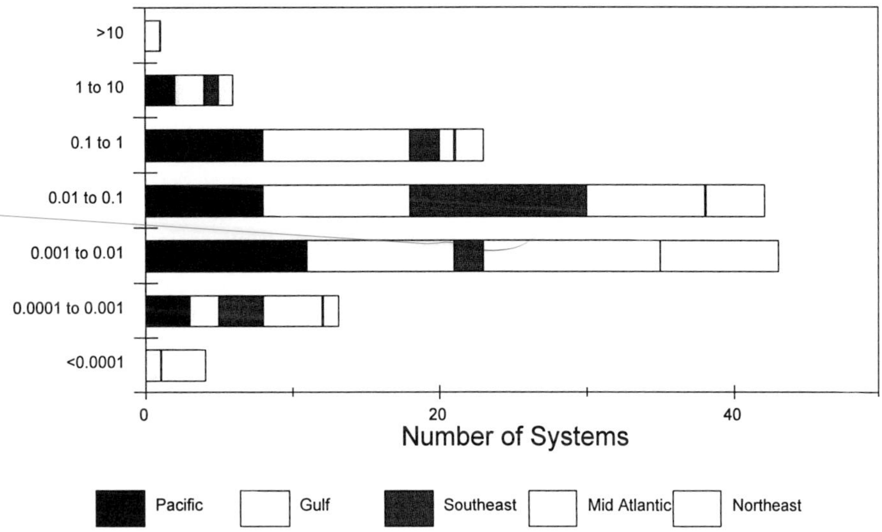

Figure 3b. Frequency diagram, by US geographic regions, of the ratio of freshwater volume to tidal volume input. From data compiled by NOAA estuarine susceptibility/eutrophication survey (see Bricker et al. 1999; P. Orlando, personal comm.), using tide gauges near mouths of estuaries. Pacific N=32, Gulf N=35, Southeast N=20, Mid Atlantic N=27, Northeast N=18.

The northeast and its macrotidal conditions appears to be skewed to Fw/tidal ratios below 1, whereas other US geographic regions have a distribution that includes ratios above 1, or even above 10 (Figure 3b). There is considerable overlap in the frequency distribution for each region; clearly the potential for both river-dominated and ocean-dominated flows exists in all regions. The significance of ocean loading as a nutrient source will vary with the offshore N concentration, which may show a general decrease with latitude. Increased atmospheric N deposition (e.g. Prospero et al. 1996, Paerl and Whithall 1999) directly to adjacent near-coastal waters could increase the role of the oceanside source of N to estuaries and embayments. These uncertainties reinforce the notion that we are still learning to quantify sources of nutrients to many coastal systems, and that source characterization is a big factor that has limited development of loading-effects relationships.

3.2. Trend of increasing N concentrations and loading to many coastal systems

Human populations along coastlines have been dramatically increasing with global population rise, as have associated anthropogenic pressures on coastal systems. Recent increases in loading to coastal systems are rather spectacular in some cases, but they are also just part of a general global increase in N circulation throughout the atmosphere, terrestrial, and aquatic ecosystems (e.g. Nixon 1995, Howarth et al. 1996, Prospero et al. 1996, Vitousek et al. 1997, other chapters in this volume)

For a number of coastal systems, it has been possible to measure or reconstruct trends of increase in N loading or *in situ* water column N concentrations over several decades or even from pre-European settlement in the US. A few examples, all with markedly increasing N trends, include: the open and coastal Baltic Sea since the 1950s and 1960s (Elmgren 1989, Cederwall and Elmgren 1990, Rosenberg et al. 1990), Narragansett Bay/Albemarle-Pamlico Sound since the 1800s (Nixon 1995), the Ythan estuary in Scotland from 1960s to 1990s (Balls et al. 1995), the Mississippi River plume from the 1960s to 1990s (Rabalais et al. 2000), and Chesapeake Bay from 1950 onward (Boynton 2000). Conley (2000) compared several of these published trends in the US and Europe; he summarized that N loads have increased by a factor of 1.5 to 4.5 over the 20th century and are presently as much as 60 times more than what might be judged as "pristine" condition loading. Certainly, we are at a point in the growth of human society where the development, use, and input to coastal systems is higher than at previous points in history. Thus comes the pressing concern to develop understanding of related, undesirable effects.

3.3. Modifiers of N loading that have consequence for expression of effects

Like other aquatic systems, coastal systems experience multiple stressors. When we look for nutrient-related effects there can be confounding problems from suspended solids, toxic contaminants, and habitat loss. But even in cases where we think we know all the N inputs and other stressors, we don't necessarily know much. There are a number of features within coastal systems that modify how and when nutrients reach biological receptors and in essence create the N "exposure." Principal among these modifiers is flushing and the residence time of water within the system; this feature is an emphasis of this review. Other physical features, like stratification, are also significant. Additionally, work of Seitzinger (2000) and Nixon et al. (1996) show that sediment microbial denitrification converts DIN to N_2 gas, removing >25% of N loading to longer residence time systems. Also, larger biological organisms modify the distribution of N forms, spatially or seasonally, or graze upon plankton and affect the way primary producers respond to nutrients. As will be discussed, many features complicate relationships between N loading and effects, in part by affecting the concentration experienced at a given loading rate in different systems.

4. LOADING-RESPONSE RELATIONSHIPS

4.1. Chlorophyll response to N loading and concentrations

Nutrient inputs (especially N) generally stimulate plankton biomass in coastal systems, and this is a first response in the sequence of related effects (Figure 1). The response is regularly measured in terms of chlorophyll *a*. Considerable evidence for stimulation exists at all levels of ecological organization and complexity (cf. Hecky and Kilham 1988). Studies note enhancement by N additions in: axenic cultures, community (bottle) assays whole-system enclosure/mesocosm experiments, and natural systems. The latter is inferred from historical trends, taking advantage of natural system "experiments" (e.g. sewage increases or diversions), and comparative trend analyses for many coastal systems (e.g. Boynton et al. 1982, Nixon 1983, Howarth 1988; Nixon et al. 1986, Nixon 1992).

In examining empirical patterns to develop quantitative relationships across whole systems (experimental or natural), an interesting challenge is the characterization of chlorophyll concentration, which has very high space and time variability. Should we be looking at peak (individual sample) concentrations, mid-summer ranges, or annual depth-integrated/spatially-averaged means? The most successful efforts to relate N and chlorophyll have been constructed using annual means and spatial averaging across a range of sites when possible. Where data are too infrequent or poorly spaced (in time or across the estuary) the estimate of a systems value may be miscast and create variability for pattern analyses. When we look across time or across systems we must constantly ask, are data summaries comparable and reliable, and how wide are the bounds of the estimate?

Increasing chlorophyll concentrations over years to decadal or greater time scales have been observed at very many sites around the world in the last half-century. In many cases a rise in benthic microalgae or phytoplankton chlorophyll has been correlated with N concentration increases. To generally summarize from many studies: when viewed across sites along enrichment gradients within or between ecosystems, a basic pattern between planktonic chlorophyll and water column DIN concentrations often emerges. Over a range of annual average DIN from <1 to >20 μM, chlorophyll tends to rise slightly less than 1 μg/L of chlorophyll with every 1 μM increase in DIN, about 0.7 to 0.8 μg Chl/μM DIN as a very rough rule. But there is variability in different areas and a greater tendency for the chlorophyll rise to fall off this trend at increasingly higher DIN levels. *Within* a given coastal system there can often be a similar trend where enrichment gradients are observable; i.e. at the upper levels of enrichment in a given system (often near sewage treatment or other strong point source of nutrients) are well below the trend for the other stations. Observations such as this have been used to suggest light limitation at very high nutrient levels (e.g. Malone 1982, Monbet 1992).

Many of these generalizations can be seen in Figure 4, which also adds a dimension of classification to the trends. The parameter range is broad enough that it has to be viewed on a log scale. The increasing general trends and variability noted above are nonetheless apparent, but for two fairly distinct classes of systems—those which have very large tidal ranges ("macrotidal") and those which have smaller ("microtidal"). Microtidal systems appear to be more sensitive to N enrichment, judging by a higher chlorophyll level observed at any given N level. Tidal energy may produce effects upon the light received by plankton by increasing vertical mixing. Destratification, sediment resuspension, and flushing may all reduce the chlorophyll a response per unit N.

A number of studies, Monbet's included, have tried to relate chlorophyll with *N loading*, not just *in situ* N concentrations, with differing degrees of success. No doubt this is due to underlying variability in the nature of different systems (such as suggested by Figure 4), as well as uncertainties in both loading and response measurements. Efforts generally have confirmed a strong correlation to N loading (vs. P) and provide evidence of a general relationship, but not necessarily a satisfyingly predictive one. Some issues of comparability and reliability inherent with empirical trends observed from cross-system correlations have been overcome by whole-system mesocosm experiments. For example, Marine Ecosystem Research Laboratory (MERL) enrichment gradient experiments (e.g. Nixon et al. 1986, Keller 1988) show an unequivocal tie between N loading and chlorophyll standing stock. Data show both a general increase in mean annual chlorophyll and an increase in the overall range of spot measurement variability within increasing nutrients. Nixon et al. (1986) showed a strong relationship between annual DIN inputs, annual average *in situ* DIN concentrations, and chlorophyll over the following ranges for DIN (~5 to 300 μM) and chlorophyll (~3 to 75 μg L^{-1}). Marine and coastal systems for which there are comparable loading and chlorophyll data somewhat the general MERL trend, with exceptions noted by Nixon (1992) and observable in Figure 5. The data overall (log-log scale) suggest a hyperbolic relationship familiar from bottle assays. Chlorophyll, although it continues to increase with additional nutrients, does not keep pace 1:1 with increasing nutrient loads or concentrations.

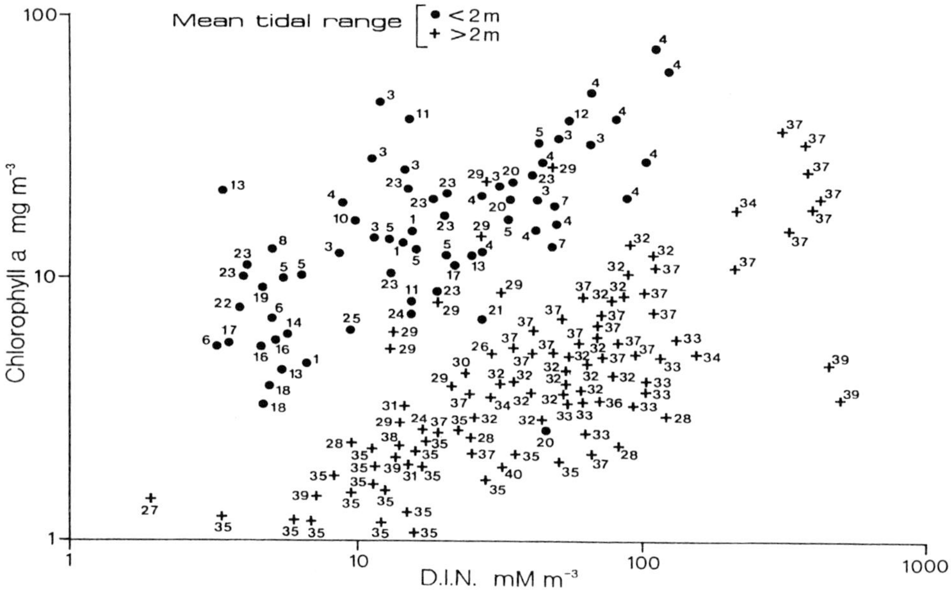

Figure 4. Mean annual concentrations of DIN and chlorophyll at multiple sites within different estuaries. From Monbet (1992). Units are equal to μM (DIN) and μg L^{-1} (chlorophyll). See original reference for coastal systems (represented by a series of similarly numbered stations) and data sources.

Nixon and co-workers' efforts to conduct experiments and compare results with natural systems have provided strong quantitative evidence of the relationship between N and chlorophyll. Scatter in the available data, however, suggest a single empirical relationship may not apply as a strong site predictive model unless we improve in normalizing for other critical variables that will also influence the response. For example, all else being equal, shorter water residence time systems (including those more energetically flushed by tides) will tend to have lower DIN concentrations for a given N loading (Kelly 1997a,b), so that the problem of chlorophyll "sensitivity" to loading (*vs. concentration*) in different systems is further complicated. Boynton and Kemp (2000) have tried a "primitive" scaling of nutrient loading (correcting areal input for hydraulic fill time and depth, as has been successful for lakes (Vollenweider 1976)). Interestingly, their significant linear regression using mean chlorophyll and "scaled nutrient loading" for various Chesapeake Bay sites (and a few others) begins to suggest that physically different systems can be better aligned along similar response trends if properly normalized.

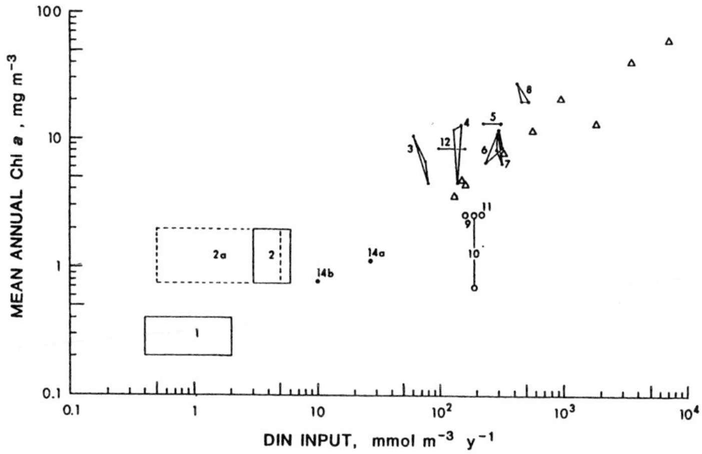

Figure 5. N loading and chlorophyll response in a mesocosm study compared to a range of coastal and marine systems. From Nixon 1992. Triangles are data from a MERL mesocosm enrichment experiment. Numbered points or polygons are natural field systems. The rectangles (1,2) are open sea systems (Sargasso Sea, N. Central Pacific). Systems 9, 10, 11 are continental shelf and upwelling areas. The remainder includes estuaries or bays (e.g. Kaneohe Bay (HA), 14a/b [before/after sewage removal]; sub-estuaries of Cheseapeake Bay, 3, 4, 6, 7, 8). See original reference for systems and data sources.

Finally, it has to be noted that the range of chlorophyll concentrations (e.g. Figures 4 and 5) is largely what would be considered mesotrophic (mean 4.7 μg L^{-1}, range 3-11) to eutrophic (mean 14.3 μg L^{-1}, range 3-78), with some hypereutrophic (range 100-150 μg L^{-1}). This judgement applies the standard lake classification (see Vollenweider 1976, Wetzel 1983, NRC 1993). Few coastal marine examples would be oligotrophic (by the lake standard: mean 1.7 μg/L, range 0.3-4.5); Kaneohe Bay (tropical, microtidal) or offshore areas of larger, open-water temperate macrotidal systems (e.g. Buzzards Bay (MA), Bay of Brest (France) might qualify. More fully marine systems, like mid-ocean gyres (Sargasso and North Pacific) as well as some continental shelves some distance from land also have annual mean chlorophyll < 2 μg L^{-1} and also qualify as oligotrophic. The relative lack of low-N/low-chlorophyll coastal marine systems, at least among those that have been more extensively studied, has implications for trying to observe effects from N loading, and this is next put into perspective through examination of productivity-nutrient loading trends.

4.2. Productivity response to N loading

It is important to derive a relationship between primary production and N loading, because *in situ* productivity in large part sets the system's organic supply and establishes a potential for metabolic effects (i.e. low DO, Figure 1); organic supply has been cast as a prime basis for establishing "eutrophication" classes (Nixon 1995). The ability to quantify the productivity–loading relationship suffers from the same difficulties faced in relating chlorophyll and N. Year to year variability in production in coastal systems can be considerable even with fairly constant loading rates, a situation that limits site-specific predictability. Variability occurs because many factors besides nutrient loading can moderate production processes and response to enrichment (e.g. cloudiness, climate, water stratification and circulation; *in situ* physico-chemical properties; as well as grazing rates and biological structure). Nonetheless, a relationship should be at least broadly evident because phytoplankton production correlates well with chlorophyll biomass. For example, Keller (1988a) provides an empirical regression between annual productivity (P_y, g C m^{-2} y^{-1}, using the ^{14}C technique) and mean annual chlorophyll biomass (B, mg Chl *a* m^{-3}) for data from about 9 natural systems and a MERL experiment:

$$P_y = 95.4\ (\pm 20.2)\ +13.0\ (\pm 1.0)\ B\text{, with n= 20, r}^2\text{=0.91, standard errors in parentheses.} \qquad (1)$$

Using MERL studies and extensive data for Narragansett Bay from 1978 to 1983, P_y was correlated to a composite parameter (Keller 1998b, following Cole and Cloern 1987) that recognizes not only the influence of B, but incorporates the influence of the depth of the photic zone (Z_p) and incident light (I_o). The resulting relationship was:

$$P_y = 25\ (\pm 10)\ +0.3\ (\pm 0.02)\ B\ Z_p I_o\text{ , with n= 32, r}^2\text{=0.92, standard errors in parentheses.} \qquad (2)$$

For the last century, we have known there is a connection between nutrient inputs and plankton productivity for marine systems (cf. Johnstone 1908 or several historical considerations of productivity [Nixon et al. 1986, 1992]). Even so, the relationship between N input and productivity has only comparatively recently been quantified, and this can best be seen in a succession of progressive efforts reported by Nixon (1983, et al. 1986, 1992, et al. 1996, 1997). When restricted to those relatively few field systems—mostly for open shelf and open or semi-enclosed seas (i.e. the Baltic)—for which there is high confidence in estimates of *total* DIN inputs (including ocean loading) and ^{14}C-based production, and combined with experimental MERL mesocosm data, Scott Nixon's analyses show a trend that would suggest a strong predictive ability (Figure 6).

Nixon's resolution of the producers' response, because it relies on some of the most complete input budgets and productivity data, confirms a strong coupling, shows a strikingly tight trend, and has been already much cited. It draws from relatively few systems, mostly more open coastal or marine, and the upper end is primarily driven by MERL results. Not all estuarine and coastal systems will strictly abide by it, a point illustrated in Figure 6, where I have added a few other systems that I believe have comparable and suitable data. The first "anomalous" example included on the plot is for outer Boston Harbor, a highly enriched shallow coastal embayment—one of the few such systems where all inputs, including offshore exchange have been estimated. Production in

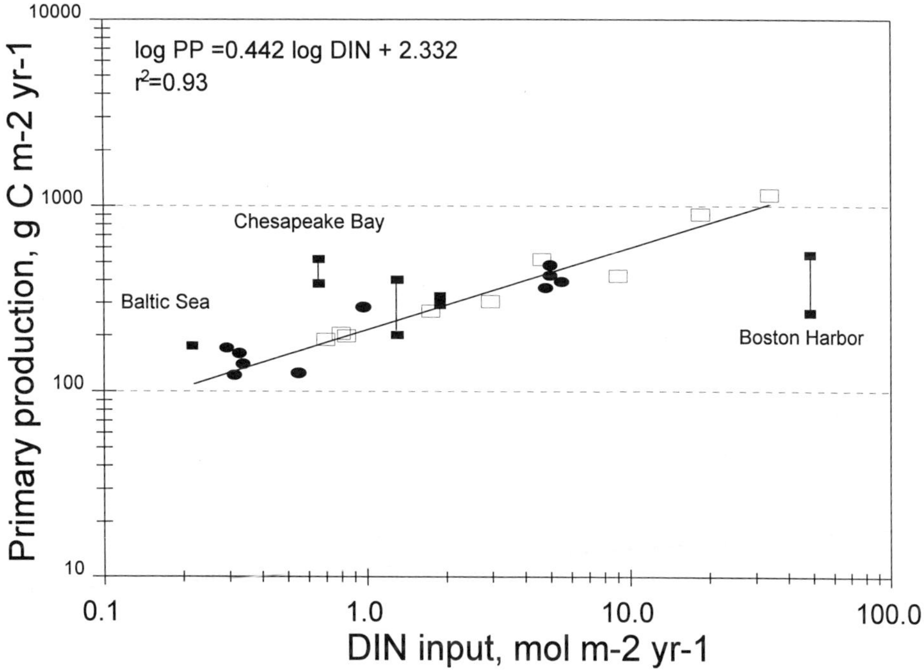

Figure 6. DIN inputs and phytoplankton primary productivity. Modified from Nixon 1997. Open squares are the first year of a MERL experiment (Nixon et al. 1986, Nixon 1992). Closed circles are various open marine systems, coastal continental shelf, or estuary systems. The regression shown from Nixon 1997 does not include data (all closed squares) which I have added to the original plot: outer Boston Harbor, Chesapeake Bay, Delaware Bay, Potomac River, and Baltic Sea. These additional areas have suitable data, including DIN loading estimates with offshore exchanges considered (Boynton et al. 1995, Nixon et al. 1996, Kelly 1998, Nixon 1992). Chesapeake uses 0.7TN to estimate DIN, from Boynton et al. 1995 (following Nixon et al. 1996). Data for Boston Harbor are for the northern outer harbor where production measurements were made (Kelly 1998). Note that with plots like this we are trying to derive the response to N as the primary stimulant; there are other nutrients and inputs occurring in both experimental and field situations.

Boston Harbor appears distinctly low compared to the prediction. Kelly and Doering (1997) suggested this might be due to light limitation or a short water residence time. If the plankton doubling times are not always shorter than the water residence time, then the plankton population will be regulated by “washout” of cells to the offshore. The physics in such a case does not allow higher production because the population level to support it simply cannot accumulate. Boston Harbor stations have low chlorophyll for their nutrient levels, which is consistent with a lower than expected cell buildup and *in situ* production rate, but data otherwise seemed to follow the basic rules of chlorophyll-production relationships that apply to most other coastal waters (Cole and Cloern

1987; Keller 1988a,b; Kelly and Doering 1997). We should indeed look at other factors to explain a situation like Boston Harbor, rather than think the anomaly contradicts the general prediction of increasing production with increasing input. On the other hand, we also know that there are upper limits on production (e.g. Bannister 1974) and the increasing trend will not go without bounds, as self-shading by bloom conditions will become a factor (Wetzel 1983). The MERL mesocosms operate with strong vertical mixing and a favorable light environment for plankton, so the upper treatments may be more productive than can be achieved in many field settings. However, treatments showing >900 g C m-2 y-1 (Figure 6) are higher than those used in the model to estimate them and seem inconsistently high compared with another measure of metabolism (Nixon 1992), so there is reason to view them with caution. There are some natural systems with apparent loading rates that exceed the upper end of the MERL experiment (Jaworski 1981), Monbet 1992), but I am unaware of production estimates for them, excepting the Boston Harbor example.

A second "anomaly" is Chesapeake Bay. It has been noted (e.g. Boynton et al. 1982, Nixon 1992) that this bay's chlorophyll and productivity ranges are relatively high for the input of N, as shown on the plot. The same appears to be true for the Baltic Sea, although less pronounced.

The main trend and deviations of Figure 6 are usefully put in another perspective (Figure 7). Following an earlier paper (Kelly and Levin 1986), I have overlain production data for freshwater systems (using P loading) with coastal and marine systems (using N loading) by rectifying the axis to a Redfield ratio (N:P =16:1, by atoms). From the previous summary, I excluded a couple lakes where production data had included macrophytes or other producers in addition to plankton. For lakes, instead of actual N input, one is simply representing the P input times 16, which is the necessary N equivalent for the average marine or freshwater plankton tissue (e.g. Schindler 1974, Hecky and Kilham 1988). This approach is preferable to using actual N inputs to freshwaters, since lakes are usually responsive to P and can make up for N-deficiencies by N-fixation, which is often not measured as a loading term (see Howarth et al. 1988).

Figure 7 also shows predictions of pelagic primary production (*PP*) based on the empirical lake model of Vollenweider (1979—see Wetzel 1983). This is of the form:

$$PP\ (\mathrm{g\ C\ m^{-2}\ yr^{-1}}) = 6.985\ (X^{0.76})/(0.29 + 0.11X^{0.76}) \qquad (3)$$

where

$$X = [P_i]/(1+\sqrt{T_w}), \qquad (4)$$

based further on P_i = average P inflow concentration and T_w = average residence time. X is the "expected" or predicted P concentration for the water body (see Wetzel 1983).

Using this empirical model, aquatic production has been forecast several water residence times (1 week to 10 years) and a standard depth of 5 m, typical of many coastal systems. A projection for a longer residence time (20 y) and depth (50 m) appropriate for some lakes and the Baltic Sea, is also shown. The *PP* prediction is actually based on X which estimates in-lake concentration, not loading. From it, one can back-calculate to units of areal loading and plot results consistent with the standard expression that has been used in marine studies (Figure 6). The formulation, within the bounds of parameters chosen, encloses most of the lake data (from which it is generally derived); use of longer residence times would include virtually all of it.

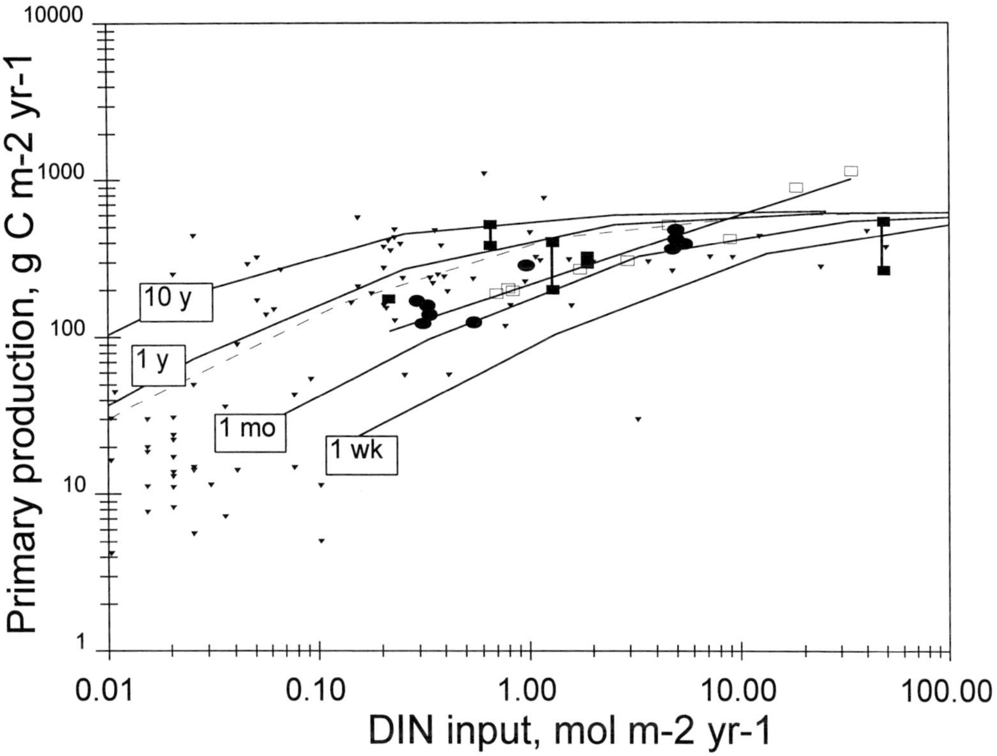

Figure 7. Production-loading response in aquatic ecosystems. Lakes (closed triangles) are overplotted on the estuarine/marine data of Figure 6, by converting P inputs to the equivalents needed by phytoplankton (Kelly and Levin 1986). Using a classic empirical model for lakes (see text), which recognizes the influence of water residence time and depth, the plot shows lines of predicted production for conditions, as a function of nutrient input on an areal basis. The family of solid curves shows different water residence times for a 5-m water depth. The unmarked dotted curve below the 1-y, 5-m projection represents a 20-y, 50-m condition, such as the Baltic Sea.

There are a number of messages to be gleaned from this exercise. One is the notion pointed out above with respect to Figures 2, 4, and 5—that coastal systems generally receive high nutrients and are not like oligotrophic lakes. The more rapid rise in production that is apparent from the least-loaded freshwater systems cannot be assessed for coastal systems because, on this scale, no such poorly loaded ones are studied. Even a prehistoric input estimate to Narragansett Bay, about 0.27-0.33 mols N m^{-2} y^{-1} (Nixon 1997), is an order of magnitude higher than equivalent loading to oligotrophic lakes.

Secondly, it has often been noted that the range of production in marine coastal systems is not very large, especially in comparison to loading (e.g. Nixon and Pilson 1983, Nixon et al. 1986, Oviatt et al. 1986, Nixon 1992). The regression of Figure 6 shows the relation is non-linear and there is only a factor of 4.4 increase for each order of magnitude increase in loading. From the perspective of Figure 7, this range of production and degree of stimulation is consistent with the fact that the coastal marine systems are biased to the upper half of the general aquatic nutrient saturation curve, a non-linearity that would be more evident if the plot were not on a log-log scale. Both lake and marine studies recognize a self-shading effect begins to limit phytoplankton at very high nutrient levels. An appreciation for the difficulty of sorting out a production increase "signal," amidst the "noise" of system differences (and potential signal modifiers) is gained, and the role of the MERL mesocosm experiment in defining an unambiguous response is recognized.

Thirdly, lakes and coastal systems may be described by the same simple rules—loading, depth, and residence time—once critical limiting nutrients are accounted for. Hecky and Kilham (1988) point out fundamental similarities in physiology between freshwater and marine algae, and perhaps it should not be surprising that the two conditions could fundamentally follow the same model. But to my knowledge, this has never been fully recognized and Figure 7 is the first suggestion this may be so. The freshwater model curve pretty well predicts Nixon's trend for conditions of a 1-month water residence time and a 5-m water column, which is basically the configuration of the MERL nutrient gradient experiment (Figures 6, 7). The model does not, however, duplicate the trend for the upper two MERL points, although it is a near-perfect hit for the very enriched outer Boston Harbor (2-d residence time, 5-m depth). The Baltic Sea, with a 20-y, 50-m condition is predicted. By the model, the Chesapeake Bay (~8-mo, 6-m depth) should indeed have higher production than the MERL trend and the match for the Chesapeake is close; it would improve if TN input (rather than DIN, ~0.7 TN) had been used.

There is undoubtedly room to review more coastal data and improve model formulations, but the principal lesson from Figure 7 involves water residence time. A linchpin of the lake model concept is that internal system physics modifies loading to produce different concentrations of nutrients maintained within the receiving water, to which biology responds. Evidence confirms a relationship between residence-time corrected inputs and *in situ* concentrations for open coastal systems and the MERL experiment (e.g. Kelly 1997a). Estuarine scientists have been slow to incorporate the concept, developed long ago for lakes (e.g. Dillon 1975). Part of the problem is that residence time in estuaries is not dictated just by freshwater throughput as it is in lakes, and it has been difficult to come to grips with this. Tidal inflow and mixing are significant and in very many cases the "freshwater residence time" is much longer than the true estuarine water residence time. Some physical oceanographers (Geyer et al. 2000) provide a perspective: "One of the most important quantities relating physics to the ecology of estuaries is residence time. A widely cited example is the work of Vollenweider (1976), who demonstrated that in lakes it is not just the nutrient loading, but rather the product of nutrient loading and residence time that determines the impact of phytoplankton production. Unfortunately, estuarine physicists have been rather unenthusiastic about attempting to quantify residence time, due in part to how easily misinterpreted a single number would be in characterizing the complex exchange processes that influence an estuary."

While "more effort should be placed in developing more accurate and sophisticated approaches to estimating residence time" (Geyer et al. 2000), there are already simple box-modeling techniques to derive estuarine water residence times useful for exercises like Figure 7 (e.g. Officer 1980, Pilson

1985, Doering et al. 1990, Asselin and Spaulding 1993, Smith 1993, Kelly 1998, Hagy et al. 2000). Such studies, along with results in some cases of complex hydrodynamic mixing models, are the basis for residence-time corrections used by Nixon et al. (1996) in assessing estuary retention of nutrients, in characterizing the anomalies of Figure 6, and later in this chapter.

4.3. Dissolved oxygen (DO) response to N

The most obvious concern of an adverse ecological effect with N enrichment is development of low DO (hypoxia) or even anoxia (no oxygen) in the water column of coastal marine systems. The fundamental conceptual model for the effect of nitrogen on DO is simple. Nitrogen stimulates primary production (i.e. it causes "eutrophication"). At some point of stimulation, the associated respiration rate of accrued autotrophic biomass begins to exceed the capacity of the water body to replenish itself by re-aeration and equilibration with the atmosphere and DO concentrations can fall to hypoxic or anoxic levels. A water column concentration of DO > 0 but <2 mg L^{-1} is the common definition of hypoxia. Most often noted in stable bottom waters of vertically stratified systems (and thus affecting sessile benthic organisms), hypoxic/anoxic levels can also occur throughout the water column, even in vertically well-mixed conditions. It is, of course, true that DO concentrations often go to zero within mm's or cm's of the surface of soft-sedimentary deposits. Benthic infauna, which live in these sediments and which cannot easily move to avoid conditions, can tolerate low DO (even hypoxia) in the overlying water column. For example, Rosenberg (1980) suggested ~2.8 mg L^{-1} as a limit noted for coastal benthic communities, and later (Rosenberg et al. 1991) lowered this limit to an overlying water exposure of 1.4 mg L^{-1} for several days to weeks, using shallow shelf organisms tested within their natural sediment environment. Many US states have long used 5 or 6 mg L^{-1} as a standard, recognizing that the lower thresholds for biological effects are higher in sensitive species and sensitive life stages (e.g., NRC 2000), including species which live within the water column, where DO concentrations are measured. Bricker et al. (1999) recognized this in the National Oceanic and Atmospheric Administration (NOAA) survey, and thus characterized hypoxia as >0 and ≤2 mg L^{-1}, with >2 to 5 mg L^{-1} characterized as "biologically stressful," in an effort to note different levels of potential DO problems. These characterizations are offered as a point of reference; it is not the goal of this review to develop estuarine/marine DO criteria, which is an ongoing effort within the US EPA. I focus on the occurrence of hypoxia/anoxia (≤2 mg L^{-1}), as a very serious condition documented in coastal systems, and explore how it may generally relate to N loading.

Hypoxia and a "dead zone" in the northern Gulf of Mexico have received recent attention in the both scientific and public sectors (e.g. Rabalais et al. 1991, 2000; CENR 2000, NRC 2000). But a DO problem has been found in many coastal systems worldwide (e.g., GESAMP 1990, Nixon 1998) and major one-time or chronic low DO events have been detected since the 1970's. Examples include the NY Bight, Chesapeake Bay, Potomac River, Baltic Sea, Scheldt River estuary, western Long Island Sound, the Venice lagoon, northern Adriatic Sea, several Alabama estuaries, Pamilco River, Providence River, and Hudson River areas (Falkowski et al. 1980, Officer et al. 1984, Oviatt et al. 1984, Larson et al. 1985, Kullenberg 1986b, Turner et al. 1987, Justic et al. 1987, Parker and O'Reilly 1991, Stanley and Nixon 1992, Nixon et al. 1996, NRC 2000).

A recent Science news article (Malakoff 1998) suggested the Gulf of Mexico hypoxia was one of more than 50 coastal regions worldwide experiencing severe oxygen decline. A citation for these 50 systems was not given, but worldwide the number of coastal systems and the total area presently affected or vulnerable to low oxygen in the near future is certainly substantial. In the US alone,

NOAA's National Estuarine Eutrophication Assessment survey (Bricker et al. 1999) categorized 42 of 121 estuaries (~35%) with sufficient information as having "moderate or high" depression of DO concentrations. The NOAA survey relied on conditions described by regional experts with extensive first-hand knowledge of each estuary. A US EPA Environmental Monitoring and Assessment Program (EMAP) statistical study (Summers 2001) has just reported results of an unbiased random sample (n = 1,133 stations) of 1,516 Atlantic (south of Cape Cod) and Gulf Coast estuaries. The study included a total of 74,744 km^2 (42 large estuaries [>250 km^2], 1464 small estuaries [2-250 km^2] and tidal portions of 10 large tidal rivers). Stations were sampled between 1990 and 1997 in late summer, when DO problems tend to be most pronounced. The spatial distribution of stations with measured hypoxia was centered among northern Gulf of Mexico estuaries and Chesapeake Bay sub-estuaries, with a sprinkling in the Florida and NY/southern NE regions. The EMAP study estimated that 4% of the represented area had hypoxic conditions and another 16% had DO concentrations between 2 and 5 mg L^{-1}. Thus, an estimated ~3000 km^2 was hypoxic and a total of ~15,000 km^2 with DO within a threshold range for biological responses. In comparison, the Gulf of Mexico hypoxic zone may cover up to an additional ~20,000 km^2 of the Louisiana continental shelf adjacent to the Mississippi and Atchafalaya River deltas (Rabalais et al. 2000).

The inherent vulnerability of systems to low DO events must vary, independent of the N delivery, because factors such as climate, river flow, tides, physical oceanography, individual bathymetry, and geomorphology have influence through constraints on flushing, stratification, and temperature (as a regulator of metabolic processes). Such processes are usually mathematically formalized in sophisticated, coupled hydrodynamic-water quality models or even in simpler DO models (e.g. Officer et al. 1984). Models may not yet fully capture some finer-scale physical processes (Kelly and Doering 1999) nor include all significant biological structure, such as benthic grazers, which potentially affect DO via food web and metabolic influences (e.g., Cloern 1982, Doering et al., 1986; Doering et al. 1989, Simenstad 2000). In principle though, sophisticated mass-balance or process-type models can link nutrient loading to DO response. Importantly, model formulations explicitly recognize that DO levels will vary with factors other than nutrient delivery or organic matter supply, and they can be useful sensitivity tools for that reason. Models are available and parameterized for a handful of coastal systems, but there are scores of coastal ecosystems for which there exists no calibrated or validated predictive model. Recognizing this lack, it may still be possible to examine time and space trends for a variety of coastal systems to make broader statements about a relationship between N levels and DO depression.

Malakoff (1998) suggested there has been a tripling of reports of dead zones in the last 30 years. At a broad scale, such reports of hypoxia/anoxia in coastal waters map principally within the northern hemisphere (US Atlantic and Gulf coasts, western and northern Europe, areas of the Mediterranean, Japan) around industrialized regions with high human populations and downstream of their associated N exports (Malakoff 1998, Nixon 1998). However, it is always a quandary to determine if an increased incidence in part arises from looking more and the connection to a specific cause can be tenuous as we try to interpret, in essence, "epidemiological" data on the basis of an observed symptom. In the US, for NOAA's (Bricker et al., 1999) study, of the 44 (of 121) systems expressing what were termed highly "eutrophic" conditions, only about 22 were included for having a "high" or " moderately high" expression of low DO symptoms. Moreover, many systems with "eutrophic" symptoms may more relate to conditions other than increases in N loading (e.g. Bricker et al. 1999).

Fortunately, there is more than epidemiological evidence. It is not common to have the intensity of monitoring information to detect signals among the noise of natural variability, but there are cases of an increased scale or intensity of low DO documented within the last half century. Some cases show DO strongly correlated to nutrient deliveries (Officer et al, 1984, Parker and O'Reilly 1991, Justic et al. 1987, Boynton and Kemp 2000).

One study, Rabalais et al. (2000), has described the Mississippi plume dynamics in the northern Gulf of Mexico. By patching together and comparing several time series, it is shown that surplus oxygen concentrations in surface water (indicator of net production from river-originated nutrients) peaks about one month after the Mississippi River flow peaks. A resultant DO minimum in bottom-water follows the surface water peak by about another month, and is thus associated with decay of recently produced organic matter settling from surface production by diatoms. Usually coincident with most pronounced hypoxia is high water column stability and stratification produced by strong

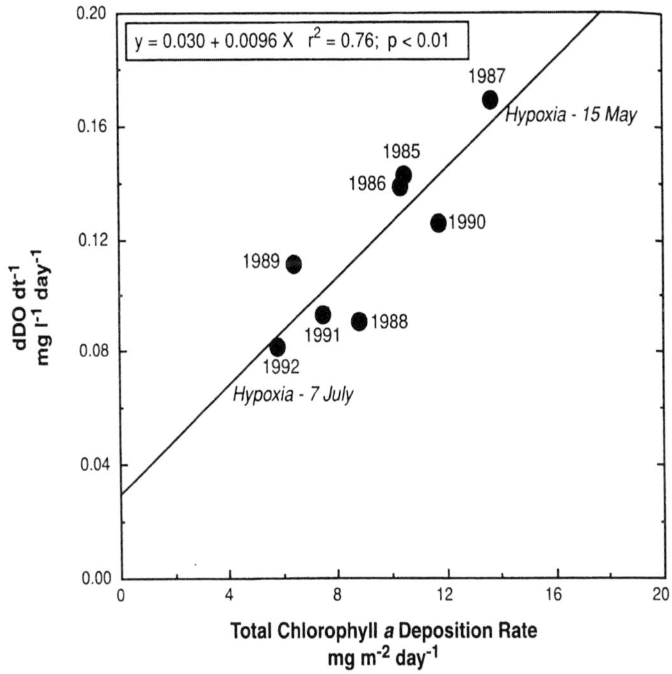

Figure 8a. Empirical relationship between spring organic matter input and summer seasonal DO decline rates in bottom water at a site in the mainstem of Chesapeake Bay. From Boynton and Kemp (2000). The date at which hypoxia (as DO <1 mg L-1) was first encountered in years with highest and lowest organic matter is indicated.

surface-to-bottom density differences. A lighter, freshwater surface plume overlying a denser, saline layer creates such stratification. Importantly, the rate of N loading and the level of diatomaceous remains (as Si) in underlying sediments appear to increase in lock-step through the century, thus indicating how an increase in diatom blooms and resultant hypoxia has arisen in the latter half of the 20th century.

Boynton and Kemp (2000) examined a lengthy time series (1985-1992) at a mesohaline site in Chesapeake Bay (Figure 8a). They were able to correlate seasonal DO decline in subpycnocline deep water to spring bloom deposition of organic matter. Chlorophyll, primary production, and organic deposition were all strongly correlated with river flow, which is a primary determinant of nutrient input to this region. Thus, in part by proxy, higher N input and lower DO are related by a series of expected connections that lead to a secondary consequence from initial plankton stimulation. In spite of strong correlations, Boynton and Kemp (2000) note that other factors may be involved, such as annual variability in temperature and stratification.

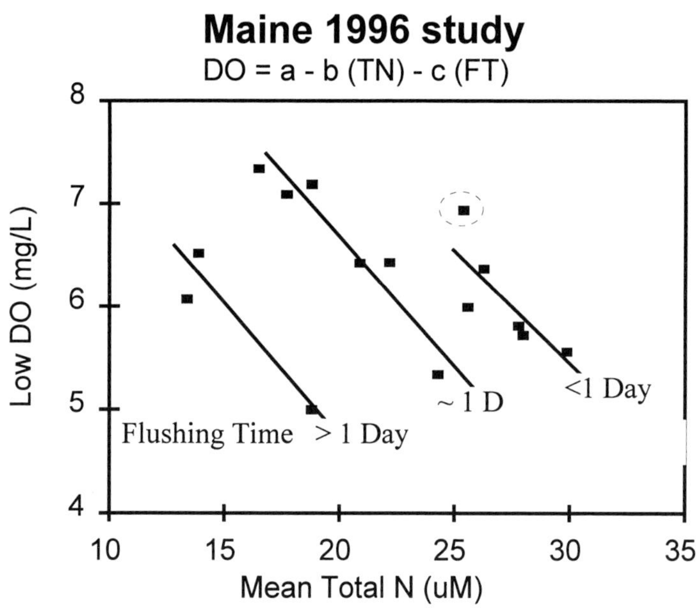

Figure 8b. Relationship observed among N concentrations, minimum DO concentrations, and flushing time for 15 small, short-residence time, tidally flushed estuaries and embayments. Adapted from Kelly 1997c, in which stepwise multiple regression analyses selected TN and flushing as first-order explanatory variables accounting for >60% of the variability. One point (circled) did not fit the trend. Flushing time is based on replacement of estuarine volume by tidal volume input every 12.42 h, and as such assumes complete mixing. Freshwater replacement time is much slower; these systems generally have low Fw/tidal volume ratios (see Figure 3a).

Some trends over space or across systems have been also noted. As the work of Boynton and Kemp (2000) and others (Jay et al. 2000, NRC 2000) tend to highlight, it requires some level of "scaling" or "classification" to apply cross-system analyses most appropriately in a search for more generalized rules. These efforts have not yet been generally extended to examination to N-induced DO effects.

Table 1: DO, nutrient loading, and other characteristics for selected coastal areas and a Marine Ecosystem Research Laboratory mesocosm enrichment experiment.

System	Area	Depth Avg (m)	Annual TN loading (mmol m^{-2})	Res. Time (mo)	[a]DO status	Vertical Mixing status	[b]Normalized TN Loading (μM)	Primary Production (g C m^{-2} y^{-1})
[c]Experimental	(m^2)							
MERL-control	2.63	5	800	0.9	OK	mixed	12	190 (100)
MERL-1X	2.63	5	1,750	0.9	OK	mixed	26	270 (115)
MERL-2X	2.63	5	2950	0.9	OK	mixed	44	305 (243)
MERL-4X	2.63	5	4,850	0.9	OK	mixed	72	515 (305)
MERL-8X	2.63	5	9,000	0.9	~H	mixed	133	420 (171)
MERL-16X	2.63	5	18,500	0.9	H	mixed	274	900 (601)
MERL-32X	2.63	5	34,000	0.9	A	mixed	503	1150 (901)
[d]Field	(km^2)							
[e]Baltic Sea	374,600	55	217	250	H/A	stratified	81	~149-170
Scheldt	277	11.2	13,400	3	H/A	??	295	?
[f,g]Chesapeake Bay	11,542	6	938	7.6	A	stratified	98	~380 to 520 (361-858)
[f]Potomac River	1,210	5.9	2,095	5	H/A	stratified	146	~290 to 325
[h]Guadalupe estuary	551	1.4	548	10	?	??	322	?
	551	1.4	2,058	1	?	??	121	?
Ochlocknee Bay	24	1	5,995	0.1	OK		49	?
Delaware Bay	1,989	9.7	1,900	4	OK	stratified	64	~200 to 400
[i]Narragansett Bay	328	8.3	1,960	0.9	OK	weak strat	17	270 to 290
[j]Providence River	24.13	3.7	13,600	0.083	H	stratified	25	?
[j,k]Providence Riv.	24.13	3.7	13,600	0.233	H	stratified	70	?
[l]Boston Harbor	103	5.5	21,600	0.266	~H	weak strat	86	?
[m]N. Outer Harbor	13	10	107,692	0.03	OK	mixed	27	263 to 546
[n]N. Gulf of Mexico	20,000	30	6500	[o]6	H/A	stratified	107	~290 to 320

Table 1: (Concluded).

Notes:

[a]H= hypoxia, A=anoxia.

[b]Volumetric TN loading is normalized for residence time to yield an "expected" or potential concentration. The value is calculated as: Annual TN Loading * Residence time (expressed in years) divided by Depth. Units are thus mmol/m3, or µM. See Kelly 1997a,b; 1998. The value is not decremented for denitrification or burial, removal processes that have greater effect on concentrations in longer residence time systems (cf. Nixon et al. 1996, Kelly 1998).

[c]See Nixon et al. 1984, Oviatt et al. 1986, Nixon 1992, Nixon et al. 1996. DIN was used to enrich treatment conditions (e.g. 1X...32X) and is represented in Figures 5, 6, and 7. TN values include input of organic forms with feedwater, which is only a substantial portion of input at the control and the low end of the enrichment gradient. Production for year 1 of experiment was extrapolated using empirical model of Keller 1988, which did not include measurements of primary production above 600 g C m^{-2} y^{-1} (Nixon 1992). These values are used in Figures 6 and 7. Parenthetical production values for year 2 are from Keller 1988. Hypoxic and anoxic events were periodic, not chronic.

[d]Except for Providence River, Boston Harbor and Gulf of Mexico, loading is TN as reported by Nixon et al. 1996. With noted exceptions for individual systems below, see Nixon (1992, 1997) for productivity references.

[e]Also see Elmgren 1989, Cederwall and Elmgren 1990, Rosenberg et al. 1990. Table value for TN loading from Nixon et al. 1996 is lower than DIN input in Nixon 1997 plot, which included N input across the halocline. Lower value is labeled in Figure 6.

[f]Also see Boynton et al. 1995, Boynton and Kemp 2000; historical Chesapeake production range (parenthetical) is from Boynton et al. 1982.

[g]Mainstem stratification, increasing anoxic extent; Officer et al. 1984, Boynton and Kemp 2000.

[h]Top line is for dry flow, bottom line is for wet flow.

[i]Only strongly stratified by freshwater at head of Bay in Providence River area, see notes j, k below. Production range is from Nixon 1997 (does not include historical pre-settlement estimate of 120-130 g C m^{-2} y^{-1}).

[j]Oviatt et al. 1984, Doering et al. 1990, Asselin and Spaulding 1993; TN loading from seaward and landward inputs, avg residence time (2.5 d), low DO in 13-15 m channel.

[k]Uses longer 7-d residence time during very low flow conditions, Asselin and Spauling 1993.

[l]TN budget includes direct estimate of ocean loading as well as land loading. Nixon et al. 1996 gave a preliminary budget; table shows improved budget of Kelly 1998. Freshwater stratification and near hypoxia/occasional hypoxia only occur in inner Harbor. See Signell and Butman 1992 for flushing estimate of whole harbor.

[m]Northern harbor section , Kelly 1998. Harbor station production of Kelly and Doering 1997.

[n]Area represents greatest measured extent of hypoxic zone. Higher production is for immediate plume (Rabalais et al. 2000). TN loading is to a 20,000-km^2 hypoxic zone only (and thus is a maximal rate) based on Mississippi/Atchafalaya input of 130 x 10^9 moles y^{-1} (Howarth et al. 1996; Turner and Rabalais 1991). Rate is consistent with long-term average (1980-1996) estimated by CENR 2000 of 1,567,900 metric tons y^{-1}.

[o]Assumed a 6-mo residence time (~seasonal turnover) *for illustration only*; if longer, then normalized concentration would increase accordingly.

There is compelling *experimental* evidence on the relationship between N loading and low DO. A MERL nutrient gradient experiment produced oxygen problems at its upper N loading levels, > 9,000 mmol TN m^{-2} y^{-1} (Oviatt et al. 1986). Low DO was also concomitant with primary production that reached at least 400 g C m^{-2} y^{-1} but the two most enriched MERL treatments, which had more severe, chronically-low DO, averaged production rates above 750 g C m^{-2} y^{-1} for the two years it was measured (Oviatt et al. 1986 and see Table 1). In contrast to many natural systems where DO problems are known, the MERL studies were conducted using (intermittently) well-mixed mesocosms with a 5-m deep water column and underlying active and functional benthic community (e.g. Nixon et al. 1986, Kelly et al. 1985).

I have brought several pieces together to explore patterns across experiments and field data (Table 1). Included are natural or experimental systems having the most complete N loading (total N) and budgets available in the literature, each with indirect or direct estimates of ocean N loading. Not all of these systems have DO problems, but many do. Summarized systems have great diversity—in latitude (Baltic to Gulf of Mexico), size (for natural systems, 10s to 100,000s of km^2), depth (1 to 55 m), estuarine residence time (days to years), and vertical stratification (well-mixed to strongly stratified). Areal TN loading rates have a wide range (217 to 107,692 mmol N m^{-2} y^{-1}). Table comparisons indicate DO problems in some low to medium loaded systems (Baltic, Chesapeake), but not necessarily all those with higher loading (e.g., Delaware Bay or Narragansett Bay), which illustrates some of the difficulty of defining directly an N loading–DO relationship.

In spite of all their differences, systems with DO problems may share a similar residence-time corrected loading, or "expected" concentration. The "expected" concentration of Table 1 is a simple correction of areal N loading for residence-time and depth, a parameter that correlates well with observed mean *in situ* N concentrations in some coastal systems (Kelly 1997a, b). This is a similar scaling concept analogous to that used in Figure 7 and explored by others (e.g. Valiela and Costa 1988; Kelly 1998; Boynton and Kemp 2000). A rough hypoxic threshold value, scanning the data of Table 1, might be on the order of 80 μM. There are several important issues of scale. First, Providence River has a DO problem compared to its parent system and also a higher "expected" (as well as measured) concentration (Table 1). In contrast, the outer Boston Harbor region itself does not itself have a DO problem and its value is lower than its whole parent system (Table 1). Both these sub-area observations support a threshold concept. Second, freshwater residence time in the Providence River strongly affects the expected value (Table 1) and it may be significant to occasional development of hypoxia/anoxia. This is a phenomenon similar to that described recently by Howarth and co-workers (in NRC 2000) for low flow conditions in the Hudson River estuary; in that case not only was residence time affected by low flow but other elements of the hypoxic recipe, stratification and primary production, both increased. Third, note that "illustration" value calculated for the northern Gulf of Mexico uses various assumptions that should be challenged. Input occurs to an area larger than the immediate hypoxic zone, so loading must also be lower; I do not know of an estimate for residence time in this open shelf situation and just assumed a seasonal turnover. Lastly, a suggested value near 80 for stratified natural systems is lower than indicated for the well-mixed conditions of the MERL experiment, where low DO was produced at values >130. If the very speculative concept were valid, it would operate mechanistically through an influence of residence time on production (as per Figure 7). It is clear that hypoxia occurs at lower primary production levels in stratified natural systems than it took in well-mixed MERL conditions. Production ranges for each system are large. If hypoxia occurs more towards the higher end of most systems' range (Table 1), we could tentatively place most hypoxia-associated conditions with production ≥300 g C m^{-2} y^{-1}. The Baltic Sea would be a distinct exception. Possibility, the very long residence time allows greater long-term accrual of organic material, so legacies of past production help promote low DO. This is unproven, as is a distinct threshold of production to produce hypoxia. The simple point is that DO problems obviously occur at different levels of (area-based) N loading, so a consideration of flushing and residence time, through effects on production, seems a logical aspect to explore.

Besides production, the strength and spatial details of stratification, among others, are factors influencing DO (e.g. Turner et al. 1987, Kelly and Doering 1999). A pattern like Figure 8b might arise in part through flushing effects on stratification. Also, temperature, turbidity or periods of cloudiness, or even shallowness itself may also be key factors. The growth of macroalgae and associated hypoxia in shallow water may occur at lower levels of N than it takes to produce hypoxia in deeper areas (see Section 4.4). In contrast, grazing by benthic filter feeders may moderate enrichment effects of chlorophyll or productivity (e.g. Cloern 1982). Ultimately, we have to recognize that N loading and productivity create only a *potential* for lowered DO; to develop quantitative relationships we need continued work to classify systems by attributes which make a DO problem be more or less likely.

4.4. Benthic primary producer response (SAV, macroalgae) to N in shallow systems

There are a number of excellent site summaries and reviews of submerged (often called submersed) aquatic vegetation (SAV) and macroalgae in coastal systems—freshwater, estuarine, and marine. SAV is a broad term that includes seagrasses (marine angiosperms) as well as freshwater macrophytes which are found in fresher regions of estuaries (e.g. Dennison et al. 1993 Studies describe many facets of SAV: the ecological importance of rooted macrophytes and seagrasses in coastal water; temporal patterns of seagrass decline, including possible relationships to nutrient loading, water quality, or other historical factors; potential for recovery from anthropogenic nutrient/sediment loads; and still other studies describe the stimulation of nuisance blooms of macroalgae by nutrients in shallow coastal systems, including coral reefs. The reader is referred to a number of examples (Thayer et al. 1975, Zieman 1982, Sand-Jensen and Borum 1991, Stevenson et al. 1993, Dennison et al. 1993, Duarte 1995, Valiela et al. 1997, Lapointe 1997, Stevenson 1988, Fourqurean and Robblee 1999, NRC 2000).

SAV is ecologically significant. It is important to waterfowl, it affects water quality by buffering turbidity in estuaries, contributes very high primary productivity and feeds a significant food chain through (mostly) detrital pathways, and offers habitat for adult fish and shellfish as well as nursery for many juvenile or larval forms. As an example, Heck et al. (1995) suggests that eelgrass habitat can support macroinvertebrate production (prey items for fish) that is disproportionately large compared to unvegetated areas (intertidal and subtidal muds). SAV can dominate overall secondary productivity of shallow estuaries even when its areal coverage is as low as 10% , and its contribution to the consumer food web can be more significant than implied by its level of contribution to primary production. Simply put, concern for SAV decline or loss focuses on the loss of all the stated functions above, especially for the food web (fish and shellfish) supported by its presence.

The evidence for SAV response to nutrients goes beyond epidemiological and anecdotal site trends, and there are a number of detailed examples, worldwide, of coastal SAV decline in the last half century, in Europe, North America, and Australia (e.g. Orth and Moore 1983, Costa 1988, Valiela et al. 1992, Dennison et al. 1993, Fourqurean and Robblee 1999). Throughout history there have been other causes of seagrass declines, but many during the last half of the 20th century have been linked to nutrients, specifically N. A variety of controlled experiments, including in mesocosms, have confirmed a link to N and the qualitative sequence of events with increasing loading (cf. Kemp et al. 1983, Twilley et al. 1985, Short et al. 1995, Taylor et al. 1995a,b). Based on various site trends, comparative analyses, and experimental evidence, Duarte (1995) determined that there was “an adequate empirical basis to formulate qualitative predictions on the direction of change in submerged vegetation upon nutrient enrichment,” but there was a lesser basis to predict recovery with lessening of nutrient loading.

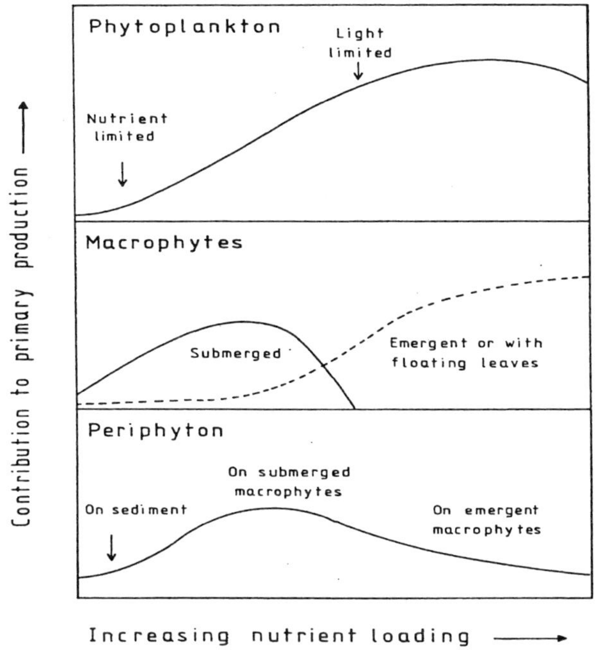

Figure 9a. General pattern of changes among primary producers with increased nutrient loading in shallow aquatic ecosystems. From Sand-Jensen and Borum (1991). Conceptual progression based on summary of data for temperate lakes.

It has been noted that changes in SAV are not gradual, but have thresholds and appear as step changes with a sudden shift in vegetation, implying both direct and indirect effects are at play. A principal mechanism for nutrient effects on SAV is uniformly recognized as a secondary consequence of enrichment of other primary producers. Hansson (1988) confirmed that under very low nutrient conditions in lakes, benthic algae can access nutrients from sediments and have a competitive advantage over planktonic algae, whose advantage grows with nutrients in the water column, due to their superiority of access to light. Similar concepts apply where principal benthic producers are rooted macrophytes or seagrasses (Figure 9a). Direct nutrient stimulation of plankton, periphyton on sediments, epiphytes on the vegetation, or other algal, emergent, or floating overgrowth induces light limitation of the seagrass or macrophyte, rooted to the bottom and thus subject to shading by unattached forms. Some studies also suggest that algal stimulation can affect root metabolism and indirectly affect SAV and there is some variability in the paradigm that may be induced by the effects of grazers on different producer forms. But the simple progression of Figure 9a, long described as a freshwater eutrophication paradigm (Wetzel 1983), appears applicable to estuarine areas (Stevenson 1988, Sand-Jensen and Borum 1991) and has been a principal conceptual foundation of studies examining SAV decline.

Figure 9a is fundamentally similar to trends established for a shallow southern Massachusetts estuary (Figure 9b), although the "emergent/floating" macrophyte forms are replaced by nuisance macroalgae (e.g. *Cladophora*, *Gracilaria* sp.) with high nutrient uptake rates. In Waquoit Bay, eelgrass was shown to decline from 1951 to 1987, from an extensive spatial coverage to restriction to a small patch near the mouth of the estuary. Duarte (1995) compared producer forms (seagrass, macroalgae, phytoplankton) in terms of various physiological properties in relation to the environment (light, nutrients) to suggest that macroalgal forms have physiological advantages over seagrasses in N-loaded systems, being more nutrient- and less light-limited. The Valiela et al. (1997b) trends in Waquoit Bay fundamentally followed the qualitative predictions of Duarte (1995). Sub-estuary data (Figure 9b) suggest that at only modest nitrogen loading enrichments, macroalgae replaced eelgrass, and increased watershed N inputs could be isotopically linked to all producers. Previously, Valiela et al. (1992) showed that enhanced macroalgal development facilitates development of anoxia, a mechanism to promote problems inherent to shallower systems that may not be captured in previous DO discussions (Table 1). Anoxia is among several ecological process/food web changes accompanying the shift to macroalgal dominance that are consequential to commercial shellfish populations (Valiela et al. 1992).

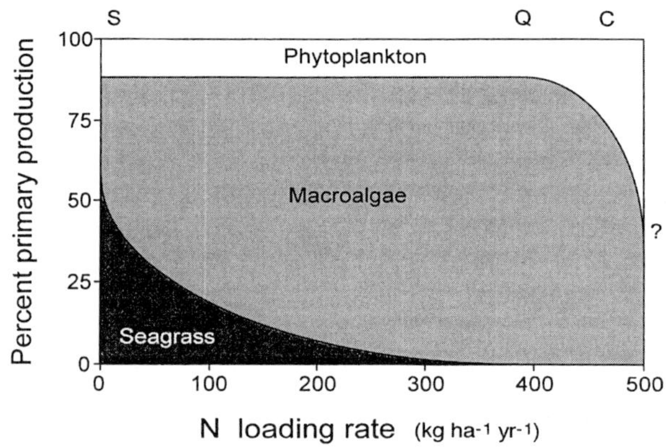

Figure 9b. Trends in primary producers with increased nutrient loading: an example from three sub-estuaries of Waquoit Bay, MA. From Valiela et al. (1997b). The position of three short-residence time sub-estuaries (S, Q, C) placed along a loading axis. Note: 1 kg ha^{-1} is about ~7 mmol N m^{-2}. Eelgrass (*Zostera marina*) decline is rapidly promoted; >50% relative reduction is >700 mmol N m^{-2}. A speculated threshold for full phytoplankton domination is ~ 3500 mmol N m^{-2}.

Consistent with a theoretical succession which shifts, essentially from nutrient to light limitation, various studies have noted that seagrasses colonize to a depth with a certain light level. Requirements are species-specific, but it tends to average about 11% of surface light (Duarte 1995). Negative relationships between nutrient concentrations and the depth limit of benthic macrophtyes have noted. Decline in seagrass beds sometimes has been observed from depth shoreward as phytoplankton and epiphytes reduce available light. Related to this effect, the bathymetry of vegetated area can affect the pace and spatial distribution of seagrass decline and resultant patchiness in different systems. This phenomenon contributes to a lack of a general quantitative relationship between SAV declines and N loading (Duarte 1995).

There is, however, some information to draw on. For example, the data of Figure 9b can be used to place some bounds on thresholds for change eelgrass and macroalgae, in response to N. A value of <700 mmol m^{-2} yr^{-1} (converted from Figure units of kg ha^{-1} y^{-1} to be consistent with other expressions in this chapter) is suggested for eelgrass decline and macroalgal dominance. Given a very short water residence time (~1-4 d) and shallow water depth (0.9 m) (Valiela et al. 1997b, Jay et al. 1997) this value would equate to a residence-time and depth normalized loading of about 4 μM (cf. data of Table 1). In a rough sense this expectation seems consistent with conditions in the main body of Waquoit Bay where there was a vestige of seagrass bed still remaining in 1987. This sub-area has higher salinity, low chlorophyll (~3-5 μg L^{-1}), little NO_3, and NH_4 concentrations averaging perhaps 2 μM or so in summer (Valiela et al. 1992). N concentrations increase upstream into fresher water and this is where extensive *Cladophora* mats are found, where water concentrations average perhaps 20 μM (Valiela et al. 1992), or roughly consistent with expectations for a 3500 mmol m^{-2} yr^{-1} threshold implied as the edge of phytoplankton dominance (Figure 9b).

Another extended example for assessing thresholds comes from Chesapeake Bay and its subareas. Brush and Hilgartner (2000) present a record of SAV in the upper Bay since the 1600s, from paleo-evidence of SAV seeds in sediments. SAV distributions have high variability in space and over time. Nonetheless a distinct threshold is suggestible in response to land use change (with sediment/nutrient loading increases) that began in the 1700-1800s and intensified in the last half of the 20th century. The number of tributaries with SAV has decreased markedly in the 1900s. In 1983, Orth and Moore (1983) detailed a major loss of SAV. Studies at that time and soon after, including controlled microcosm and field experiments, demonstrated a connection between SAV, turbidity, and N (Kemp et al. 1983).

Subsequent studies have used an understanding of light requirements to define conditions where water quality is sufficient to support SAV (Dennison et al. 1993). By surveying nutrients, chlorophyll, turbidity, and light extinction at different depths and areas, studies established where SAV was present, or transplants were able to survive. Using an experimental field study, Stevenson et al. (1993) transplanted plugs of living plants (*Ruppia maritima*, *Potamogeton perfoliatus* and *Potamogeton pectinatus*) to different areas in the Choptank River, and assessed survival. The following water quality thresholds for survival were indicated: ~15-20 mg L^{-1} total suspended solids, 15 μg L^{-1} chlorophyll, DIN below 10 μM, and PO_4 below 0.35 μM. Stevenson et al. (1993) emphasize that survival may occur at lower levels than that which would instigate declines. Concurrent work of Dennison et al. (1993) in higher salinity areas of the Bay (York River) indicated a similar range for patterns of eelgrass, *Zostera marina.*

Boynton (2000) provides a final Chesapeake Bay example, for the Patuxent River. He shows a precipitous seagrass decline concomitant with increased chlorophyll and decreased light penetration from 1960 to 1970 and to 1980. During this time, chlorophyll rose from <10 to almost 30 μg L^{-1}. TN loading increased from ~ 0.91 to 1.73 X 10^6 kg N yr^{-1} from 1963 to 1985-86 (Boynton et al. 1995). Using dimensions from Boynton et al. (1995) and the estuarine residence time recently estimated by Hagy et al. (2000), one can calculate the TN loading range as: ~515 to 980 mmol m^{-2} yr^{-1}. With a mean depth of about 4.8 m and median estuarine residence time of 25 d, calculations suggest an increase in residence–time corrected, volumetric loading from ~7 to 14 μM over the period (see above, and Table 1). These values coincide with DIN concentrations near 10 μM in

1969 (cf. Nixon and Pilson 1983, Boynton et al. 1982), increasing to an average DIN value of ~15 μM in 1985-86 (Boynton et al. 1995). Interestingly, these values all surround the survival conditions suggested by Dennison et al. (1993) and Stevenson et al. (1993).

There are other examples of trends, both response and the recovery, for both seagrass decline and macroalgal problems. For example, with lowering of chlorophyll levels to a target 8 $\mu g\ L^{-1}$, seagrass recovery appears to be proceeding, after a lag, in Tampa Bay (NRC 2000). More examples should be used to develop appropriate comparisons of threshold levels of loading for different systems, SAV species, and geographic regions, but the few examples here suffice to put the problem in some quantitative perspective in relation to other effects (Section 5).

Before leaving SAV effects, it is worth considering in concept possible differences across systems. Valiela et al. (1997b) discuss two factors. First, they hypothesize that presence of fringing salt marsh may intercept groundwater and surface flows and lead to denitrification along the flow path; thus variations in the area of tidal salt marshes in an estuary could affect its vulnerability by affecting the eventual loading to the estuarine receiving waters. Secondly, Waquoit Bay has a short water residence time (~1-2.5 d); phytoplankton have less ability to respond to nutrients at these very short residence times. This may exacerbate the ability of macroalgae to replace seagrasses in this estuary, compared to those with longer residence times, where phytoplankton may more easily dominate and shade both seagrass and macroalgae, at relatively lower input rates.

4.5. Phytoplankton species response to N, stimulation of "harmful algal blooms"

There are a variety of nuisance algal blooms (such as blue-green algae) which cause aesthetic and other problems, generally in only the oligohaline portions of estuaries (salinity of 0-5 PSU) (Paerl 1988). Of more concern to this review are saline forms, which characteristically include dinoflagellates. There are nearly two dozen noted genera of phytoplankton that produce potent toxins, including ones historically called "red tide" dinoflagellates (Anderson and Garrison 1997). There are species-specific toxins, which include those named for their symptomology in human consumers: paralytic, neurotoxic, amnesic, and diarrhetic shellfish poisoning (respectively, PSP, NSP, ASP, and DSP). There are endotoxins that accumulate through the food chain (and thus to commercially sought fish and shellfish species) and there are exotoxins that are exuded in the water. But not all "red tides" or dinoflagellates are harmful, not all toxic species are dinoflagellates (e.g. cyanobacteria of the genus *Trichodesmium*, diatoms of the genus *Pseudo-nitzchia*, prymnesiophytes of the genus *Phaeocystis* —various references in Anderson and Garrison 1997, such as Turner and Tester 1997), and not all problem species discolor the water at all (or are "brown tides"). Blooms also can disrupt normal filter feeding or grazing, change the food chain, foul beaches, or through rapid accumulation and decay cause acute DO problems. Thus, the term "harmful algal bloom(s),"or HAB(s), was coined to include species-level growth that is toxic, hypoxia-inducing, or food-web disrupting. Many HABs are elusive in the sense that they exist in some type of resting stage (such as a cyst), which can lay dormant in sediments, until it "excysts" and provides a seed population in favorable conditions. The triggers for this action are not well understood, but cysts are a mechanism for remaining in a location for a long time once advected or carried there (in ballast water?) and established. The various known mortality modes and impact mechanisms of HABs are summarized in Anderson and Garrison 1997 (cf. Smayda 1997). In broadest use, HABs includes both microalgae and macroalgae; the latter has been included in the previous Section (4.4.). Many microplankton HABs problems occur in slightly deeper coastal waters, so there is often a physical separation in potential SAV/macroalgal and HAB effects, whereas DO effects can occur in both shallow and deep systems.

ECOHAB (1995) and Anderson and Garrison (1997) offer excellent summaries of the problem. Concerns for HABs have heightened principally because the types of observed problems (numbers of newly-identified problems and problem species), the spatial extent or new locations of cases, and the incidence of reported occurrence all have expanded in the past few decades. This seems especially true in Western Europe and North America, where N loading increases are particularly notable. As an example, Paerl and Whitall (1999) examine the case for open coastal systems of the North Atlantic Ocean (Europe and North America), where new atmospheric inputs, in particular, have increased and form a substantial portion of the external N input. Concurrence of HAB events with high atmospheric N loading is part of the epidemiological evidence that has been compiled to suggest a linkage with increasing N inputs. Earlier, Smayda (1990) suggested a global epidemic of "novel" (~harmful) blooms and summarized evidence for increased spatial occurrence around the world. There was (as remains the case) provocative epidemiological evidence, but strong direct linkages to N loading have not been confirmed and, certainly, no cross-system comparisons can be developed to suggest that a certain critical N load is involved.

There are a couple sites with long-term data sets sites showing an increased occurrence frequency coincident with a temporal increase in nutrient loading (e.g. Tolo Harbour, Hong Kong; Smayda 1990). NRC (2000) cites another example from the inland Sea of Japan. Burkholder and Glasgow (1997) make an argument for a recently identified "phantom" dinoflagellate (with encysting form), *Pfiesteria*. They suggest that nutrients may foster outbreaks of these organisms, which can kill fish and also cause human health effects. Of course, there are areas of the world with increasing nutrient loading which do not have an increased occurrence of HAB species. In all, the challenge to epidemiological and time series evidence is that increasing reports could be due to increased attention and detection ability.

Smayda (1990) speculated that changes in observed N/Si/P ratios in some coastal areas (with increased N loading) over the last few decades may be promoting growth of forms that have low (or no) Si requirements (e.g., dinoflagellates, *Phaeocystis*) over more favorable bloom diatom species (expected from tenets of Officer and Ryther 1980, Ryther and Officer 1991). Rabalais and Turner (2000) show N/Si ratio changes in the Mississippi and changes in the mix of diatom species. They note that some harmful forms (e.g. *Pseudo-nitzchia* spp. and maybe others) are more recently observed, but a wholesale shift to HAB forms has not been observed. Interestingly, various MERL mesocosm enrichment studies have never noted a shift to HABs or extensive HAB species development even though plankton biomass and productivity climb (Oviatt et al. 1986, Doering et al. 1989). Moreover, there is a contrasting case. Keller and Rice (1989) noted that a brown tide organism (*Aureococcos anophagegefferens*) was present at a MERL experiment's start (from Narragansett Bay feedwater), in which nutrient levels and N/Si ratios were subsequently altered. After a brief response to initial enrichment, populations declined, appearing to be out-competed by diatoms; the organism did best in initial low nutrient conditions. Perhaps the simple message is that species-level response predictions are exceedingly difficult in complex ecosystems.

In all, we do know that relative increases of N might selectively favor some phytoplankton forms. It is possible that HABs could increase as part of a general increase in the phytoplankton community biomass and production associated with higher N loads, as is now somewhat described (Section 4.1., 4.2.). It is far more controversial, as there seems meager evidence, to suggest that HAB species are being selectively stimulated. The contrasting physical (and chemical, ecological) conditions favorable to diatoms, dino-and micro-flagellates as *groups* have been outlined for marine systems (e.g. Pingree et al. 1975, Margalef 1978, Demers et al. 1986, Legendre and Le Fevre 1989). However, we cannot easily predict when any particular phytoplankton species among the community will flourish. In sum, a major concern exists, and HABs have been expanding according to available records, but a quantitative linkage to N for individual harmful species has not yet been confirmed.

5. A CONCLUDING SUMMARY AND SPECULATION ON PROGRESSIONS WITH INCREASING ENRICHMENT

The evidence demonstrating a variety of effects of N on coastal systems is strong, and found at many levels of investigation. By most accounts, the scales of the problems have been growing rapidly throughout the 20th century. As summarized, we now have some general rules relating chlorophyll and production responses to N loading. By and large, the chlorophyll and production trends have strong similarity to those established for lakes. Chlorophyll and production increases are precursors to adverse secondary effects of concern (Figure 1), but even for these primary effects we do not yet have site-specific predictability. I would like to conclude with three brief summary topics related to this.

One recurrent theme in this review has been the significance of physics, specifically water residence time. Recognition of the importance of residence time has long been woven into coastal studies, but only as specks of color here and there; it needs to be a dominant hue in the fabric of eutrophication research. We know, for example, the water residence time can affect how coastal systems remove N loading via dentrification losses (e.g. Nixon et al. 1996) and how it can influence the expression of benthic grazers on overlying plankton (e.g. Simenstad et al. 2000). We recognize its influence on water quality/biological dynamics and role in determining vulnerability to enrichment effects. Residence time has been a cornerstone of the concept of lake eutrophication, where general predictive relationships with residence-time normalized loading have been developed for chlorophyll, secchi depth, primary production, hypolimnetic oxygen depletion, and fish yield (e.g. Jones and Lee 1986). This summary suggests that residence time plays a very similar role in coastal estuarine/marine production.

The second topic focuses on our understanding, both qualitative and quantitative, of the primary and secondary effects of N enrichment (e.g. Figure 1). In at least a handful of systems, enrichment progressions have been noted and linked to increasing nutrients, such as depicted in Figure 9 (a, b) for shallow systems. Observation or historical reconstruction of change shows subtle-to-dramatic algal increases, sharp food web shifts from benthic producers to planktonic producers and associated higher trophic-level organisms, and mortality from anoxia. From these, management strategies for specific systems have been formulated; there are examples where target reduction goals have been set which have helped with the problem (e.g. NRC 2000).

There are, however, different levels of confidence in our general ability to link the response to N loading for the different categories of effects reviewed. Considering confidence, quantification, and generality of findings, I believe it reasonable to rank our overall understanding of effects (in decreasing order) as:

Chlorophyll > Primary production > DO > SAV ~ Macroalgae >> HABs (phytoplankton)

Some will argue the exact order, the middle being the contentious ranking. The ranking is not to imply we have strong predictive capability for any effects. For even the best of the derived quantitative relationships, one always seems to be able to find new, outlier systems, as examples in this chapter illustrate.

The ranking, of course, suggests that the closer the effect is to the cause, the greater our ability to couple the two. With secondary effects, indirect mechanisms, and specific population responses, each highly dependent on many confounding factors, the requirements for details about the character, history, and structure of the system grow. Intensive studies and uniquely tailored

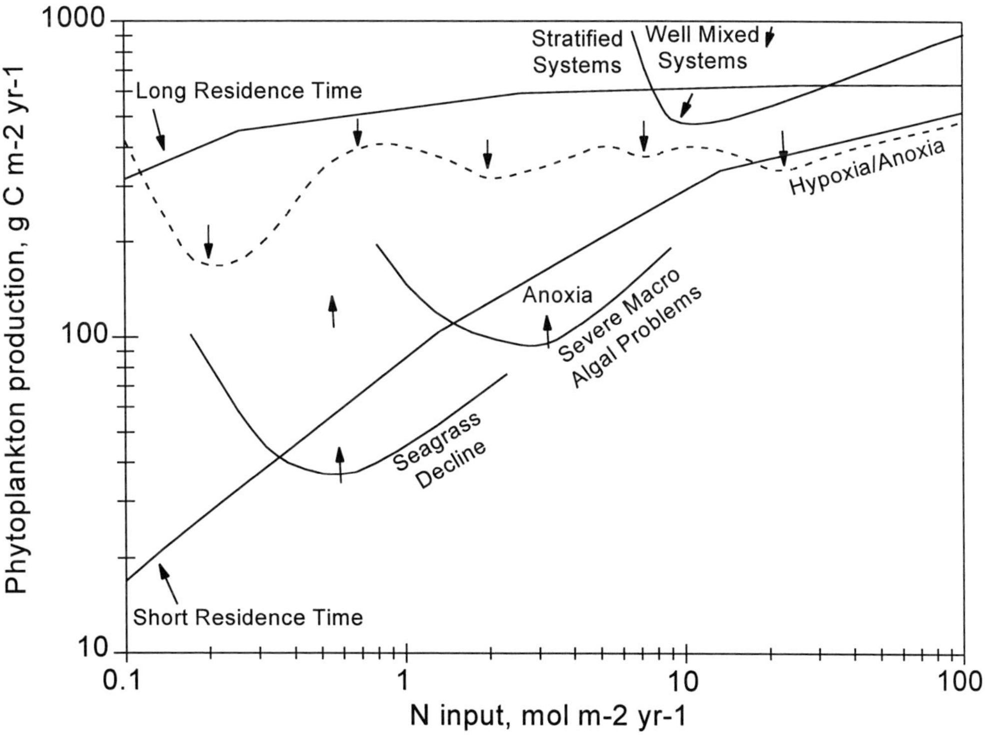

Figure 10. A speculative concept of the progression and thresholds for effects of N loading. Wavy lines or partial curves note possible thresholds. The sequence of arrows along a line suggesting hypoxia, represent (from left to right): Baltic Sea, Chesapeake Bay, Potomac River, Mississippi plume, and Providence River/Boston Harbor (Table 1). The partial curves for seagrass and macroalgae are based on description for Waquoit Bay and the Patuxent River (see test). The figure concept is borrowed from Figure 7, but ranges for axes have been narrowed to reflect the ranges observed in coastal systems.

simulation models should convince us that resolving such effects with some level of predictive confidence is not impossible, but takes considerable effort. The demands and precision of research must be balanced against a coarser level of guidance that can be effective for management action, where setting targets within a range can help be protective.

With this thought, I have compiled information to illustrate where we are with respect to development of quantitative thresholds (Figure 10). Increasing nutrients cause various algal changes affecting SAV, followed or accompanied by macroalgal changes (in shallow systems), with hypoxia/anoxia therefore one of the later effects. I have crudely attempted to map some boundaries for effects over the pattern and range of N loading, water residence time, and productivity observed for the bulk of coastal systems. These "thresholds" should, at best, be a speculative hypothesis to be tested and improved. Productivity data for systems with a DO problem are from Table 1. For SAV and macroalgae, the domains are suggested without productivity data, i.e. only from measured N input and residence times. It is clear that a given effect, such as hypoxia, occurs across a range of areal loading rates, probably being manifest more by productivity and affected by residence time. The array of north temperate systems in which DO problems have been noted (Table 1) includes many with productivity >300-400 g C m^{-2} y^{-1}. But it is also clear that only the potential for an effect is suggested and not all at these levels have hypoxia (thus a wavy dotted line, Figure 10). High loading and very high productivity were necessary to induce a chronic DO problem in the well-mixed MERL mesocosms.

Some have wondered whether the suggested growing global incidence of hypoxia might be our figurative "canary in the coal mine." This may only be true only in the sense of presently providing warning of the increasing scale of coastal problems. At least for shallow systems, evidence suggests that before situations actually advance to a low DO problem arising from loading and plankton productivity, effects such as SAV loss and problem macroalgae blooms, will indeed appear. Based on chlorophyll data for the systems used to suggest SAV loss or macroalgal bloom thresholds, one would expect associated plankton productivity to be very low, as is suggested (Figure 10). We don't know this to actually be the case in all systems and the figure only begins to illustrate how variations in physics may modify a sequence of effects at a given loading rate. Overall, the illustration indicates a great deal of complexity still to be resolved, probably through further classification of systems and their responses to enrichment. Importantly, though, the illustration reinforces the notion it would help to have greater study of systems in different physical settings, and especially, more at the lower end of the coastal loading range. Many coastal areas being observed at their high present day levels of loading probably have passed already through a succession of changes.

With progressive enrichment comes consequential species change, SAV being our best example. We have the least information on the general topic of species compositional change, and have little to guide us as to whether there is any threshold stimulation point for a specific biological change, such as HABs. This raises the third related topic. Food webs and fisheries are a fundamental societal concern, but they are ecologically removed from the direct effects of nutrient loading. The world is not lacking for evidence of fish kills, but it is fascinating that, with hypoxia and benthic mortality documented at a huge scale in the northern Gulf of Mexico, analyses have difficulty showing the effect on total fish catch even though loss decline in important species (e.g. brown shrimp) have been noted (CENR 2000). There are reasons that this could be so, including the difficulty of obtaining data on fisheries that reflect the actual conditions of the stock that assist detection. There may also be time lags for expression of effects in longer-lived species. Unlike infaunal benthos, fish and epifaunal organisms (adult shrimp) can move to avoid hypoxia, but with such a large benthic food base affected, the concerns are large for the long-term sustainability of the fishery and fundamental shifts in the nature of the fish consumers in the food web (Caddy 1993). In Caddy's view (Figure 11), there are consumer food-web changes across the loading regime (often to

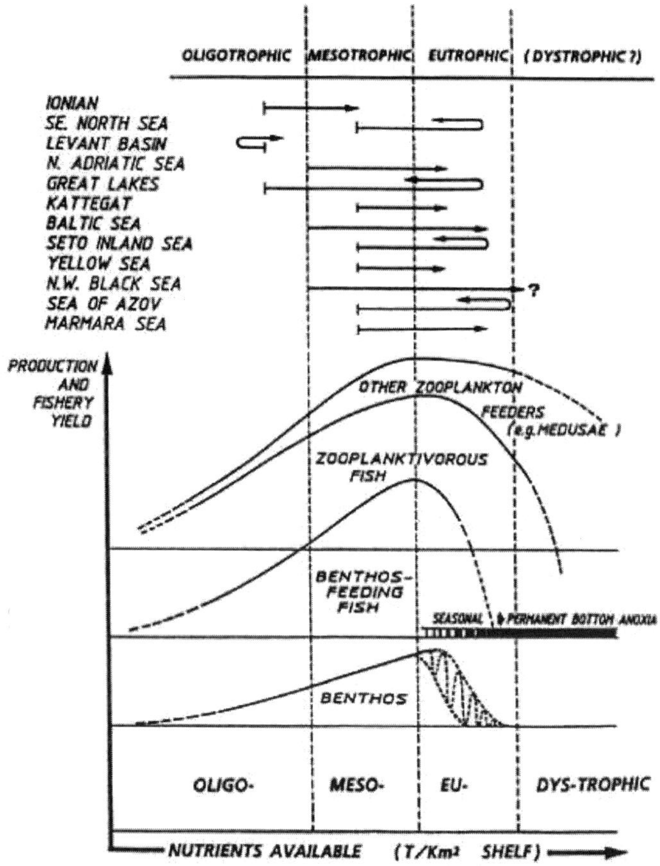

Figure 11. A speculative concept of fisheries change with nutrient enrichment. From Caddy (1993). A qualitative progression was suggested from a review of patterns in different enclosed and semi-enclosed seas. The recent trajectory of different systems with respect to trophic status is indicated at the top. The bottom suggests a progression of change in the structure of food webs and composition of fisheries prior to a dramatic loss of yield with permanent bottom anoxia.

less desirable, commercially sought species), many of which become more consequential at high loading. The progression to anoxia, even in deeper, unvegetated systems, begins a decoupling of the functional connection of pelagic and benthic food webs (cf. Pearson and Rosenberg 1978, Oviatt et al. 1986). Sediments become uninhabitable and only selected organisms thrive (less "choice" for some commercial fish species). Eventually, the benthos is lost totally, and further pelagic consumer food web changes follow, sometimes with lags typical of longer-lived species.

Finally, Caddy's image of eventual collapse can be compared with some other noted trends. There appears a fundamental relationship between primary production in the water column and fisheries yield of different marine areas, as well as lakes (Nixon 1998). Interestingly, with increasing production (such as is stimulated by higher nutrients), the efficiency of conversion to fish appears to increase, not decrease. It has been suggested that this could relate to the nutritional quality (higher N for protein) of the phytoplankton (Iverson 1990, Nixon 1992). The trend compiled by Nixon does not in any way suggest a fisheries collapse at high production, although his summary does not include eutrophic/hypereutrophic areas with sustained productivity above 500 g C m^{-2} y^{-1}. Judging from this level compared to Figure 10, perhaps protecting against anoxia will generally prevent wholesale fisheries collapse due to eutrophication, but it will not prevent shifts in fish and shellfish, nor the loss of some species that are most valued by humans. The next generation of N enrichment studies should maintain a sharp focus on these subtler food web effects of N enrichment.

6. ACKNOWLEDGMENTS

I thank John Morrice, Jim Latimer, Mary Ann Starus, and Al Batterman for technical and editorial reviews of this long chapter. I thank the volume editors, Ron Follett and Jerry Hatfield, for their patience. Thanks also to Scott Nixon, Candace Oviatt, and Walter Boynton, who responded to my request for recent papers and pre-prints; this review has drawn heavily from their body of work and insights on coastal eutrophication.

This document has been reviewed in accordance with U. S. Environmental Protection Agency policy and approved for publication. Approval does not signify that contents reflect the views of the Agency. Mention of trade names or commercial products does not constitute endorsement or recommendation for use.

REFERENCES

Anderson, D.M. and D.J.Garrison (eds.). 1997. The ecology and oceanography of harmful algal blooms. Limnol. Oceanogr 42(5, part 2):1009-1305.

Asselin, S. and M.L. Spaulding. 1993. Flushing times for the Providence River based on tracer experiments. Estuaries 16(4):830-839.

Balls, P.W., A. Macdonald, K. Pugh, and A.C. Edwards. 1995. Long-term nutrient enrichment of an estuarine system: Ythan, Scotland (1958-1993). Environ. Poll. 90(3):311-321.

Bannister,T.T. 1974. Production equation in terms of chlorophyll concentration, quantum yield, and upper limit to production. Limnol. Oceanogr. 19:1-12.

Boynton, W.R. 2000. Impact of nutrient inflows on Chesapeake Bay. In: A.N. Sharpley (ed.), Agriculture and Phosphorus Management. Lewis Publishers, Boca Raton, pp. 23-40.

Boynton, W.R., W.M. Kemp, and C.W. Keefe. 1982. A comparative analysis of nutrients and other factors influencing estuarine phytoplankton production. In: V.S. Kennedy (ed.), Estuarine Comparisons. Estuarine Research Federation Biennial Conference, Academic Press, New York, pp. 69-90.

Boynton, W.R., J.H. Garber, R. Summer, and W.M. Kemp. 1995. Inputs, transformations and transport of nitrogen and phosphorus in Chesapeake Bay and selected tributaries. Estuaries 18(1B):285-314.

Boynton, W.R. and W.M. Kemp. 2000. Influence of river flow and nutrient loads on selected ecosystem processes: a synthesis of Chesapeake Bay data. In: Hobbie, J. E. (ed.), Estuarine Science: A Synthetic Approach to Research and Practice. Island Press, Washington DC, pp. 269-298.

Bricker, S.B. and J.C. Stevenson. 1996. Nutrients in coastal waters: A chronology and synopsis of research. Estuaries 19(2B):337-341.

Bricker, S.B., C.G. Clement, D.E. Pirhalla, S.P. Orlando, and D.R.G. Farrow. 1999. National estuarine eutrophication assessment: Effects of nutrient enrichment in the nation's estuaries. NOAA, National Ocean Service, Special Projects Office and the National Centers for Coastal Ocean Service. Silver Spring, MD: 71 pp.

Brush, G.S. and W.B. Hilgartner. 2000. Paleoecology of submerged macrophytes in the upper Chesapeake Bay. Ecol. Monogr. 70(4):645-667.

Burkholder, J. M. and H.B. Glasgow, Jr. 1997. *Pfiesteria piscidida* and other *Pfiesteria*-like dinoflagellates: Behavior, impacts, and environmental controls. Limnol. Oceangr. 42(5, part 2):1052-1075.

Caddy, J.F. 1993. Toward a comparative evaluation of human impacts on fishery ecosystems of enclosed and semi-enclosed seas. Reviews in Fisheries Science 1(1):57-95.

Carpenter, E.J. and D.G. Capone. 1983. Nitrogen in the Marine Environment. Academic Press, New York. 900 pp.

Cederwall, H. and R. Elmgren. 1990. Biological effects of eutrophication in the Baltic Sea, particularly the coastal zone. Ambio 19(3):109-112.

CENR. 2000. Integrated assessment of hypoxia in the northern Gulf of Mexico. National Science and Technology Council Committe on Enviornment and Natural Resources, Washington, DC.

Chesapeake Bay Program. 1983. Chesapeake Bay: A Framework for Action. US EPA, Region 3 Philadelphia, PA, 186 pp + appendices.

Cloern, J.E. 1982. Does the benthos control phytoplankton biomass in South San Francisco Bay? Mar. Ecol. Prog. Ser. 9:191-202.

Cole, B.E. and J.E. Cloern. 1987. An empirical model for estimating phytoplankton productivity in estuaries. Mar. Ecol. Prog. Ser. 36:299-305.

Conley, D.J. 2000. Biogeochemical nutrient cycles and nutrient management strategies. Hydrobiologia 410:87-96.

Costa, J.J. 1988. Distribution, production, and historical changes in abundance of eelgrass (*Zostera marina*) in southeastern Massachusetts. Ph.D. Thesis. Boston University, 352 pp.

Dillon, P.J. 1975. The phosphorus budget of Cameron Lake, Ontario: the importance of flushing rate to the degree of eutrophy in lakes. Limnol. Oceanogr. 20:28-39.

Demers, S., Legendre, L., and J.C. Therriault. 1986. Phytoplankton response to vertical tidal mixing. In: J. Bowman, C.M. Yentsch, and W.T. Peterson (eds.), Tidal Mixing and Plankton Dynamics. Springer, Berlin, pp. 1-40.

Dennison, W.C., R.J. Orth, K.A. Moore, J.C. Stevenson, V. Carter, S. Kollar, P.W. Bergstrom, and R. Batiuk. 1993. Assessing water quality with submersed aquatic vegetation. Bioscience 43:86-94.

Doering, P.H., C.A. Oviatt, and J.R. Kelly. 1986. The effects of the filter-feeding clam *Mercenaria mercenaria* on carbon cycling in experimental marine mesocosms. J. Mar. Res. 44:839-861.

Doering, P.H., C.A. Oviatt, L.L. Beatty, V.F. Banzon, R. Rice, S.P. Kelly, B.K. Sullivan, and J.B. Frithsen. 1989. Structure and function in a model coastal ecosystem: silicon, the benthos, and eutrophication. Mar. Ecol. Prog. Ser. 52:287-299.

Doering, P.H., C.A. Oviatt, M.E.Q. Pilson. 1990. Control of nutrient concentrations in the Seekonk-Providence River region of Narragansett Bay, Rhode Island. Estuaries 13(4):418-430.

Doering, P.H., C.A. Oviatt, B.L. Nowicki, E.G. Klos, and L.W. Reed. 1995. Phosphorus and nitrogen limitation of primary production in a simulated estuarine gradient. Mar. Ecol. Prog. Ser. 124:271-287.

Duarte, C. 1995. Submerged aquatic vegetation in relation to different nutrient regimes. Ophelia 41:87-112.

ECOHAB. 1995. The ecology and oceanography of harmful algal blooms, a national research agenda. ECOHAB Workshop Rep. Woods Hole Oceanogr. Inst., Woods Hole MA, 66 pp.

Elliott, M. and J-P. Ducrotoy (eds.). 1991. Estuaries and coasts: Spatial and temporal intercomparisons. ECSA19 Symposium, University of Caen, FR. International Symposium Series, Olsen and Olsen Publishing, Fredensborg, Denmark, 390 pp.

Elmgren, R. 1989. Man's impact on the ecosystem of the Baltic Sea: energy flows today and at the turn of the century. Ambio 18(6):326-332.

Elser, J.J., E.R. Marzolf, and C.R. Goldman. 1990. Phosphorus and nitrogen limitation of phytoplankton growth and the freshwaters of North America: A review and critique of experimental enrichments. Can J. Fish. Aquat. Sci. 47:1468-1477.

Falkowski, P.G., T.S. Hopkins, J.J. Walsh. 1980. An analysis of factors affecting oxygen depletion in the New York Bight. J. Mar. Res. 38(3):479-506.

Fourqurean, J.W. and M.B. Robblee. 1999. Florida Bay: A history of recent ecological changes. Estuaries 22(2B):345-357.

Garside, C., G. Hull, and C.S. Yentsch. 1976. Coastal source waters and their role as a nitrogen source for primary production in an estuary in Maine. In: M.L. Wiley (ed.), Estuarine Interactions. Academic Press, New York, pp. 565-575.

GESAMP. 1990. The state of the marine environment. UNEP Regional Seas Reports and Studies No. 115 (IMO/FAO/UNESCO/WMO/WHO/IAEA/UN/UNEP, Joint Group of Experts on scientific aspects of marine pollution). UNEP 1990.

Geyer, W.R., J.T. Morris, F.G. Prahl, and D.A. Jay. 2000. Interaction between physical processes and ecosystem structure: a comparative approach. In: Hobbie, J. E. (ed). 2000. Estuarine Science: A Synthetic Approach to Research and Practice. Island Press, Washington DC, pp. 177-210.

Hagy, J.D., W.R. Boynton, and L.P. Sanford. 2000. Estimation of net physical transport and hydraulic residence time for a coastal plain estuary using box models. Estuaries 23(3):328-340.

Hansson, L.A. 1988. Effects of competitive interactions on the biomass development of planktonic and periphytic algae in lakes. Limnol. Oceanogr. 33(1):121-128.

Heck, K.L. Jr., K.W. Able, C.T. Roman, and M.P. Fahy. 1995. Composition, abundance, biomass, and production of macrofauna in a New England estuary: comparisons among eelgrass meadows and other nursery habitats. Estuaries 18(2):379-389.

Hecky, R.E. and P. Kilham. 1988. Nutrient limitation of phytoplankton in freshwater and marine environments: A review of recent evidence on the effects of nutrient enrichment. Limnol. Oceanogr. 33:796-822.

Hobbie, J. E. (ed). 2000. Estuarine Science: A Synthetic Approach to Research and Practice. Island Press, Washington DC.

Howarth, R. W. 1988. Nutrient limitation of net primary production in marine ecosystems. Ann. Rev. Ecol. Sys. 19:89-110.

Howarth, R.W., R. Marino, J. Lane, and J.J. Cole. 1988. Nitrogen fixation in freshwater, estuarine, and marine ecosystems: rates and importance. Limno. Oceanogr. 33:669-687.

Howarth, R.W., G. Billen, D. Swaney, A. Townsend, N. Jaworski, K. Lajtha, J.A. Downing, R. Elmgren, N. Caraco, T. Jordan, F. Berendse, J. Freney, V. Kudeyarov, P. Murdoch, and Zhu Zhao-Liang. 1996. Regional nitrogen budgets and riverine N and P fluxes for the drainages to the North Atlantic Ocean: Natural and human influences. Biogeochemsitry 35:75-139.

Iverson, R. 1990. Control of marine fish production. Limnol. Oceangr. 35:1593-1604.

Jaworski, N.A. 1981. Sources of nutrients and the scale of eutrophication problems in estuaries. In: B.J. Nielson and L.E. Cronin (eds.), Estuaries and Nutrients. Humana Press, Clifton, NJ, pp. 83-110.

Jay, D.A., R.J. Uncles, J. Largier, W.R. Geyer, J. Vallino, and W.R. Boynton. 1997. A review of recent developments in estuarine scalar flux estimation. Estuaries 20(2):262-280.

Jay, D.A., W.R. Geyer, and D.R. Montgomery. 2000. An ecological perspective on estuarine classification. In: Hobbie, J. E. (ed), Estuarine Science: A Synthetic Approach to Research and Practice. Island Press, Washington DC, pp. 149-176.

Johnstone, J. 1908. Conditions of Life in the Sea. Cambridge University Press. Reprinted from Arno Press, New York. 1977. 332 pp.

Jones, R.A. and G.F. Lee. 1986. Eutophication modeling for water quality management: an update of the Vollenweider-OECD model. WHO Water Quality Bulletin 11(2):67-74.

Justic, D., T. Legovic, and L. Rottini-Sandrini. 1987. Trends in the oxygen content 1911-1984 and occurrence of benthic mortality in the northern Adriatic Sea. Est. Coast. Shelf Sci. 25:435-445.

Keller, A.A. 1988a. An empirical model of primary productivity (14C) using mesocosm data along a nutrient gradient. J. Plankton Res. 10(4):813-834.

Keller, A.A. 1988b. Estimating phytoplankton productivity from light availability and biomnass in the MERL mesocosms and Narragansett Bay. Mar. Ecol Prog. Ser. 54:159-168.

Keller, A.A. and R.L. Rice. 1989. Effects of nutrient enrichment on natural populations of the brown tide phytoplankton *Aureococcus anophagefferens* (Chrysophyceae). J. Phycol. 25:636-646.

Kelly, J.R. 1997a. Nitrogen flow and the interaction of Boston Harbor with Massachusetts Bay. Estuaries 20(2):365-380.

Kelly, J.R. 1997b. Nutrients and human-induced change in the Gulf of Maine–"One, if by land, and two, if by sea." In: Wallace, G.T. and E.F. Braasch (eds.), Proceedings of the Gulf of Maine Ecosystem Dynamics Scientific Symposium and Workshop. RARGOM Report, 97-1. Hanover, NH: Regional Association for Research on the Gulf of Maine, 1997, pp. 169-181.

Kelly, J.R. 1997c. Dissolved oxygen in Maine estuaries and embyaments. Final report

(DEPLW97-23) to Wells National Estuarine Research Reserve (WNERR) Maine Department of Environmental Protection, and Casco Bay Estuary Project. WNERR, Wells, ME 04090. 18+ pp.

Kelly, J.R. 1998. Quantification and potential role of ocean nutrient loading to Boston Harbor, Massachusetts, USA. Mar. Ecol. Prog. Ser. 173:53-65.

Kelly, J.R., V.M. Berounsky, S.W. Nixon, and C.A. Oviatt. 1985. Benthic-pelagic coupling and nutrient cycling across an experimental eutrophication gradient. Mar. Ecol. Prog. Ser. 26:207-219.

Kelly, J.R. and S.A. Levin. 1986. A comparison of aquatic and terrestrial nutrient cycling and production processes in natural ecosystems, with reference to ecological concepts of relevance to some waste disposal issues. In: G. Kullenberg (ed.), The Role of the Oceans as a Waste Disposal Option. D. Reidel Publishing Co, Dordrecht, pp. 165-203.

Kelly, J.R. and P.H. Doering. 1997. Monitoring and modeling primary production in coastal waters: studies in Massachusetts Bay 1992-1994. Mar. Ecol. Prog. Ser. 148:155-168.

Kelly, J.R. and P.H. Doering. 1999. Seasonal deepening of the pycnocline in a shallow shelf ecosystem and its influence on near-bottom dissolved oxygen. Mar. Ecol. Prog. Ser. 178:151-168.

Kemp, W.M., R.R. Twilley, J.C. Stevenson, W.R. Boynton, and J.C. Means. 1983. The decline of submerged vascular plants in upper Chesapeake Bay: summary of results concerning possible cause. Marine Technol. Soc. J. 17(2):78-89.

Kennedy, V.S. (ed.). 1984. The Estuary as a Filter. Estuarine Research Federation, Biennial Conference, Academic Press, Orlando. 511 pp.

Ketchum, B.H. 1969. Eutrophication of estuaries. In: NAS, Eutrophication: Causes, Consequences, Correctives. Proceedings of an International Symposium on Eutrophication. University of Wisconsin, 1967. NAS Printing and Publishing Office, Washington D.C., pp. 197-209.

Krause-Jensen, D. and K. Sand-Jensen. 1998. Light attenuation and photosynthesis of aquatic plant communities. Limnol. Oceanogr. 43(3):396-407.

Krom, M.D., N. Kress, S. Brenner, and LI Gordon. 1991. Phosphorus limitation of primary production in the eastern Mediterranean. Limnol. Oceanogr. 36:424-432.

Kullenberg, G. (ed.). 1986a. The Role of the Oceans as a Waste Disposal Option. D. Reidel Publishing Co, Dordrecht.

Kullenberg, G. 1986b. The Baltic Sea: conditions and options of management. In: Kullenberg, G. (ed.), The Role of the Oceans as a Waste Disposal Option. D. Reidel Publishing Co, Dordrecht, pp. 325-345.

Lapointe, B. E. 1997. Nutrient thresholds for bottom-up control of macroalgal blooms on coral reefs in Jamaica and southeast Florida. Limnol. Oceanogr. 42:1119-1131.

Larsson U., R. Elmgren, and F. Wulff. 1985. Eutrophication and the Baltic Sea: causes and consequences. Ambio 14(1):9-14.

Lauff, G.H.(ed) 1967. Estuaries. Amer. Assoc. Adv. Sci. Pub. No. 83. Washington DC. 757 pp.

Legendre, L. and J. Le Fevre. 1989. Hydrodynamic singularities as controls of recycling versus export production in oceans. In: W.H. Berger, V.S. Smetacek, and G. Wefer (eds.), Productivity of the Oceans: Present and Past. John Wiley and Sons, Ltd., Dahlheim Konferensen, pp. 49-63.

Likens, G.E.(ed.) 1972. Nutrients and eutrophication: The limiting-nutrient controversy. Limnol. Oceanogr. Special Symposia, Vol. 1.

Malakoff, D. 1998. Death by suffocation in the Gulf of Mexico. Science 281:190-192.
Malone, T.C. 1982. Factors influencing fate of sewage derived nutrients in lower Hudson Estuary and New York Bight. In: G.F. Meyer (ed.), Ecological Stress and the New York Bight, Science and Management. Estuarine Research Federation, Columbia, SC, pp. 389-400.
Malone, T.C., D.J. Conley, T.R. Fisher, P.M. Glibert, L.W. Harding, Jr., and K.G. Sellner. 1996. Scales of nutrient limited phytoplankton productivity in Chesapeake Bay. Estuaries 19:371-385.
Margalef, R. 1978. Life-forms of phytoplankton as survival alternatives in an unstable environment. Oceanol. Act. 1:493-509.
Monbet, Y. 1992. Control of phytoplankton biomass in estuaries: a comparative anlaysis of microtidal and macrotidal estuaries. Estuaries 15(4):563-571.
National Academy of Sciences (NAS). 1969. Eutrophication: Causes, Consequences, Correctives. Proceedings of an International Symposium on Eutrophication. University of Wisconsin, 1967. NAS Printing and Publishing Office, Washington D.C.
National Research Council (NRC). 1993. Managing wastewater in coastal urban areas. National Academy Press, Washington DC.
National Research Council (NRC). 2000. Clean coastal waters: Understanding and reducing the effects of nutrient pollution. National Academy Press, Washington DC.
Nielson, B.J. and L.E. Cronin (eds.). 1981. Estuaries and Nutrients. Humana Press, Clifton, NJ. 643 pp.
Nixon, S.W. 1981. Remineralization and nutrient cycling in coastal marine ecosystems. In: B.J. Nielson and L.E. Cronin (eds.), Estuaries and Nutrients. Humana Press, Clifton, NJ, pp. 111-138.
Nixon, S.W. 1983. Estuarine ecology—a comparative and experimental analysis using 14 estuaries and the MERL microcosms. Final Report to U.S. Environmental Protection Agency, Chesapeake Bay Program, Annapolis, MD.
Nixon, S.W. 1988. Physical energy inputs and the comparative ecology of lake and marine ecosystems. Limnol. Oceanogr. 33(4, part 2):1005-1025.
Nixon, S.W. 1992. Quantifying the relationship between nitrogen inputs and the productivity of marine ecosystems. Proc. Adv. Mar. Tech. Conf. 5:57-83.
Nixon, S.W. 1995. Coastal marine eutrophication: A definition, social causes and future concerns. Ophelia 41:199-219.
Nixon, S.W. 1997. Prehistoric nutrient inputs and productivity in Narragansett Bay. Estuaries 20:253-261.
Nixon, S.W. 1998. Enriching the sea to death. Scientific American VOL: 48-53.
Nixon, S.W., J.R. Kelly, B.N. Furnas, C.A. Oviatt, and S.S. Hale. 1980. Phosphorus regeneration and the metabolsim of coastal marine bottom communities. In: K.R. Tenore and B.C. Coull (eds.), Marine Benthic Dynamics. Univ. of South Carolina Press, Columbia, S.C., pp. 219-243.
Nixon, S.W. and M.E.Q. Pilson. 1983. Nitrogen in estuarine and coastal marine ecosystems. In: Carpenter, E.J. and D.G. Capone (eds.), Nitrogen in the Marine Environment. Academic Press, New York. pp. 565-648.
Nixon, S.W., M.E. Q. Pilson, C.A. Oviatt, P. Donaghay, B. Sullivan, S. Seitzinger, D. Rudnick, and J. Frithsen. 1984. Eutrophication of a coastal marine ecosystem—an experimental study using the MERL microcosms. In: M.J.R. Fasham (ed.), Flows and Energy and Materials in Marine Ecosystems. Plenum Press, London, pp.105-135.
Nixon, S.W., C.A. Oviatt, J. Frithsen, and B. Sullivan. 1986. Nutrients and the productivity of

estuarine and coastal marine ecosystems. J. Limnol. Soc. Sth. Afr. 12(1/2):43-71.
Nixon, S.W., S.L. Granger, and B.L. Nowicki. 1995. An assessment of the annual mass balance of carbon, nitrogen, and phosphorus in Narragansett Bay. Biogeochemistry 31:15-61.
Nixon, S.W., J.W. Ammerman, L.P. Atkinson, V.M. Berounsky, G. Billen, W.C. Boicourt, W.R. Boynton, T.M. Church, D. M. Ditoro, R. Elmgren, J.H. Garber, A.E. Giblin, R.A, Jahnke, H.J.P. Owens, M.E.Q. Pilson, and S.P. Seitzinger. 1996. The fate of nitrogen and phosphorus at the land-sea margin of the North Atlantic Ocean. Biogeochemistry 35: 141-180.
Officer, C.B. 1980. Box models revisited. In: Hamilton, P and K. Macdonald (eds.), Estuarine and wetland processes with emphasis on modeling. Plenum Press, New York, pp. 65-114.
Officer, C.B. and J.H. Ryther. 1980. The possible importance of silicon in the marine environment. Mar. Ecol. Prog. Ser. 3:83-91.
Officer, C.B., R.B. Biggs, J.L. Taft, L.E. Cronin, M.A. Tyler, and W.R. Boynton. 1984. Chesapeake Bay anoxia: origin, development, and significance. Science 223:22-27.
Oviatt, C.A., M.E. Q. Pilson, S.W. Nixon, J.B. Frithsen, D.T. Rudnick, J.R. Kelly, J.F. Grassle, and J.P. Grassle. 1984. Recovery of a polluted estuarine system: a mesocosm experiment. Mar. Ecol. Prog. Ser. 16:203-217.
Oviatt, C.A. A.A. Keller, P.A. Sampou, L.L. Beatty. 1986. Patterns of productivity during eutrophication: a mesocosm experiment. Mar. Ecol. Prog. Ser. 28:69-80.
Oviatt, C., P. Doering, B. Nowicki, L. Reed, J. Cole, and J. Frithsen. 1995. An ecosystem level experiment on nutrient limitation in temperate coastal marine environments. Mar. Ecol. Prog. Ser. 116:171-179.
Orth, R.J. and K.A. Moore. 1983. Chesapeake Bay: An unprecedented decline in submerged aquatic vegetation. Science 222:51-53.
Parker, C.A. and J.E. O'Reilly. 1991. Oxygen depletion in Long Island Sound: a historical perspective. Estuaries 14:248-264.
Pearson, T.H. and R. Rosenberg. 1978. Macrobenthic succession in relation to organic enrichment and pollution of the marine environment. Oceanogr. Mar. Biol. Ann. Rev. 163:229-311.
Paerl, H.W. 1988. Nuisance phytoplankton blooms in coastal, estuarine, and inland waters. Limnol. Oceanogr. 33(4, part 2):823-847.
Paerl, H.W. and D.R. Whitall. 1999. Anthropogenically-derived atmospheric nitrogen deposition, marine eutrophication and harmful algal bloom expansion: is there a link? Ambio 28(4):307-311.
Pilson, M.E. Q. 1985. On the residence time of water in Narragansett Bay. Estuaries 8:2-14.
Pingree, R. P.Pugh, P. Holligan, and G. Forster. 1975. Summer phytoplankton blooms and red tides along fronts in the approaches to the English channel. Nature 258:672-677.
Prospero, J.M., K. Barrett, T. Church, F. Dentener, R.A. Duce, J.N. Galloway, H. Levy II, J. Moody, and P. Quinn. 1996. Atmospheric deposition of nutrients to the North Atlantic Basin. Biogeochemistry 35:27-73.
Rabalais, N.N., R.E. Turner, W.J. Wiseman, Jr., and D.F. Boesch. 1991. A brief summary of hypoxia on the northern Gulf of Mexico continental shelf: 1985-1988. In: Tyson, R.V. and T.H. Pearson (eds.), Modern and Ancient Continental Shelf Anoxia. Geological Society Special Publication No. 58, The Geological Society, London, 470 pp.
Rabalais, N.N., R.E. Turner, D. Justic, Q. Dortch, W. Wiseman, Jr., and B. K. Sen Gupta. 2000. Gulf of Mexico biological system responses to nutrient changes in the Mississippi River. In: Hobbie, J. E. (ed.), Estuarine Science: A Synthetic Approach to Research and Practice. Island

Press, Washington DC, pp. 241-268.
Rosenberg, R. 1980. Effect of oxygen deficiency on benthic macrofauna. In: Freeland, J.H., D.M. Farmer, D.M. and C.D. Levings (eds.), Fjord oceanography. Plenum Publ. Corp., New York, pp. 499-514.
Rosenberg, R. 1985. Eutrophication – the future marine coastal nuisance? Mar. Poll. Bull. 16:227-231.
Rosenberg, R. R. Elmgren, S. Fleischer, P. Jonsson, G. Persson, and H. Dahlin. 1990. Marine eutrophication case studies in Sweden. Ambio 19(3):102-108.
Rosenberg, R., B. Hellman, B. Johansson. 1991. Hypoxic tolerance of marine benthic fauna. Mar. Ecol. Prog. Ser. 79:127-131.
Ryther, J. H. 1954. The ecology of phytoplankton blooms in Moriches Bay and Great South Bay, Long Island, New York. Biological Bulletin 106:198-209.
Ryther, J.H. and W.M. Dunstan. 1971. Nitrogen, phosphorus, and eutrophication in the coastal marine environment. Science 171:1008-1013.
Ryther, J.H. and C.B. Officer. 1981. Impact of nutrient enrichment on water uses. In: Nielson, B.J. and L.E. Cronin (eds.), Estuaries and Nutrients. Humana Press, Clifton, NJ, pp.247-261.
Sand-Jensen, K. and J. Borum. 1991. Interactions among phytoplankton, periphyton, and macrophytes in temperate freshwaters and estuaries. Aquat. Bot. 41: 137-175.
Schindler, D.W. 1974. Eutrophication and recovery in experimental lakes: implications for lake management. Science 184:897-899.
Seitzinger, S. P. 2000. Scaling up: Site specific measurements to global-scale estimates of denitrification. In : Hobbie, J. E. (ed.), Estuarine Science: A Synthetic Approach to Research and Practice. Island Press, Washington DC, pp. 211-240.
Short, F.T. 1987. Effects of sediment nutrients on seagrasses: literature review and mesocosm experiment. Aquatic Botany 27(1):41-57.
Short, F.T., D.M. Burdick, and J.E. Kaldy III. 1995. Mesocosm experiments quantify the effects of eutrophication on eelgrass, Zostera marina. Limnol. Oceanogr. 40(4):740-749.
Signell, RP and B. Butman. 1992. Modeling tidal exchange and dispersion in Boston Harbor. J. Geophys Res 97:15191-15606.
Simenstad, C.A., S.B. Brandt, A. Chalmers, R. Dame, L.A. Deegan, R. Hodson, and E.D. Houde. 2000. In: Hobbie, J. E. (ed.), Estuarine Science: A Synthetic Approach to Research and Practice. Island Press, Washington DC, pp. 427-455.
Smith, N.P. 1993. Tidal and nontidal flushing of Florida's India River lagoon. Estuaries 16(4):739-746.
Smayda, T.J. 1990. Novel and nuisance phytoplankton blooms in the sea: evidence for global epidemic. In: Graneli, E., B. Sundstrom, R. Edler, and D.M. Anderson (eds.), Toxic Marine Phytoplankton. Elsevier, New York, pp.29-40.
Smayda, T.J. 1997. What is a bloom? A commentary. Limnol. Oceanogr. 42(5, part 2):1132-1136.
Stanley, D.W., and S.W. Nixon. 1992. Stratification and bottom-water hypoxis in the Pamlico River estuary. Estuaries 15(3):270-281.
Stevenson, J.C. 1988. Comparative ecology of submersed grass beds in freshwater, etsuarine, and marine environments. Limnol. Oceanogr. 33(4, part 2): 867-893.
Stevenson, J.C., L.W. Staver, and K.W. Staver. 1993. Water quality associated with survival of submersed aquatic vegetation along an estuarine gradient. Estuaries 16(2):346-361.

Summers, J.K. 2001. Ecological condition of the estuaries of the Atlantic and Gulf coasts of the United States. Environ. Tox. and Chem. 20(1):99-106.

Sutcliffe, W.H., Jr. 1972. Some relations of land drainage, nutrients, particulate material, and fish catch in two eastern Canadian bays. J. Fish. Res. Bd. Canada 29:357-362.

Sutcliffe, W.H. Jr. 1973. Correlations between seasonal river discharge and local landings of American lobster (*Homarus americanus*) and Atlantic halibut (*Hippoglossus hippoglossus*) in the Gulf of St. Lawrence. J. Fish Res. Bd. Canada 30:856-859.

Sutcliffe, W.H., Jr., K. Drinkwater, and B. S. Muir. 1977. Correlations of fish catch and environmental factors in the Gulf of Maine. J. Fish Res. Bd. Canada 34:19-30.

Sutcliffe, W.H., Jr., R. H. Loucks, K. Drinkwater, and A.R. Coote. 1983. Nutrient flux onto the Labrador Shelf from Hudson Strait and its biological consequences. Can. J. Fish. Aq. Sci. 40:1692-1701.

Taylor, D., S.Nixon, S.Granger, and B.Buckley. 1995a. Nutrient limitation and the eutrophication of coastal lagoons. Mar. Ecol. Prog. Ser. 127:235-244.

Taylor, D. S.W. Nixon, S.L. Granger, B.A. Buckley, J.P McMahon, and H.J. Lin. 1995b. Response of coastal lagoon plant communities to different forms of nutrient enrichment—a mesocosm experiment. Aquatic Botany 52:19-34.

Thayer, G.W., D.A. Wolfe, and R.B. Williams. 1975. The impact of man on seagrass systems. Amer. Sci. 63:289-296.

Twilley, R.R., W.M. Kemp, K.W. Staver, J.C. Stevenson, and W.R. Boynton. 1985. Nutrient enrichment of estuarine submerged vascular plant communities: I. Algal growth and effects on production of plants and associated communities. Mar. Ecol. Prog. Ser. 23:179-191.

Turner, J.T and P. A Tester. 1997. Toxic marine phytoplankton, zooplankton grazers, and pelagic food webs. Limnol. Oceanogr. 42(5, part 2):1203-1214.

Turner, R.E., W.W. Schroeder, and W.J. Wiseman, Jr. 1987. The role of stratification in the deoxygenation of Mobile Bay and adjacent shelf bottom waters. Estuaries 10(1):13-19.

Turner, R.E. and N.N. Rabalais. 1991. Changes in Mississippi River water quality this century. BioSci. 41(3):140-147.

Valiela, I. 1984. Marine Ecological Processes. Springer-Verlag, New York. 546 pp.

Valiela, I. and J. Costa. 1988. Eutrophication of Buttermilk Bay, a Cape Cod coastal embayment: Conentration of nutrients and watershed nutrient budgets. Environ. Manage. 12:539-551.

Valiela, I., K. Foreman, M. Lamontagne, D. Hersh, J. Costa, P. Peckol, B. DeMeo-Andreson, C. D'Avanzo, M. Babione, C. H. Sham, J. Brawley, and K. Lajtha. 1992. Couplings of watersheds and coastal waters: sources and consequences of nutrient enrichment in Waquoit Bay, Massachusetts. Estuaries 15(4):443-457.

Valiela, I., G. Collins, J. Kremer, K. Lajtha, M. Gesit, B. Seely, J. Brawley, and C.H. Sham. 1997a. Nitrogen loading from coastal watersheds to receiving estuaries: new method and application. Ecol. Appl. 7(2):358-380.

Valiela., I., J. McClelland, J. Hauxwell, P.J. Behr, D. Hersh, and K. Foreman. 1997b. Macroalgal blooms in shallow estuaries: controls and ecophysiological and ecosystem consequences. Limnol. Oceanogr. 42(5, part 2):1105-1118.

Vitousek, P.M., J. Aber, S.E. Bayley, R.W. Howarth, G.E. Likens, P.A. Matson, D.W. Schindler, W.H. Schlesinger, and G.D. Tilman. 1997. Human alteration of the global nitrogen cycle: causes and consequences. Ecol. Issues 1:1-15.

Vollenweider, R.A. 1976. Advances in defining critical loading levels of phosphorus in lake eutrophication. Mem. Ist. Ital. Idrobiol. 33:53-83.

Vollenweider, R.A. 1979. (Cited in Wetzel, 1983). Das Nahrsstoffbelastungskonzept als Grundlage fur den externen Eingriff in den Eutrophierungsprozess stehender Gewasser und Talsperren. Z. Wasser- u. Abwasser-Forschung 12:46-56.

Vollenweider, R.A., R. Marchetti, and R. Viviani (eds.). 1992. Marine Coastal Eutrophication, Elsevier, Mew York, 1310 pp.

Wetzel, R.G. 1983. Limnology. 2nd Edition. Saunders College Publishing, Fort Worth.

Zieman, J.C. 1982. The ecology of seagrasses of south Florida: a community profile. United States Fish and Wildlife Service FWS/OBS/82-25.

Section 3:

Atmospheric Effects

Chapter 10. Gaseous Nitrogen Emissions from Livestock Farming Systems

Oene Oenema[1], André Bannink[2], Sven G. Sommer[3] and Gerard L. Velthof[1]

[1]) Alterra, Wageningen University and Research Center, P.O. Box 47, NL-6700 AA Wageningen, The Netherlands, Email O.Oenema@alterra.wag-ur.nl

[2]) Institute for Animal Science and Health (ID), Wageningen University and Research Center P.O. Box 65, 8200 AB Lelystad, The Netherlands,

[3]) Danish Institute of Agricultural Sciences (DIAS), P.O. Box 536, 8700 Horsens, Denmark

ABSTRACT

On a global scale, livestock farming systems contribute about 70% to the total anthropogenic emission of ammonia (NH_3) and about 30% to the total anthropogenic emission of nitrous oxide (N_2O) into the atmosphere. This chapter discusses the origin and controlling factors of these emissions, the uncertainty in the estimates, and possible measures that may be taken to decrease these emissions.

Basically, livestock farming systems transform carbohydrates and protein from plants into milk, meat and eggs. Usually not more than 5 to 30% of the plant protein is transformed into animal protein, depending on animal type and management. The remaining 70 to 95% is excreted via urine and dung as organically bound nitrogen (N). Following its deposition on the floor of animal housing systems or in pastures, a major fraction of the organic N is rapidly hydrolyzed into ammonium (NH_4^+). The NH_4^+ in urine and dung is conducive to volatilization as NH_3. The NH_4^+ is also substrate for nitrifying bacteria upon aeration of the manure (dung and urine). The nitrifying bacteria convert NH_4^+ into nitrate (NO_3^-) which than can be converted subsequently into dinitrogen (N_2) by denitrifying bacteria under anoxic conditions. During the nitrification of NH_4^+ and the denitrification of NO_3^-, nitrogen oxide (NO) and nitrous oxide (N_2O) may escape into the atmosphere, together also with the gaseous N_2 from denitrification. The total loss of NH_3, NO, N_2O and N_2 from animal housing systems and manure storage systems ranges from about 10% of the excreted N in dung and urine from dairy cattle up to more than 30% for pigs and poultry in intensive livestock operations. Another 10 to 50% of total N in the manure may escape as NH_3, NO, N_2O and N_2 from the soil following application to agricultural land. The uncertainty in these estimates is large, because there are many possible sites for gaseous N compounds to escape from the livestock farming system, and there are also many different types of farm animals, livestock farming systems and manure management systems, while the number of measurements of gaseous N losses is still limited.

It is suggested that the emissions of NH_3, NO, N_2O and N_2 from livestock farming systems will continue to increase because of the increasing quest of animal protein by the growing human population, unless effective mitigation measures are implemented in practice.

1. INTRODUCTION

Livestock farming systems produce milk and meat for human consumption. The milk and meat provide important nutritional compounds, especially protein. Nomadic derive their nutrition almost completely from milk and meat from ruminant animals, but for most people, animal protein supplements a predominantly vegetable diet with essential amino acids and vitamins that are in short supply in vegetarian diets. Consumption of animal proteins is related to human culture but also to the level of prosperity; the higher the standard of living the larger the consumption of animal protein. The increase in human population, together with the increase in the standard of living in developed countries, are the main reasons for the rapid growth in numbers of farm animals worldwide and also for the increased impacts of livestock farming systems on the biosphere during the last few decades.

The farm animals utilize only a fraction of the plant nutrients contained in the animal feed. For nitrogen (N), only 5 to 30% of the amount of N in feed is transformed into animal products. The remainder is excreted via feces and urine, and conducive to gaseous N losses. Livestock farming systems are important sources of gaseous N compounds in the atmosphere. They emit about 22 (range 20-60) Tg NH_3-N per year into the atmosphere, and contribute as much as 50 % of the total global emission of ammonia (NH_3) into the atmosphere (Dentener and Crutzen, 1994). Further, they are estimated to contribute about 15 % (equivalent to about 2 Tg N_2O-N) of the total global emission of nitrous oxide (N_2O) into the atmosphere (Berges and Crutzen, 1996; Mosier et al., 1998).

Recent estimates suggest that about 10 to 40% of the N in livestock feces and urine is lost from animal housing and animal manure storage systems via emissions of gaseous nitrogen (Poulsen & Kristensen, 1997; Oenema et al., 2000a). Losses are related to animal type, composition of the animal feed (mainly protein content), housing system, manure storage system and manure management. In addition, up to 50% of the N in manure may be lost following the application of the manure to agricultural land. Next, about 10% of the N in feces and urine deposited in pastures by grazing animals is volatilized into the atmosphere. Evidently, the N cycle of livestock farming systems is a leaky cycle, with many opportunities for the release of gaseous N forms into the atmosphere (see e.g. Jarvis & Pain, 1997 and references therein).

The purpose of this chapter is to discuss the origin, importance and controls of gaseous N emissions from livestock farming systems. Following a brief description of the N cycle of livestock farming systems, we proceed with a discussion of the N transformation during feed digestion and of the relationship between feed composition and the composition of feces and urine. Thereafter, we describe the emissions from feces and urine deposited on pastures by grazing animals, and the emissions from manure in animal housing systems and storage systems. Then, gaseous N losses from manure applied to agricultural land are discussed.

2. NITROGEN CYCLING IN LIVESTOCK FARMING SYSTEMS

2.1. Livestock farming systems

Livestock farming systems comprise domesticated and farmed animals and their feeding, housing and management, including the management of animal excrements. There is a wide variety in types of farm animals. Most important animal categories in terms of numbers and animal protein production are cattle, pigs (swine), poultry, sheep and goats. The production of milk, beef, pork and poultry have become highly specialized and concentrated geographically in various parts of the world. Here, the animal products are produced for the (global) market, which is almost free. Competition forces farmers to specialize and to decrease the cost of production. Modern technology and transport facilities have contributed to the introduction of large automated housing systems that rely on imported animal feed. These developments have led to strong increases in labor productivity, low prices of animal products, and to a segregation of crop (animal feed) production systems from animal production systems. There is growing concern about animal welfare in modern livestock operations, about the consequences of modern biotechnology (hormones, antibiotics, genetically modified animals), and about the large scale transfer of plant nutrients with animal feed.

Conveniently, four major compartments are distinguished in whole livestock farming systems, i.e. livestock, manure, land and crop (animal feed) (Fig. 1). Nutrients cycle through these compartments, but there are costs associated with the transfer of matter from one compartment to the other. Animals utilize only a fraction (5 to 30%) of the N in the feed for the production of milk, meat, eggs and offspring (animal products) exported from the system. The greater part is excreted via feces and urine, which is stored and managed for some time in various types of manure storage systems, or deposited directly on pastoral land and allowed to lie there unmanaged. The manure from manure storage systems will be applied to agricultural land as fertilizer to nourish the growing crop. However, only about 30 to 60% of the manure N will be utilized by growing crops for the production of plant protein, and only the protein in the harvested fraction of the crop will feed the livestock. Hence, only a minor fraction (usually less than 10%) of the N from manure will be exported from the farm in animal products; the greater part will have dissipated into the wider environment.

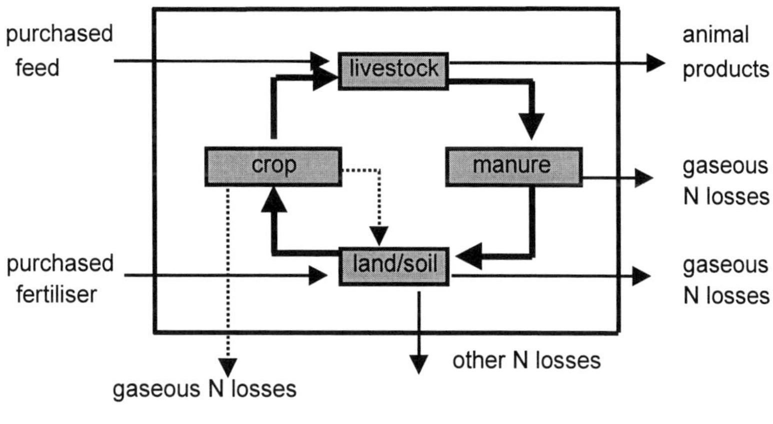

Figure 1. Simplified diagram of a whole livestock farming system, showing the four main compartments livestock, manure, land/soil and crop. The land/soil and crop compartments can be separated physically from the livestock and manure compartments, as is the case in specialized livestock farming systems without land. Arrows indicate the flows of N in products and as gaseous N losses.

Various modifications exist to the whole livestock farming system depicted in Fig. 1. In extensively managed rangeland systems, cattle graze on unfertilized swards. Usually only 10-20% of the above-ground biomass herbage is consumed in one year, due to the low stock density (Coleman et al., 1977), except in situations with overgrazing. Total annual N input via atmospheric deposition and biological N fixation in and total annual N output via animal products from these extensively managed rangeland systems is less than 10 kg per ha per year (Woodmansee, 1979). By contrast, intensively managed mixed farming systems purchase fertilizers to boost crop production and purchase animal feed (cereals, concentrates, residues from food processing industry) to supplement the ration of the livestock and to increase the animal production. Total annual N input via purchased fertilizers and animal feed in intensively managed dairy and beef farming systems is in the range of 100 to 500 kg per ha per year, and total annual N output via animal products ranges from 40 to 80 kg per ha per year (e.g. Goh and Williams, 1999; Brouwer and Hellegers, 1997). Specialized livestock operations that are without land, purchase all animal feed and export all animal manure. These farms consists of just two compartments, i.e. livestock and manure storage. Total annual N input via purchased animal feed into these farms and total annual N output via animal products and animal manure from these farms can be very large. Brouwer and Hellegers (1997) provide N balances for these so called granivore farms in the European Union. They indicate that the mean difference between N inputs and N outputs exceed 1000 kg N per ha per year. Evidently, the potential for N losses differ greatly between livestock farming systems.

2.2. Gaseous N losses from livestock farming systems

There are many opportunities and places for N to escape from livestock farming systems (Fig. 1). Significant losses of gaseous N compounds may occur via volatilization of NH_3 and via emissions of nitric oxide (NO), nitrous oxide (N_2O) and dinitrogen (N_2) from nitrification and denitrification processes. In addition, N may be lost from soil via leaching and runoff. The gaseous N compounds may escape from feces and urine during storage in manure storage systems (lagoons, pits, manure heaps, manure silos), after deposition on pastures and paddocks by free ranging animals and after application of manure and fertilizers to agricultural land. Crops, especially well-fertilized grasslands and ensiled forages may also emit NH_3 into the atmosphere, but this amount is considered to be small. Likewise, livestock may emit NH_3 during rumination and flatulation, but these N losses are consider to be small as well, and are therefore also not discussed here further.

Volatilization of NH_3 occurs at an early stage in the sequence of processes following the excretion of feces and urine (Fig. 2). The total loss of NH_3 is related to the amount of easily hydrolyzable ammoniacal N in the dung and urine, environmental conditions, i.e. temperature, rainfall and wind, and to manure management. A significant fraction of the N in feces and urine hydrolyzes rapidly (within hours to days) into NH_4^+. The concomitant production of bicarbonate (HCO_3^-) contributes to deprotonation of NH_4^+ and to the volatilization of NH_3 upon exposure to the atmosphere. Central to the control of NH_3 volatilization is the composition of the animal feed and the storage and management of the manure, as further explained below.

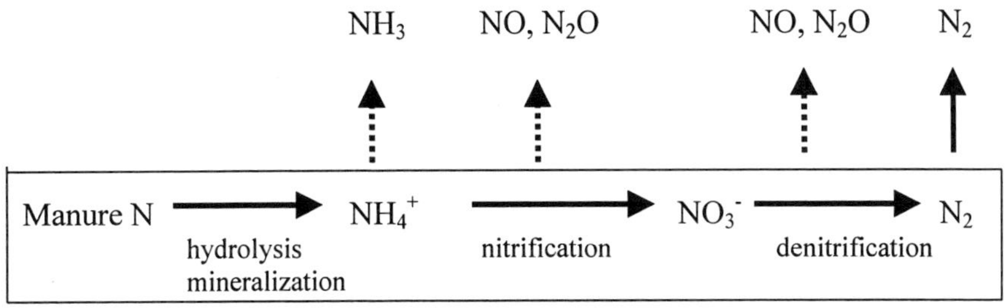

Figure 2. Sequence of N transformation processes, and the release of gaseous N compounds from the N in animal manure.

Exposure to the atmosphere and treatment (aeration) of the initially anoxic feces and dung provide nitrifying bacteria the opportunity to utilize the energy in NH_4^+ and to transform NH_4^+ to nitrite (NO_2^-) and nitrate (NO_3^-). The NO_2^- and NO_3^- may subsequently act as electron acceptor for denitrifying bacteria active in anoxic environments. Various intermediates and side products, such as NO and N_2O may escape to the atmosphere in this sequence of processes. The emissions of NO, N_2O and N_2 from livestock farming systems are related to this sequence of nitrification and denitrification processes (Fig. 2).The release of NO, N_2O and N_2 depends on a complex of interacting factors, in which animal nutrition, animal housing system, manure management and environmental conditions all play a key role, as further discussed below.

3. FEED DIGESTION AND NITROGEN EXCRETION

Animals require energy, protein, water, various nutrients including trace elements, and vitamins for their nutrition. The value of animal feed is usually defined by the quantity of energy and protein that can be metabolized by the animal after the digestion of feed in the gastrointestinal tract. The protein value of a diet is estimated by the fraction of crude protein (N content $\times$ 6.25) that is absorbed from the gastrointestinal tract (Fig. 3). For pig and poultry diets, the protein value is also defined by the content of individual amino acids in absorbed protein in order to identify the amino acid most limiting protein deposition in animal products.

3.1. Feed digestion and N utilization

The digestion of feed by farm animals is a multi-step process that involves dynamic interactions among the diet, microbial populations and the animal itself. In ruminants, the primary digestion of feed occurs by microbial fermentation in the rumen. The unfermented feed components and the microbial matter synthesized in the rumen are subsequently degraded in the small intestine by digestive enzymes secreted by the animal. Finally, a significant fermentation may occur in the large intestine, depending on the diet, before the feces is excreted. The rumen acts as an imperfectly stirred, continuous-flow reactor, whereas the small intestine acts like a plugged-flow reactor (e.g. Mertens, 1993; Nolan, 1993).

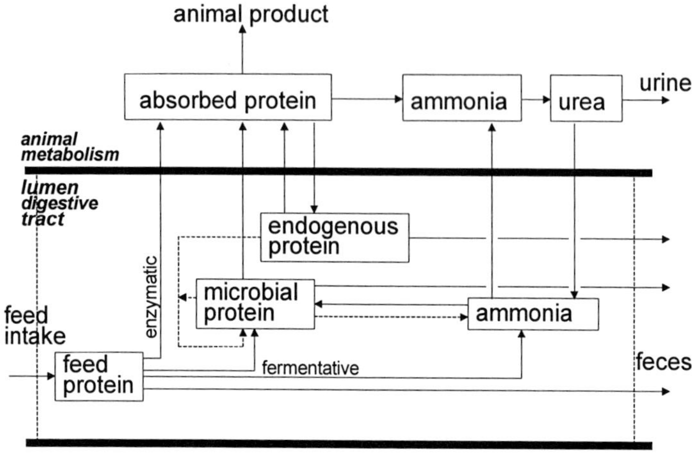

Figure 3. A schematic representation of the N flows in an animal. A distinction has been made between feed digestion by enzymes secreted by the gastrointestinal tract wall and fermentative digestion by microbial activity. In ruminants, the majority of the organic matter in feed is converted into fermentation end products (microbial matter and volatile fatty acids) because of microbial fermentation in the rumen. The importance of fermentative digestion in the small intestine is relatively small for ruminants as well as for monogastric animals, whereas it may become significant again in the large intestine, depending on the diet.

The amount of protein that becomes available for animal use is defined by the amount of protein ingested minus the amount retrieved at the end of the small intestine. The true digestibility exceeds the apparent digestibility, because there is also a considerable influx of protein to the lumen of the gastrointestinal tract. This influx of protein to the lumen is caused by the secretion of digestive enzymes and by cells that are sloughed from the tract wall with passage of digesta (Fig. 3). Therefore, the net rate of protein disappearance in the small intestine is the net result of the rate of absorption minus the rate of endogenous loss.

The majority of the protein in the diet of *ruminants* is hydrolyzed by extracellular microbial proteases to peptides and amino acids in the rumen and then incorporated into microbial protein or deaminated intracellularly to volatile fatty acids and ammonia. This ammonia is available for assimilation and synthesis into protein by other micro-organisms, but when the diet is rich in protein, the ammonia formed from deamination of excess protein is absorbed across the gut wall, converted to urea ($CO(NH_2)_2$) in the liver and subsequently excreted in the urine. Evidently, when the diet is rich in protein, animals use the protein inefficiently; a surplus of digestible protein ends up almost completely in urea. Therefore, rumen fermentation of ingested feed affects the composition of urine. In addition, extensive fermentation in the large intestine affects the composition of both feces and urine. When more protein enters the large intestine and becomes fermented and utilized as an energy source for microbial synthesis, more NH_4^+ will be formed. This NH_4^+ will be absorbed to blood and after conversion to urea in the liver, excreted via urine, or it will be used by micro-organisms as a source of N for protein synthesis (Fig. 3). Conversely, fermentation of feed low in protein in the large intestine may result in a net uptake of urea from blood, to be used as an additional source of N for microbial protein synthesis. The uptake of urea from blood shifts the excretion of the N from urea in urine towards microbial protein in feces.

3.2. Nutritional effects on N excretion

Dairy cows deposit about 10 dung pats and 10 urine puddles per day. The N deposition ranges from 20 to 80 g m^{-2} in urine puddles and from 50 to 200 g m^{-2} in dung pats, depending on animal nutrition and on the volume and spreading of the dung and urine (Lantinga et al., 1987; Haynes and Williams, 1993). Table 1 presents ranges of the N content of urine and feces of dairy cows, pigs and chicken. The fraction of total N excretion via urine ranges from less than 40% to more than 80% in ruminants. The variation in N concentration in the urine is even larger because of the effects of electrolyte concentration on volume and frequency of urination. Typically over 70% of the N in urine is present as urea and the rest consists of amino acids and peptides (Bristow et al., 1992). The bulk of the N in feces is in organic form. The water soluble organic N compounds in dung hydrolyze rapidly, but the remaining organic N is resistant and it may take months to years before these compounds are mineralized (Castellanos and Pratt, 1981). Evidently, animal nutrition has a large influence on the total N content and the degradability of the various N fractions in feces and urine.

Lowering of the protein content in the diet by changing the ratio of energy to protein is one of the most important dietary measures to improve the efficiency of the deposition of dietary protein into animal products. Total N excretion by dairy cows depends on protein content of the diet, animal production level and the size of the cow (Table 2). The protein content of the diet of dairy cattle may range from less than 15% to more than 20% (Ketelaars and Van der Meer, 1999), though 16% seems to be the minimum for a highly productive herd. The average protein content of the diet of dairy cows in intensive grassland farming systems is near 20%, indicating that the total N excretion ranges from 100 to 160 kg N per cow per year, depending on animal size and milk production level (Ketelaars and Van der Meer, 1999).

Table 1. Composition of feces and urine from dairy cattle, fattening pigs and chicken. Range of values as observed in literature (after Oenema et al., 2000a)

Animal category	Dry matter g per kg	Total N g per kg feces/urine	Urea % of total N	Uric acid % of total N	Protein-N, % of total N	Ammonium % of total N
Dairy cattle						
- Feces	100-175	10-17	0	0	90-95	1-4
- Urine	30-40	4-10	60-95	0-2	0	1
Finishing pigs						
- Feces	200-300	7-15	0		90-95	1-7
- Urine	10-50	2-10	30-90		10-20	5-65
Chicken	200-300	10-20	5-8	35-50	30-50	6-8

Table 2. Relationship between protein content of the diet and the total N excretion by dairy cows.The N excretion varies with the size of the dairy cow and with milk production level (adapted from Ketelaars and Van der Meer, 1999).

		Nitrogen excretion, kg per cow per year					
		Small cow (450 kg)			Big cow (650 kg)		
Protein	Nitrogen	Milk production, kg per year			Milk production, kg per year		
%	%	4000	5000	6000	6000	7000	8000
15	2,38	68	76	82	98	104	112
16	2,52	73	81	89	106	113	121
17	2,66	79	87	96	113	121	130
18	2,80	84	93	102	121	130	139
18	2,94	89	99	109	129	138	148
19	3,08	94	104	116	137	147	157
20	3,22	100	111	122	143	154	166

Supplementing the protein-rich grass diet with protein-poor (silage) maize can strongly decrease the N excretion, without much effect on dairy production. Small dairy cows excrete less N per kg milk produced compared to big cows, but big cows produce more milk. Hence, less N is excreted by one big cow producing 8000 kg of milk, compared to two small cows producing 4000 kg of milk each, when similar diets are provided (Table 2). In the case of pigs and poultry nutrition, lowering the protein content of the diet may be combined with supplementing synthetic amino acids, that limit animal production. This allows for an even further reduction of dietary protein content without compromising animal performance. A lower dietary protein content results in a lower ratio of urea and uric acid N to organic N in the excrements.

Effects of nutritional measures on fermentation in the large intestine of ruminants are relatively small because of the extensive fermentation in the rumen. Nutritional measures which increase the fermentation in the large intestine are particularly relevant in *pig* nutrition when substantial quantities of non-starch polysaccharides or structural carbohydrates are included in the diet. Bakker and Dekker (1998) offered a diet with 65% of the maize starch replaced by raw potato starch to growing pigs and obtained a 33% decrease in urea excretion via urine and a 140% increase of the amount of N in feces. Potato starch is less digestible in the small intestine, which delivered more fermentable carbohydrate to the large intestine. A larger quantity of dry matter disappeared from the large intestine and a disappearance of N changed into a net appearance of N. Evidently, the diet determines whether the balance between protein fermentation and ammonia absorption, and the balance between urea influx and microbial protein synthesis, result in a net appearance or disappearance of N in the large intestine (Van der Meulen et al., 1997).

Inclusion of structural carbohydrates in the form of soluble pectin in the diet also stimulates microbial fermentation in the gastrointestinal tract of *poultry* (Langhout, 1999). Protein digestibility decreased in conventional chicks but not in germ free chicks, which indicates that the observed effects of structural carbohydrates have to be attributed to the activity of micro-organisms. The inclusion of structural carbohydrates in the diet will cause more N to be excreted with feces, less to be absorbed and metabolized by the animal, and probably less N to be excreted with urine. Although effects on total N excretion remain small because urine and feces are excreted together by poultry, the relative contribution of fecal and urinary N may change. For example, the exchange of feed ingredients in an experiment of Carré et al. (1995) increased the dietary content of polysaccharides with 1%, and decreased fecal N digestibility with 2%.

3.3. Nutritional effects on NH_3 emissions

The concentration of urea in the urine depends on the amount of urea (uric acid in case of poultry) generated by animal metabolism and on the total volume of urine excreted by the animal. As explained before, the more NH_4^+ absorbed from the gastrointestinal tract, the larger the quantity of urea excreted via urine. Urine volume depends mainly on the quantity of electrolytes (cations and anions) in the diet (Fig. 4). The concentrations of sodium (Na), potassium (K) and N in the urine explain most of the variation in the volume of urine excreted by lactating dairy cows; the more salts and N in the urine, the larger the total urine volume excreted.

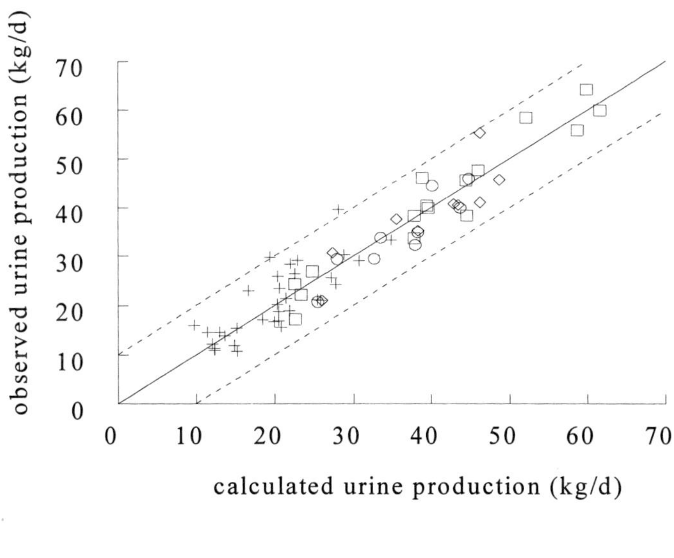

Figure 4. Comparison between measured and calculated urine production by lactating dairy cows. Calculated values were obtained from regression analysis, according to: urine production (kg/d) = 0.134 (0.010) × urinary excreted sodium (g/d) + 0.061 (0.005) × urinary excreted potassium (g/d) +0.024 (0.007) × urinary excreted N (g/d) [R^2 =0.898, SE = 4.22 kg/d]. (Bannink et al., 1999).

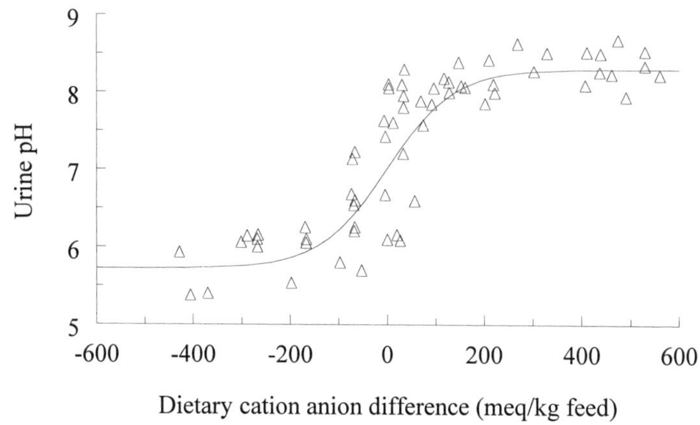

Figure. 5. The relationship between dietary cation-anion balance (Na+K-Cl-2×S in meq/kg feed dry matter) and the pH of urine excreted by dairy cows (Bannink and Van Vuuren, 1998). The following relationship was derived by regression of a logistic equation to observed data: Urine pH = 5.72 (0.17)+ 2.57 (0.29) / [1+ exp (-0.015(0.004) x (cation-anion balance] $R^2 = 0.692$; SE = 0.612

Increasing the volume of urine by increasing the amount of electrolytes in the diet, decreases the concentration of urea in the urine, and increases the frequency of urination and also increases the volume of a single urination. Such multiple and interactive effects complicate the prediction of the effects of changes in animal's diet on N excretion and especially on NH_3 volatilization (Smits et al., 1995).

The pH of freshly excreted urine mainly depends on the dietary content of electrolytes (Fig. 5). Though the pH will eventually rise towards alkaline values during the hydrolysis of urea, irrespective of initial pH, it are the initial pH and the pH buffering capacity of urine which determine the rate of ammonia volatilization immediately after urination.

Volatilization of NH_3 from animal manure is closely related to urinary excretion of urea, which is rapidly hydrolyzed by the enzyme urease into NH_4^+/NH_3 and HCO_3^- according to:

$$CO(NH_2)_2 + 2H_2O \rightarrow (NH_4)_2CO_3 \leftrightarrow 2NH_4^+ + CO_3^{2-} \leftrightarrow NH_3\uparrow + NH_4^+ + HCO_3^-$$

Grazed pastures and the floors in animal houses that are fouled with fecal matter have a high urease activity, and the hydrolysis of urea into NH_3/NH_4^+ is generally completed within a few hours or days. The hydrolysis of urea is associated with an increase in pH because of the

production of (bi)carbonate(s). This combined with the high pK_a value (9.15) of the NH_3/NH_4^+ equilibrium make that urea puddles have a high potential for ammonia volatilization.
A low pH of the urine decreases the urease activity and also affects the NH_4^+ / NH_3 equilibrium and the NH_3 partial pressure.

Cahn et al. (1998b) showed that the pH of fresh urine decreased by more than two units and the volatilization of NH_3 from slurry decreased by up to 50% following the introduction of acidifying salts in the diet of growing pigs. Stimulation of fermentation in the large intestine also lowers the pH of slurry because it will contain a higher concentration of volatile fatty acids as end products of microbial fermentation. In controlled experiments with growing pigs, Cahn et al. (1998a) obtained a 45% decrease in NH_3 volatilization from pig slurry when starch-rich tapioca constituting 15% of dietary dry matter was replaced by pressed sugar beet pulp rich in structural carbohydrates. Though such experimental diets are not representative for the diets used in practice, it shows that fermentation in the large intestine is an important determinant for the site of N excretion in pigs and also for NH_3 volatilization.

4. GASEOUS N LOSSES FROM DUNG AND URINE IN PASTURES

Grazed grassland are aggregations of grazed and ungrazed areas, urine patches, dung patches, compacted hoof holes, camping areas and mixtures of these areas. At the end of the growing season of intensively grazed grassland, there is a mosaic of fresh and old, and of single, overlapping and possibly mixed urine and dung patches covering up to 40 % of the total area (Lantinga et al., 1987). The spatial distribution over the field of patches containing fecal and urinary N is known to be very heterogeneous (Afzal and Adams, 1992). The urine and dung patches have a much higher potential for emitting NH_3, NO, N_2O and N_2 into the atmosphere than the urine and dung free areas. As a consequence, the spatial variability of gaseous N losses in grazed grasslands is extremely high (e.g. Colbourn, 1992; 1993; Velthof et al., 1996a; 1996b; Van Cleemput et al., 1994; Oenema et al., 1997).

4.1. Volatilization of NH_3 from pastures

The loss of NH_3 from grazed pastures is related to pasture productivity, grazing intensity and environmental conditions. There are strong seasonal variations due to variations in weather conditions, grazing periods and fertilizer applications (e.g. Denmead et al., 1974; Ball and Ryden, 1984; Jarvis and Pain, 1990; Bussink, 1992; 1994; Petersen et al., 1998). Plantaz (1998) showed that net emission of NH_3 occurred during the grazing season and a net deposition of NH_3 to grassland during the non-grazing winter season. He also showed that NH_3 emission via the stomata of plants was a way of NH_3 emission from grassland towards the atmosphere. Similar observations have been made by Harper et al. (1996).

Enclosure measurements of NH_3 volatilization from single urine patches indicate that NH_3 losses range from 4 to 52 percent of the urine N (Lockyer and Whitehead, 1990; Sherlock and Goh, 1984; Petersen et al., 1998; Vertregt and Rutgers, 1988; Vallis et al., 1982; Whitehead and Raistrick, 1991; 1993). More than 50 percent of the NH_3 volatilization from urine patches occurred during the process of urea hydrolysis in the first few days after urine deposition. Largest NH_3 losses from urine treated soil were found at high urea-N concentrations (Petersen et al., 1998), high temperatures (Lockyer and Whitehead, 1990), relatively dry conditions (Whitehead and Raistrick, 1991), and a relatively low cation exchange capacity of the soil (Whitehead and Raistrick, 1993). Whitehead et al. (1989) showed that the composition of urine has a strong influence on NH_3 volatilization; volatilization of the five major components of urine decreased in the order urea > allantoin > creatinine > creatine > hippuric acid. However, the NH_3 volatilization from a mixture of hippuric acid and urea was higher than from urea only, probably because hippuric acid increases soil pH. The NH_3 volatilization from dung pats are (much) smaller than from urine patches and range from less than 1 percent (Petersen et al., 1998; Ryden et al., 1987a) to 13 percent of the dung N (Vertregt and Rutgers, 1988).

An integral estimate of the NH_3 loss from grazed pastures, i.e. from both the dung and urine contaminated areas and from the dung and urine free grazed and ungrazed areas can be obtained from micro-meteorological studies. Table 3 provides an overview of results obtained with micro-meteorological methods. Total annual NH_3 volatilization from grazed grasslands ranges from 1 tot 42 kg N per ha per year or 3.3 to 14.4 percent of the total N excreted. Losses of NH_3 tend to be lower from sheep grazed grasslands than from cattle grazed grasslands, possibly because of the lower N content in sheep urine and the smaller volume of a single urination (Bristow et al., 1992).

The NH_3 volatilization from grazed grassland is related to N fertilizer input. Fertilizer N increases herbage production and the N content of the herbage. As a consequence, more animals can graze the pasture and more N is excreted by the grazing animals. Further, more N will be excreted via urine when the protein content of the herbage increases. Misselbrook et al. (2000) derived the following relationship between NH_3 losses from grazed grasslands and the input of mineral fertilizer N, using data of measured NH_3 volatilization of complete grazing seasons in the UK, The Netherlands and New Zealand:

$$NH_3\ loss = 2.27 + 0.0683 * fertilizer\ N\ input$$

Where *NH_3 loss* is the loss in g N per live weight unit per day (1 live weight unit = 500 kg), and *fertilizer N input* is the fertilizer application rate in kg N per ha per year. Because the stocking density increases with an increase in fertilizer N input, the equation indicates that the NH_3 loss increases more than proportionally with an increase in N input.

Table 3. Annual emissions of NH_3 from grazed grasslands. The emitted amount of NH_3-N is expressed in % of the total N or the urinary N excreted by the grazing animal.

Site	Type of animal	Fertilizer input, kg N $ha^{-1}\ yr^{-1}$	NH_3-N emission kg $ha^{-1}\ yr^{-1}$	NH_3-N emission % of total excreted N	NH_3 –N emission % of urinary N	Reference
UK	Beef cattle	210	10	6.7	11.2	Jarvis et al., 1989
UK	Beef cattle	420	25	9.0	12.1	Jarvis et al., 1989
UK	Beef cattle	0 (clover)	7			Jarvis et al., 1989
UK	Sheep	420	9			Jarvis et al., 1991
UK	Sheep	0	4			Jarvis et al., 1991
UK	Sheep	0 (clover)	1			Jarvis et al., 1991
NL	Dairy cows	550	39 - 42	7.7 - 8.5		Bussink, 1992
NL	Dairy cows	250	4 – 9	3.3 - 5.3		Bussink, 1992
NL	Dairy cows	400	12 - 27	6.9 - 13.9		Bussink, 1994
NL	Dairy cows	550	15 - 33	6.9 - 14.4		Bussink, 1994
NZ	Dairy cows	0 (clover)	15	3.5		Ledgard et al., 1996

[1]) UK: United Kingdom, NL: the Netherlands, NZ: New Zealand

4.2. Volatilization of NO, N_2O and N_2 from pastures

Freshly excreted urine and dung contain energy-rich and chemically-reduced carbon and nitrogen compounds, which provide substrates for consortia of autotrophic and heterotrophic bacteria. Upon partial aeration, autotrophic nitrifiers oxidize the NH_4^+, mineralized from urine and dung, to NO_2^- and subsequently to NO_3^-. The rate of nitrification depends on the NH_4^+ and oxygen (O_2) contents, pH and temperature. Usually, it takes weeks before all NH_4^+ in urine patches has been converted into NO_2^- and NO_3^-. The high osmotic pressure and high NH_3 partial pressure in urine patches may partly inhibit the nitrification process (Darrah et al., 1985; Monaghan and Barraclough, 1992). Further, nitrite-oxidizing bacteria are more rapidly inhibited than the ammonia-oxidizing bacteria. As a consequence, a temporary accumulation of NO_2^- may develop in urine and dung patches, which promotes the production and release of NO and N_2O from the urine and dung (Monaghan and Barraclough, 1993). Low O_2 concentrations in the soil may also promote the temporary accumulation of NO_2^- during nitrification and hence, the release of NO and N_2O.

Heterotrophic denitrifiers use the NO_3^- and NO_2^- as electron acceptors; they chemically reduce these N compounds to NO, N_2O and N_2. The rate of denitrification and the release of NO, N_2O and N_2 into the atmosphere are controlled by (i) NO_2^- and NO_3^- contents, (ii) availability of organic C as substrate for the heterotrophic denitrifiers, (iii) O_2 content, (iv) pH, and (v) temperature (Tiedje, 1988). Urine and dung are sources of both NO_3^- and available organic C and, therefore, the denitrification activity can be very high in urine and dung affected soil (Ryden, 1986). Treading and trampling by grazing animals also contribute to denitrifying activity because of soil compaction (Naeth et al., 1990; Warren et al., 1986). Soil compaction retards water infiltration rate and O_2 diffusivity in soil, enhancing both N_2O production during nitrification and denitrification activity (Hansen and Bakken, 1993; Torbert and Wood, 1992).

Results from experiments dealing with N_2O emission and denitrification losses from dung and urine on pastures are shown in Table 4. The emission of N_2O from soils is generally measured using enclosures (Mosier, 1989). Denitrification is generally determined using the acetylene inhibition technique with intact soil cores (Ryden et al., 1987b). In some studies ^{15}N labeling was used to quantified gaseous N losses (e.g. Monaghan and Barraclough, 1993). The N_2O emission from dung pats ranges from 0.1 to 0.7 percent. However, the amount of mineralizable organic N in dung pats is large, which can make these pats conducive to N_2O release via nitrification and denitrification over a long period (Yamulki et al., 1998). These findings suggest that the data presented so far underestimate the N_2O emission from dung, because most measurement periods have been relatively short (Table 4). Emissions of N_2O from urine patches range from 0.1 to 5.0%. This wide range has been attributed to variations in urine composition, soil type and environmental conditions, which all can have large effects (Sherlock and Goh, 1983; Monaghan and Barraclough, 1993; Allen et al., 1996; Yamulki et al. 1998). Denitrification losses from urine affected soils range from less than 1% of the urine N under dry conditions up to 65% of the urine N under moist conditions (Table 4). Monaghan and Barraclough (1993) showed large losses of N_2 during the first days after urine application, which was attributed to an increased amount of available carbon in soil, an increased microbial activity and a decreased oxygen content in the soil.

Only few studies have examined the integral effect of grazing, i.e. the combined effects of urine patches, dung pats, soil compaction and grazing on N_2O emission and denitrification, corrected for background level and N fertilizer derived N_2O emissions. Grazing derived N_2O emissions range from 0.2% of total excreted N for a silty loam in New Zealand to 9.8% of the

Table 4. Emission of N_2O from animal dung and urine deposited in grassland; a compilation of published and unpublished data. The emitted amount of N_2O-N is expressed in % of the amount of N excreted by the grazing animal.

Country[1]	Soil	Object	Period, days	N_2O emission, % of N excreted N	Denitrification % of excreted	Reference
UK	clay loam	dung	66-417	0.1-0.7		Yamulki et al., 1998
G	loess	dung	77	0.5		Flessa et al., 1996
G	sand	dung	365	0.4		Poggemann et al., 1995
NL	sand	dung	184	0.7		Velthof, unpublished data
USA	clay loam	urine	300	0.6		Mosier and Parton, 1985
UK	clay loam	urine	30	1.0-5.0		Monaghan and Barraclough, 1993
UK	clay loam	urine	60-417	0.1-1.4		Yamulki et al., 1998
B	sand loam	urine	19-35	0.1-2.4		Vermoesen et al., 1997
NZ	peat	urine	100	<1.0		Clough et al., 1996
NZ	silt loam	urine	100	1.5-3.0		Clough et al., 1996
NZ	-	urine	42	< 0.5		Sherlock and Goh, 1983
G	loess	urine	77	3.8		Flessa et al., 1996
NL	clay	urine	28	0.5		Velthof and Oenema ,1994
G	sand	urine	365	0.4-1.3		Poggemann et al., 1995
UK	clay loam	grazing	7	8.0		Velthof et al., 1996a
NL	sand	grazing	730	1.5		Velthof et al., 1996c
NL	clay	grazing	730	3.3		Velthof et al., 1996c
NL	peat I	grazing	730	2.3-9.8		Velthof et al., 1996c
NZ	silt loam	grazing	730	0.2		Carran et al., 1995
NZ	silt loam	grazing	730	1.0		Carran et al., 1995
NL	sand	urine	14		18.0	De Klein and Logtestijn,'94
NL	peat	urine	31		2.2	Koops et al., 1997
UK	clay loam	urine	30		30-65	Monaghan and Barraclough, 1993
UK	silt loam	urine	40		0.6 – 3.2	Colbourn, 1992
NZ	clay	urine	31		0.33[2]	Clough and Ledgard, 1997
NZ	peat	urine	31		2.35[2]	Clough and Ledgard, 1997
NZ	sandy loam	urine	31		1.35[2]	Clough and Ledgard, 1997
NZ	slity loam	urine	31		0.65	Clough and Ledgard, 1997

[1]B: Belgium; G: Germany, NL: the Netherlands, NZ: New Zealand, UK: United Kingdom
[2]N_2 loss + N_2O loss

total excreted N for an intensively managed grassland on peat soil in the Netherlands (Table 4). Measured denitrification losses from grazed grasslands range from 3 to 19 kg N per ha per year in New Zealand (Ledgard et al., 1996; Ruz-Jerez et al., 1994), 3 to 108 kg N per ha in the growing season in the Netherlands (De Klein and Logtestijn, 1994; Koops et al., 1996; 1997) and 8 to more than 100 kg N per ha per year in the UK (Garret et al., 1992). However, it is not always clear which fraction of the denitrification losses comes from fertilizer N, biological N fixation, urine N and dung N. Ryden (1986) showed that denitrification losses were 1.7 to 6 times higher from grazed grasslands than from a comparable cut grassland. High denitrification rates from grazed grasslands are found at the end of the grazing season, when N uptake by the grass ceases and soil water and mineral N contents and soil temperature are still at an elevated level.

Only few studies have quantified the NO emission from urine and dung patches. Studies of Williams et al. (1998) and Lovell and Jarvis (1996) indicate that the emission of N_2O-N is higher than that of NO-N from urine treated soil; the ratio between NO-N to N_2O-N soil ranged from <0.001 to 0.011 in laboratory studies and from 0.005 to 0.263 in a field study. Also Colbourn et al. (1987) showed that NO-N losses from urine patches are small. The NO in urine patches is suggested to be mainly produced by nitrifiers (Colbourn et al., 1987; Williams et al., 1998).

5. GASEOUS N LOSSES FROM STABLES AND MANURE STORAGE SYSTEMS

Housed animals deposit feces and urine in the housing system on either litter, concrete floors, or fully or partially slatted floors. There is a wide variety of animal housing systems, ranging from simple shelters where animals find protection against sun, rain, cold and or wind, to climate controlled and mechanically vented large housing systems for e.g. poultry in battery cages and finishing hogs. Manure storage and manure management differs widely among these housing systems, ranging from paddocks and unpaved feedlots in dry climates where the manure is allowed to dry until it is periodically removed, to cubicle houses with slurry storage underneath slatted floors, to the manure belt system underneath battery cages in mechanically vented poultry housing systems with forced drying of the poultry manure.

Animal manure collected in housing systems has to be stored for some time inside or outside the housing system until timely spreading of the manure on the field, i.e. during the growing season when the crop will be able to utilize the plant nutrients. In animal housing systems with (partially) slatted floors, the urine and feces are mixed and stored as slurry in pits and channels underneath the slats. When the storage capacity inside the housing system is small, slurry will be stored outside in silos, tanks and lagoons. In tie stalls with litter, feces mixed with litter and liquid manure are collected daily and stored outside in separate storage systems. In deep litter systems in organic farming, feces and urine are absorbed by straw and compacted by the loose housed animals. The stacked manure will be removed from the deep litter system only two to three times a year and transported to a manure heap outside for another storage period, or it is directly applied to agricultural land. In dry climates, animals may be kept on unpaved feedlots where the manure and urine is allowed to dry until it is periodically removed and spread on agricultural land. In some dry areas, sun dried dung cakes are collected in the field or in paddocks and burned for fuel. The total storage period of slurries and manure may range from a few weeks to more than 9 months (e.g. Bloxham and Svoboda, 1996). Because of the differences in housing system, manure management and storage period, there are large differences in N losses via volatilization of NH_3, NO, N_2O and N_2.

5.1. Volatilization of NH_3, NO, N_2O and N_2 from slurry in animal housing systems

The volatilization of NH_3 from the urine and feces in slurry inside the animal housing system is related to the NH_4^+ concentration, pH and surface area of the slurry and to the temperature and ventilation in the housing system. The potential for NH_3 volatilization from slurry is large, because of the abundance of NH_4^+ and the relatively high pH of the slurry.

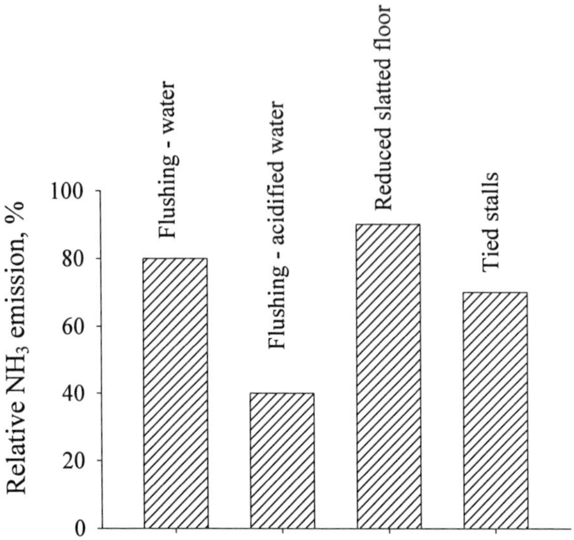

Figure 6. Potential for decreasing losses of NH_3 from animal houses of four measures relative to a fully slatted floor (after Monteny and Erisman, 1998).

Generally, the total N content of slurry varies from 3 to 5 g per kg and 40 to 75% of this N is NH_4^+, depending on animal type and the protein content of the animal feed (Table 1; Safley et al., 1986; Petersen et al., 1998a; 1998b). Ammonia is emitted from both the (slatted) floors fouled with urine and feces and from the slurry channels under slats. The larger the area fouled by the animals, the larger the NH_3 loss. The ventilation of the housing system determines the air exchange between the housing system and the outside atmosphere and thereby also the NH_3 loss; the larger the ventilation the larger the loss (e.g. Aarnink, 1997; Groot Koerkamp, 1994). Thus, emission may be mitigated by decreasing the fouled area either by decreasing the slatted area, by tying the animals (Fig. 6) or by decreasing the ventilation. Frequent cleaning of the floor also decreases NH_3 losses (Fig. 6).

Mean NH_3 losses are larger from pig housing systems than from dairy cattle housing systems, because of differences in the amount of NH_4^+ in the slurry and temperature. The loss of NH_3 from cattle housing systems with slatted floors in Denmark is estimated at about 8 % of the total-N in the slurry, and losses from pig housing systems with slatted floors are estimated at 15 % of the total-N in the slurry (Poulsen and Kristensen, 1997). Estimated losses of NH_3 from dairy cattle housing systems with slatted floors in The Netherlands range from 2% when the cattle are housed for 180 days per year in tie stalls to about 15% of the total-N in the cattle slurry when the cattle are housed all year round in cubicle houses (summer feeding). This wide range is caused by the large difference in the areas of fouled floor between tie stalls and cubicle houses, and by the difference in housing period. Estimated NH_3 losses from pig housing systems with slatted floors range from 17% of total N for piglets to 29% of total N for rearing pigs (Monteny and Erisman, 1998; Oenema et al., 2000a). Large losses of up to 50% of the total initial N content have been measured in poultry housing systems with partial drying of the poultry manure (Groot Koerkamp et al., 1995; 1998). By contrast, small NH_3 losses of less than 5% occur in mechanically vented low-emission housing systems in which the area fouled with urine and feces is minimal and where most of the NH_3 is scrubbed from the exhaust air via chemical or biological scrubbers.

Slurry stored in pits and canals underneath slatted floors is not a significant source of N_2O, NO or N_2, because very little NH_4^+ from the slurry is oxidized in the highly anoxic environment. However, the fouled surface of slats, with a large interface between air and slurry, may be a source of N_2O. This was shown in the study of Thelosen et al (1993), who measured a total annual emission of 0.2 kg N_2O-N per pig place. Very large N_2O losses may occur following acidification of the slurry with nitric acid (HNO_3) to decrease the volatilization of NH_3 and to increase the N fertilizer value of the slurry (Oenema and Velthof, 1993; Oenema et al., 1993).

5.2. Volatilization of NH_3, NO, N_2O and N_2 from slurry in tanks, silos and lagoons

Ammonia emission from slurry in open tanks, silos and lagoons ranges from 6 to 30% of the total N in stored slurry (Sommer, 1997; Harper et al., 2000). The NH_3 loss is related to environmental conditions (temperature and wind), slurry composition and surface area. Losses are larger from pig slurry than from cattle slurry, due to differences in NH_4^+ content. Further, losses tend to be twice as large from slurry that has been fermented in a biogas plant than from unfermented slurry, because fermented slurry has a higher pH and NH_4 content (Sommer et al., 1993; Sommer, 1997).

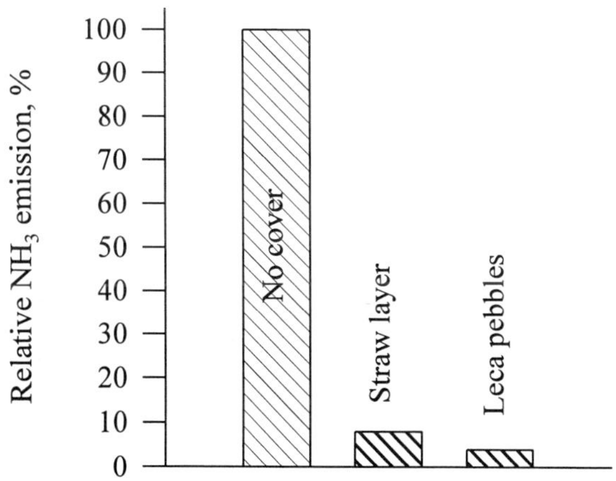

Figure 7. Decrease in NH_3 losses following the covering of slurry storage systems, in percent (after Sommer, 1997).

A cover on the slurry significantly decreases NH_3 loss. The cover may be a natural surface crust formed by solids floating on the surface, a cover of straw, peat or floating expanded clay particles, or a roof. Covers greatly decrease the air exchange rate between the surface of the slurry and the atmosphere by creating a stagnant air layer above the slurry through which NH_3 has to be transported by the slow process of diffusion. This decreases the NH_3 losses to less than 10% of those from uncovered slurry (Fig. 7).

Because stored slurry is anaerobic, there will be little or no nitrification. As a result, little NO, N_2O and N_2 will be lost (Sommer, 1997). However, under drying conditions a mosaic of anaerobic and aerobic sites may emerge in the porous crust, creating an environment where N_2O is produced (Hüther et al., 1997; Sommer et al., 2000). Then, emissions of N_2O may go up to 25 mg N_2O-N m^{-2} h^{-1}.

5.3. Volatilization of NH_3, NO, N_2O and N_2 from solid manure in animal housing systems

Litter is added to animal housing systems for animal well-being and for absorbing the liquid excrements of the animals. The litter can be hay, straw, chopped or unchopped, heathy plants, woody snips and saw dust or simply sand, depending on the housing system and the availability of litter. The fraction of litter to dung and urine varies widely, depending again on housing system. Further, the dung and urine mixed with litter can be removed from the housing system on a daily basis or just once or a few times a year. In tie stalls with a small channel behind the animals, manure mixed with a little litter will be scraped to the manure heap outside on a daily basis. Losses of NH_3, NO, N_2O and N_2 from tie stalls with daily removal of the manure are only a few percent of the amount of N excreted; the time period for N losses in the stable is simply too short for large losses. By contrast, the stacked manure in deep litter systems and in paddocks and drylot systems, will be removed only a few times a year. Such long storage times create conditions for volatilization of NH_3, NO, N_2O and N_2.

In deep litter systems, the composition of the stacked manure is related to the animal type and animal nutrition, straw admixture, downward urine transport, and to fermentation and microbial transformation processes in the manure. A significant fraction of the NH_4^+ mineralized from the easily metabolizable N fractions in urine and dung can be absorbed by the straw and transformed into organically bound N by micro-organisms (Henriksen et al., 2000). This would suggest that the potential for N losses via volatilization of NH_3, NO, N_2O and N_2 from deep litter systems is small, because of the immobilization of NH_4^+. However, animals walk around and foul the litter in the surface layer with fresh urine and feces. Further, molecular oxygen (O_2) diffuses into the porous surface layer, using straw as channels. Fermentation processes increase the temperature and induce an upward current of air. As a result, NH_3 losses from deep litter systems are up to 10% of the N that is excreted and collected in the straw litter (Rom and Hendriksen, 2000). Aerobic microbial activity in deep litter may cause a temperature increase to about 40-50°C at 10 cm depth. In this layer, O_2 is depleted. The NO, N_2O and N_2 production in deep litter is related to the partial aeration of the top layer, which is a function of litter addition and animal activity. When little litter is added, nitrification is inhibited by a combination of low oxygen partial pressure, high temperature and a high NH_3 concentration (Henriksen et al., 2000). Mixing of the top layer once a week may increase total N losses via nitrification and denitrification to 47% of the N excreted (Thelosen et al., 1993). Then, losses of N_2O are in the range of 15 to 21% and losses of NH_3 range from 9 to 17% (Groenestein et al., 1993; Groenestein and Van Faassen, 1996).

5.4. Volatilization of NH_3, NO, N_2O and N_2 from manure heaps

Nitrogen losses from manure heaps via volatilization of NH_3, NO, N_2O and N_2 depend on the composition and stacking of the manure and on storage conditions. Manure heaps store manure collected from dry lots, feed lots and deep litter stables once in a few months. There are also manure heaps that accumulate fresh manure from stables on a daily basis, either via additions on top or via intrusion from the bottom. When fresh manure is added daily on top of the heap, there is a constant source of fresh urea and a near constant flux of NH_3 into the atmosphere, but there is little chance for nitrification and denitrification processes. By contrast, when fresh manure is added via intrusion from the bottom, volatilization of NH_3 is small because of lack of fresh urea, but surface layers become partly aerobic and than nitrifying bacteria may transform NH_4^+ into NO_3^-. This provides denitrifying bacteria the opportunity to denitrify NO_3^- when it moves to anoxic micro sites in the manure, for example when rain water leaches the NO_3^- from the surface layers downwards. This makes these heaps conducive to NO, N_2O and N_2 losses.

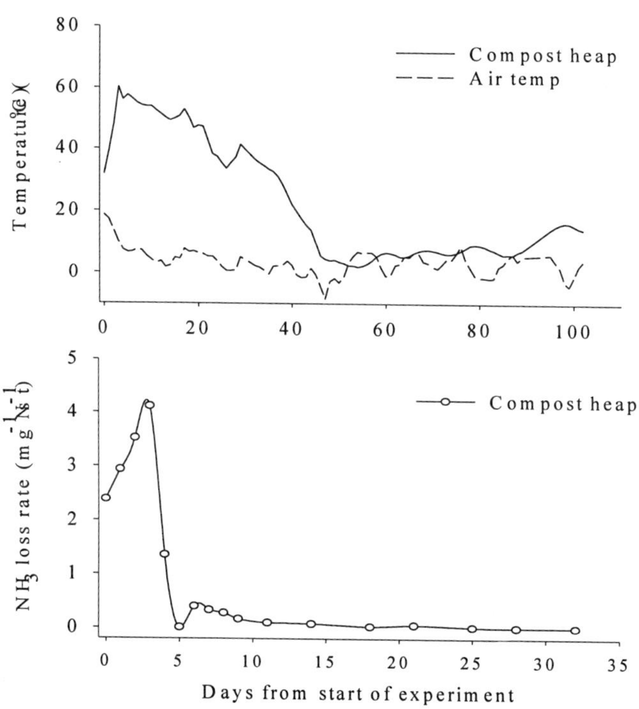

Figure 8. Air temperature and manure temperature (upper panel) and the NH_3 loss from the manure (lower panel). Note differences in scale of X-axes: upper panel 0-100 days and lower panel 0-35 days. The increase in NH_3 loss during the first four days (lower panel) coincides with the rise in temperature of the manure (upper panel).

Usually, the manure lies on a concrete floor surrounded with concrete walls, but manure heaps on bare agricultural land without any provision for the collection of drainage water are also common. Some heaps may be covered to decrease NH_3 ammonia volatilization and to prevent the infiltration of rain in the manure and the leaching of solutes from the manure. Evidently, storage conditions vary widely, as do gaseous N losses from manure heaps.

During the formation of a manure heap, the temperature inside the heap may increase to 70°C due to aerobic microbial metabolism, i.e. composting (Fig. 8). Composting generates an upward airflow in the heap and, as a consequence, fresh air from the atmosphere will enter through the lower section of the heap. Further, composting causes an increase in pH, which increases the NH_3 fraction relative to NH_4^+. As a result, volatilization of NH_3 from composting solid manure and deep litter may be high. Losses of 25-30% of the total-N in stored pig manure and deep litter have been recorded (Petersen et al., 1998; Karlsson and Jeppson, 1995). Losses can be lowered by decreasing the convection of air through the heap with a cover of for example tarpaulin or through compaction of the litter. In solid manure with low straw content, the diffusion rate of O_2 is low and composting nearly absent (Forshell, 1993).

During the composting phase of solid manure little N_2 and N_2O is produced, because nitrifying and denitrifying micro organisms are generally not thermophilic (Hellman et al., 1997). After the temperature declines in compost heaps, N_2O concentration in the stores generally increases, due to the co-existence of aerobic zones with nitrification and anoxic zones with denitrification (Petersen et al., 1998). Emissions of N_2O from composting manure are in the range of 0.1–0.3% of the total-N (Czepiel et al., 1996; Petersen et al., 1998), depending also on the compaction, water content of the manure and the environmental conditions (e.g. Sommer, 2000; Brown et al., 2000).

6. GASEOUS N LOSSES FROM SLURRY AND MANURE APPLIED TO SOIL

Animal slurry and manure collected in animal houses are valuable sources of plant nutrients for crop production. However, farmers in many developed countries have had little incentive for applying manure timely and accurately to the soil during the last decades, because of the availability of cheap mineral fertilizers. This development has contributed to large N losses from slurry and manure during storage but also during and after application to soil. Current governmental legislation on the proper use and management of slurries and manure has reversed this trend in many countries from about 1985 onwards.

Application of animal slurry to soil induces a sequence of reactions. During the first few hours, the NH_4^+ concentration in the surface soil is high and may even increase further due to evaporation of water, despite losses via NH_3 volatilization (Sommer and Sherlock, 1996). Next, the concentration of NH_4^+ in the soil surface decreases (Comfort, et al., 1990; Kirchmann and Lundvall, 1993), but this decrease can not be explained fully by the measured N loss via NH_3 volatilization and the measured increase in NO_3^- concentration. The decrease in total inorganic N in the surface layer is likely caused by a combination of NH_3 volatilization, coupled nitrification-denitrification and immobilization of inorganic N in microbial biomass. Animal slurry has a high content of easy degradable carbon in the form of volatile fatty acids (VFA), and these VFA are rapidly metabolized (Kirchmann and Lundvall, 1993; Paul and Beauchamp, 1989). The transformation of these easily degradable VFA may contribute to immobilization of inorganic N and to the development of anoxic micro sites, which subsequently may lead to increased emissions of NO, N_2O and N_2.

6.1. Volatilization of NH_3 from slurry and manure applied to soil

Rates of NH_3 volatilization as high as 12 kg N ha^{-1} per hour have been measured, immediately after slurry application (Pain et al. 1989). This high initial loss rate (see Fig. 9) is related to both the initial high NH_4^+ concentration in the soil surface, and to the rise in pH in the soil surface. The pH in the soil slurry mixture increases, because the volatilization of CO_2 is faster than the volatilization of NH_3 and because of the degradation of VFA (Sommer and Sherlock, 1996). The cumulative NH_3 loss increases hyperbolic with time as shown in Fig. 9. Generally, the rate of NH_3 volatilization is very low after a few days, because the concentration of dissolved NH_4^+ in the soil surface decreases rapidly due to volatilization, immobilization, infiltration and nitrification (Van der Molen et al., 1990; Brunke et al., 1988; Oenema et al., 1993). Hence, 50% of the total NH_3 loss occurs within 4 to 12 h after slurry application (Pain et al., 1989; Moal et al., 1995).

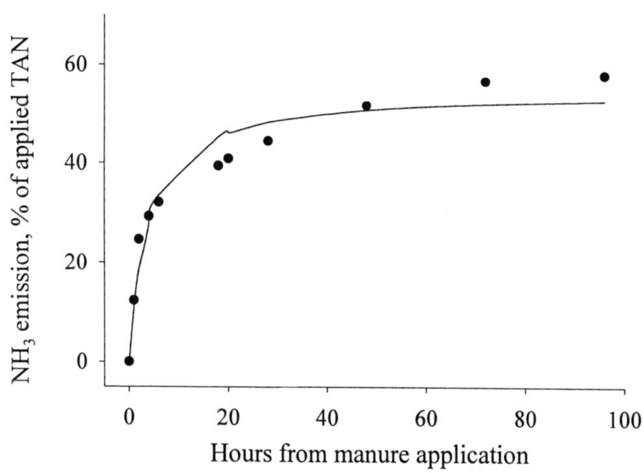

Figure 9. Pattern of NH_3 volatilization following surface application of pig slurry to arable land on top of stubble, in % of the total amount of ammonium N (TAN) in the slurry. (after Bless et al., 1991).

The rate of NH_3 volatilization from *slurry* applied to soil is related to temperature and solar radiation (Brunke et al,. 1988; Moal et al.,1995); the higher the temperature the larger and faster the NH_3 loss. At low temperature and on frozen soil, volatilization may continue for a long time and result in a large cumulative NH_3 loss too. In this case, the large losses are explained by either low rates of infiltration of NH_4^+ in the soil, or by low rates of immobilization and nitrification (Sommer Incorporating slurry into the soil is a most effective way of decreasing NH_3 volatilization. Shallow direct injection of slurry can decrease losses by about 70%, while deep injection will stop losses completely. Incorporation of slurry by ploughing or by rotary harrow and immediate ploughing of the soil following surface application of slurry decreases NH_3 losses by 80% (Pain et al., 1991). Furthermore, application of slurry with trailing hoses on the soil beneath the plant canopy may decrease NH_3 volatilization with more than 50%; the efficiency of this technique increases with increasing leaf area and height of the crop (Sommer et al., 1997).

Losses of NH_3 from *solid manure* applied to soil are still poorly understood. The volatilization of NH_3 from solid manure follows a pattern over time that is different from that of animal slurry. The initial loss is low, but the volatilization continues for a long period, probably because the NH_4^+ from the manure does not infiltrate the soil at the same rate as NH_4^+ from slurry (Sommer and Christensen, 1990; Chambers et al., 1997). The few studies dealing with NH_3 emission from soild manure applied to soil indicate that about 50% of the loss occurs within 24 h and that the volatilization may continue for about 10 days. Incorporation of solid manure into the soil by ploughing decreases NH_3 losses as effectively as ploughing does after application of animal slurry.

6.2. Volatilization of NO, N_2O and N_2 from slurry and manure applied to soil

Application of slurry and manure increases the loss of N from soil via volatilization of NO, N_2O and N_2 produced via nitrification and denitrification processes. The application increases the contents of NH_4^+, and of easily mineralizable N and C in the topsoil, and thereby activates microbial respiration and biological oxygen consumption in the soil. This in turn may increase nitrification and subsequently denitrification locally.

Organic compounds from slurry and manure provide readily available substrate for denitrifiers (Comfort et al., 1990; Dendooven et al., 1998a; 1998b; Ellis et al., 1998; Paul and Zebarth, 1997). Paul and Beauchamp (1989) and Dendooven et al. (1998a) showed that volatile fatty acids (VFA) may be used by denitrifiers as substrate and thereby increase N losses via denitrification. This is a short-term effect because VFA are only present during a few days after application of slurry to soil (Kirchman and Lundvall, 1993). However, there are many interacting factors that control nitrification and denitrification processes and the patterns of N losses and emission of N_2O are complex. Results presented in Fig. 10 show that the N_2O emission following the application of pig slurry to bare soil was high initially (on the day of application), than dropped to background level and increased again after 20 days. The short N_2O flux immediately after slurry application is probably related to the addition of easily available C which increased denitrification of nitrate initially present in the soil. The increase after 20 days is probably related to nitrification and subsequently denitrification of NO_3^- from applied slurry N. In contrast, N_2O emissions from soil treated with ammonium nitrate fertilizer increased rapidly during the first week and than dropped during the second week to background level. These patterns are strongly influenced also by initial soil moisture content and rainfall.

Nitrogen losses via denitrification in slurry amended soil vary from less than 10% of total N applied in relatively dry, sandy soils (Comfort et al., 1990; De Klein et al., 1996; Egginton and Smith, 1986a; 1986b; Van der Weerden et al., 1994; Velthof et al, 1997) to more than 10% of total N applied in clay soil late in the growing season (Thompson et al., 1987). Relatively low losses were found for spring application of cattle slurry to grassland (Velthof et al., 1997). In the growing season, the N uptake of a crop is high, especially for grassland, and little applied slurry N will be available for denitrifiers. Relatively high losses are associated with autumn/winter application of slurries (Thompson et al., 1987; Pain et al., 1990). In the winter, the N uptake by the crop is negligible, so that slurry N will be available for denitrifiers. Moreover, wet conditions prevail during winter in areas with a temperate climate, which may enhances denitrification activity whenever nitrification is not rate limiting for the availability of NO_3^-. It has been shown that the addition of a nitrification inhibitor to slurry and manure may strongly decrease denitrification losses (De Klein et al., 1996; Pain et al., 1990).

Application of animal slurries to soils increases the emission of N_2O. Slurry-derived N_2O-emissies range from less than 0.1% of total N applied up to about 1% of total N applied (Chadwick et al., 2000; Comfort et al., 1990; Egginton and Smith, 1986b; Ferm et al., 1999; Misselbrook et al., 1998; Velthof and Oenema, 1993; Velthof et al., 1997; Weslien et al., 1998). Both nitrification and denitrification may be sources of N_2O. Dendooven et al. (1998b) concluded that about 33 percent of the N_2O produced after pig slurry application was derived from nitrification.

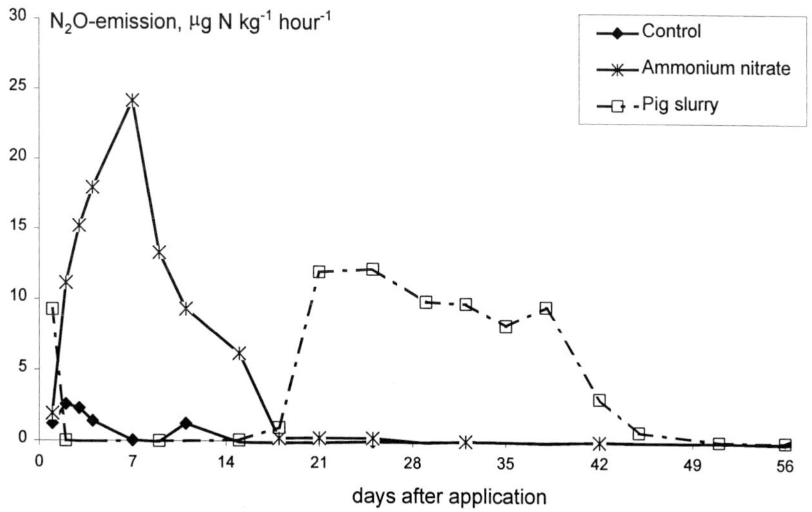

Figure 10. Patterns of N_2O emission following application of ammonium nitrate fertilizer and of pig slurry at a rate of 200 mg N per kg soil (Velthof et al., unpublished results).

Table 5. N_2O-emission from animal slurries and manure applied to grassland in April, July and October (Chadwick et al., 2000).

Application time	Manure	Total N application rate kg N ha^{-1}	N_2O-emission % of N applied
April	Dairy slurry	125	0.97
	Pig slurry	97	0.44
July	Dairy slurry	110	0.12
	Pig slurry	295	0.12
October	Pig slurry	300	0.24
	Dilute dairy cow effluent	35	0.38
	Pig farm yard manure	310	0.05
	Beef farm yard manure	315	0.20
	Layer manure	416	0.05

So far, most studies suggest that the N_2O emission from animal slurries applied to grassland is less than the N_2O emission from an equivalent amount of nitrate-based N fertilizer (e.g. Egginton and Smith, 1986a; 1986b; Velthof and Oenema, 1993; Velthof et al., 1997). This suggests that the NH_4^+ from slurry is taken up by the grass or by soil microbial biomass before it can be transformed into gaseous N oxides.

The composition of the animal manure can have a large effect on the emission of N_2O (Table 5). Large emissions are associated with slurries with high contents of inorganic N, and easily mineralizable N and C (i.e. VFA). Animal nutrition affects the composition of the manure and thereby also N and C transformations and losses during storage and after soil application (Kreuzer et al., 1998; Misselbrook et al., 1998; Mroz et al., 1995; Paul et al., 1998). Adjustment of animal nutrition is thus a tool to decreases gaseous N losses from animal manure.

7. DISCUSSION AND CONCLUDING REMARKS

Evidence for losses of gaseous N compounds that decrease the fertilizer value of animal manure dates back to about 1850 (e.g. Russel, 1912; Bussink and Oenema, 1998 and references therein), but it is only during the last 3 decades that such losses have been associated with environmental effects. It is now well-recognized that emission of gaseous N compounds into the atmosphere may contribute to a cascade of environmental effects, in both the atmosphere and biosphere (e.g. Galloway, 1998), though all cause-effect relationships are not well-understood yet. Further, it is recognized that livestock farming systems have a relatively large share in the total atmospheric burden of NH_3 and N_2O (e.g. Dentener and Crutzen, 1994; Mosier et al., 1998), and that this share may increase further during next decades if human population and animal protein consumption continue to increase. However, the uncertainty in the estimated NH_3, NO and N_2O losses from livestock farming systems is still large. More interestingly, there is potential for mitigation of these losses, as suggested also by various authors cited in this chapter.

7.1. Uncertainties in estimated gaseous N losses from livestock farming systems

The uncertainty in the estimated contributions of livestock farming systems to the total emissions of NH_3, NO and N_2O into the atmosphere stems in part from the paucity in measurement data of gaseous N losses from animal housing systems and manure storage systems, especially also for livestock farming systems in the developing countries. The number of farm animals is larger in developing countries than in developed countries, while the number of measurements of the emissions of NH_3, NO, N_2O and N_2 from livestock farming systems is much larger in developed countries than in developing countries. Further, there are more data about N losses associated with the application of slurry and manure to agricultural land than about N losses from animal housing systems and manure storage systems, while N losses from housing systems, manure storage systems and from slurry application to land may be equally large. To some extent, gaseous N losses from housing systems and manure storage systems seem to have been neglected or simply considered as unavoidable.

The uncertainty in the estimates stems in part also from the complexity of the N cycle of livestock farming systems and from the many interacting factors that control the sources of NH_3, NO and N_2O in these systems. There are still many unknowns and there is a wide variety in animal nutrition, animal housing systems, manure storage systems and manure management. Environmental conditions also have a large impact on gaseous N losses, though the effects of environmental conditions are in part entangled with the livestock farming systems itself. These interacting factors complicate an accurate assessment of the total gaseous N losses from livestock farming systems. Further, data statistics on animal housing systems, manure storage and manure management are poor. As discussed in the previous sections of this chapter, subtle differences in manure storage and management can have major effects on the emissions of NH_3, NO, N_2O and N_2.

The availability of measurement data about gaseous N losses from animal manure decrease in the order $NH_3 > N_2O > NO > N_2$. There are very few direct measurements of NO emissions and there are no direct measurements of N_2 losses, because of technical and analytical difficulties. The measurement of gaseous losses from animal housing systems is complex, especially for natural vented housing systems with numerous sites of air exchange. Small enclosures inside the housing system appear to provide biased estimates (Monteny and Erisman, 1998) and the measurement of NH_3, N_2O, and NO concentrations in the inside atmosphere, in combination with a tracer gas such as for example SF_6 requires expensive equipment that only recently have become available. Because of specialization and the funding driven research interests, there are only very few combined or simultaneous measurements of NH_3, N_2O, NO and N_2 emissions from livestock farming systems or from a compartment of a livestock farming system (e.g Harper et al., 2000). Fortunately, techniques and instrumentation have greatly improved during the last decades and it has become easier to measure NH_3, NO and N_2O emissions. Combined measurements would improve our understanding of the factors that control NH_3, N_2O, and NO emissions from livestock farming systems and would help to consider the interactive nature of the gaseous N losses.

Simulation models have become common tools to improved our understanding of the N transformations in animals, animal manure and soil. Our knowledge of the microbiology and biogeochemistry of NH_3, NO, N_2O, and N_2 formation processes is quite extensive, and the number of possible biophysicochemical controls incorporated into simulation models is considerable, but there is still a great need for calibration and validation of these models. There is also a need for linking and integrating of models as most models have been developed separately for animals, manure, housing system, soil and crop. Models can also help to explain the gap that is often observed between laboratory and field. Our knowledge of the managerial effects on NH_3, NO, N_2O, and N_2 emissions from livestock farming systems in practice seem to be still inadequate.

Improving the accuracy of the estimates of gaseous N losses from livestock farming systems requires also improved characterization of livestock farming systems and improved inventories of animal housing systems and manure management. Monitoring of N balances of animal housing systems and especially of the manure compartment (e.g. Aarts et al., 2000), may provide indirect estimates of the total gaseous N losses from manure, especially when the N flows are quantified relative to an element that is not conducive to gaseous N loss, such as phosphorus or potassium (e.g. Oenema et al., 2000a). Establishing N/P balances of animal housing systems and manure storage systems requires proper sampling procedures and accurate chemical analyses of animal feed and animal manure, but do not require expensive equipment and complicated calculations. When combined with direct measurements of NH_3, N_2O, and NO emissions, N/P balances can provide an independent check of the accuracy of the direct measurements. Evidently, monitoring N/P balances can be practiced also in remote areas, without much complications.

7.2. Mitigation of gaseous N losses from livestock farming systems

Measures to decreases NH_3 emissions from animal manure have been researched from the end of the eighteenth century onwards (e.g. Bussink and Oenema, 1998), so as to improve the N fertilizer value of the manure. This type of research received new impetus when the emphasis in the research on N in agricultural systems shifted from agronomic to environmental effects nearly one century later. It is now well-recognized that a whole farming systems approach is needed to examine measures that may decrease N losses and more specifically gaseous N emissions. A systems approach shows how intervening in one part may affect losses in another part of the system. Because of the interactive nature of the processes involved, it is important that any changes that are imposed are based on a mechanistic understanding of the N cycle and N transformations (e.g. Jarvis, 1997; Aarts et al., 2000).

As the quest for food and more specifically animal protein will increase concomitant with the increase in global human population and standard of living, the quest for measures to decrease gaseous N emissions from livestock farming systems will increase too. Apart from social, economical, political and cultural implications, it is a major challenge to transfer the message and the available insights and techniques to decreases gaseous N emissions from livestock farming systems to the stakeholders, i.e. the farmers. Education and implementation of best management practices are the first steps, but financial incentives will be needed when the necessity for decreasing gaseous N emissions increases. Bans and controls (do's and don'ts) will be effective only when accompanied by education, extension, and financial incentives.

The N balance of the system is a good overall indicator for both the N use efficiency and total N losses. Therefore, targets for N surpluses for individual farms may facilitate the implementation of technical measures to decrease N losses and more specifically gaseous N emissions. Improving the productivity and N use efficiency of the system, in combination with a lowering of the N input into the system seems to be the best strategy for decreasing gaseous N emissions from livestock farming systems (e.g. Oenema et al., 2000b). Improving the productivity and N use efficiency of the system shows up in the N balance, in the difference between total inputs and total outputs. Further, N balances for the different compartments of the livestock farming systems may indicate the weakest chain in the system and the potentials for improvement. Improving the productivity of the herd genetically, improving the composition of the animal feed, and improving the storage and management of manure and the utilization of the N in the manure for the production of high yielding crops, all contribute to both the agronomic and environmental performances of the farming system. Measures like lowering the protein intake of animals, shifting N excretion from urine to feces, regular flushing of the floor in animal houses with acidified water, covering manure storage systems, direct injection of slurry in the soil, and proper timing of manure application may all greatly decrease NH_3 emissions from livestock farming systems. Nevertheless, total N losses will remain the same when these measures are not accompanied with a decrease in total N input into the system or with an increase in total output via animal products.

While many technical measures have been developed and tested in practice for the mitigation of NH_3 emissions from animal manure in some countries, there is still little information about measures to decrease NO, N_2O, and N_2 emissions from livestock farming systems in practice. The formation of these gases is related to nitrification and denitrification and hence to the partial aeration of the initial anoxic manure. Addition of litter and drying facilitates the partial aeration of manure, and so do animals that tread, snout and scratch on and in the manure, as in deep litter stables. Hence, emissions of NO, N_2O, and N_2 are related to the type of animal housing system and to manure management; emissions from slurries are low and emissions from solid manure with litter and from manure that is subject to drying-wetting fluctuations or partial aeration tend to be high. Nitrification is the rate limiting step here. Though nitrifying bacteria are slow growers, there can be a steady transformation of NH_4^+ into NO_2^- and NO_3^- concomitant with a small but steady loss of NO and N_2O during the aeration phase. This phase can be followed by a pulse of NO, N_2O, and N_2 from rapid denitrification of the accumulated NO_3^-, with highly varying ratios of NO to N_2O to N_2 versus time. The ratios seem to depend on site specific environmental conditions. Evidently, mitigation measures for emissions of NO, N_2O, and N_2 require further considerations of the effects of manure management and environmental conditions in practice.

Acknowledgements

We would like to acknowledge the EU (project FEEDMAN) and the ministries of Agriculture, Nature Management and Fisheries in The Netherlands and Denmark for funding this study.

REFERENCES

Aarnink, A.J.A. 1997. Ammonia emission from houses for growing pigs as affected by pen design, indoor climate and behaviour. PhD Thesis, Wageningen Agricultural University, Wageningen, The Netherlands, 175 pp.

Aarts, H.F.M., B. Habekotté, and H. van Keulen. 2000. Nitrogen management in the "De Marke" dairy farming system. Nutri. Cycl. in Agroecosyst. 56:231-240.

Afzal, M., and W.A. Adams. 1992. Heterogeneity of soil mineral nitrogen in pasture grazed by cattle. Soil Sci. Soc. Am. J. 56:1160-1160.

Allen, A.G., S.C. Jarvis, and D.M. Headon. 1996. Nitrous oxide emissions from soils due to inputs of nitrogen from excreta return by livestock on grazed grassland in the UK. Soil Biol. Biochem. 28:297-607.

Bakker, G.C.M., and R.A. Dekker. 1998. Effect of source and amount of non-starch polysaccharides on the side of excretion of nitrogen in pigs. J. Animal Sci. 76, Suppl. 1, 665.

Ball, P.R., and J.C. Ryden. 1984. Nitrogen relationships in intensively managed temperate grasslands. Plant Soil 76:23-33.

Bannink, A., and A.M. van Vuuren. 1998. Cation anion difference in dairy cow rations as a measure to influence urine pH and ammonia emission. ID-DLO report 98.032, Lelystad, NL.

Bannink, A., H. Valk, and A.M. van Vuuren. 1999. Intake and excretion of sodium, potassium, and nitrogen and the effects on urine production by lactating dairy cows. J. Dairy Sci. 82:1008-1018.

Beauchamp, E.G., G.E. Kidd, and B. Thurtell. 1978. Ammonia volatilization from sewage sludge applied in the field. J. Environ. Qual. 7:141-146.

Beauchamp, E.G., G.E. Kidd, and B. Thurtell. 1982. Ammonia volatilization from liquid dairy cattle manure in the field. Can. J. Soil Sci. 62:11-19.

Berges, M.G.M., and P.J. Crutzen. 1996. Estimates of global N_2O emissions from cattle, pig and chicken manure, including a discussions of CH_4 emissions. J. Atm. Chem. 24:241-269.

Bless, H.G., R. Beinhauer, and B. Sattelmacher. 1991. Ammonia emission from slurry applied to wheat stubble and rape in North Germany. J. Agric. Sci., Cambridge 117:225-231.

Bloxham, P.F., and I.F. Svoboda. 1996. Management of livestock manures in the U.K. Ingénieries: Eau-Agriculture Territoires. Special Issue. 39-49.

Brown, H.A., C. Wagner-Riddle, and G.W. Thrtell. 2000. Nitrous oxide flux from solid manure in storage as affected by water content and redox potential. J. Environ. Qual. 29:630-638.

Brunke, R., P. Alvo, P. Schuepp, and R. Gordon. 1988. Effect of meteorological parameters on ammonia loss from manure in the field. J. Environ. Qual. 17:431-436.

Bristow, A.W., D.C. Whitehead and J.F. Cockburn. 1992. Nitrogenous constituents in the urine of cattle, sheep and goats. J. Sci. Food Agric. 59:387-394.

Brouwer, F., and P.H. Hellegers. 1997. Nitrogen flows at farm level across European Union Agriculture. p. 11-26. *In* E. Romstad, J. Simonsen, and A. Vatn (eds.) Controlling Mineral Emissions in European Agriculture; Economics, Policies and Environment. CAB International, Wallingford, UK.

Bussink, D.W. 1992. Ammonia volatilization from grassland receiving nitrogen fertilizer and rotationally grazed by dairy cattle. Fert. Res. 33:257 – 265.

Bussink, D.W. 1994. Relationships between ammonia volatilization and nitrogen fertilizer application rate, intake and excretion of herbage nitrogen by cattle on grazed swards. Fert. Res. 38:111-121.

Bussink, D.W., J.F.M. Huijsmans, and J.J.M.H. Ketelaars. 1994. Ammonia volatilization from nitric-acid-treated cattle slurry surface applied to grassland. Neth. J. of Agric. Sci. 42:293-309.

Bussink, D.W., and O. Oenema. 1996. Ammonia volatilization from dairy farming systems in temperate areas; a review. Nutri. Cycl. in Agroecosyst. 51:19-33.

Cahn, T.T., J.W. Schrama, A.J.A. Aarnink, M.W.A. Verstegen, C.E. van 't Klooster, and M.J.W. Heetkamp. 1998a. Effect of dietary fermentable non-starch polysaccharides from pressed sugar beet pulp silage on ammonia emission from slurry of growing-finishing pigs. Animal Sci. 67:583-590.

Cahn, T.T., A.J.A. Aarnink, Z. Mroz, A.W. Jongbloed, J.W. Schrama, and M.W.A. Verstegen. 1998b. Influence of electrolyte balance and acidifying calcium salts in the diet of growing-finishing pigs on urinary pH, slurry pH and ammonia volatilisation from slurry. Livestock Production Sci. 56:1-13.

Carran, R.A., P.W. Theobald, and J.P. Evans. 1995. Emission of nitrous oxide from some grazed pasture soils in New Zealand. Aust. J. Soil Res. 33:341-352.

Carré, B., J. Gomez, and A.M. Chagneau. 1995. Contribution of oligosaccharide and polysaccharide, and excreta losses of lactic acid and short chain fatty acids, to dietary metabolisable energy values in broiler chickens and adult cockerels. British Poultry Sci. 36:611-629.

Castellanos, J.Z., and P.F. Pratt, P.F. (1981) Mineralisation of manure nitrogen – correlation with laboratory indexes. Soil Sci. Soc. Am. J. 45:354-357.

Chadwick, D.R., B.F. Pain, and S.K.E. Brookman. 2000. Nitrous oxide and methane emisisons following application of animal manure to grassland. J. Environ. Qual. 29:277-287.

Chambers, B.J., K. Smith, and T.J. van der Weerden. 1997. Ammonia emissions following the land spreading of solid manure. p. 275-280. *In* S.C. Jarvis, and B.F. Pain (Eds.) Gaseous nitrogen emissions from grasslands. CAB International, Wallingford, UK.

Clough,T.J., R.R. Sherlock , K.C. Cameron, and S.F. Ledgard. 1996. Fate of urine nitrogen on mineral and peat soils in New Zealand. Fert. Res. 178:141-152.

Clough, T.J., and S.F. Ledgard. 1997. Denitrification losses from urine-affected soils. p. 71-73 *In* S.C. Jarvis, and B.F. Pain (eds.) Gaseous nitrogen emissions from grasslands. CAB International, Wallingford, UK.

Colbourn, P., J.C. Ryden, and G.J. Dollard. 1987. Emissions of NOx from urine-treated pasture Environ. Pollut. 46:253-261.

Colbourn, P. 1993. Limits to denitrification in two pasture soils in a temperate maritime climate. Agric. Ecosyst. Environ. 43:49-68.

Colbourn, P. 1992. Denitrification and N_2O production in pasture soil: the influence of nitrogen supply and moisture. Agric. Ecosyst. Environ. 39:267-278.

Coleman, D.C., R. Andrews, J.E. Ellis, and J.S. Singh. 1977. Energy flow and partitioning in selected man-managed and natural ecosystems. Agroecosystems 3:45-54.

Comfort, S.D., K.A. Kelling, D.R. Keeney, and J.C. Converse J. C. 1990. Nitrous oxide production from injected liquid dairy manure. Soil Sci. Soc. Am. J. 54:421-427.

Czepiel, P., E. Douglas, R. Harriss, and P. Crill, P. 1996. Measurements of N_2O from composted organic wastes. Environ. Sci. Technol. 30:2519–2525.

Darrah, P.R., R.E. White, and P.H. Nye. 1985. Modelling growth responses of soil nitrifiers to additions of ammonium sulphate and ammonium chloride. Plant Soil 86:424-439.

De Klein, C.A.M., and R.S.P. van Logtestijn. 1994. Denitrification and N_2O emission from urine-affected grassland soil. Plant Soil 163:235-242.

De Klein, C.A.M., R.S.P. van Logtestijn, H.G. van der Meer, and J.H. Geurink. 1996. Nitrogen losses due to denitrification from cattle slurry injected into grassland soil with and without a nitrification inhibitor. Plant Soil 183:161-170.

Dendooven, L., E. Bonhomme, R. Merckx, and K. Vlassak. 1998a. Injection of pig slurry and its effect on dynamics of nitrogen and carbon in a loamy soil under laboratory conditions. Biol. Fert. Soils 27:5-8.

Dendooven, L., E. Bonhomme, R. Merckx, and K. Vlassak. 1998b. N dynamics and sources of N_2O production following pig slurry application to a loamy soil. Biol. Fert. Soils 26:224-228.

Denmead, O.T., J.R. Simpson, and F.R. Freney. 1974 Ammonia flux into the atmosphere from a grazed pasture. Science 185: 609-610.

Dentener, F.J., and P.J. Crutzen. 1994. A three dimensional model of the global ammonia cycle. J. Atmos. Chem. 19(4):331-369.

Egginton, G.M., and K.A. Smith. 1986a. Nitrous oxide emission from a grassland soil fertilized with slurry and calcium nitrate. J. Soil Sci. 37:59-67.

Egginton, G.M., and K.A. Smith. 1986b. Losses of nitrogen by denitrification from a grassland soil fertilized with cattle slurry and calcium nitrate. J. Soil Sci. 37:69-80.

Ellis, S., S. Yamulki, E. Dixon, R. Harrison, and J.C. Jarvis. 1998. Denitrification and N_2O emissions from a UK pasture soil following the early spring application of cattle slurry and mineral fertiliser. Plant Soil 202:15-25.

Ferm, M., A. Kasimir-Klemedtsson, P. Weslien, and L. Klemedtsson. 1999. Emission of NH_3 and N_2O after spreading of pig slurry by broadcasting or band spreading. Soil Use Manage. 15:27-33.

Flessa, H., P. Dörsch, F. Beese, H. König, and A.F. Bouwman. 1996. Influence of cattle excrements on nitrous oxide and methane fluxes in pasture land. J. Environ. Qual. 25:1366-1370.

Forshell, L.P. 1993. Composting of cattle and pig manure. J. Vet. Med. B. 40:634-640.

Galloway, J.N. 1998. The global nitrogen cycle: changes and consequences. Environ. Pollut. 102:15-24.

Garret, M.K., C.J. Watson, C. Jordan, R.W.J. Steen, and R.V. Smith. 1992. The nitrogen economy of grazed grassland. Proceedings Fertiliser Society No. 326, The Fertiliser Society, London.

Goh, K.M., and P.H. Williams. 1999. Comparative nutrient budgets of temperate grazed pastoral systems. P. 265-294. *In* E.M.A. Smaling, O.Oenema, and L.O. Fresco (eds.) Nutrient Disequilibria in Agroecosystems; Concepts and Case-studies. CAB-International, Wallingford, UK.

Groenestein, C.M., and H.G. van Faassen. 1996. Volatilization of ammonia, nitrous oxide and nitric oxide in deep-litter systems for fattening pigs. J. Agric. Eng. Res. 65:269-274.

Groenestein, C.M., J. Oosthoek, and H.G. van Faassen. 1993. Microbial processes in deep-litter systems for fattening pigs and emission of ammonia, nitrous oxide and nitric oxide. p. 307-312. *In* M.W.A. Verstegen et al., (eds.) Nitrogen flow in pig production and environmental consequences. EAAP Publication 69, Pudoc Scientific Publishers, Den Haag, NL.

Groot Koerkamp, P.W.G. 1994. Review on emissions of ammonia from housing systems for laying hens in relation to sources, processes, building design and manure handling. J. Agric. Eng. Res. 59:73-87.

Groot Koerkamp, P.W.G., A. Keen, T.G.C.M. van Niekerk, and S. Smit. 1995. The effect of manure and litter handling and indoor climatic conditions on ammonia emissions from a battery cage and an aviary housing system for laying hens. Neth. J. Agric. Sci. 43:351-373.

Groot Koerkamp, P.W.G., J.H.M. Metz, G.H. Uenk, V.R. Phillips, M.R. Holden, R.W. Sneath, J.L. Short, R.P. White, J. Hartung, J. Seedorf, M. Schroder, K.H. Linkert, S. Pedersen, H. Takai, J.O. Johnson, and C.M. Wather. 1998. Concentrations and emissions of ammonia in livestock buildings in Northern Europe. J. Agric. Eng. Res. 70:79-95.

Hansen, S., and L.R. Bakken. 1993. N_2O, CO_2 and O_2 concentrations in soil air influenced by organic and inorganic fertilizers and soil compaction. Norwegian J. Agric. Sci. 7:1-10.

Harper, L.A., D.W. Bussink, H.G. van der Meer, and W.J. Corré. 1996. Ammonia transport in a temperate grassland: I. Seasonal transport in relation to soil fertility and crop management. Agron. J. 88:614-621

Harper, L.A., R.R. Sharpe, and T.B. Parkin. 2000. Gaseous nitrogen emissions from anaerobic swine lagoons: Ammonia, nitrous oxide and dinitrogen gas. J. Environ, Qual. 29:1356-1365.

Haynes, R.J., and P.H. Williams. 1993. Nutrient cycling and soil fertility in the grazed pasture ecosystem. Advances Agron. 49:119-199.

Hellmann, B., Zelles, L., Palojärvi, A., Bai, Q. 1997. Emission of climate-relevant trace gases and succession of microbial communities during open-windrow composting. Appl. Environ. Microbiol. 63, 1011–1018.

Henriksen, K., T. Olesen, T., and H.B. Rom. 2000. Carbon and nitrogen transformations in deep litter in cattle housing. p. 14-21. *In* S.G. Sommer, and J. Eriksen (eds.) Husdyrgødning og kompost. Forskningscenter for Økologisk Jordbrug, Tjele, Denmark.

Hüther, L., F. Schuchardt, and T. Willke. 1997. Emissions of ammonia and greenhouse gases during storage and composting of animal manures. p. 327-333. *In* J.A.M. Voermans and G.J. Monteny (eds.) Ammonia and odour emissions from animal production facilities. NVTL AB Rosmalen, The Netherlands.

Jarvis, S.C. 1997. Emission processes and their interactions in grassland soil. p. 1-18. *In* S.C. Jarvis, and B.F. Pain (eds.) Gaseous Nitrogen Emissions from Grassland. CAB International, Wallingford, UK.

Jarvis, S.C., D.J. Hatch, and D.R. Lockyer. 1989. Ammonia fluxes from grazed grassland: annual losses from cattle production systems and their relation to nitrogen input. J. Agric. Sci. Cambridge 113:99-108.

Jarvis, S.C., and B.F. Pain. 1990. Ammonia volatilisation from agricultural land. Proceedings Fertiliser Society 298, The Fertiliser Society, London.

Jarvis, S.C., D.J. Hatch, R.J. Orr, and S.E. Reynolds. 1991 Micrometeorological studies of ammonia emission from sheep grazed swards. J. Agric. Sci. Cambridge 117: 101-109.

Jarvis, S.C. and B.F. Pain (eds.). 1997. Gaseous Nitrogen Emissions from Grassland. CAB International, Wallingford, UK, 452 pp.

Karlsson, S., and K.H. Jeppson. 1995. Deep Litter in Livestock Buildings and Field Storage. JTI Report 204, Swedish Institute of Agricultural Eng., Ultuna Uppsala, Sweden.

Ketelaars, J.J.M.H., and H.G. van der Meer. 1999. Establishment of criteria for the assessment of the nitrogen content of animal manures. Final report of the second phase of ERM project 5765. AB-DLO, Wageningen, ERM London, 68 p.

Kirchmann, H., and A. Lundvall. 1993. Relationship between N immobilization and volatile fatty acids in soil after application of pig and cattle slurry. Biol. Fertil. Soils 15:161-164.

Koops, J.G., O. Oenema, and M.L. van Beusichem. 1996. Denitrification in the top and sub soil of grassland on peat soils. Plant Soil 184:1-10.

Koops, J.G., M.L. van Beusichem, and O. Oenema. 1997. Nitrous oxide production, its source and distribution in urine patches on grassland on peat soil. Plant Soil 191:57-65.

Kreuzer, M., A. Machmüller, M.M. Gerdemann, H. Hanneken, and M. Wittman. 1998. Reduction of gaseous nitrogen loss from pig manure using feeds rich in easily-fermentable non-starch polysaccharides. Anim. Feed Sci. Technol. 73:1-19.

Langhout, D.J., J.B. Schutte, P. van Leeuwen, J. Wiebenga, and S. Tamminga. 1999. Effect of dietary high- and low-methylated citrus pectin on the activity of the ileal microflora and morphology of the small intestinal wall of broiler chicks. British Poultry Sci. 40:340-347.

Lantinga, E.A., J.A. Keuning, J. Groenwold, and P.J.A.G. Deenen. 1987. Distribution of excreted nitrogen by grazing cattle and its effects on sward quality, herbage production and utilization. p. 103-117. *In* H.G. van der Meer, R.J. Unwin, T.A. van Van Dijk, and G.C. Ennik, G.C. (eds.) Animal Manure on Grassland and Fodder Crops, Martinus Nijhoff Publishers, Dordrecht, Boston and Lancaster.

Ledgard, S.F., M.S. Sprosen, G.J. Brier, E.K.K. Nemaia, and D.A. Clark. 1996. Nitrogen inputs and losses from New Zealand dairy farmlets, as affected by nitrogen fertilizer application: year one. Plant Soil 181:65-69.

Lockyer, D.R., and D.C. Whitehead. 1990. Volatilization of ammonia from cattle urine applied to grassland. Soil Biol. Biochem. 22:1137-1142.

Lovell, R.D. and S.C. Jarvis. 1996. Effects of urine on soil microbial biomass, methanogenesis, nitrification and denitrification in grassland soils. Plant Soil. 186:265-273.

Mertens, D.R. 1993. Rate and Extent of Digestion. p. 13-51. *In* J.M. Forbes, and J. France (eds.) Quantitative Aspects of Ruminant Digestion and Metabolism. CAB International Wallingford, UK.

Misselbrook, T.H., D.R. Chadwick, B.F. Pain, and D.M. Headon. 1998. Dietary manipulation as a means of decreasing N losses and methane emissions and improving herbage N uptake following application of pig slurry to grassland. J. Agric. Sci. 130:183-191.

Misselbrook, T.H., T.J. van der Weerden, B.F. Pain, S.C. Jarvis, B.J. Chambers, K.A. Smith, V.R. Phillips, and T.G.M. Demmers. 2000. Ammonia emission factors for UK agriculture. Atmos. Environ. 34:871-880.

Moal, J.F, J. Martinez, F. Guiziou, and C.M. Coste. 1995. Ammonia volatilization following surface-applied pig and cattle slurry in France. J. Agric. Sci. Cambridge 125:245-252.

Monaghan, R.M. and B. Barraclough. 1992. Some chemical and physical factors affecting the rate and dynamics of nitrification in urine-affected soil. Plant Soil 143:11-18.

Monaghan, R.M., and B. Barraclough. 1993. Nitrous oxide and dinitrogen emissions from urine-affected soil under controlled conditions. Plant Soil 151:127-138.

Monteny, G.J., and J.W. Erisman. 1998. Ammonia emission from dairy cow buildings: a review of measurement techniques, influencing factors and possibilities for reduction. Neth. J. Agric. Sci. 46:225-247.

Mosier, A.R. 1989. Chamber and isotope techniques. p. 175-187. *In* M.O. Andrea, and D.S. Schimel (eds.) Exchange of Trace Gases between Terrestrial Ecosystems and the Atmosphere. Life Sciences Research Report 47, John Wiley, Chichester.

Mosier, A.R., and W.J. Parton. 1985. Denitrification in a shortgrass prairie: A modelling approach. p. 441-452. *In* D.E. Caldwell, J.A. Brierley, and C.L. Brierley (eds.) Planetary Ecology. Van Nostrand Reinhold Co, New York.

Mosier, A., C. Kroeze, C. Nevison, O. Oenema, S. Seitzinger, and O. van Cleemput. 1998. Closing the global N_2O budget: nitrous oxide emissions through the agricultural nitrogen cycle. Nutri. Cycl. in Agroecosyst. 52:225-248.

Mroz, Z., A.W. Jongbloed, N. Lenis, and K. Vreman. 1995. Water in pig nutrition: physiology, allowances and environmental implications. Nutrition Res. Rev. 8:137-164.

Naeth, M.A., D.J. Pluth, D.S. Chanasyk, A.W. Bailey, and A.W. Fedkenheuer. 1990. Soil compaction impacts of grazing in mixed prairie and fescue grassland ecosystems of Alberta. Can. J. Soil Sci. 70:157-167.

Nolan, J.V. 1993. Nitrogen Kinetics. p. 123-143. *In* J.M. Forbes and J. France (eds.) Quantitative Aspects of Ruminant Digestion and Metabolism. CAB International Wallingford, UK.

Oenema, O. and G.L. Velthof. 1993. Denitrification in nitric-acid-treated cattle slurry during storage. Neth. J. Agric. Sci. 41:63-80.

Oenema, O., G.L. Velthof, and D.W. Bussink. 1993. Emissions of ammonia, nitrous oxide and methane from cattle slurry. p. 419-433. *In* R.S. Oremland (ed.) Biogeochemistry of Global Change: Radiatively Active Trace Gases. Chapman & Hall, New York.

Oenema, O., G.L. Velthof, S. Yamulki, and S.C. Jarvis. 1997. Nitrous oxide emissions from grazed grassland. Soil Use Manage. 13:288-295.

Oenema, O., G.L. Velthof, N. Verdoes, P.W.G. Groot Koerkamp, A. Bannink, H.G. van der Meer, and K.W. van der Hoek. 2000a. Calculated losses of gaseous nitrogen compounds from livestock manure in stables and manure storage systems (in Dutch with 15 page

English summary). Alterra-report 107, Wageningen University and Research Center, Wageningen, 185 pp.

Oenema, O., G.L. Velthof, and P.J. Kuikman. 2000b. Technical and policy aspects of strategies to decrease greenhouse gas emissions from agriculture. Nutri. Cycl. in Agroecosys. (in press.)

Pain, B.F., V.R. Phillips, C.R. Clarkson, and J.V. Klarenbeek. 1989. Loss of nitrogen through NH_3 volatilisation during and following application of pig or cattle slurry to grassland. J. Sci. Food Agric. 47:1-12.

Pain, B.F., R.B. Thompson, Y.J. Rees, and J.H. Skinner. 1990. Reducing gaseous losses of nitrogen from cattle slurry applied to grassland by the use of additives. J. Sci. Food Agric. 50:141-153.

Pain, B.F., V.R. Phillips, J.F.M. Huijsmans, and J.V. Klarenbeek. 1991. Anglo-Dutch experiments on odour and ammonia emission following the spreading of piggery wastes on arable land. Rapport 91-9, IMAG-DLO, Wageningen, 28 pp.

Paul J.W., E.G. Beauchamp. 1989. Effect of carbon constituents in manure on denitrification in soil. Can. J. Soil Sci. 69:49-61.

Paul, J.W. and B.J. Zebarth. 1997. Denitrification and nitrate leaching during the fall and winter following dairy cattle slurry application. Can. J. Soil Sci. 77:231-240.

Paul, J.W., N.E. Dinn, T. Kannangara, and L.J. Fisher. 1998. Protein content in dairy cattle diets affects ammonia losses and fertiliser nitrogen value. J. Environ. Qual. 27:528-534.

Petersen, J. 1996. Animal manure – a source of nutrients. SP report nr. 11. Danish Institute of Agricultural Sciences, Tjele, Denmark.

Petersen, S.O., S.G. Sommer, O. Aaes, and K. Søegaard. 1998a. Ammonia losses from urine and dung of grazing cattle: effect of N intake. Atmos. Environ. 32:295-300.

Petersen, S.O., A.M. Lind, and S.G. Sommer. 1998b. Nitrogen and organic matter losses during storage of cattle and pig manure. J. Agric. Sci. Cambridge. 130:69–79.

Plantaz, M.A.H.G. 1998. Surface/atmosphere exchange of ammonia over grazed pasture. PhD thesis Wageningen University, 199 p.

Poggemann, S., F. Weissbach, and U. Künzel. 1995. N2O-Freisetzung von Exkrementflecken auf dem Weideland, Abstract VDLUFA, Garmisch-Partenkirchen 1995, pp. 275.

Poulsen, H.D., and V.F. Kristensen (eds.). 1997. Standard Values for Farm Manure. An evaluation of the Danish standard values concerning the nitrogen, phosphorus and potassium content of manure. DIAS Report Animal Husbandry no. 7. Tjele, Denmark, 160 pp.

Rom, H.B., and K. Henriksen. 2000. Nitrogen loss from cattle housed on deep litter. p. 5-13. *In* S.G. Sommer, and J. Eriksen (eds.) Husdyrgødning og kompost. Forskningscenter for Økologisk Jordbrug, Tjele, Denmark.

Russell, E.J. 1912. Soil Conditions and Plant Growth. Longman, London, UK.

Ruz-Jerez, B.E., R.E. White, and P. Roger Ball. 1994. Long-term measurement of denitrification in three contrasting pastures grazed by sheep. Soil Biol. Biochem. 26:29-39.

Ryden, J.C. 1986. Gaseous losses of nitrogen from grassland. p. 59-73. *In* H.G. van der Meer, J.C. Ryden, and G.C. Ennik (eds.) Nitrogen Fluxes in Intensive Grassland Systems. Martinus Nijhoff Publishers, Dortrecht, NL.

Ryden, J.C., D.C. Whitehead, D.R. Lockyer, R.B. Thompson, J.H. Skinner, E.A. Garwood. 1987a. Ammonia emission from grassland and livestock production systems in the UK. Environ. Pollution 48:173-184.

Ryden, J.C., J.H. Skinner, and D.J. Nixon. 1987b Soil core incubation system for the field measurement of denitrification using acetylene-inhibition. Soil Biol. Biochem. 19:753-757.

Safley, L.M., P.W. Westerman, and J.C. Barker. 1986. Fresh dairy manure characteristics and barnlot nutrient losses. Agri. Wastes 17, 203-215.

Sherlock, R.R., and K.M. Goh. 1983. Initial emission of nitrous oxide from sheep urine applied to pasture soil. Soil Biol. Biochem. 15:615-617.

Sherlock, R.R., and K.M. Goh. 1984. Dynamics of ammonia volatilization from simulated urine patches and aqueous urea applied to pasture. I. Field experiments. Fert. Res. 5:181-195.

Smits, M.C.J., H. Valk, A. Elzing, and A. Keen. 1995. Effect of protein nutrition on ammonia emission from a cubicle house for dairy cattle. Livestock Production Sci.44:147-156.

Sommer, S.G. 1997. Ammonia volatilization from farm tanks containing anaerobically digested animal slurry. Atmos. Environ. 31:863-868.

Sommer, S.G. 2000. Effect of composting on nutrient loss and nitrogen availability of cattle deep litter. Europ. J. Agron. (Submitted).

Sommer, S.G., and B.T. Christensen. 1990. NH_3 volatilization from solid manure and fresh, fermented and separated slurry after surface application, injection, incorporation into the soil and irrigation. Tidsskrift for Planteavl. 94:407-417. (in Danish, with English summary).

Sommer, S. G., J.E. Olesen, and B.T. Christensen. 1991. Effects of temperature, wind speed and air humidity on NH_3 volatilization from surface applied cattle slurry. J. Agric. Sci. Cambridge 117:91-100.

Sommer, S.G., B.T. Christensen, N.E. Nielsen, and J.K. Schjørring. 1993. Ammonia volatilization during storage of cattle and pig slurry: Effect of surface cover. J. Agric. Sci. Cambridge. 121:63-71.

Sommer, S.G., and R.R. Sherlock. 1996. pH and buffer component dynamics in the surface layers of animal slurries. J. Agric. Sci. Cambridge 127:109-116.

Sommer, S.G., E. Friis, A.B. Bak, and J.K. Schjørring. 1997. Ammonia volatilization from pig slurry applied with trail hoses or broadspread to winter wheat: Effects of crop developmental stage, microclimate, and leaf ammonia absorption. J. Environ. Quality. 26: 1153-1160.

Sommer, S.G., S.O. Petersen, and H.T. Søgaard. 2000. Emission of greenhouse gases from stored livestock slurry. J. Environ. Qual. 29:744-751.

Thelosen, J.G.M., B.P. Heitlager, and J.A.M. Voermans. 1993. Nitrogen balances of two deep litter systems for finishing pigs. p. 318-323. *In* M.W.A. Verstegen et al. (eds.) Nitrogen flow in pig production and environmental consequences. EAAP Publication 69, Pudoc Scientific Publishers, Den Haag, NL.

Thompson, R.B., J.C. Ryden, and D.R. Lockyer. 1987. Fate of nitrogen in cattle slurry following surface application or injection to grassland. J. Soil Sci. 38:689-700.

Tiedje, J.M. 1988. Ecology of Denitrification and dissimilatory nitrate reduction to ammonium. p. 179-224. *In* A.J.B. Zehnder (ed.) Biology of anaerobic microorganisms.

Torbert, H.A., and C.W. Wood. 1992. Effects of soil compaction and water-filled pore space on soil microbial activity and N losses. Commun. Soil Sci. Plant Anal. 23:1321-1331.

Vallis, I., L.A. Harper, V.R. Catchpoole, and K.L. Weier. 1982.Volatilization of ammonia from urine patches in a subtropical pasture. Aust. J. Agric. Res. 33:97-107.

Van Cleemput, O., A. Vermoesen, C.J. de Groot, and K. van Ryckeghem. 1994. Nitrous oxide emission out of grassland. Environ. Monitoring Assessment 31:145-152.

Van der Meulen, J., G.C.M. Bakker, J.G.M. Bakker, H. de Visser, A.W. Jongbloed, and H. Everts. 1997. Effect of resistant starch on net portal-drained viscera flux of glucose, volatile fatty acids, urea, and ammonia in growing pigs. J. Animal Sci. 75:2697-2704.

Van der Molen, J., A.C. Beiljaars, W.J. Chardon, W.A. Jury, and H.G. van Faassen. 1990. Ammonia volatilization from arable land after application of cattle slurry. 1. Field estimates. Neth. J. Agric. Sci. 38:145-158.

Van der Weerden, T.J., T.H. Misselbrook, and B.F. Pain. 1994. Ammonia emission and denitrification following surface-spreading and injection of cattle slurry to grassland. p.

471-474. *In* L. 't Mannetje and J. Frame (eds.) Grassland and Society, Wageningen Pers, Wageningen.

Velthof, G.L., and O. Oenema. 1993. Nitrous oxide flux from nitric-acid treated cattle slurry applied to grassland under semi-controlled conditions. Neth. J. Agric. Sci. 41:81-93.

Velthof, G.L., and O. Oenema. 1994. Effect of nitrogen fertilizer type and urine on nitrous oxide flux from grassland in early spring. p. 458-462. *In* L. 't Mannetje and J. Frame (eds.) Grassland and Society, Wageningen Pers, Wageningen.

Velthof, G.L., S.C. Jarvis, A. Stein, A.G. Allen, and O. Oenema. 1996a. Spatial variability of nitrous oxide fluxes in mown and grazed grasslands on a poorly drained clay soil. Soil Biol. Biochem. 28:1215-1225.

Velthof, G.L., J.G. Koops, J.H. Duyzer, and O. Oenema. 1996b. Prediction of nitrous oxide fluxes from managed grassland on peat soils using a simple empirical model. Neth. J. Agric. Sci. 44:339-356.

Velthof, G.L., A.B. Brader, and O. Oenema. 1996c. Seasonal variations in nitrous oxide losses from managed grassland in The Netherlands. Plant Soil 181:263-274.

Velthof, G.L., O. Oenema, R. Postma, and M.L. van Beusichem. 1997. Effects of type and amount of applied nitrogen fertilizer on nitrous oxide fluxes from intensively managed grassland. Nutri. Cycl. Agroecosys. 46:257-267.

Vermoesen, A., O. van Cleemput, and G. Hofman. 1997. Contribution of urine patches to the emission of N2O. p. 189-194. *In* S.C. Jarvis, and B.F. Pain (eds.) Gaseous Nitrogen Emissions from Grassland. CAB International, Wallingford, UK.

Vertregt, N., and B. Rutgers. 1988. Ammonia volatilization from grazed pastures. CABO-report 84, Dutch Priority Programme on Acidification, Wageningen, 37 p.

Warren, S.D., M.B. Nedvill, W.H. Blackburn and N.E. Garza. 1986. Soil response to trampling under intensive rotation grazing. Soil Sci. Soc. Am. J. 50:1336-1341.

Weslien, P., L. Klemedtsson, L. Svensson, B. Galle, A. Kasimir-Klemedtsson, and A. Gustafsson. 1998. Nitrogen losses following application of pig slurry to arable land. Soil Use Manage. 14:200-208.

Whitehead, D.C., D.R. Lockyer, and N. Raistrick. 1989. Volatilization of ammonia from urea applied to soil: influence of hippuric acid and other constituents of livestock urine. Soil Biol. Biochem. 21:803-808.

Whitehead, D.C., and N. Raistrick. 1991. Effects of some environmental factors on ammonia volatilization from simulated livestock urine applied to soil. Biol. Fertil. Soils 11:279-284.

Whitehead, D.C., and N. Raistrick. 1993. The volatilization of ammonia from cattle urine applied to soils as influenced by soil properties. Plant Soil 148:43-51.

Williamson, J.C., S.C. Jarvis. 1997. Effect of dicyandiamide on nitrous oxide flux following return of animal excreta to grassland. Soil Biol. Biochem. 29:1575-1578.

Williams, P.H., S.C. Jarvis, and E. Dixon. 1998. Emission of nitric oxide and nitrous oxide from soil under field and laboratory conditions. Soil Biol. Biochem. 30:1885-1893.

Woodmansee, R.G. 1979. Factors influencing input and output of nitrogen in grassland. p. 117-134. *In* N.R. French (ed.) Perspectives in Grassland Ecology. Springer Verlag, New York. I

Yamulki, S., R.M. Harrison, K.W.T. Goulding, and C.P. Webster. 1995. Effect of fertilizer application on NO and N2O fluxes from agricultural fields. J. Geophys. Res. 100:25923-25931.

Yamulki, S., S.C. Jarvis, and P. Owen. 1998. Nitrous oxide emissions from excreta in a simulated grazing pattern. Soil Biol. Biochem. 30:491-500.

Nitrogen in the Environment: Sources, Problems, and Management
R.F. Follett and J.L. Hatfield (Eds)
Published by Elsevier Science B.V.

Chapter 11. Exchange of Gaseous Nitrogen Compounds between Terrestrial Systems and the Atmosphere

A. R. Mosier

U.S. Department of Agriculture/Agricultural Research Service, P.O. Box E Fort Collins, Colorado, 80522, USA

Food production systems are important contributors to local, regional and global NH_3, NO_x ($NO + NO_2$) and N_2O budgets. Emissions of NH_3 and NO_x (which are biologically and chemically active) into the atmosphere serve to redistribute fixed N to local and regional aquatic and terrestrial ecosystems that otherwise may be disconnected from the sources of the N gases. The emissions of NO_x also contribute to local elevated ozone concentrations while N_2O emissions contribute to global greenhouse gas accumulation and to stratospheric ozone depletion.

Ammonia is the major gaseous base in the atmosphere and serves to neutralize about 30% of the hydrogen ions in the atmosphere. 50 to 75% of the ~ 55 Tg N y^{-1} NH_3 from terrestrial systems is emitted from animal and crop-based agriculture from animal excreta and synthetic fertilizer application. About half of the ~50 Tg N y^{-1} of NO_x emitted from the earth surface annually arises from fossil fuel combustion and the remainder from biomass burning and emissions from soil. The NO_x emitted, principally as nitric oxide (NO), reacts rapidly in the atmosphere and in a complex cycle with light, ozone and hydrocarbons, and produces nitric acid and particulate nitrate. These materials can interact with plants and the soil locally or be transported from the site and interact with atmospheric particulate to form aerosols. These salts and aerosols return to fertilize terrestrial and aquatic systems in wet and dry deposition. A small fraction of this N may be biologically converted to N_2O. About 5% of the total atmospheric greenhouse effect is attributed to N_2O from which ~70% of the annual global anthropogenic emissions come from animal and crop production.

The coupling of increased population with a move of a large sector of the world population to diets that require more energy and N input, will lead to continued increases in anthropogenic input into the global N cycle. This scenario suggests that emissions of NH_3, NO_x and N_2O from agricultural systems will continue to increase and impact global terrestrial and aquatic systems, even those far removed from agricultural production, to an ever growing extent, unless N resources are used more efficiently or food consumption trends change.

1. GLOBAL TRENDS IN HUMAN-INDUCED ADDITIONS OF REACTIVE NITROGEN TO EARTH'S BIOGPHERE

During the past four decades we have seen historically unprecedented additions of reactive nitrogen (N) into terrestrial systems (Vitousek et al. 1997). These inputs of reactive N (termed here as NH_x (NH_3 and NH_4^+), NO_x ($NO + NO_2$), and nitrate and nitrite) are through increased biological fixation of atmospheric N_2 in crop production, combustion of fossil fuels and release of NO_x, and production of synthetic fertilizer nitrogen (Galloway et al. 1994; 1995). There is an additional release of reactive N from stable soil organic matter as a result of soil disturbance for crop production that is not readily quantifiable.

"Natural" fixation of atmospheric N_2 primarily by lightning and biological processes is estimated to be about 100 Tg N, globally (Galloway et al. 1995), and is assumed to have remained relatively constant during the past century. A new global estimate (Cleveland et al. 1999) suggests that background terrestrial biological N fixation is higher, on the order of 100 to 290 Tg N y^{-1}. Imposed upon the natural conversion of atmospheric N_2 to usable forms of N is the human-induced production of reactive N. Anthropogenic production has increased from ~40 Tg N y^{-1} in 1961 to ~160 Tg in 1995 (Smil, 1999). Holland and Lamarque (1997) estimated NO_x produced during fossil fuel combustion to be ~11 Tg in 1961 while biological N fixation in crops such as pulses, soybeans, ground nuts and forage legumes and by algae and other microorganisms in rice fields was about 18 Tg N (Galloway et al. 1995); and synthetic N fertilizer production was 12.9 Tg N (FAO, 1999). In 1995 these inputs had increased to 25, 48 and 87 Tg N, respectively. Galloway et al. (1995) discussed that we can account for only about 60 Tg of this anthropogenic N fixed in 1990, through accumulation of N_2O in the atmosphere (~3 Tg), movement into coastal waters by river flow (~41 Tg) and atmospheric deposition into the open oceans (~18 Tg). The remaining 80 Tg N y-1 must either be retained in terrestrial systems in groundwater, soils, vegetation or denitrified to N_2. Although the exact values for each area of N accumulation is uncertain, Galloway et al. (1995) concluded that reactive N is accumulating in terrestrial systems.

We also know that much of the reactive N that is produced from combustion, legume crop N fixation and synthetic N-fertilizer is distributed to terrestrial and aquatic systems in a variety of ways. Gaseous transport of N compounds is one such mechanism. Much of the newly fixed N is directed to food production. The reprocessing of N through animals leads to the distribution of N through volatilization of ammonia and runoff and leaching from soils. Part of the N applied to soils used for crop production is lost from the field through leaching of nitrate into ground water and/or runoff from fields into surface water systems. The N consumed by humans is mostly released in waste processing systems and returned to aquatic and soil systems. These loss processes lead to problems of interactions with other biological systems down wind or down stream.

2. GLOBAL FOOD PRODUCTION AND NITROUS OXIDE

In most agricultural soils, biogenic formation of N_2O is enhanced by an increase in available mineral N which, in turn increases nitrification and denitrification through which N_2O is produced. N-fertilization, therefore, directly results in additional N_2O formation in the field in which the N is applied. The IPCC 1997 National Inventory Guidelines for N_2O from

Agriculture (IPCC, 1997; Mosier et al. 1998) refer to N_2O emissions directly from agricultural fields as direct emissions. In addition, these N inputs may lead to indirect formation of N_2O after N leaching or runoff, or following gaseous losses and deposition of NO_x and NH_3 and following human consumption of agricultural products and waste processing in sewage systems, termed indirect emissions (IPCC, 1997). These National N_2O Inventory Guidelines consider a variety of sources of N in agricultural systems as anthropogenic; including synthetic fertilizers, animal manures (urine and feces), N derived from enhanced biological N-fixation through N_2-fixing crops, crop residue returned to the field after harvest and human sewage sludge application. Some part of the animal manure N, crop residue and sewage may have come from previous application of synthetic fertilizer. However, the re-entry of this N back into the soil system renders it again susceptible to microbial processes which produce N_2O (IPCC, 1997; Mosier et al. 1998). In addition to the direct and indirect N_2O emissions the IPCC (1997) Guidelines estimate N_2O emissions that result from animal waste management systems separately from animal manures that are used as fertilizers. The animal component includes N_2O emissions from animal waste deposition in pastures and animal confinements.

2.1. Total Fertilizer N Input into Agricultural Soils

According to the IPCC (1997) Guidelines N input into agricultural systems includes N input into cropping systems from synthetic fertilizer N, biological N fixation, return of crop residue and animal manures. Using FAO data and human population estimates to back calculate N input into crop production, assuming linear changes in crop and livestock production, 1970 FAO data was used as the baseline to calculate N used in crop production are estimated for 1800 to 1961 (Fig. 1). From 1961, the year that the FAO data bases begin, the FAO (1999) data are used. Before 1960, animal and human wastes were the major fertilizer source (Kroeze et al. 1999). In 1950 synthetic fertilizer N input comprised about 7% of a total N input of ~56 Tg and 1996 synthetic N input was ~43% of a total input of ~190 Tg N. Animal waste used as fertilizer was an estimated 37 Tg in 1950 compared to 65 Tg N in 1996.

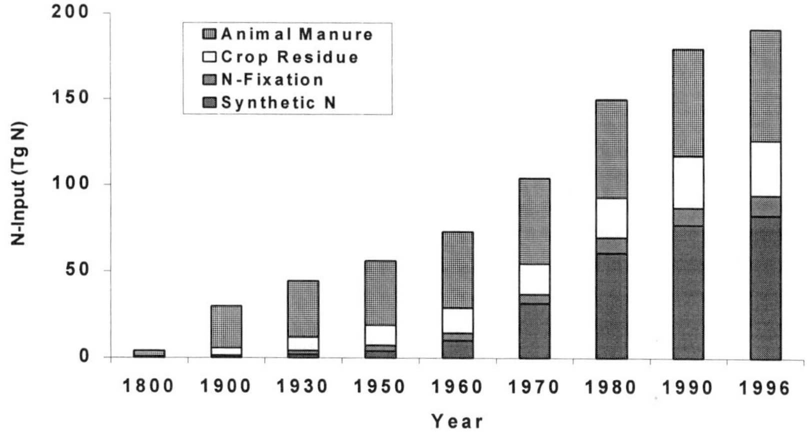

Figure 1. Estimate of global annual N input into crop production from synthetic N, biological N-fixation, crop residue return and animal manures (Kroeze et al. 1999; Mosier, 2000).

2.2. Global annual N_2O emissions estimates

Using the N input data shown in Fig. 1, estimates of global N_2O emissions from food production systems from 1800 until 1996 are shown in Fig. 2. Before 1950 <50% of the estimated N_2O emission resulted directly or indirectly from N input into crop production, with the greater portion coming from animal production. As a result of increased N input and increased crop production by 1996 about one third of the N_2O originated from each of the three segments of the N_2O estimate. In 1950 total N_2O from global food production systems totaled ~3 Tg N compared to ~6 Tg in 1996 (Kroeze et al. 1999). Note that these N_2O estimates present a

very different picture from the IPCC 1990 and 1992 projections. In those estimates agricultural N input was considered only synthetic fertilizer and the livestock waste management segment was not included (Cole et al. 1996). The estimates presented here and by Mosier et al. (1998) and Kroeze et al. (1999) indicate that when a more complete picture of the agricultural N cycle is depicted that N_2O emissions resulting from food production have been a significant part of the global budget for centuries.

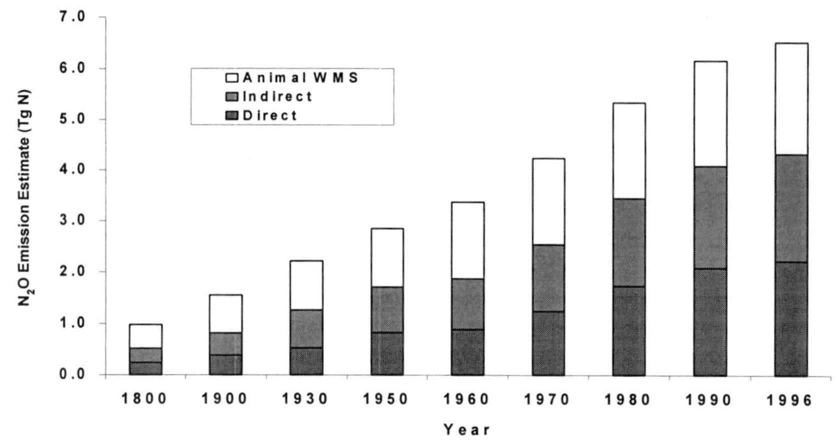

Figure. 2. Estimates of global N_2O emissions as a result of crop and livestock production from 1800 to 1996 (Kroeze et al. 1999); Direct = N_2O emitted directly from cropped fields, AnimalWMS = N_2O emitted from livestock excreta where emissions are related to the waste management system used, and Indirect = N_2O emitted from N that was used in crop production after it was leached or eroded from the site of application (Mosier, 2000).

3. NUTRIENT REDISTRIBUTION BY SOIL-ATMOSPHERE EXCHANGE OF NITROGEN COMPOUNDS

3.1. NH_3 Exchange between the Soil, Plants and the Atmosphere

Plant-Atmosphere NH_3 Exchange. Plants can serve as both a sink and a source of atmospheric NH_3 (Farquhar et al. 1980; Harper et al. 1996; Hutchinson et al. 1972; Schjorring and Mattsson, 2000). Whether NH_3 is emitted or absorbed by plants depends upon the NH_3 compensation point of the plant. Ammonia is absorbed when the mole fraction of NH_3 in the atmosphere above the plant leaf is greater than the NH_3 concentration in the atmosphere around the mesophyll cells in the stomatal cavity. Ammonia is emitted when the atmospheric NH_3 concentration is lower than the leaf mesophyll cell concentration (Farquhar et al. 1980).

The impact of plant NH_3 exchange on field crop N balance and exchange with the regional ecosystem is not yet clear. The net NH_3 exchange at a location is likely dependent upon several interacting factors, thus general simplifying conditions to describe NH_3 exchange are not known. Wetselaar and Farquhar (1980) showed that N loss from wheat (*Triticum aestivum*) was increased with increasing amount of N-fertilizer application. They showed that the N-content of plant leaves typically declined between anthesis and maturity in most field crops, including wheat, rice (*Oryza sativa*) rye (*Secale cereale*), corn (*Zea mays*) and barley (*Hordeum vulgare*). They attributed much of this N loss to foliar volatilization of NH_3 described by Hutchinson et al. (1972). Both laboratory (Farquhar et al. 1980; Parton et al. 1988) and field studies (Harper et al. 1987; Harper and Sharpe, 1995) show that NH_3 compensation point can vary with plant type, temperature, phenological growth stage, time of day and soil N availability. Because of the continuous exchange of NH_3 between the atmosphere and the plant, Francis et al. (1997) concluded that determining the true loss or gain in between corn plants and the atmosphere is difficult.

Schjorring and Mattsson (2000) note that the role of crop NH_3 emissions in agroecosystem N balance is not clear. Schjorring (1995) and Schjorring and Mattsson (2000) did not observe a general relationship between N-fertilizer application rate and net plant-atmosphere NH_3 exchange. On the other hand, Harper et al. (1987) suggested a direct relationship between plant N status, fertilization and NH_3 loss. They reported a relatively large loss between corn anthesis and harvest (~15 kg NH_3-N ha^{-1}) compared to the 1-2 kg N ha^{-1} $season^{-1}$ reported by Schjorring (1995). Schjorring and Mattsson (2000) observed that N loss from crop vegetation was important only when the N harvest index at maturity (ratio between grain and shoot N content) was less than 0.63. Net NH_3 losses from spring barley that had N harvest indices at maturity of 0.8 were only 0.5 to 1.5 kg N ha^{-1}.

Net NH_3 exchange within plants, because NH_3 can be both absorbed and emitted, can influence fertilizer ^{15}N balance studies (Harper and Sharpe, 1995; Francis et al. 1997). Plants grown with ^{15}N-enriched fertilizer tend to loose $^{15}NH_3$ and gain $^{14}NH_3$ even if the net NH_3 flux is zero (Sharpe and Harper, 1997). The actual impact of this exchange on N-fertilizer studies is not always clear, although the NH_3 exchange would suggest that N-fertilizer losses estimated by ^{15}N-balance would be overestimated (Francis et al. 1997). Schjorring and Mattsson (2000) point out that overestimates of N-loss by isotopic techniques can also be due to leaf drop before flowering, NH_3 emissions from decaying leaves, or NH_3 exchange within the crop canopy. They conclude that crops are typically a net source of NH_3 to the atmosphere on a seasonal basis of up to 5 kg NH_3-N ha^{-1}. The amount of NH_3 lost from crop vegetation ranges between one and four per cent of fertilizer N applied and between one and four per cent of the N present in the crop (Schjorring and Mattsson, 2000). As a result, agroecosystems play an important role in regional atmospheric N gas composition and may be the dominant feature where NH_3 emissions from livestock waste is small (see Oenema et al. this volume).

NH_3 Absorption/Volatilization by Soil. Uptake of atmospheric NH_3 by soils has been documented for more than 150 years (references and discussion in Hanawalt, 1969). In many areas of the world, atmospheric NH_3 concentrations are several times higher than in relatively pristine environments and in these areas the atmosphere-soil exchange of NH_3 is a main supplier of ecosystem N (Vitousek et al. 1997). Generally speaking, NH_3 volatilization to the atmosphere is from soils that have insufficient sorption capacity to hold ammonium from NH_x deposition or from ammonium fertilizer application (Terman, 1979). Soil characteristics and basic chemical factors control the soil/atmosphere exchange of NH_3. Soil NH_3 retention is controlled by chemical equilibria in soil water. As a basic gas, NH_3 reacts readily with protons, metals and acidic compounds to form ions. Ammonium formed can be ionically or physically bound to soil particles and the interaction between these processes control the soil-atmosphere exchange of NH_3. These interactions are fully described by Freney et al. (1983) and will not be described in detail here. Briefly, the various reactions which regulate NH_3 exchange with soil can be represented as follows:

$$\textbf{Absorbed } NH_4^+ \leftrightarrow NH_4^+{}_{(solution)} \leftrightarrow NH_{3\,(solution)} \leftrightarrow NH_{3\,(gas)} \leftrightarrow NH_{3\,(gas\ in\ atm.)} \qquad \textbf{(Eq. 1)}$$

Adsorption of NH_3 and fixation of NH_4^+ onto clay particles is typically regulated by soil organic matter content and the type of clay minerals in the soil. Adsorption is related to the surface area of the sorbing material and generally clay soils sorb more NH_3 than do sandy soils (Terman, 1979). The rate of NH_3 volatilization may be controlled by the rate of removal and dispersion of NH_3 into the atmosphere by changing the concentration of NH_4^+ or NH_3 in the soil solution or by displacing any of the equilibria in Eq. 1. The driving force for NH_3 volatilization from soil solution is the difference in the NH_3 partial pressure between that in equilibrium with the liquid phase and that in the ambient atmosphere. The equilibrium vapor pressure of NH_3 is controlled by the NH_3 concentration in adjacent solutes which, in the absence of other ionic species, is affected by NH_4^+ ion concentration and pH. At pH 9.2 a solution contains approximately equal amounts of solution NH_4^+ and solution NH_3. At pH 7.2 the solution contains approximately 99% solution NH_4^+ and 1% solution NH_3. Thus, NH_3 emissions are typically higher in more basic soils. Chemical equilibria dictate that an aqueous solution will hold less NH_3 with increasing temperature, so temperature affects soil-atmosphere NH_3 exchange as well (Freney et al. 1983).

Ammonia must be transported to the soil surface before it can be lost to the atmosphere. As a result, NH_3 losses are typically reduced by subsurface application of fertilizers (Terman, 1979). Ammonia transport can be accomplished by movement in liquid or gaseous phases and their relative importance depends on soil water content. Volatilization of NH_3 from solution at the soil surface occurs in response to a difference in vapor pressure between solution and ambient air. Increasing wind speed increases the rate of volatilization by permitting more rapid transport of NH_3 away from the water surface (Freney et al. 1983).

Factors which influence soil-atmosphere NH_3 exchange include soil cation exchange capacity (CEC), soil pH, soil buffer capacity and calcium carbonate content. As NH_4^+ is positively charged it readily reacts with the soil cation exchange complex. The disassociation of ammonium ion releases a proton in addition to NH_3. Consequently as NH_3 loss proceeds the solution becomes acidified and as the fraction of ammoniacal N is reduced an equilibrium is reached. For this reason NH_3 volatilization is generally not important in soils that have a high base saturation (Freney et al. 1983). A strong correlation between NH_3 loss and calcium carbonate content of soils has been observed (Fenn et al. 1981). The apparent stimulation of NH_3 volatilization has been related to soil clay-sized calcium carbonate content and the formation of calcium fluoride, sulfate and phosphate precipitates and ammonium bicarbonate (Fenn et al. 1981; Terman, 1979).

NH_3 Volatilization from Surface Residues. Since NH_3 emissions are readily reduced by placing ammonium-based fertilizers below the soil surface, surface application of fertilizers can lead to increased NH_3 losses (Terman, 1979). Increased use of no-till management could contribute to this increase if new fertilizer-N management tools are not developed. An example of the impact of crop residue management on NH_3 loss is the practice of trash retention following green cane harvesting in sugar cane production. As practiced in north Queensland, Australia, cane trash is left on the field following harvest to retain soil moisture, prevent weed growth and to minimize soil erosion (Freney et al. 1992). Surface application of urea onto green cane trash was a common practice which typically resulted in large NH_3 loss. These losses are driven by addition of small amounts of water from dewfall, light rain and condensation of evaporated soil water. This water dissolves the urea, permits urea hydrolysis to NH_3 and allows NH_3 to be lost as the water evaporated (Freney et al. 1992). Efforts to decrease these NH_3 losses included fertilizer banding, supplemental irrigation and delaying fertilization until cane canopy was developed. Freney et al. (1991) found that banding urea increased NH_3 loss while irrigating with 16 mm of water decreased NH_3 loss by about 50%. The water addition dissolved part of the urea and washed it into the soil. A one meter high crop canopy absorbed about 20% of the NH_3 emitted from the residue surface, thus decreasing net NH_3 loss. Substituting ammonium sulfate for urea reduced NH_3 loss to $< 1.8\%$ of the N applied (Freney et al. 1992).

NH_3 Volatilization from Flooded Rice. Wetland rice is grown on approximately 125 million ha globally (Neue, 1992) and is frequently limited by soil N supply (Simpson et al. 1984). Fertilizer N application typically, greatly increases rice yield but it is costly to Asian farmers because of low N-use-efficiency. Recoveries of applied N can be as low as 10% and rarely exceeds 50% (Freney et al. 1990). The main cause of fertilizer inefficiency is gaseous emissions of NH_3 or denitrification. The greatest N losses are reported to occur when fertilization leads to high concentrations of ammoniacal N in flood water and thus NH_3 volatilization (Freney et al. 1990). One method of reducing NH_3 emissions is to incorporate the fertilizer into the soil after application. Although this practice generally reduces NH_3 emissions it does not always lead to decreased total N loss (Cai et al. 1986). Freney et al. (1990) found that when urea was broadcast into the flood water at rice transplanting, NH_3 loss varied from 10% to 56% of applied N at four different locations in the Philippines. Losses were greater where temperature and wind speed were greatest. Ammonia loss was reduced by 7% to 56% by harrowing after fertilizer application. Total N loss from the basal urea application ranged from 59% to 71% of N applied. Incorporating the urea by harrowing had little effect on total N loss. Denitrification losses ranged from 3% to 50% of N applied. The denitrification losses were low when NH_3 losses were high and vice versa (Freney et al. 1990).

In China, ammonium bicarbonate and urea are the main N-fertilizer sources (Bouwman et al. 1997). Surface application of ammonium bicarbonate can result in losses of 30% to 70% of N applied by NH_3 volatilization (Cai et al. 1986). Application of urea or ammonium bicarbonate to calcareous soil in north-central China at rice transplanting resulted in 33% and 39% volatilization as NH_3, respectively. Denitrification losses were also large, 33% and 30%, respectively. Incorporation of fertilizer into drained soil substantially increased N utilization by the rice crop. Fertilizer N-use-efficiency was improved mainly through the reduction in NH_3 loss (Zhu et al. 1989).

3.2. Global Emission Sources of NO_x and NH_3

NH_3. Global NH_3 emissions from earth surface to the atmosphere in 1990 total about 54 Tg N (Bouwman et al. 1997). These emissions arise from four main sources (Table 1) which include excreta from domesticated animals (21.6 Tg N), synthetic fertilizers (9 Tg N), biomass burning (5.9 Tg N) and crops and crop decomposition (3.6 Tg N). These data indicate that about 65% of the total global NH_3 is emitted from agricultural systems. In animal production much of N contained in organic compounds in animal excreta is rapidly converted to NH_3 which can be volatilized directly from the animal production system or when the excreta is applied to the soil. Synthetic fertilizers applied in the form of urea (about 55% of global fertilizer N production, FAO, 1999) or ammonium bicarbonate (about 11% of global fertilizer N production) are susceptible to significant NH_3 losses when applied to the soil surface. Urea is typically converted to NH_4^+ within a few days of application to the soil by the enzyme urease (Bouwman et al. 1997) thereby making NH_3 volatilization a problem for surface applied urea.

Table 1.
Total global NH_3 emission estimates adapted from Bouwman et al. (1997) for 1990 (Mosier, 2000).

Source	NH_3-N Emission	
	Estimate	Uncertainty of Estimate
	------------ Tg N	-------------------------
Excreta from domestic animals	21.6	10-30
Excreta of wild animals	0.1	0-1
Synthetic fertilizers	9.0	4.5-1
Biomass burning	5.9	3-7.7
Soils under natural vegetation	2.4	0-10
Oceans	8.2	3-16
Fossil fuel combustion	0.1	0-0.2
Industrial processes	0.2	0.1-0.3
Human excreta & pets	2.6	1.3-3.9
Crops & crop decomposition	3.6	1.4-5.0
Total Emission	54	40-70

When wood, crop residue or fossil fuels are burned, a portion of the N contained in these fuels is converted to NH_3 which is emitted to the atmosphere. Crops can either take up NH_3 from the atmosphere or emit NH_3 to the atmosphere, depending upon atmospheric concentrations and plant N status. When plants begin to senesce they typically lose N content through NH_3 volatilization (Bouwman et al. 1997).

Ferm (1998) estimated for 1990 that 74, 12 and 5.5% of the NH_3 emitted from western Europe came from livestock, synthetic fertilizers and crops, respectively. He also presented data indicating that NH_3 emissions from Europe started to increase dramatically following the second world war. Extending Ferm's information to 1990 suggests that a slowing of this rate of increase in NH_3 emissions has occurred, during the past two decades, probably as a result of the decline in cattle population and decrease in fertilizer use in western Europe during this time.

The rate of increase in NH_3 emissions globally also should have slowed during the past decade. Using FAO data for global domestic animal populations and synthetic fertilizer use and emissions calculations from Bouwman et al. (1997), global NH_3 emission estimates from domestic animal excreta and synthetic fertilizers from 1961 to 1994 are shown in Fig.3. During these three decades NH_3 from animal production increased 54% from 14.2 to 22.1 Tg N y^{-1} at a relatively constant rate of 1.7% y^{-1}. During this time NH_3 from use of synthetic fertilizer increased 540%, 1.4 Tg N y^{-1} to a maximum of 9 Tg N y^{-1} in 1990. The rate of increase between 1961 and 1980 was 21.4% y^{-1} but this rate of increased slowed to 1.5% from 1980 to 1994, mirroring the rate of annual increase from animal production.

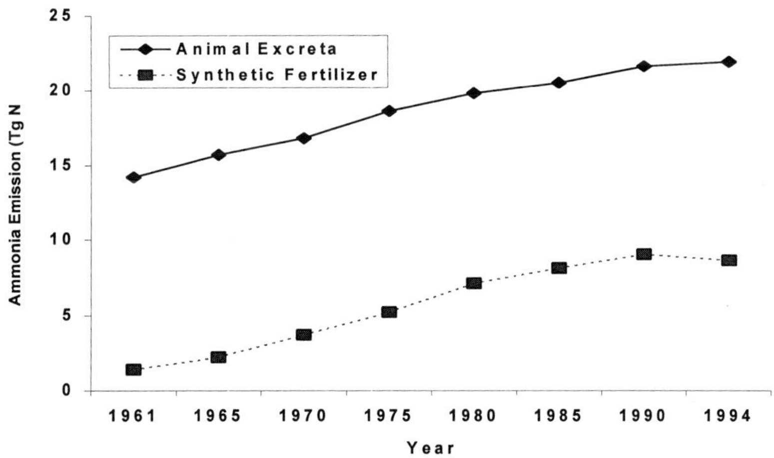

Figure 3. Estimates of global NH_3 emissions from animal excreta and synthetic fertilizer (Bouwman et al. 1997; Mosier, 2000).

NO_x. The release of nitrogen oxides ($NO_x = NO + NO_2$) has accelerated during the last few decades through, primarily the increase in fossil fuel combustion (Galloway et al., 1995; Holland and Lamarque, 1997). With this increase in emissions from ~ 5 Tg N in 1940 to ~ 25

Tg N in 1995, combustion of fossil fuels account for about 50% of the total global NO_x emissions (Table 1), for 1990. Of the anthropogenic sources, fossil fuel, aircraft, biomass burning and part of the soils emission are most important. Although the estimates are also similar for the soil source; 6.2 Tg (Holland et al. 1997) and 5.5 from Delmas et al. (1997), Davidson and Kingerlee (1997) estimated, from field flux measurements, total global soil NO_x emissions to be about 21 Tg N.

When plant uptake of NO_x at the place of NO emission is considered, the total soil-based emission of NO_x are at least 13 Tg N. NO_x emissions from the soil are primarily a result of NO production by the microbial oxidation of ammonium (nitrification) (Williams et al. 1992). NO production in the soil also occurs through the microbial reduction of nitrate (denitrification). This reaction generally occurs only in water-saturated soils where little NO is released from the soil to the atmosphere (Conrad, 1996). Davidson and Kingerlee (1997) estimate that about 5 Tg of NO_x-N is emitted annually from cultivated soils globally. Through an analysis of published NO_x flux measurements in agricultural fields Veldkamp and Keller (1997) estimated that about 0.5% of fertilizer N applied to agricultural fields was emitted to the atmosphere as NO. The total global N input into agricultural soils in 1990 was about 120 Tg. This amount includes N from: (1) synthetic fertilizer (~78 Tg); (2) animal excreta used as fertilizer (~50 Tg); (3) biological N-

Table 2.
Global sources of NO_x emissions to the atmosphere. Compiled from Holland et al. (1997) and Delmas et al. (1997), from Mosier (2000).

Source	Holland et al.[1]		Delmas et al.	
	Tg N			
Fossil fuel combustion	21.1	(20-22.4)	22	(15-29)
Lightning	6.0	(3-10)	2	(1-4)
Soils	6.2	(5-10)	5.5	(3.3-7.7)
Aircraft	0.4	(0.23-0.6)	0.4	0.5-0.6)
Biomass burning	7.0	(4.4-10)	7	(3-10.4)
Stratospheric injection	0.3	(0.2-0.6)	0.5	(0.4-0.6)
Total	41.0	(35-48.8)	38.2	(23.7-53.8)

[1]Numbers presented are the mean of emission estimates from five different 3-dimensional chemical transport models, numbers in parentheses are the range of model output values.

fixing crops [includes pulses and soybeans but not forage crops] (~10 Tg); and, (4) crop residues returned to agricultural fields (~28 Tg) (Mosier et al. 1998). Using the 120 Tg as N input into cropped soils the fertilizer-induced NO_x emissions total 0.6 Tg N y^{-1}. The remainder of the soil sources come from natural systems (Davidson and Kingerlee, 1997), all of which now have received some amount of anthropogenic N input through atmospheric deposition in precipitation and dry fall. Including the additional soil NO_x source estimated by Davidson and Kingerlee (7 Tg N) total global NO_x emissions in 1990 were approximately 48 Tg N using the Holland et al. (1997) numbers (Table 2).

3.3. Fate of NH_3 and NO_x in the atmosphere (NH_x and NO_y deposition)

NH_3. When NH_3 evolves from the earth's surface it is either redeposited near the source of emission or, since it is highly water soluble, dissolved in atmospheric water (Fig. 4). When NH_3 is dissolved in water, ammonium ions (NH_4^+) and hydroxyl ions (OH^-) are formed. The reaction is reversible and the solubility increases if acidic species are also dissolved. Since NH_3 is the dominant basic compound in the atmosphere it reacts readily with acidic gases and particles to form hygroscopic salts containing ammonium sulfate and ammonium nitrate. These two reactions occur very rapidly, a few hours to a few days in the troposphere, so NH_3 is rapidly removed from the atmosphere through deposition as NH_3 and NH_4^+ in precipitation particulates (Ferm, 1998), or through foliar absorption of NH_3 (Schjorring and Mattsson, 2000).

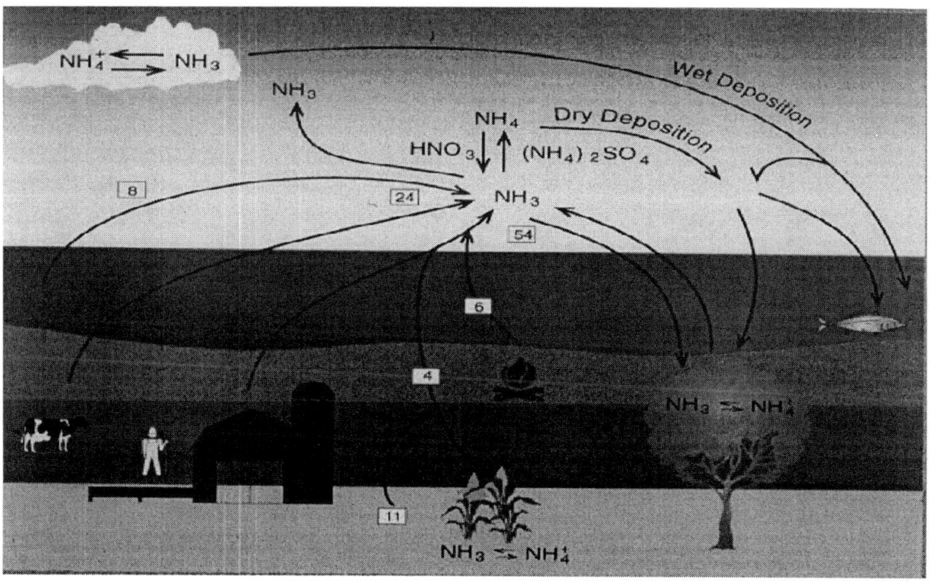

Figure 4. A cartoon depiction of the terrestrial-atmosphere exchange of NH_3 and deposition of NH_x (adapted from Langford et al. 1992; Mosier, 1998).

Fowler et al. (1998) showed that atmospheric NH_3 concentrations drop from ~ 90 ug NH_3 m^{-3} at a poultry production facility to ~ 10 ug NH_3 m^{-3} 100 m downwind. At 300 m distance the NH_3 concentration was near ambient at ~ 0.5 ug NH_3 m^{-3} (Fig. 5). Although redeposition and atmospheric dispersion rapidly decrease NH_3 concentrations as distance from a livestock facility increases, total deposition within 1-km radial distance was less than 10% of the NH_3 emitted (Fig. 5). Pitcairn et al. (1998) extended this work and looked at relationships between NH_3 emissions from livestock facilities and N-deposition onto nearby woodlands. They found that foliar N concentration of a number of plant species were large close to the livestock buildings and declined with distance from them. Within the woodlands ectohydric moss tissue N concentration increased from ~ 1% at ambient atmospheric NH_3 concentration to ~3.2% at NH_3 concentrations of > 20 ug m^{-3}. Pitcairn et al. (1998) also found that moss tissue N content continued to increase to ~ 3.7% at NH_X deposition rates of 40 kg N ha^{-1} y^{-1}. Since most of NH_3 evolved from terrestrial systems is redeposited within 20 km of the emission point, NH_x deposition near cattle production operations can be substantial. For example, NH_3 from a 90,000 head cattle feedlot collected in acid traps located 2 km from the feedlot was 30 kg N ha^{-1} y^{-1} greater than that collected 15 km from the feedlot (Hutchinson and Viets, 1969). The NH_3 evolved from the feedlot provided about 50 kg N ha^{-1} y^{-1} to agricultural soils within a few km of the feedlot. The atmospheric deposition was not considered in fertilization budgets for crops within the area.

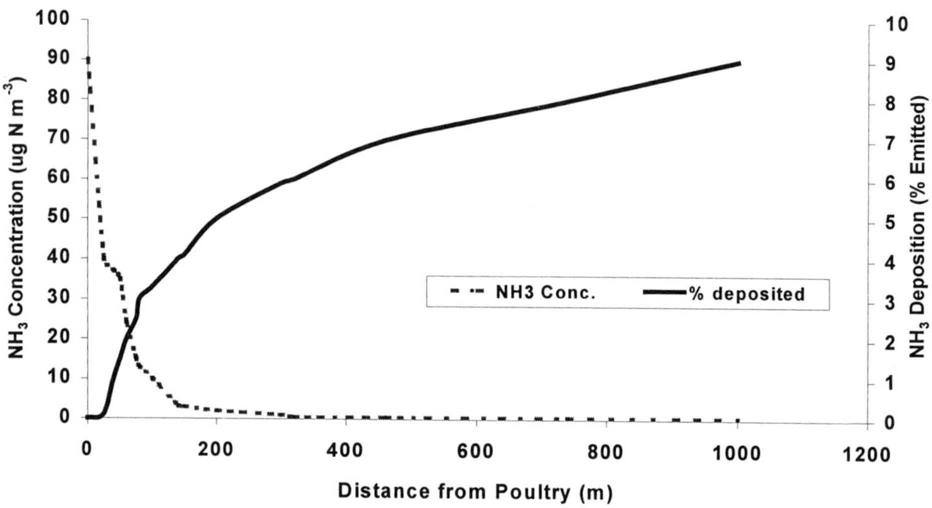

Figure 5. Distance from poultry production buildings and atmospheric NH_3 concentration and NH_x deposition (adapted from Fowler et al. 1998) from Mosier (2000).

NH_x can be transported long distances in the atmosphere. NH_x and other highly water soluble compounds are mainly dispersed in the lower troposphere in the so-called mixing layer. The transport distances within the mixing layer depends upon wind speed and the deposition rate. Model estimates indicate that about 50% of emitted NH_3 is redeposited to earth surface systems within 50 km of the emission source (Ferm, 1998). When gaseous NH_x reaches the top of the mixing layer (about 500 m in winter and 1500 m in summer in northern Europe) (Fig. 6) the NH_x can be transported long distances. When clouds within the mixing layer contain water, the NH_x concentration is halved every 10 km from the source (or every 30 min.). If there is no precipitation, the NH_x concentration is halved every 1400 km (or every third day) (Ferm, 1998). Ferm also estimated that, over Europe, the half life of NH_x is 3 to 6 hours over a distance of 65-130 km. Overall, the average transport distance of NH_x depends upon the atmospheric concentration of SO_2 and NO_x (Ferm, 1998). Atmospheric residence times and transport distances of NH_x are sufficiently long for significant interaction with terrestrial and aquatic systems far removed from the NH_3 sources.

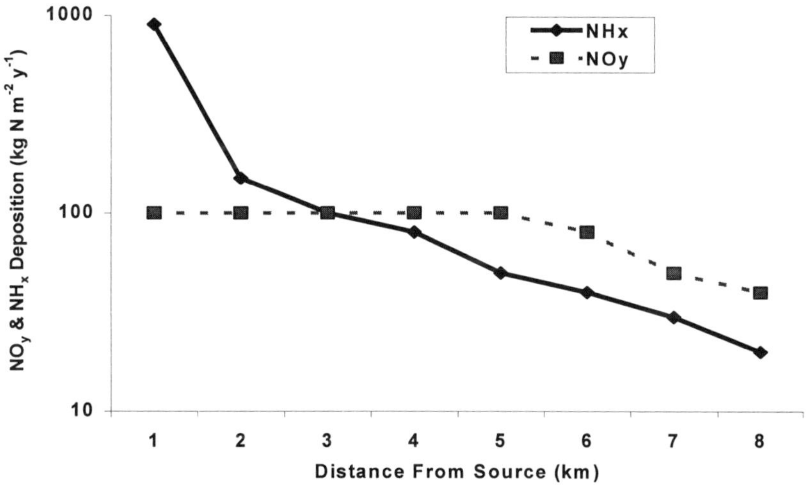

Figure 6. Estimated deposition patterns of NO_y and NH_x emitted from Western Europe (adapted from Ferm, 1998), Mosier (2000).

When NH_x is deposited back onto soil it reenters the terrestrial N cycle. If it is not taken up by plants directly as NH_3 or NH_4^+, it contributes to ecosystem acidification as protons are released when NH_4^+ is oxidized (nitrified). Reactive N deposition can also decrease soil consumption of atmospheric methane (Steudler et al. 1989).

NO_x. Upon its emission into the atmosphere NO_x plays a central role in tropospheric chemistry (Williams et al. 1992). NO_x interacts in a series of concentration dependent cycles with atmospheric oxidants, ozone (O_3), hydroxyl radicals (OH) during the oxidation of carbon monoxide (CO), methane (CH_4) and nonmethane hydrocarbons. The oxidizing capacity of the atmosphere is regulated by NO_x through regulation of OH and O_3 production and loss (Holland et al, 1997). Increasing NO_x concentrations are contributing substantially to observed increases in O_3 concentrations in the Northern Hemisphere (Chameides et al. 1994). Detailed descriptions of the complex interactive cycles of NO_x in the atmosphere can be found in many references (e.g., Ehhalt et al. 1992; Holland and Lamarque, 1997; Liu et al. 1992 and Williams et al. 1992).

As a general simple description (Fig. 7), when NO enters the atmosphere it reacts rapidly with O_3 to form NO_2. In the sunlight, part of NO_2, in a complex set of reactions that involve hydrocarbons, ozone and CO, dissociates to reform NO and O_3. In the presence of NO_x, CO is oxidized to produce CO_2 and O_3. NO_2 is further oxidized to a mixture of products referred to as NO_y (NO_x + organic nitrates, inorganic nitrates and nitrites). During the daytime this set of reactions takes place within a few hours when the sun is bright but may require a few days when clouds block the sun (Williams et al. 1992). At night NO_2 accumulates. At very low NO_x concentrations (<0.08 ppbv) there is net O_3 consumption. As NO_x concentrations increase to 0.08 ppbv to about 1 ppbv, OH and O_3 production increases proportionately with increasing NO_x. At NO_x concentrations above 1.1 ppbv there is a decline in OH and O_3 production (Holland and Lamarque, 1997). Nitric acid production becomes the dominant end product. Nitric acid and peroxynitric acid (HO_2NO_2) are soluble and are removed from the atmosphere by precipitation. Estimates of NO_y deposition (Table 3) indicate that a balance between emission of NO_x and deposition of NO_y exists.

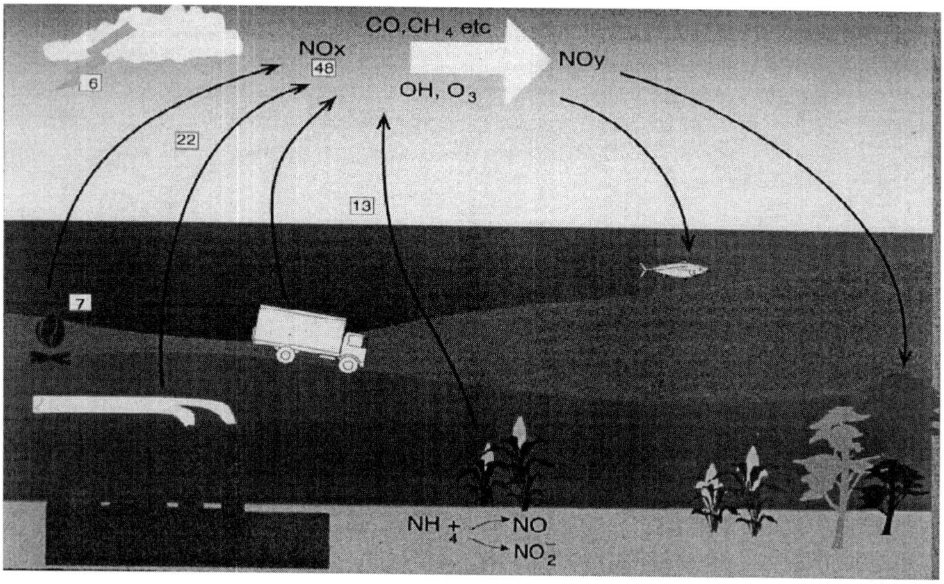

Figure 7. A cartoon depiction of the terrestrial-atmosphere exchange of NO_x and deposition of NO_y (adapted from Williams et al. 1992), Mosier (1998).

Although the lifetime of NO_x in the atmosphere is short (hours to a few days) it can be deposited back to earth long distances from its point of origin. Ferm (1998) estimated that dry

NO_y deposition is rather constant the first 800 km from the emission source then is halved approximately every 600 km (Fig. 5). This long distance transport of NO_y serves to fertilize the global oceans and terrestrial systems. Prospero et al. (1996) estimated NO_y deposition to the open ocean portion of the North Atlantic Ocean to be about 5 Tg N y^{-1}, or about 25% of the NO_x emission from North America and Europe. Globally, about 11 Tg of NO_y-N are returned to agricultural land annually (Table 3) according to model estimates (Holland et al. 1997). This NO_y deposition was equivalent to 14% of the synthetic N fertilizer used globally in 1990 and is not generally included in N budget balance sheets in agricultural systems.

Table 3.
Total NO_y-N deposition on the earth's surface. Data presented are the mean and range of deposition quantities determined by five, 3-dimensional chemical transport models and values in parenthesis are the range of the five estimates (Holland et al. 1997), from Mosier (1998).

Deposition Location	NO_y-N Deposited	
	Mean	Range
	---------Tg N -----------	
Oceans	16.2	11.7-25.8
Land	22.9	15.4-27.6
ice or deserts	1.3	1.2-1.4
agricultural areas	10.6	6.6-14.1
natural vegetation	10.9	7.7-13.8
forests	3.1	2.2-4
unforested land	7.8	5.4-9.8
Total Global Deposition	39.1	35.3-41.2

4. PERSPECTIVES ON SOIL-ATMOSPHERE EXCHANGE OF GASEOUS N COMPOUNDS

Reactive N species are highly mobile through the emissions mainly of NO_x and NH_3 via the atmosphere and release of nitrate into ground and surface waters. These compounds may cross national boundaries and be deposited as NO_y and NH_X or nitrate far from the source of N release (Fig. 4 & Fig. 7) (Aber et al. 1995; Matson et al. 1997; Vitousek et al. 1997). Additional production of N_2O from N deposition and release is part of the anthropogenic N_2O global budget (IPCC 1997). The redistribution of N from industrial, automobile and agricultural sources has and is having profound impacts on the quality of terrestrial and aquatic ecosystems on our atmosphere. The recent estimates noted herein indicate that globally, about 150 Tg of biologically and chemically active forms of N are emitted to the atmosphere annually (Smil, 1999). The emission of these compounds have increased exponentially since the end of World

War II and can be attributed most directly to fossil fuel combustion, agricultural production and biomass burning.

Deposition of N can stimulate productivity in N-limited grasslands, forests and aquatic systems (Galloway et al. 1995; Holland et al. 1997). Excessive N deposition, however, can and has led to acidification and eutrophication of aquatic systems (Aber et al. 1995) and to forest decline a decrease in plant species diversity and soil acidification (Ferm, 1998; Pitcairn et al. 1998). NO_x emissions also contribute to the increase of tropospheric O_3 which is damaging crop production in some areas of the globe (Holland and Lamarque, 1997). Ozone in the troposphere is also a greenhouse gas. N deposition onto formerly pristine areas also contribute to increased radiative forcing through increased N_2O emissions (Mosier et al. 1998) and decreased soil consumption of atmospheric CH_4 (Steudler et al. 1989). A large portion of this N deposition occurs on agricultural lands (10-20 Tg annually) that needs to be accounted for in fertilizer recommendations, an aspect of the agricultural N cycle that has been neglected to date.

Because of the above characteristics of gaseous N emissions, we continue to be faced with the challenge of limiting their emissions from food production systems while sustaining the systems themselves. Technical advances are allowing accessability of field scale analytical devices to a larger number and variety of research groups to permit larger scale assessment of NH_3, NO_x and N_2O emissions. A host of such studies are needed to assess the impact of changes in management in crop production such as tillage, crop varieties, fertilization timing and livestock production. Particularily interesting is the role of plants in the balance of NH_3 and NO_x in the atmosphere. Schjorring and Mattsson (2000) conclude that crops can represent a significant input of NH_3. Davidson and Kingerlee (1997) note that NO_x emissions from fertilized fields can be large but that the resorption of NO_2 by plant foliage prevents an unknown and likely variable amount of NO_x from exiting the plant canopy. All of the problems, environmental and economic, associated with the loss of N from the soil and its redistribution, reinforces the need to more tightly couple the crop and livestock parts of food production.

REFERENCES

Aber, J.D., A. Magill, S.G. McNulty, R.D. Boone, K.J. Nadelhoffer, M. Downs, and `R. Hallett. 1995. Forest biogeochemistry and primary production altered by nitrogen saturation. Water Air and Soil Poll. 85:1665-1670.

Bouwman, A.F., D.S., Lee, W.A.H. Asman, F.J. Dentener, K.W. Van der Hoek, and J.G.J. Olivier. 1997. A global high-resolution emission inventory for ammonia. Global Biogeochem. Cycl. 51:561-587.

Cai, G.X., Z.L. Zhu, A.C.F. Trevitt, J.R. Freney, and J.R. Simpson. 1986. Nitrogen loss from ammonium bicarbonate and urea fertilizers applied to flooded rice. Fert. Res. 10:203-215.

Chameides, W. L., P.S. Kasibhatla,J. Yienger, and H. Levy II. 1994. Growth of continental scale metro-agro-plexes, regional ozone pollution, and world food production. Science. 264:74-77.

Cleveland, C.C., A.R. Townsend, D.S. Schimel, H. Fisher, R.W. Howarth, L.O. Hedin, S.S. Perakis, E.F. Latty, J.C. Von Fischer, A. Elseroad, and M.F. Wasson. 1999. Global patterns of terrestrial biological nitrogen (N2) fixation in natural ecosystems. Global Biogeochem. Cycl. 13:623-645.

Cole, C.V., C. Cerri, K. Minami, A.R. Mosier, N. Rosenberg, and D. Sauerbeck. 1996. Chapter 23. Agricultural options for mitigation of greenhouse gas emissions. pp 745-771. *In* R.T. Watson, M.C. Zinyowera and R.H. Moss (eds.). Climate Change 1995. Impacts, Adaptations and Mitigation of Climate Change: Scientific Technical Analyses. Published for the Intergovernmental Panel on Climate Change. Cambridge University Press, Cambridge, UK.

Conrad, R. 1996. Soil microorganisms as controllers of atmospheric trace gases (H_2, CO, CH_4, OCS, N_2O and NO. Microbiological Reviews. 60:609-640.

Davidson, E.A., and W. Kingerlee. 1997. A global inventory of nitric oxide emissions from soils. Nut. Cycl. in Agroecosys. 48:37-50.

Delmas, R., D. Serca, and C. Jambert. 1997. Global inventory of NO_x sources. Nut. Cycl. in Agroecosys. 48:51-60.

Ehhalt, D.H., F. Rohrer, and A. Wahner. 1992. Sources and distribution of NO_x in the upper troposphere at northern mid-latitudes. J. Geophys. Res. 97(D4):3725-3738.
FAO United Nations Food and Agricultural Organization. 1999. FAOSTAT: Agricultural Data, are available on the world wide web: (http://www.apps.fao.org/cgi-bin/nph-db.pl?subset=agriculture).

Farquhar, G.D., P.M. Firth, R. Wetselaar, and B. Weir. 1980. On the gaseous exchange of ammonia between leaves and the environment: Determination of the ammonia compensation point. Plant Physiol. 66:710-714.

Fenn, L.B., J.E. Matocha, and E. Wu. 1981. A comparison of calcium carbonate precipitation and pH depression on calcium-reduced ammonia loss from surface-applied urea. Soil Sci. Soc. Am. J. 45:1128-1131.

Ferm, M. 1998. Atmospheric ammonia and ammonium transport in Europe and critical loads - A review. Nut. Cycl. in Agroecosys. 51:5-17.

Fowler, D., C.E.R. Pitcairn, M.A. Sutton, C. Flechard, B. Loubt, M. Coyle, and R.C. Munro. 1998. The mass budget of atmospheric ammonia in woodland within 1 km of livestock buildings. Environ. Poll. 102: 343-348.

Francis, D.D., J.S. Schepers, and A.L. Sims. 1997. Ammonia exchange from corn foilage during reproductive growth. Agron. J. 89:941-946.

Freney, J.R., J.R. Simpson, and O.T. Denmead. 1983. Volatilization of ammonia. *In* J.R. Freney and J.R. Simpson (eds.) pp. 1-32. Gaseous Loss of Nitrogen from Plant-Soil Systems. Kluwer Academic Publishers, The Hague.

Freney, J.R., A.C.F. Trevitt, S.K. DeDatta, W.N. Obcemea, and J.G. Real. 1990. The interdependence of ammonia volatilization and denitrification as nitrogen loss processes in flooded rice fields in the Philippines. Biol. Fertil. Soils 9:31-36.

Freney, J.R., O.T. Denmead, P.G. Saffigna, A.W. Wood, L.S. Chapman, and A.P. Hurney. 1991. Ammonia loss from sugar cane fields as affected by fertilizer placement, irrigation and canopy development. Proc. of Aust. Soc. Sugar Cane Techs. pp. 38-43.

Freney, J.R., O.T. Denmead, A.W. Wood, P.G. Saffigna, L.S. Chapman, G.J. Ham, A.P. Hurney, and R.L. Stewart. 1992. Factors controlling ammonia loss from trash covered sugarcane fields fertilized with urea. Fert. Res. 31:341-349.

Galloway, J.N., W.H. Schlesinger, H. Levy II, A. Michaels, and J.L. Schnoor. 1995. Nitrogen fixation: Anthropogenic enhancement-environmental response. Global Biogeochem. Cycl. 9:235-252.

Galloway, J.N., H. Levy II, and P.S. Kasibhatla. 1994. Year 2020: consequences of population growth and development on the decomposition of oxidized nitrogen. Ambio. 23:120-123.

Hanawalt, R.B. 1969. Environmental factors influencing the sorption of atmospheric ammonia by soils. Soil Sci. Soc. Am. Proc. 33:231-234.

Harper, L.A., R.R. Sharpe, G.W. Langdale and J.E. Giddens. 1987. Nitrrogen cycling in a wheat crop: Soil plant and aerial nitrogen transport. Agron. J. 79:965-973.

Harper, L.A., and R.R. Sharpe. 1995. Nitrogen dynamics in irrigated corn: Soil-plant nitrogen and atmospheric ammonia transport. Agron. J. 87:669-675.

Harper, L.A., and R.R. Sharpe. 1998. Atmospheric ammonia: Issues on transport and nitrogen isotope measurement. Atmos. Environ. 32:273-277.

Harper, L.A., D.W. Bussink, H.G. van der Meer. and W. J. Corrre. 1996. Ammonia transport in a temperate grassland: I. Seasonal transport in relation to soil fertility and crop management. Agron. J. 88:614-621.

Holland, E.A., and J.F. Lamarque. 1997. Modeling bio-atmospheric coupling of the nitrogen cycle through NO_x emissions and NO_y deposition. Nut. Cycl. in Agroecosys. 48:7-24.

Holland, E.A., B.H. Braswell, J.F. Lamarque, A. Townsend, J.M. Sulzman, J.F. Muller, F. Dentener, G. Brasseur, H. Levy II, J.E. Penner, and G. Roelofs. 1997. Variations in the predicted spatial distribution of atmospheric nitrogen deposition and their impact on carbon uptake by terrestrial ecosystems. J. Geophys. Res. (Atmospheres). 102(D13), 15849-15866.

Hutchinson, G.L., and F.G. Viets. 1969. Nitrogen enrichment of surface water by absorption of ammonia volatilized from cattle feedlots. Science. 166:514-515.

Hutchinson, G.L., R.J. Millington, and D.B. Peters. 1972. Atmospheric ammonia: Absorption by plant leaves. Science. 175:771-772.

IPCC. 1997. Intergovernmental Panel on Climate Change Guidelines for National Greenhouse Gas Inventories. Chap. 4. Agriculture: nitrous oxide from agricultural soils and manure management. OECD, Paris.

Kroeze, C., A.R. Mosier, and A.F. Bouwman. 1999. Closing the global N_2O budget: A retrospective analysis 1500-1994. Global Biogeochem. Cycl. 13:1-8.

Liu, S.C., M. Trainer, M.S. Carroll, G. Hubler, D.D. Montzka, R.B. Norton, B.A. Ridley, J.G. Walega, E.L. Atlas, B.G. Heides, B.J. Huebert, and W. Warren. 1992. A study of the photochemistry and ozone budget during the Mauna Loa observatory photochemistry experiment. J. Geophys. Res. 97(D10):100463-10471.

Matson, P.A., W.J. Parton, A.G. Power, and M.J. Swift. 1997. Agricultural intensification and ecosystem properties. Science 277:504-509.

Mosier, A.R. 1998. Nutrient redistribution by soil-atmosphere exchange of nitrogn compounds. pp 672-684. *In* O. Van Cleemput et al. (eds.). Fertilization for Sustainable Plant Production and Soil Fertility. Int'l. Sci. Centre of Fert. (CIEC). Braunschweig/Budapest/Vienna.

Mosier, A.R. 2000. Exchange of gaseous nitrogen compounds between agricultural systems and the atmosphere. Plant and Soil (In Press).

Mosier, A.R., C. Kroeze, C. Nevison, O. Oenema, S. Seitzinger, and O. Van Cleemput. 1998.

Closing the global atmospheric N_2O budget: nitrous oxide emissions through the agricultural nitrogen cycle. Nut. Cycl. in Agroecosys. 52:225-248.

Neue, U.H. 1992. Agronomic practices affecting methane fluxes from rice cultivation. Ecol. Bull. (Copenhagen) 42:174-182.

Parton, W.J., J.A. Morgan, J.M. Altenhofen, and L.A. Harper. 1988. Ammonia volatilization from spring wheat plants. Agron. J. 80:419-425.

Pitcairn, C.E.R., I.D. Leith, I.J. Sheppard, M.A. Sutton, D. Fowler, R.C. Munro, S. Tang, and D. Wilson. 1998. The relationship between nitrogen deposition, species composition and foliar nitrogen concentrations in woodland flora in the vicinity of livestock farms. Environ. Poll. 102:41-48.

Prospero, J.M., K. Barrett, T. Church, F. Dentener, R.A. Duce, J.N. Galloway, H. Levy II, J. Moody, and P. Quinn. 1996. Atmospheric deposition of nutrients to the North Atlantic basin. Biogeochem. 35:27-73.

Schjorring, J.K. 1995. Long-term quantification of ammonia exchange between agricultural cropland and the atmosphere. I. Evaluation of a new method based on passive flux samplers in gradient configuration. Atmos. Environ. 29:885-893.

Schjorring, J.K., and M. Mattsson. 2000. Long-term quantification of ammonia exchange between agricultural cropland and the atmosphere. II. Measurements over oilseed rape, wheat, barley and pea. Plant & Soil (In Press).

Sharpe, R.R. and L.A. Harper. 1997. Apparent atmospheric nitrogen loss from hydroponically grown corn. Agron. J. 89:605-609.

Simpson, J.R., J.R. Freney, R. Wetselaar, W.A. Muirhead, R. Leuning, and O.T. Denmead. 1984. Transformations and losses of urea nitrogen after application to flooded rice. Aust. J. Agric. Res. 35:189-200.

Smil, V. 1999. Nitrogen in crop production: An account of global flows. Global Biogeochem. Cycl. 13:647-662.

Steudler, P.A., R.D. Bowden, J.M. Melillo, and J.D. Aber. 1989. Influence of nitrogen fertilization on methane uptake in temperate forest soils. Nature. 341:314-316.

Terman, G.L. 1979. Volatilization losses of nitrogen as ammonia from surface-applied fertilizers, organic amendments, and crop residues. *In* N.C. Brady (ed.). Adv. in Agron. 31:189-223.

Veldkamp, E, and M. Keller. 1997. Fertilizer-induced nitric oxide emissions from agricultural soils. Nut. Cycl. in Agroecosys. 48:69-77.

Vitousek, P.M., J. Aber, R.W. Howarth, G.E. Likens, P.A. Matson, D.W. Schindler, W.H. Schlesinger, and D.G. Tilman. 1997. Human alteration of the global nitrogen cycle: Causes and Consequences. Issues in Ecology. 1:1-15.

Wetselaar, R. and G.D. Farquhar. 1980. Nitrogen losses from tops of plants. *In* N.C. Brady (ed.). Adv. in Agron. 33:263-302.

Williams, E.J., G.L. Hutchinson, and F.C. Fehsenfeld. 1992. NO_x and N_2O emissions from soil. Global Biogeochem. Cycl. 6:351-388.

Zhu, Z.L., G.X. Cai, J.R. Simpson, S.L. Zhang, D.L. Chen, A.V. Jackson and J.R. Freney. 1989. Processes of nitrogen loss from fertilizers applied to flooded rice fields on a calcareous soil in north-central China. Fert. Res. 18:101-115.

Nitrogen in the Environment: Sources, Problems, and Management
R.F. Follett and J.L. Hatfield (Eds)

Chapter 12. The Impacts of Nitrogen Deposition on Forest Ecosystems

K. J. Nadelhoffer

The Ecosystems Center, Marine Biological Laboratory, Woods Hole, Massachusetts, 02543 USA

Forest plants rely mainly on ammonium (NH_4^+) released from decomposing litter and humus, and often on nitrate (NO_3^-) produced by ammonium oxidizing bacteria ("nitrifiers"), to meet their N uptake demands (Waring and Running 1998). Although some forest species are capable of taking up amino acids from soils (Näsholm et al. 1998), most N assimilated by tree roots is inorganic. Forest litter and soil N pools are maintained, in turn, by organic inputs originating as plant materials. The total amount of N in forest floor and soil pools typically exceeds the amount in vegetation and is at least an order of magnitude greater than the amount of N taken up annually by forest vegetation (Johnson and Curtis 2001). Low mineralization and turnover rates of soil and litter N pools combine with the pivotal role of N as a component of photosynthetic and respiratory enzymes, to limit annual primary productivity (plant biomass production per unit area) in forests. This is particularly true in temperate and boreal forests, but is less so in tropical forests where primary production is often limited by other elements, often phosphorus (Matson et al. 1999). Therefore, primary production in most temperate and boreal forests is considered "N limited" at yearly time scales (Vitousek and Howarth 1991).

Forest litter and soils typically interact with vegetation to form "closed" N cycles in which annual rates of plant N uptake per unit area approximately balance annual N returns to the forest floors and soil as litter and root material. This closed pattern of N cycling contrasts with the more "open" cycles of agricultural systems subjected to fertilizer N inputs (Frink et al. 1999). It also differs from the open cycles of many wetland, freshwater and marine ecosystems in which inputs from surrounding landscapes, biological N fixation, or upwelling bottom waters function to supply much of the N used to support primary production (Jahnke 1990, Mitsch and Gosselink 1986, Vollenweider 1968). Increased N deposition on forests and other ecosystems with relatively closed cycles has the potential to drive forest N cycles to the more open states by increasing the importance of N inputs and outputs relative to rates of internal N cycling

The increases in N emission and deposition that have occurred in industrialized regions of the Northern Hemisphere, and which are projected to continue into the future across more of Earth's surface (Galloway et al. 1994, Galloway et al. 1995, Holland et al. 1999a), could serve to diminish the degree of N limitation of forest growth, to alter forest community composition, to change plant-soil interactions, and to open forest N cycles. These responses to N deposition have implications not only for forests, but also for the atmosphere and for ecosystems that receive nutrient outputs from forests. This chapter summarizes research addressing the influences of N deposition on forest N cycles and on forest ecosystem structure and function.

1. NITROGEN INPUTS

Although N mineralized from decomposing litter and humus supplies most N taken up by forest plants, forest N cycles are not completely closed. In mid-successional and mature forests that are not subject to elevated atmospheric N deposition or to recent physical disturbance, N inputs are small (<2 to 4 kg N ha^{-1} yr^{-1}) relative to the ~40 to 130 kg N ha^{-1} yr^{-1} that cycles between vegetation and soils in temperate and boreal forests (Aber et al. 1985, Bonan and Van Cleve 1991, Reich et al. 1997). Longer growing seasons in tropical and subtropical regions can lead to higher annual rates of N exchange between forest vegetation and soils (Vitousek and Matson 1988) with little or no increase in N inputs (except where N fixers are active, below). Biological N fixation (the enzymatic reduction of inert N_2 gas to NH_4^+) and atmospheric N deposition, although usually of minor importance to plant N uptake on an annual basis under pristine conditions (no anthropogenic N deposition), are the primary processes responsible for N accretion in forests.

1.1. Biological N fixation

The presence of certain plant species (e.g. *Robinia, Acacia, Alnus*, *Myrica*) capable of hosting N-fixing microbial symbionts (*Rhizobia*, *Frankia*) in root nodules can increase N fixation and the reliance of vegetation on this process to meet N demands. Symbiotic N fixation is of greater importance when host plant species are abundant and when other elements (e.g. phosphorus, calcium, molybdenum) required by N fixers are in sufficient supply (Binkley et al. 1992, Högberg and Alexander 1994, Liu and Deng 1991). Nitrogen fixation by free-living microbes in soils and litter can also function to add N to forests, but generally at much lower rates than are required for plant growth (Johnson and Curtis 2001). Nitrogen fixation, both by symbiotic and free-living organisms, is typically more important early in primary succession than in mid- or late succession, on N-deficient than more N-rich soils, and in tropical and sub-tropical regions than in temperate and boreal regions soils (Binkley et al. 1992, Vitousek and Sanford 1986 , Zou et al. 1995).

1.2. N deposition

Before the industrial age, rates of atmospheric N deposition on forests were exceedingly low and contributed insignificantly to N uptake by vegetation (Galloway et al. 1982, Hedin et al. 1995, Likens et al. 1987). Most of these inputs were as NO_3^- (generated by lighting-driven N_2 oxidation) and particulate organic N that is not immediately available to plants. Increases in fertilizer use, animal husbandry and fossil fuel combustion, especially during the latter half of the 20th century, have served increased emissions of NH_x (= NH_3 and NH_4^+) and NO_x (= NO and NO_2). Transport and transformations of NH_x and NO_x in the atmosphere have led to elevated deposition of biologically active N (mostly NH_x and NO_3^-) on forests, particularly in populated temperate areas in eastern North America, central and western Europe, and far eastern Asia (Galloway et al. 1995, Holland et al. 1999a).

Atmospheric N deposition rates on forests vary widely across industrialized regions in the North Temperate Zone. Total inorganic N deposition (wet plus dry) ranges from about 3 to 32 kg N ha^{-1} yr^{-1} in North American forests (Fenn et al. 1998, Ollinger et al. 1993) and from about 1 to 75 kg N ha^{-1} yr^{-1} in Europe (Dise and Wright 1995). The highest deposition rates typically occur near agricultural "hot spots" of intensive fertilization or dense animal concentrations where ammonia volatilization occurs, near point sources of NO_x emission, and at high elevations subject to fog and cloud droplet formation (Bytnerowicz and Fenn 1996, Glatzel 1990, Lovett and Kinsman 1990, van Breemen and van Dijk 1988). Interpolations of N deposition measurements across regions, such as by Ollinger et al. (Ollinger et al. 1993), illustrate the influences of elevation and distances from emission sources on the spatial pattern of N deposition (Figure 1).

Because their plant canopies are taller and more irregular than either cropland and pasture canopies, forests typically intercept more inorganic N ions (and other pollutants)

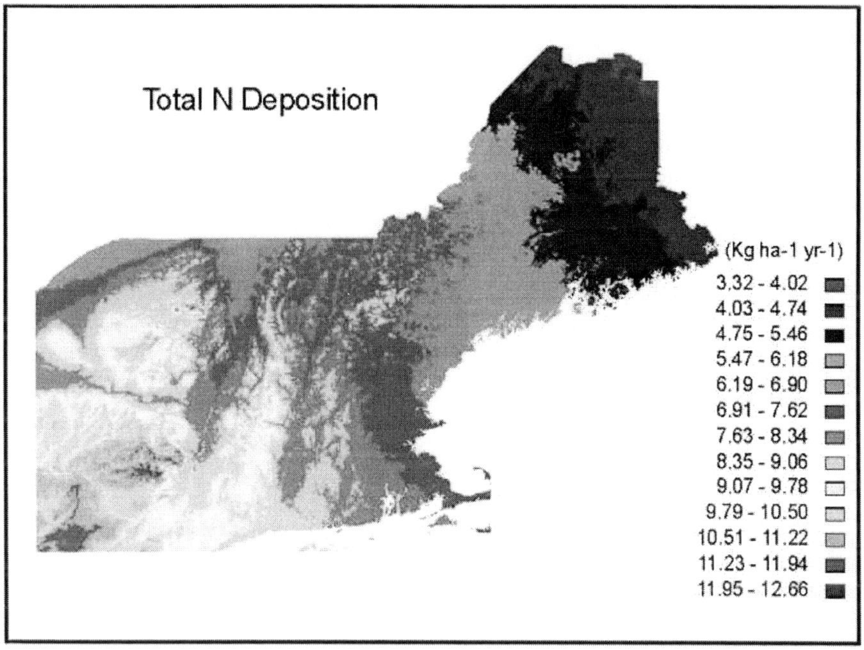

Figure 1. Patterns of N deposition in the northeastern US. Higher rates occur near to sources of N emission in the Midwest and Mid-Atlantic regions (from the southwest) and at higher elevations. Modifed from Ollinger et al. (1993). Courtesy of S. Ollinger

than do adjacent ecosystems of lower stature (Fowler et al. 1999). Given this, and the presence of point sources of N emissions that can contribute high N deposition in small areas (Fowler et al. 1998a, Pitcairn et al. 1998), many forest stands are subject to higher N deposition rates than predicted by regional models (Fowler et al. 1998b, Sutton et al. 1998).

2. EFFECTS ON FOREST STRUCTURE AND FUNCTION

Elevated N deposition was not recognized as an important stressor relative to other atmospheric pollutants such as sulfuric acid and ozone until the mid 1980s (Aber 1992). This initial lack of attention to N deposition as a potential pollutant resulted from recognition that tree growth and primary production in forests are often limited by N availability, which in turn is largely dependent on decomposition and N mineralization processes. In fact, a major focus of forest nutrient cycling research before the 1980s was on the long-term implications of forest product removal for N cycling (Aber et al. 1979). In the early 1980s, however, reports of soil acidification and "forest decline" in European regions subjected to highly elevated ammonium inputs led to speculation that elevated N deposition could eventually lead to forest productivity declines, soil acidification and stream water nitrate contamination (Nihlgård 1985).

Nihlgård's 1985 "ammonium hypothesis", together with reports increasing nitrate levels in forest streams along N deposition gradients (Driscoll et al. 1987, Grennfelt and Hultberg 1986), led to various investigators to propose "Nitrogen Saturation" as the possible end point of forest responses to chronically elevated inputs of reactive N from the atmosphere (Aber et al. 1989, Ågren and Bosatta 1988, Skeffington and Wilson 1988). Although differing in some respects, these definitions all addressed the possibility that elevated N deposition could lead to perturbations of the N cycle and the biological and physical processes that otherwise serve to assimilate and retain N inputs in forests. In short, these conceptual models suggested that elevated N deposition could lead to the transition of forest N cycles from closed to open states. At the end-point of N Saturation, N outputs from forests might eventually approach or even exceed N inputs. Moreover, these N Saturation models proposed various mechanisms and interactions to explain how "N-limited" ecosystems such as forests might "leak" biologically reactive N forms. The proposed effects of N deposition on key biogeochemical proposes have been and are subjects of active research, the results of which provide a basis for assessing how forest structure and function are influenced by N deposition.

One view of N saturation is that of stages of responses to chronically elevated N deposition (Aber et al. 1995, Aber et al. 1998, Aber et al. 1989). In this view, plant and soil processes undergo transitions from closed to open N cycles resulting from biogeochemical responses to N inputs (Figure 2). This model posits that initial responses of N-limited forests to elevated deposition (*Stage 1*) are small. However, increases in foliar biomass and primary production would likely be below the detection limits of field measurements. The most likely "indicators" of forest response to N deposition are likely to be increased N tissues such as leaves and fine roots rather than growth responses. Higher N concentrations in these tissues would allow for greater resource uptake (carbon by leaves and nutrients by fine roots) and possibly promote tree growth. Soil responses in during *Stage 1* could include increases in N mineralization due to greater N inputs as leaf litter and fine root death. Higher N mineralization rates, together with increased N deposition, would increase ammonium availability to nitrifiers and thereby increase nitrification (net NO_3^- production). Increased nitrification would (1) change the ratio of ammonium to nitrate available to tree roots; (2) stimulate the production of N_2O, NO and N_2 gasses (denitrification), and (3) lead to NO_3^- leaching losses. In summary, the possible initial responses to elevated N deposition

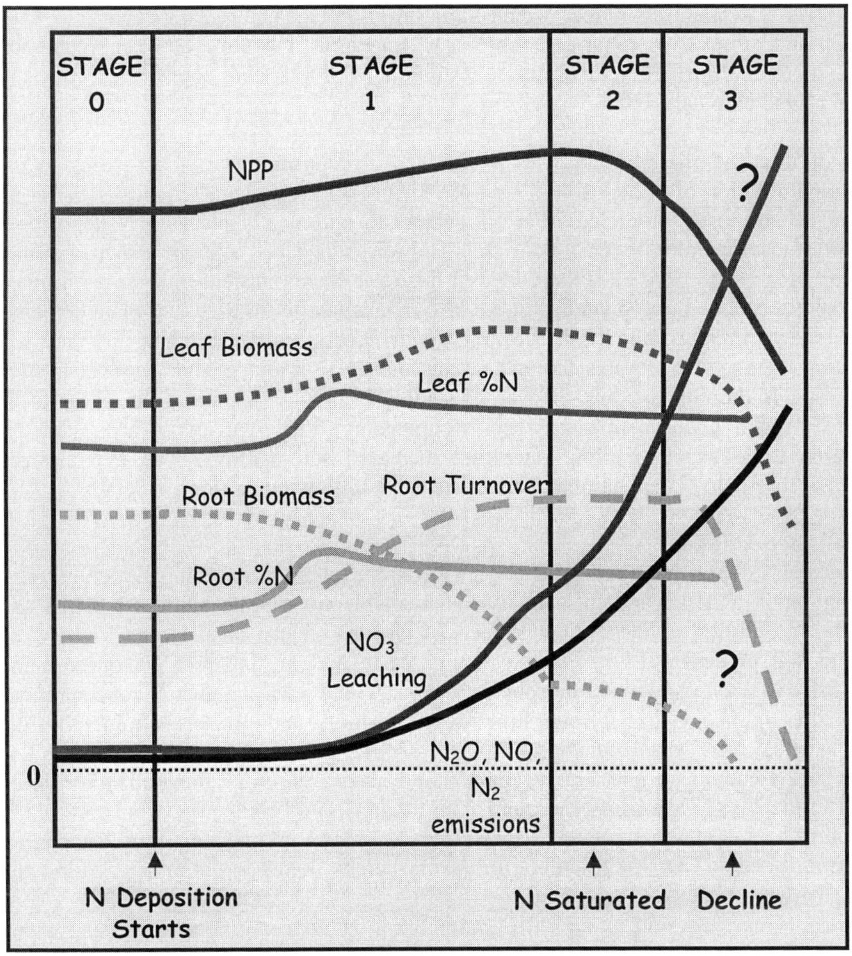

Figure. 2. Conceptual model of forest biogeochemical responses to chronically elevated N deposition. Stage 0 represents pristine conditions. Stage 1 follows the onset of elevated N deposition. Where N limits growth, increases in foliar and fine root N could leato increased N concentrations and turnover rates in resource leaves and fine roots, and higher net primary production (NPP). These increases, however, might be difficult to detect, particularly at low rates of N input. Nitrate losses above background levels can occur, but they are not large. If forests advance to Stage 2, nitrate losses increase and biomass responses are more easily detected. If forest decline (Stage 3) occurs, tree mortality is high, NPP decreases, and exports of nitrate and oxidized N gasses increase. Variations of lines along the *y*-axis indicate qualitative changes for the individual processes or properties as elevated N deposition progresses. Adapted from Aber et al. (1989, 1998) and Nadelhoffer 2000.

include higher N availability, higher N concentrations in foliage and roots, greater production and turnover of foliar and root tissue, increased N cycling between vegetation and soils, increased nitrification and nitrate availability to roots, and increased gaseous and dissolved nitrate outputs.

The duration of *Stage 1* responses to elevated N deposition varies according to the intensity and duration of N deposition, land-use history and soil characteristics (Aber et al. 1995) and is not easily predicted. Forests subject to chronically elevated N deposition, particularly those on infertile, base-poor soils (low exchangeable calcium and magnesium levels) can progress to *Stage 2* (Figure 2). At this later stage, nutrient imbalances in plant tissues (below) can lead to growth declines. Also, N availability in excess of plant demands can lead to even greater rates of nitrification, denitrification, nitrate leaching loss and soil acidification. *Stage 4* represents the end of this progression and is characterized by tree mortality and N outputs equaling or even exceeding N inputs. This stage is likely to be followed by major changes in plant species composition and a period of ecosystem re-equilibration to greater N cycling rates and increased soil acidity. The hypothesized responses of forests to N deposition are discussed in the following sections in detail.

2.1. Plant processes

Elevated N deposition directly influences forest N cycles by increasing input N:internal N cycling ratios and the amounts reactive N available for uptake by forest plants and microbes. In addition to these direct effects, N deposition can alter feedbacks between plants and soil processes. For example, increases in leaf or fine root N concentration occurring due to plant uptake of atmospheric N inputs could lead to higher N concentrations in plant litter inputs to soils, faster litter decomposition, and greater N mineralization (Hendricks et al. 2000, McClaugherty et al. 1985, Melillo et al. 1982). Such changes in leaf and root litter chemistry and any accompanying increases in N mineralization could feedback to increase N availability to plants more than would be expected from elevated N inputs alone. Likewise, increased uptake of N inputs by decomposers (*microbial immobilization*) could increase microbial turnover, thereby increasing N mineralization and N cycling between plants and soils.

Forest N cycling studies typically show foliar N concentrations increasing along atmospheric N deposition gradients and in response to experimental N additions (Burton et al. 1993, Magill et al. 2000, Magill et al. 1996, McNulty et al. 1991, McNulty et al. 1990). Although N concentrations often increase with N inputs, reports of decreased foliar Mg or Ca and increase foliar Al concentrations (Hutchinson et al. 1998, Schulze 1989, van der Eerden et al. 1998, Wilson and Skeffington 1994) suggest N deposition can lead to nutrient imbalances in trees. More recently, free amino acids (arginine) and polyamines (e.g. putrescine) indicative of plant stress have been shown to increase in pine and oak foliage in response to chronic N additions (Calanni et al. 1999, Minocha et al. 2000). Such nutrient imbalances are possible reasons why increases in primary production under elevated N deposition are sometimes short-lived or not detectable (below).

2.2. Soil processes

As with foliage, N concentrations also increase (or C:N decreases) with N deposition or fertilizer N additions to forests. (Emmett et al. 1998, McNulty et al. 1991). These decreases in forest floor C:N ratios could result from higher N concentrations in litter and senesced root inputs, from greater consumption of labile C in litter when N demands of microbes are met by increased deposition, or both. N deposition can also influence N processing and the acid-base relations as described below.

Mineralization and nitrification- Elevated N deposition can, at least initially, stimulate net rates of N mineralization nitrification, or both (Kjønaas et al. 1998, McNulty et al. 1990). However, long-term studies suggest that N mineralization rates can eventually decline in response to N deposition, whereas nitrification rates can continue to increase or remain elevated above initial conditions (Aber et al. 1995). Increases in N mineralization and nitrification, whether transient or of long duration, probably result from feedbacks caused by inputs to soils of N-enrich plant and microbial tissues. Recent work suggests that the long-term decreases in N mineralization in response to elevated inputs might result from increases in carbon limitation of microbial activity and suppression of microbial production of humus-degrading enzymes (Berg 2000, Berg and Matzner 1997, Fog 1988). Increases in nitrification have important implications for soil acid-base relations, the availability of other nutrient ions to forest plants, and nitrogen outputs to the drainage water and the atmosphere.

Soil acid-base relations- Mineralization of one unit of organically bound litter or humus N consumes one unit of acidity or H^+ (Figure 3). If the NH_4^+ produced by mineralization is taken up and re-assimilated into plant (or microbial) biomass, a unit of H^+ is produced, resulting in no net change in acidity of the system. If, however, the mineralized NH_4^+ is nitrified (oxidized by nitrifiers), two units of H^+ are produced, yielding one unit H^+ for the combined mineralization and nitrification of a single organic N unit. If NO_3^- produced by nitrification is taken up and assimilated into biomass, the H^+ produced from mineralization pluse nitrification is consumed and the acidity of the system remains balanced. Likewise, if soil NO_3^- is denitrified, acidity is consumed. When NO_3^- is not assimilated into biomass or denitrified, the H^+ generated during mineralization and nitrification increases forest soil acidity.

Because nitrate ions and clay-humus particle surfaces in most temperate forests are negatively charged, NO_3^- is readily leached with drainage water if it is not taken up by organisms or denitrified (Johnson and Lindberg 1992). The strong affinity of H^+ ions for negatively charged surfaces of soil particles, can lead to the displacement of cations such as Ca^{+2}, Mg^{+2}, and K^+ from particle exchange surfaces. These displaced "base cations" are then leached together with unassimilated NO_3^- to maintain the charge balance of water moving from forest soil profiles. This can lead to partial depletion of base cations from soils, followed by increasing exports of H^+ and Al^{+3} ions (Johnson et al. 1991, Reuss and Johnson 1989, van Breemen and van Dijk 1988).

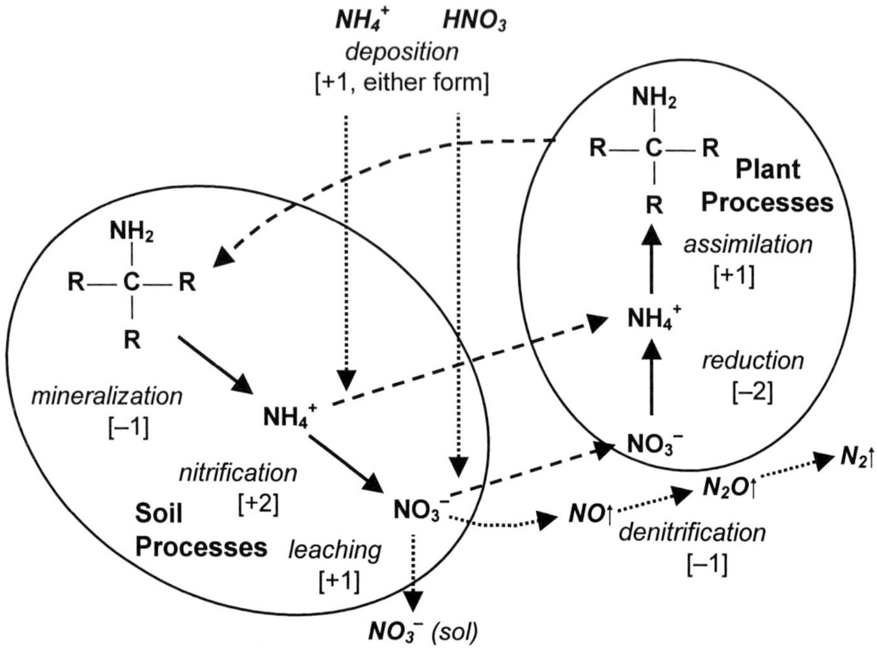

Figure 3. A simplified representation of forest N cycling influences on soil acidity. Inputs from the atmosphere and outputs to drainage waters are shown in *italics.* Ovals enclose soil processes (left) and plant processes (right). Dashed lines indicate soil-plant exchanges (plant N uptake or organic N return to soil). Solid lines show processes within soils or plants. Dotted lines show fluxes into or out of forests. Values in brackets refer to net consumption [–] or production [+] of 1 mol H^+ associated with processesing of 1 mol N. When forest N cycles are closed (small N inputs and outputs), the sum of H^+ consumed and produced by soil and plant processes is zero and no acidity is generated. When 1 mol of organic N is mineralized (1 mol H^+ consumed) and subsequently nitrified (2 mol H^+ produced), 1 mol H^+ remains to acidify soil or drainage water if nitrate is not removed from soil and converted to organic form by plants. Denitrification to any of three gaseous products consumes 1 mol H^+. Direct inputs of acidity can also result from ammonium and nitrate deposition. See Kennedy (1986) for details.

2.3. Drainage water quality

Although plant growth and primary production in temperate and boreal forests is typically N-limited, NO_3^- exports to ground water or streams can increase along N deposition gradients (Driscoll et al. 1987, Gundersen et al. 1998, Nodvin et al. 1995, Riggan et al. 1985) and in response to experimental N additions (Kahl et al. 1993, Magill et al. 1997, Tietema et al. 1998). Therefore, elevated nitrate in drainage water is considered an indicator or the onset of N saturation. Exports of nitrate and acidity from N saturated forests can have toxic effects on freshwater organism (Baker et al. 1996, Stoddard 1994) and can contribute to coastal eutrophication (Jaworski et al. 1997).

Stoddard (1994) linked stream nitrate concentrations in forested catchments to N saturation stages as described by Aber et al. (1989). He showed that nitrate concentrations in streams draining N limited forests subject to low, background levels of N deposition (Stage 0, *sensu* Aber et al. 1989) would be near zero throughout an annual cycle (Figure 4). Low nitrate concentrations at Stage 0 result from complete uptake of N deposition by vegetation and soil microbes during growing seasons and small amounts nitrate export during winter thaw events and spring runoff. After the onset of elevated N deposition, or Stage 1, the N cycle becomes more open and nitrate losses during spring increase. As the duration and magnitude of N inputs increases and forests progress to Stage 2, smaller proportions of the nitrate generated by nitrification and nitrate inputs are removed from soil solution by plants and microbes and greater proportions are exported to groundwater. This results in increased nitrate exports to streams during storm events and baseflow. If catchment forests progress to Stage 3, N cycles become completely open as biological sinks for nitrate decrease and nitrate exports increase.

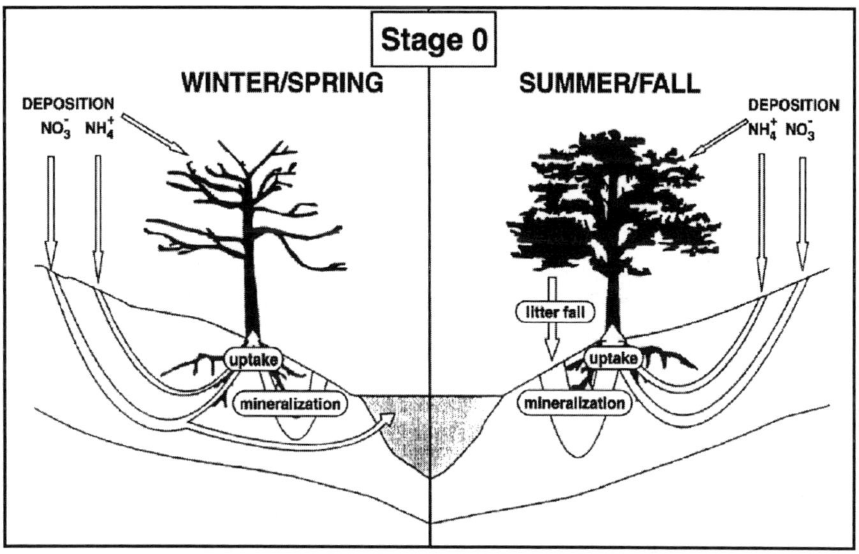

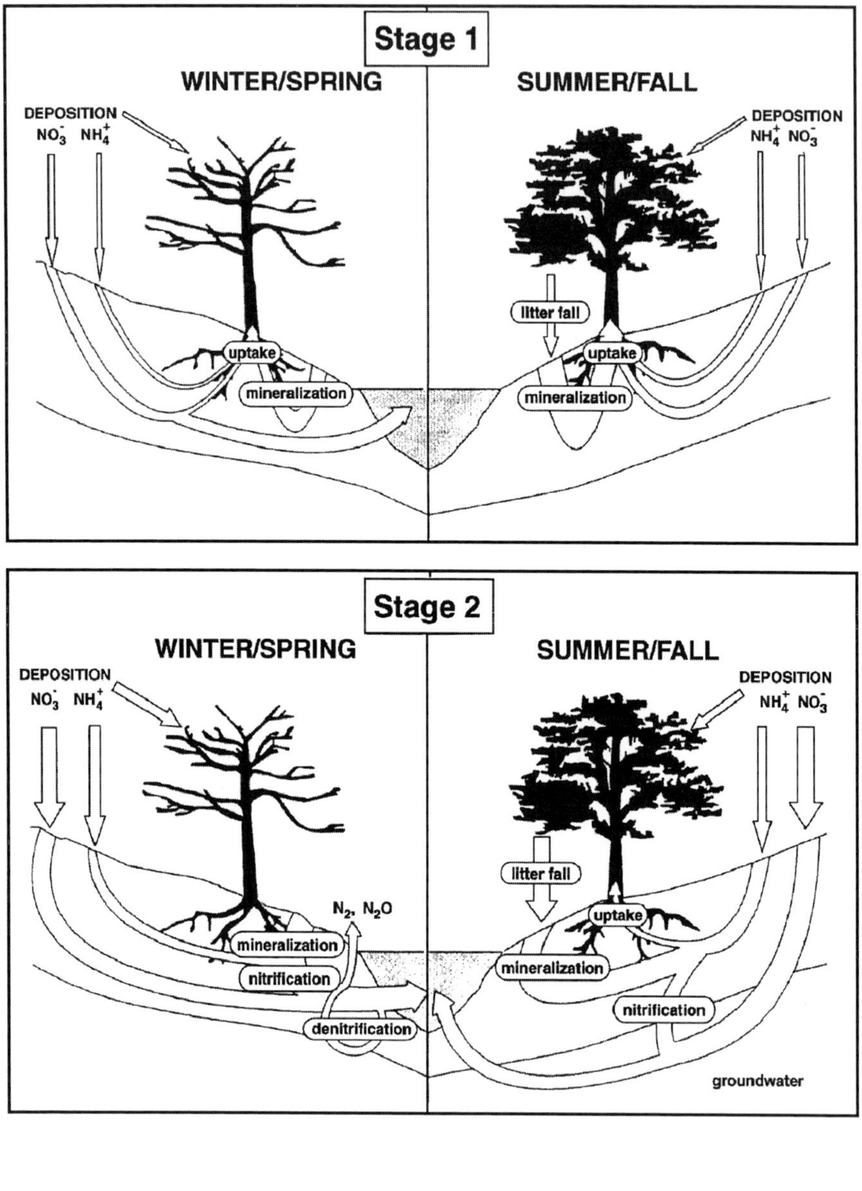
Stage 1
WINTER/SPRING
SUMMER/FALL
DEPOSITION
NO3- NH4+
DEPOSITION
NH4+ NO3-
uptake
mineralization
litter fall
uptake
mineralization
Stage 2
WINTER/SPRING
SUMMER/FALL
DEPOSITION
NO3- NH4+
DEPOSITION
NH4+ NO3-
N2, N2O
mineralization
nitrification
denitrification
litter fall
uptake
mineralization
nitrification
groundwater

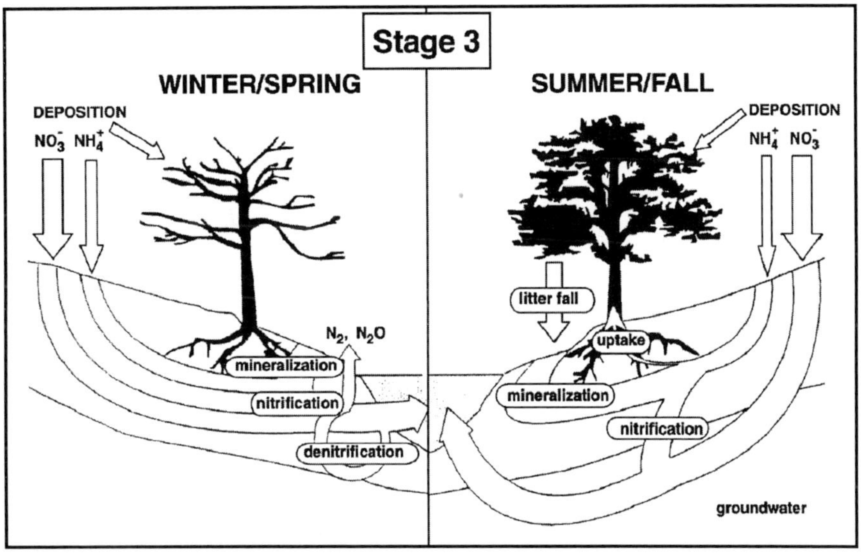

Figure 4. Diagrams for Stages 0 through 3 (above) are schematic representations of changes in nitrate concentrations of stream water associated with shifts from N-limited to increasingly N-saturated conditions of forests within catchments. Arrow thickness shows the magnitude of processes during winter-spring and summer-fall. See text for details. Adapted from Stoddard (1994)

Stoddard's 1994 analysis provided quantitative examples of forest streams at all stages of N saturation in the US. His analysis suggested that nitrate losses from North American forests contribute to episodic acidification in regions subject to elevated N deposition. It further suggested that chronic acidification of freshwaters, as reported in European catchments subjected to higher levels of N deposition is common. For the US, in contrast, chronic acidification of streams due to nitrate exports from forests is uncommon.

2.4. Atmospheric feedbacks

Although ammonia can be released from non-acidic soils as ammonia gas from alkaline soils and animal manure, forest soils are typically too acidic for ammonia volatilization (Schlesinger and Hartley 1992, Sutton et al. 1993). Therefore, the major process capable of returning N to the atmosphere from forests is denitrification, the production of NO, N_2O or N_2 gases. Although NO and N_2O can be produced as by products of intermediate steps of nitrification, most NO and N_2O released from forest soils is the result of NO_3 reduction in moisture saturates soils or aggregates (Firestone and Davidson 1989). Denitrification, particularly N_2O production, has been shown to increase with N additions to forests (Bowden et al. 1991, Firestone and Davidson 1989). Although the total N exports via denitrification are typically small compared to N inputs and plant N uptake in most forests (Bowden et al. 1991, Castro et al. 1994), N_2O losses of up to 20 kg N ha^{-1} yr^{-1} have been reported in Dutch forests subject to very high (~60 kg N ha^{-1} yr^{-1}) deposition rates (Tietema et al. 1991). Even though N losses via denitrification are often small components of total forest N budgets, increased emissions of N_2O due to elevated N deposition have the potential to contribute to the rising atmospheric concentration of this greenhouse gas (Holland et al. 1999b, Lashof and Ahuja 1990, Skiba et al. 1998).

Forest soils are normally a sink for methane (CH_4), another effective greenhouse gas that is increasing in the atmosphere. N deposition, however, can decrease methane consumption, apparently due to inhibition of methanotrophy (microbial methane oxidation) by increased ammonium availability (Goulding et al. 1998, Saari et al. 1997, Steudler et al. 1989).

In summary, elevated N deposition has the potential to increase atmospheric N_2O and CH_4 concentrations by stimulating production of the former and inhibiting consumption of the latter gas in forest soils. As with solution losses of nitrate from forests, the effects of N deposition on exchanges of these two greenhouse gases are likely to increase as forests progress to later stages of N saturation. Developing reliable quantitative estimates of the influences of N deposition on forest-atmosphere N_2O and CH_4 balances is an important challenge to environmental science and global biogeochemistry.

2.5. Community composition

Changes in the availabilities of growth limiting resources such as inorganic N can ultimately lead to changes in plant community composition by altering competitive relations among species (Wilson and Tilman 1995). It is likely, therefore, that increases in N cycling rates and nitrate availability that can result from elevated N deposition can lead to changes in forest species composition. This has been reported for ground flora and lichens in European forests (Buecking 1993, van der Eerden et al. 1998, van Dobben et al. 1999) and under experimental N additions in the US (Rainey et al. 1999). Increased turnover of fine roots associated with increased N cycling (Fig. 2) and changes in soil chemistry and plant nutrient uptake associate with N deposition can also lead to declines in mycorrhizal fungi (Arnolds 1991, Ruehling and Tyler 1991, Wallenda and Kottke 1998). Changes in mycorrhizal species and abundances could have important implications for carbon storage and N retention in forest soils (Aber et al. 1998).

The progression of forests through various stages of N saturation is likely to lead to changes in forest tree species composition as well as in microbial and ground flora species. The most likely shifts are from conifer species with high N use efficiencies to deciduous tree species with lower N use efficiencies and higher N uptake requirements (McNulty et al. 1996). Results from long-term studies of forest responses to N deposition will be required to further identify how long-lived species such as dominate most forests will respond to N deposition.

3. PRESENT AND FUTURE FOREST ECOSYSTEM RESPONSES TO ELEVATED N DEPOSITION

Most temperate and high latitude forests, even in regions with relatively high rates of N deposition, have not progressed to the advanced stages of N saturation where growth declines occur and N cycles become open. For example, Binkley and Höberg (1997) showed that although N deposition has increased above background levels in Sweden, only isolated stands near the southwest coast (where N deposition is highest) show evidence of elevated nitrate leaching. Elsewhere in Sweden, forest growth has increased due to changes in forest management, successional status, and possibly other factors such as climate change and N deposition. Likewise, forest biomass in eastern North America (Birdsey 1992, Brown et al. 1999) and Europe (Kauppi et al. 1992) has increased in recent decades despite elevated N deposition.

Increases in forest growth accompanying increases in N deposition in north temperate regions suggest that atmospheric N inputs might be fertilizing forests by partially alleviating N limitations to growth. Recent modeling studies suggested that such a fertilization effect could contribute significantly to CO_2 uptake by forests in the North Temperate Zone if most N deposition is taken up into plants and not into soil pools with long turnover times (Holland et al. 1998, Townsend et al. 1996). However, experiments simulating atmospheric N deposition by applying ammonium or nitrate fertilizers to forests in small increments across annual cycles typically do not show increased tree growth when N additions are less than about 75 kg ha^{-1} yr^{-1}, the upper limit of N deposition reported in the temperate regions (Emmett et al. 1998, Johnson 1992, Magill et al. 2000, Magill et al. 1997). These studies, together with ^{15}N tracer experiments conducted in combination with low-level N additions (Nadelhoffer et al. 1999) suggest that most N deposition is immobilized in soils rather than taken up by trees. Therefore, evidence suggests that increases in forest biomass and the consequent uptake of CO_2 is more likely due to increase in forested areas in the region, forest management practices, and climate change than to N deposition. It should be noted, that although experimental evidence does not implicate N deposition as a factor increasing forest tree growth, elevated N inputs and increased litter N content have been shown to decrease humus decay (Berg 2000, Berg and Matzner 1997, Fog 1988). Thus, even though N deposition often increases decomposition rates of fresh forest litter, it could be contributing to soil C accumulation in forests and other terrestrial ecosystems by stimulating humus formation and slowing humus turnover.

Although most forests are not presently N saturated and nitrate outputs from forests at large regional scales are not closely related to N deposition rates (Van Miegroet et al. 1992), N saturation has been shown to occur in sensitive areas. Moreover, input-output analyses of forests exporting nitrate above low background levels show correlations between rates of atmospheric N deposition and nitrate leaching (Johnson and Lindberg 1992, Tietema et al. 1997). Other factors such as forest floor C:N ratios (Emmett et al. 1998), forest stand age and successional status (Vitousek and Reiners 1975), and land use history (Aber et al. 1997, Magill et al. 1997) explain much of the variation in nitrate leaching not attributable to N deposition rate. In general, late stage N saturation with open N cycles is more likely when N deposition is high, forest stands are either old or late-successional, soils are poorly buffered (low levels of exchangeable Ca^{+2}, Mg^{+2}, K^{+}) and previous management practices have not removed large amounts of soil N from ecosystems. High elevation, steep topography, short growing seasons, and exposure to other stresses (e.g. elevated ozone) can also contribute to nitrate losses (Fenn et al. 1998). A major question is whether forests receiving low to intermediate levels of N deposition that are not old or late-successional and are not characterized by low forest floor C:N ratios or poorly buffered soils will eventually enter the later stages of N saturation as have some forests with high rates of N input (~20 to >60 kg N ha^{-1} yr^{-1}).

A major question regarding the long-term effects of elevated N deposition on forests is whether forests receiving low to intermediate levels of N deposition will eventually become N saturated with N cycles that undergo transition from closed to more open states such as has occurred in some forests subject to high levels of N inputs (~20 to >60 kg N ha^{-1} yr^{-1}). Answering this question will require continued long-term study of forest responses to N deposition and improvements in understanding of controls on N movements between plant, microbial and non-living components of forest ecosystems.

Although anthropogenic N deposition has influenced temperate and high latitude forest thus far, social and economic factors are expanding the global distribution of elevated N deposition to sub-tropical and tropical forests occupying lower latitudes (Galloway 1998). As tropical forest growth is often limited by elements other than N, increased N inputs could lead to more rapid increases in nitrate leaching and denitrification in these systems than have been observed in higher latitude forests (Matson et al. 1999). Recent work by Hall and Matson (1999) has suggests that P-limited forests in tropical regions have much higher potentials to release N oxides following the onset of elevated inputs than to N-limited forests. The shift in distribution of N deposition to lower latitudes where forest growth is not typically N-limited, together with a paucity of studies on the effects of N deposition on biogeochemical and ecological processes, limits our ability to predict the long-term effects of increased N deposition on tropical forests.

REFERENCES

Aber, J. D. 1992. Nitrogen cycling and nitrogen saturation in temperate forest ecosystems. *Trends in Ecology & Evolution [TRENDS ECOL. EVOL.]* 7: 220-224.

Aber, J. D., D. B. Botkin, and J. M. Melillo. 1979. Predicting the effects of different harvesting regimes on productivity and yield in northern hardwoods. *Canadian Journal of Forest Research [Can. J. For. Res.]* 9: 10-14.

Aber, J. D., A. Magill, S. G. McNulty, R. D. Boone, K. J. Nadelhoffer, M. Downs, and R. Hallett. 1995. Forest biogeochemistry and primary production altered by nitrogen saturation. *Water, Air and Soil Pollution* 85: 1665-1670.

Aber, J. D., W. McDowell, K. Nadelhoffer, A. Magill, G. Berntson, M. Kamakea, S. McNulty, W. Currie, L. Rustad, and I. Fernandez. 1998. Nitrogen Saturation in Temperate Forest Ecosystems: Hypotheses Revisited. *Bioscience* 48: 921-934.

Aber, J. D., J. M. Melillo, K. J. Nadelhoffer, C. McClaugherty, and J. Pastor. 1985. Fine root turnover in forest ecosystems in relation to quantity and form of nitrogen availability. A comparison of two methods. *Oecologia* 66: 317-321.

Aber, J. D., K. J. Nadelhoffer, P. Steudler, and J. M. Melillo. 1989. Nitrogen saturation in northern forest ecosystems. *Bioscience* 39: 378-286.

Aber, J. D., S. V. Ollinger, and C. T. Driscoll. 1997. Modeling nitrogen saturation in forest ecosystems in response to land use and atmospheric deposition. *Ecological Modelling [Ecol. Model.]* 101: 61-78.

Ågren, G. I., and E. Bosatta. 1988. Nitrogen saturation of terrestrial ecosystems. *Environmental Pollution* 54: 185-197.

Arnolds, E. 1991. Decline of ectomycorrhizal fungi in Europe. *Agriculture, Ecosystems & Environment [AGRIC. ECOSYST. ENVIRON.]* 35: 209-244.

Baker, J. P., J. Van Sickle, C. J. Gagen, D. R. DeWalle, W. E. Sharpe, R. F. Carline, B. P. Baldigo, P. S. Murdoch, D. W. Bath, W. A. Kretser, H. A. Simonin, and J. Wigington P.J. 1996. Episodic acidification of small streams in the northeastern United States: effects on fish populations. *Ecological Applications* 6: 422-437.

Berg, B. 2000. Litter decomposition and organic matter turnover in northern forest soils. *Forest Ecology and Management* 133: 13-22.

Berg, B., and E. Matzner. 1997. Effect of N deposition on decomposition of plant litter and soil organic matter in forest systems. *Environmental Reviews* 5: 1-25.

Binkley, D., and P. Hogberg. 1997. Does atmospheric deposition of nitrogen threaten Swedish forests? *Forest Ecology and Management FOR. ECOL. MANAGE.* 92: 119-152.

Binkley, D., P. Sollins, R. Bell, D. Sachs, and D. Myrold. 1992. Biogeochemistry of adjacent conifer and alder-conifer stands. *Ecology* 73(6): 2022-2033.

Birdsey, R. A. 1992. Carbon storage and accumulation in United States forest ecosystems. United States Department of Agriculture Forest Service.

Bonan, G. B., and K. Van Cleve. 1991. Soil temperature, nitrogen mineralization, and carbon source-sink relationships in boreal forests. *Canadian Journal of Forest Research* 22: 629-639.

Bowden, R. D., J. M. Melillo, P. A. Steudler, and J. D. Aber. 1991. Effects of nitrogen additions on annual nitrous oxide fluxes from temperate forest soils in the northeastern United States. *Journal of Geophysical Research* 96: 9321-9328.

Brown, S. L., P. Schroeder, and J. S. Kern. 1999. Spatial distribution of biomass in forests of the eastern USA. *Forest Ecology and Management* 123: 81-90.

Buecking, W. 1993. Nitrogen emmission as a new site factor in forest communities - new developments in forests of south-west Germany. *Phytocoenologia* 23: 65-94.

Burton, A. J., K. S. Pregitzer, and N. W. MacDonald. 1993. Foliar nutrients in sugar maple forests along a regional pollution-climate gradient. *Soil Science Society of America Journal* 57: 1619-1628.

Bytnerowicz, A., and M. E. Fenn. 1996. Nitrogen deposition in California forests: a review. *Environmental Pollution* 92: 127-146.

Calanni, J., E. Berg, M. Wood, D. Mangis, R. Boyce, W. Weathers, and H. Sieviring. 1999. Atmospheric nitrogen deposition at a conifer forest: response of free amino acids in Engelmann spruce needles. *Environmental Pollution [Environ Pollut]* 105: 78-79.

Castro, M. S., W. T. Peterjohn, J. M. Melillo, and P. Steudler. 1994. Effects of nitrogen fertilization on the fluxes of N_2O, CH_4, and CO_2 from soils in a Florida slash pine plantation. *Canadian Journal of Forest Research [Can. J. For. Res.]* 24: 9-13.

Dise, N. B., and R. F. Wright. 1995. Nitrogen leaching from European forests in relation to nitrogen deposition. *Forest Ecology and Management [FOR. ECOL. MANAGE.]* 71: 153-161.

Driscoll, C. T., C. P. Yatsko, and F. J. Unangst. 1987. Longitudinal and temporal trends in the water chemistry of the North Branch of the Moose River. *Biogeochemistry* 3: 37-62.

Emmett, B. A., D. Boxman, M. Bredemeier, P. Gundersen, O. J. Kjønaas, F. Moldan, P. Schleppi, A. Tietema, and R. F. Wright. 1998. Predicting the effects of atmospheric nitrogen deposition in conifer stands: Evidence from the NITREX ecosystem-scale experiments. *Ecosystems* 1: 352-360.

Fenn, M. E., M. A. Poth, J. D. Aber, J. S. Baron, B. T. Bormann, D. W. Johnson, A. D. Lemly, S. G. McNulty, D. F. Ryan, and R. Stottlemyer. 1998. Nitrogen excess in North American ecosystems: Predisposing factors, ecosystem responses, and management strategies. *Ecological Applications [Ecol. Appl.]* 8: 706-733.

Firestone, M. K., and E. A. Davidson. 1989. Microbiological basis of NO and nitrous oxide production and consumption in soil. Pages 7-21 in M. O. Andreae and D. S. Schimel, eds. *Exchange of Trace Gases between Terrestrial Ecosystems and the Atmosphere*. John Wiley & Sons, New York.

Fog, K. 1988. The effect of added nitrogen on the rate of decomposition of organic matter.

Fowler, D., J. N. Cape, M. Coyle, C. Flechard, J. Kuylenstierna, K. Hicks, D. Derwent, C. Johnson, and D. Stevenson. 1999. The global exposure of forests to air pollutants.

Fowler, D., C. E. R. Pitcairn, M. A. Sutton, C. Flechard, B. Loubet, M. Coyle, and R. C. Munro. 1998a. The mass budget of atmospheric ammonia in woodland within 1 km of livestock buildings. *Environmental Pollution* 102: 343-348.

Fowler, D., M. A. Sutton, R. I. Smith, C. E. R. Pitcairn, M. Coyle, G. Campbell, and J. Stedman. 1998b. Regional mass budgets of oxidized and reduced nitrogen and their relative contribution to the nitrogen inputs of sensitive ecosystems. *Environmental Pollution* 102: 337-342.

Frink, C. R., P. E. Waggoner, and J. H. Ausubel. 1999. Nitrogen fertilizer: Retrospect and prospect. *PNAS* 96: 1175-1180.

Galloway, J. N. 1998. The global nitrogen cycle: changes and consequences. *Environmental Pollution* 102: 15-24.

Galloway, J. N., H. Levy, II, and P. S. Kasibhatla. 1994. Year 2020: Consequences of population growth and development on deposition of oxidized nitrogen. *Ambio. Stockholm [AMBIO]* 23: 120-123.

Galloway, J. N., G. E. Likens, W. C. Keene, and J. M. Miller. 1982. The composition of precipitation in remote areas of the world. *Journal of Geophysical Research* 87: 8711-8786.

Galloway, J. N., W. H. Schlesinger, H. Levy II, A. Michaels, and J. L. Schnoor. 1995. Nitrogen fixation: anthropogenic enhancement-environmental response. *Global Biogeochemical Cycles* 9: 235-252.

Glatzel, G. 1990. The nitrogen status of Austrian forest ecosystems as influenced by atmospheric deposition, biomass harvesting and lateral organomass exchange. *Plant and Soil [PLANT SOIL.]* 128: 67-74.

Goulding, K. W. T., N. J. Bailey, N. J. Bradbury, P. Hargreaves, M. Howe, D. V. Murphy, P. R. Poulton, and T. W. Willison. 1998. Nitrogen deposition and its contribution to nitrogen cycling and associated soil processes. *New Phytologist [New Phytol.]* 139: 49-58.

Grennfelt, P., and H. Hultberg. 1986. *Effects of nitrogen deposition on the acidification of terrestrial and aquatic ecosystems*.

Gundersen, P., B. A. Emmett, O. J. Kjoenaas, C. J. Koopmans, and A. Tietema. 1998. Impact of nitrogen deposition on nitrogen cycling in forests: a synthesis of NITREX data. *Forest Ecology and Management [For. Ecol. Manage.]* 101: 37-55.

Hall, S. J., and P. A. Matson. 1999. Nitrogen oxide emissions after nitrogen additions in tropical forests. *Nature* 400: 152-155.

Hedin, L. O., J. J. Armesto, and A. H. Johnson. 1995. Patterns of nutrient loss from unpolluted, old-growth temperate forests: Evaluation of biogeochemical theory. *Ecology* 76: 493-509.

Hendricks, J. J., J. D. Aber, K. J. Nadelhoffer, and R. D. Hallett. 2000. Nitrogen Controls on Fine Root Substrate Quality in Temperate Forest Ecosystems. *Ecosystems* 3: 57-69.

Högberg, P., and I. J. Alexander. 1994. Roles of root symbioses in African woodland and forest: evidence from 15N abundance and foliar analysis. *Journal of Ecology* 82: 1-8.

Holland, E. A., B. H. Braswell, J.-F. Lamarque, A. Townsend, J. Sulzman, J.-F. Muller, F. Dentener, G. Brasseur, H. Levy, II, J. E. Penner, and G.-J. Roelofs. 1998. Variations in the predicted spatial distribution of atmospheric nitrogen deposition and their impact on carbon uptake by terrestrial ecosystems. *Journal of Geophysical Research* 102: 15,849-15,866.

Holland, E. A., F. J. Dentener, B. H. Braswell, and J. M. Sulzman. 1999a. Contemporary and pre-industrial global reactive nitrogen budgets. *Biogeochemistry* 46: 7-43.

Holland, E. A., G. P. Robertson, J. Greenberg, P. M. Groffman, R. D. Boone, and J. R. Gosz. 1999b. Soil CO_2, N_2O, and CH_4 exchange. Pages 185-201 in G. P. Robertson, D. C. Coleman, C. S. Bledsoe, and P. Sollins, eds. *Standard Soil Methods for Long-term Ecological Research.* Oxford University Press, Inc., New York.

Hutchinson, T. C., S. A. Watmough, E. P. S. Sager, and J. D. Karagatzides. 1998. Effects of excess nitrogen deposition and soil acidification on sugar maple (Acer saccharum) in Ontario, Canada: an experimental study. *Canadian Journal of Forest Research/Revue Canadienne de Rech [Can. J. For. Res./Rev. Can. Rech. For.]* 28: 299-316.

Jahnke, R. A. 1990. Ocean flux studies: a status report. *Reviews of Geophysics* 28: 381-398.

Jaworski, N. A., R. W. Howarth, and L. J. Hetling. 1997. Atmospheric deposition of nitrogen oxides onto the landscape contributes to coastal eutrophication in the Northeast United States. *Environmental Science and Technology* 31: 1995-2004.

Johnson, D. W. 1992. Nitrogen retention in forest soils. *Journal of Environmental Quality* 21: 1-12.

Johnson, D. W., M. S. Cresser, S. I. Nilsson, J. Turner, B. Ulrich, D. Binkley, and D. W. Cole. 1991. Soil changes in forest ecosystems: Evidence for and probable causes. *Proceedings of the Royal Society of Edinburgh* 97B: 81-116.

Johnson, D. W., and P. S. Curtis. 2001. Effects of forest management on soil C and N storage: meta analysis. *Forest Ecology and Management* 140: 227-238.

Johnson, D. W., and S. E. Lindberg. 1992. *Atmospheric Deposition and Forest Nutrient Cycling*. Springer-Verlag, New York.

Kahl, J. S., S. A. Norton, I. J. Fernandez, K. J. Nadelhoffer, C. T. Driscoll, and J. D. Aber. 1993. Experimental inducement of nitrogen saturation at the watershed scale. *Environmental Science and Technology* 27: 565-568.

Kauppi, P. E., Mielik,,inen, and K. Kuusela. 1992. Biomass and carbon budget of European forests, 1971-1990. *Science* 256: 70-74.

Kennedy, I. R. 1986. *Acid Soil and Acid Rain.* John Wiley & Sons, Inc., New York.

Kjønaas, O. J., A. O. Stuanes, and M. Huse. 1998. Effects of weekly nitrogen additions on N cycling in a coniferous forest catchment, Gårdsjön, Sweden. *Forest Ecology and Management [For. Ecol. Manage.]* 101: 227-249.

Lashof, D. A., and D. R. Ahuja. 1990. Relative contributions of greenhouse gas emissions to global warming. *Nature* 344: 529-531.

Likens, G. E., W. C. Keene, J. M. Miller, and J. N. Galloway. 1987. Chemistry of precipitation from a remote, terrestrial site in Australia. *Journal of Geophysical Research* 92: 299-313.

Liu, G., and T. Deng. 1991. Mathematical model of the relationship between nitrogen-fixation by black locust and soil conditions. *Soil Biology and Biochemistry [SOIL BIOL. BIOCHEM.]* 23: 1-7.

Lovett, G. M., and J. D. Kinsman. 1990. Atmospheric pollutant deposition to high-elevation ecosystems. *Atmospheric Environment* 24A: 2767-2786.

Magill, A. H., J. D. Aber, G. M. Berntson, W. H. McDowell, K. J. Nadelhoffer, J. M. Mellilo, and P. Steudler. 2000. Long-term nitrogen additions and nitrogen saturation in two temperate forests. *Ecosystems* 3: 238-253.

Magill, A. H., J. D. Aber, J. J. Hendricks, R. D. Bowden, J. M. Melillo, and P. A. Steudler. 1997. Biogeochemical response of forest ecosystems to simulated chronic nitrogen deposition. *Ecological Applications [ECOL. APPL.]* 7: 402-415.

Magill, A. H., M. R. Downs, K. J. Nadelhoffer, R. A. Hallett, and J. D. Aber. 1996. Forest ecosystem response to four years of chronic nitrate and sulfate additions at Bear Brooks Watershed, Maine, USA. *Forest Ecology and Management [FOR. ECOL. MANAGE.]* 84: 29-37.

Matson, P. A., W. H. McDowell, A. R. Townsend, and P. M. Vitousek. 1999. The globalization of N deposition: Ecosystem consequences in tropical environments. *Biogeochemistry* 46: 67-83.

McClaugherty, C. A., J. Pastor, J. D. Aber, and J. M. Melillo. 1985. Forest litter decomposition in relation to soil nitrogen dynamics and litter quality. *Ecology* 66: 266-275.

McNulty, S. G., J. D. Aber, and R. D. Boone. 1991. Spatial changes in forest floor and foliar chemistry of spruce-fir forests across New England. *Biogeochemistry* 14: 13-29.

McNulty, S. G., J. D. Aber, T. M. McLellan, and S. M. Katt. 1990. Nitrogen cycling in high elevation forests of the northeastern US in relation to nitrogen deposition. *Ambio* 19: 38-40.

McNulty, S. G., J. D. Aber, and S. D. Newman. 1996. Nitrogen saturation in a high elevation New England spruce-fir stand.

Melillo, J. M., J. D. Aber, and J. F. Muratore. 1982. Nitrogen and Lignin Control of Hardwood Leaf Litter Decomposition Dynamics. *Ecology* 63: 621-626.

Minocha, R., S. Long, A. H. Magill, J. Aber, and W. H. McDowell. 2000. Foliar free polyamine and inorganic ion content in relation to soil and soil solution chemistry in two fertilized forest stands at the Harvard Forest, Massachusetts.

Mitsch, W. J., and J. G. Gosselink. 1986. *Wetlands*. Van Nostrand Reinhold, New York.

Nadelhoffer, K. J. 2000. The potential effects of nitrogen deposition on fine root production in forest ecosystems. New Phytologist 137:131-139.

Nadelhoffer, K. J., B. A. Emmett, P. Gundersen, O. J. Kjoenaas, C. J. Koopmans, P. Schleppi, A. Tietema, and R. F. Wright. 1999. Nitrogen deposition makes a minor contribution to carbon sequestration in temperature forests. *Nature* 398: 145-148.

Näsholm, T., A. Ekblad, A. Nordin, R. Giesler, M. Högberg, and P. Högberg. 1998. Boreal forest plants take up organic nitrogen. *Nature* 392: 914-916.

Nihlgård, B. 1985. The ammonium hypothesis: an additional explanation to the forest decline in Europe. *Ambio* 14: 2-8.

Nodvin, S. C., H. Van Miegroet, S. E. Lindberg, N. S. Nicholas, and D. W. Johnson. 1995. *Acidic deposition, ecosystem processes, and nitrogen saturation in a high elevation Southern Appalachian watershed.*

Ollinger, S. V., J. D. Aber, G. M. Lovett, S. E. Millham, R. G. Lathrop, and J. M. Ellis. 1993. A spatial model of atmospheric deposition for the northeastern U.S. *Ecological Applications* 3: 459-472.

Pitcairn, C. E. R., I. D. Leith, L. J. Sheppard, M. A. Sutton, D. Fowler, R. C. Munro, S. Tang, and D. Wilson. 1998. The relationship between nitrogen deposition, species composition and foliar nitrogen concentrations in woodland flora in the vicinity of livestock farms. *Environmental Pollution* 102: 41-48.

Rainey, S. M., K. J. Nadelhoffer, W. L. Silver, and M. R. Downs. 1999. Effects of chronic nitrogen additions on understory species in a red pine plantation. *Ecological Applications [Ecol. Appl.]* 9: 949-957.

Reich, P. B., D. F. Grigal, J. D. Aber, and S. T. Gower. 1997. Nitrogen mineralization and productivity in 50 hardwood and conifer stands on diverse soils. *Ecology* 78: 335-347.

Reuss, J. O., and D. W. Johnson. 1989. *Acid Deposition and the Acidification of Soils and Waters*. Springer-Verlag, New York.

Riggan, P. J., R. N. Lockwood, and E. N. Lopez. 1985. Deposition and processing of airborne nitrogen pollutants in Medditerranean-type ecosystems of southern California. *Environmental Science and Technology* 19: 781-789.

Ruehling, A., and G. Tyler. 1991. Effects of simulated nitrogen deposition to the forest floor on the macrofungal flora of a beech forest. *Ambio. Stockholm [AMBIO.]* 20: 261-263.

Saari, A., P. J. Martikainen, A. Ferm, J. Ruuskanen, W. De Boer, S. R. Troelstra, and H. J. Laanbroek. 1997. Methane oxidation in soil profiles of Dutch and Finnish coniferous forests with different soil texture and atmospheric nitrogen deposition. *Soil Biology and Biochemistry [Soil Biol. Biochem.]* 29: 1625-1632.

Schlesinger, W. H., and A. E. Hartley. 1992. A global budget for atmospheric NH3. *Biogeochemistry* 15: 191-211.

Schulze, E.-D. 1989. Air pollution and forest decline in a spruce (*Picea abies*) forest. *Science* 244: 776-783.

Skeffington, R. A., and E. J. Wilson. 1988. Excess nitrogen deposition: Issues for consideration. *Environmental Pollution* 54: 159-184.

Skiba, U., L. Sheppard, C. E. R. Pitcairn, I. Leith, A. Crossley, S. van Dijk, V. H. Kennedy, and D. Fowler. 1998. Soil nitrous oxide and nitric oxide emissions as indicators of elevated atmospheric N deposition rates in seminatural ecosystems. *Environmental Pollution* 102: 457-461.

Steudler, P. A., R. D. Bowden, J. M. Melillo, and J. D. Aber. 1989. Influence of nitrogen fertilization on methane uptake in temperate forest soils. *Nature* 341: 314-316.

Stoddard, J. L. 1994. Long-term changes in watershed retention of nitrogen: Its causes and aquatic consequences. Pages 223-284 in L. A. Baker, ed. *Environmental Chemistry of Lakes and Reservoirs*. American Chemical Society, Washington.

Sutton, M. A., C. Milford, U. Dragosits, C. J. Place, R. J. Singles, R. I. Smith, C. E. R. Pitcairn, D. Fowler, J. Hill, H. M. ApSimon, C. Ross, R. Hill, S. C. Jarvis, B. F. Pain, V. C. Phillips, R. Harrison, D. Moss, J. Webb, S. E. Espenhahn, D. S. Lee, M. Hornung, J. Ullyett, K. R. Bull, B. A. Emmett, J. Lowe, and G. P. Wyers. 1998. Dispersion, deposition and impacts of atmospheric ammonia: quantifying local budgets and spatial variability. *Environmental Pollution* 102: 349-361.

Sutton, M. A., C. E. R. Pitcairn, and D. Fowler. 1993. The exchange of ammonia between the atmosphere and plant communities. Pages 301-393 in M. Begon and A. H. Fitter, eds. *Advances in Ecological Research*. Academic Press, London.

Tietema, A., C. Beier, P. H. B. De Visser, B. A. Emmett, P. Gundersen, O. J. Kjønaas, and C. J. Koopmans. 1997. Nitrate leaching in coniferous forest ecosystems: the European field-scale manipulation experiments NITREX (nitrogen saturation experiments) and EXMAN (experimental manipulation of forest ecosystems). *Global Biogeochemical Cycles* 11: 617-626.

Tietema, A., W. Bouten, and P. E. Wartenberg. 1991. Nitrous oxide dynamics in an oak-beech forest ecosystem in the Netherlands. *Forest Ecology and Management* 44: 53-61.

Tietema, A., A. W. Boxman, M. Bredemeier, B. A. Emmett, F. Moldan, P. Gundersen, P. Schleppi, and R. F. Wright. 1998. Nitrogen saturation experiments (NITREX) in coniferous forest ecosystems in Europe: a summary of results. *Environmental Pollution* 102: 433-437.

Townsend, A. R., B. H. Braswell, E. A. Holland, and J. E. Penner. 1996 . Spatial and temporal patterns in terrestrial carbon storage due to deposition of fossil fuel nitrogen. *Ecological applications : a publication of the Ecological Society of America.* v. 6: p. 806-814.

van Breemen, N., and H. F. G. van Dijk. 1988. Ecosystem effects of atmospheric deposition of nitrogen in the Netherlands. *Environmental Pollution* 54: 249-274.

van der Eerden, L., W. de Vries, and H. van Dobben. 1998. Effects of ammonia deposition on forests in The Netherlands. *Atmospheric Environment [ATMOS ENVIRON]* 32: 525-532.

van Dobben, H. F., C. J. F. ter Braak, and G. M. Dirkse. 1999. Undergrowth as a biomonitor for deposition of nitrogen and acidity in pine forest. *Forest Ecology and Management* 114: 83-95.

Van Miegroet, H., G. M. Lovett, and D. W. Cole. 1992. Nitrogen chemistry, deposition, and cyclin in forests: summary and conclusions. Pages 202-207 in D. W. Johnson and S. E. Lindberg, eds. *Atmospheric deposition and forest nutrient cycling, Ecological Studies 91*. Springer-Verlag, New York.

Vitousek, P. M., and R. W. Howarth. 1991. Nitrogen limitation on land and sea: how can it occur? *Biogeochemistry* 13: 87-115.

Vitousek, P. M., and P. A. Matson. 1988 . Nitrogen transformations in a range of tropical forest soils. *Soil biology and biochemistry.* v. 20: p. 361-367.

Vitousek, P. M., and W. A. Reiners. 1975. Ecosystem succession and nutrient retention: A Hypothesis. *BioScience* 25: 376-381.

Vitousek, P. M., and R. L. J. Sanford. 1986 . Nutrient cycling in moist tropical forest. *Annual review of ecology and systematics.* v. 17: p. 137-167.

Vollenweider, R. A. 1968. Scientific fundamentals of the eutrophication of lakes and flowing waters, with particular reference to nitrogen and phosphorus as factors in eutrophication. Organization for Economic Cooperative Development, Paris.

Wallenda, T., and I. Kottke. 1998. Nitrogen deposition and ectomycorrhizas. *New Phytologist [New Phytol.]* 139: 169-187.

Waring, R. H., and S. W. Running. 1998. *Forest ecosystems: analysis at multiple scales.* Academic Press., San Diego.

Wilson, E. J., and R. A. Skeffington. 1994. The effects of excess nitrogen deposition on young Norway spruce trees. Part II The vegetation. *Environmental Pollution [ENVIRON. POLLUT.]* 86: 153-160.

Wilson, S. D., and D. Tilman. 1995. Competitive responses of eight old-field plant Species in four environments. *Ecology* 76: 1169-1180.

Zou, X., D. Binkley, and B. A. Caldwell. 1995. Effects of dinitrogen-fixing trees on phosphorus biogeochemical cycling in contrasting forests. *Soil Science Society of America Journal* 59: 1452-1458.

Section 4:

Emerging Prediction and Management Technologies

Chapter 13. On-Farm Technologies and Practices to Improve Nitrogen Use Efficiency

N.R. Kitchen[a] and K.W.T. Goulding[b]
[a] USDA Agricultural Research Service, 240 Agricultural Engineering Bldg., University of Missouri, Columbia, Missouri, United States.

[b] Soil Science Department, IACR-Rothamsted, Harpenden, Herts. AL5 2JQ, UK.

Nitrogen from soil, fertilizer, and manure sources is inefficiently used (30 to 60%) in most crop production systems. As a consequence, unused inorganic N can move off crop fields and contaminate surface and ground water resources. Governments have responded with guidelines, standards, regulations, and in some cases fines when off-field losses of N have not been reduced. Numerous technologies and time-proven practices are available for producers to employ that will result in improved crop N use efficiency. We review in this chapter tools currently available, or that are soon to be available, to help producers make better N management decisions. Soil and crop measurements are considered and compared. Nonetheless, "on-farm" implies that producers will be at the center of implementing change, and "change" means N management options will motivate producers to action. Prerequisites for grower adoption require that technologies and practices be reliable, incur minimal additional expense (time and equipment), and integrate with ease into current operations. When these criteria cannot be met, external incentives (e.g., regulation, private or government cost sharing programs) may be necessary.

1. INTRODUCTION

Modern agriculture has come to embrace the concepts of environmental stewardship as a necessary component of crop production. The stories and studies that have documented agricultural nutrients moving into and impairing ground and surface waters and the environment in general (e.g., see Vitousek, 1994), an outcome which has often been followed by stepped-up governmental regulation, have compelled producers to consider nutrient management as more than an economic decision. As a nutrient, nitrogen (N) is the main fertilizer with global environmental effects. In most agricultural settings, soil N is insufficient for healthy non-leguminous crop growth; consequently yield enhancement with N fertilizer typically ranges from 10 to 200 %. The visual and subsequent yield response to inexpensive N fertilizer reinforces growers' reliance on it for profitable production. However, because of the inherent chemical properties of N, it plays a major role in dynamic, climate-mediated biological processes. Nitrogen transformation and transport in soil and water is complex, making *efficient* management of N in the food, forage, and fiber production system complicated. Nevertheless, our hope is to do better. Developing more efficient N management systems for agriculture should be a quest pursued by producers, agribusiness, and researchers around the globe.

Crop N needs--including N source, amount, and timing issues--are difficult to anticipate because of spatial (within and between fields) and temporal (within and between growing seasons) variability. As a consequence of this variability, N management strategies have shown different levels of effectiveness in meeting crop needs and minimizing environmental losses. Seldom will a single N management plan used over multiple years result in optimal N use for each of those years, at least when considering both the crop and the environment. Nitrogen use efficiency of crops varies greatly both between years and between different crops. It rarely exceeds 70 % (Pierce and Rice, 1988) and more often ranges from 30-60 % (Bock, 1984; Pierce and Rice, 1988) for many crops.

In essence, the N cycle is leaky (Figure 1). Losses to water and the atmosphere are part of the natural global N cycle. However, the conversion of stable atmospheric and organic N into reactive forms by energy production, fertilizer production, cultivation of legumes, plowing old grasslands, forest burning and land clearance, and the drainage of wetlands is reckoned to have doubled the amount of reactive N in the environment (Goulding *et al.*, 1998). New reactive N when mobilized can be readily transported in solution or *via* the. atmosphere so that local increases spread regionally and globally. The ultimate fate of this extra reactive N is uncertain. Much of it, as with much of the extra carbon dioxide, is 'missing' (i.e., current measurements and calculations cannot account for it). It could be denitrified to N_2 or be accumulating in the atmosphere, soils, groundwater, land vegetation, oceans and marine sediments, changing ecosystems through eutrophication and acidification.

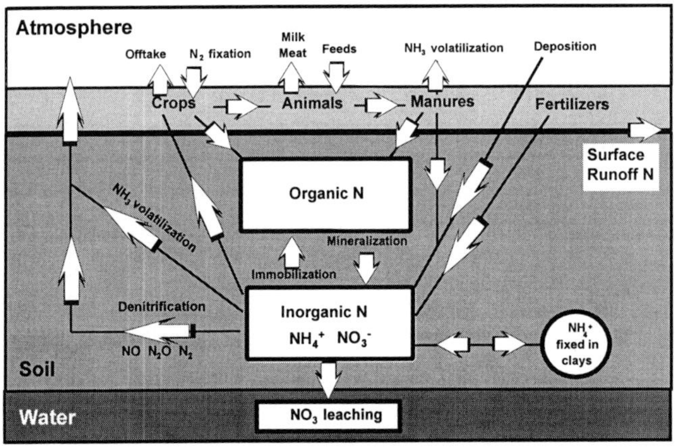

Figure 1. A simplified nitrogen cycle

Given the evasiveness of N and leakiness in the soil-plant system, many producers and researchers have asked, "Can we really do better?" New technologies, such as Global Positioning Systems (GPS), are enabling more spatially accurate measurement, and thus management, of soil and crop spatial variability. Consequently, a new generation of management options being developed hold promise for improving N use efficiency in crop production systems. Future management systems will rely upon a combination of new and improved technological tools along with time-proven practices that together are jointly responsive to the N dynamics in the crop-soil environment.

This chapter will focus on describing the technologies and practices that either have helped producers use N more efficiency or have shown promise in doing so. The phrase "N use efficiency" (NUE) is widely used in agricultural and ecological studies. However, it connotes various explicit meanings that depend on what measurements and calculations are made (Bock, 1984; Pierce and Rice, 1988). Since no standard NUE definition is available for the myriad of practices and technologies discussed here, the phrase will be used in this chapter to mean a general concept of crop uptake and utilization of soil and fertilizer N.

The chapter provides an overview of the situation primarily in Europe and North America. Those wanting more detail of the North American position can find this in Hargrove (1988), Follett *et al.*, (1991), and Havlin and Jacobsen (1994). Those wanting to learn more about the European situation are directed to Romstad *et al.* (1997) or, for the UK alone, Davies (2000). For an analysis of N management under irrigated agriculture see Rauschkolb and Hornsby, 1994. The issues of fertilizers and the environment are dealt with in other chapters of this publication, as well as other recent works (Howarth, 1998; Rengel, 1998; Laegrid *et al.*; 1999).

2. TRIED AND TRUE PRACTICES

The application of N fertilizer to agricultural crops is generally very cost-effective, i.e. the cost of applying it is far outweighed by the extra value of crop obtained. This has encouraged farmers to apply plenty of N to ensure they get the most economic return, usually close to the maximum yield. This in turn creates a surplus of inputs compared to outputs in product, which and leaves N at risk of loss to the environment. Figure 2 shows a graph of crop yield and quantity of N leached against each amount of N fertilizer applied. Applying more N than is needed for optimum yield greatly increases the potential for losses from the crop-soil system.

Nitrogen surpluses vary. Generally, the efficiency of conversion of N inputs into products for arable crops can be 60-70% or even more, but for livestock systems a 20% efficiency is good. Table 1 shows *average* N surpluses for some countries in the European Union and the U.S. expressed on an area basis. Those countries with the highest intensity of livestock production have the largest average surpluses, but the averages mask big differences between farms. Some farms have N surpluses of > 1000 kg ha^{-1} yr^{-1}. Surpluses for UK arable crops have been reducing because of environmental and economic pressures and improved technologies (Figure 3).

Factors that control N use efficiency under Northwest European conditions have been examined for the UK (Davies, 2000). The weather dominates N loss through the impact of rainfall and temperature on drainage, crop growth, and N utilization. For livestock systems, the problem of the relative inefficiency of the animal in utilizing N is not easily overcome, and our understanding of N efficiency is far from complete. However, it is clear that the better utilization of legumes and manures can have a major impact. Manipulation of diets also holds some promise.

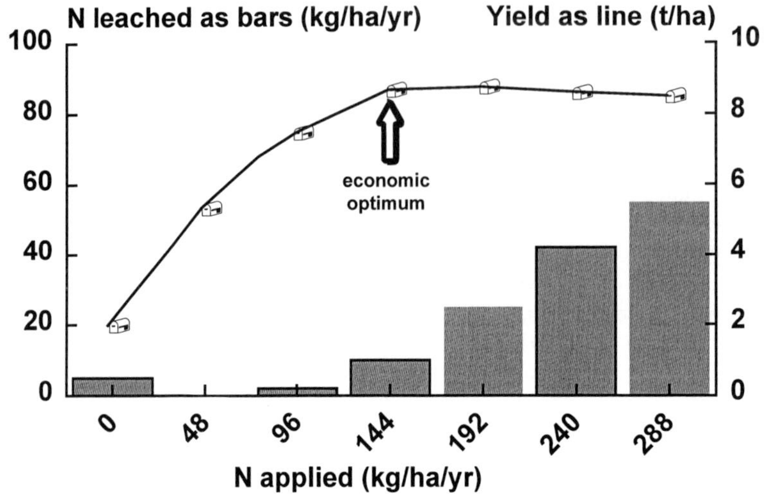

Figure 2. A nitrogen response curve and corresponding leaching losses from the Broadbalk experiment at Rothamsted.

Table 1. Country-wide N surpluses (annual fertilizer + manure applied - crop N removal in grain) for some EU countries and the U.S. (kg ha^{-1} yr^{-1}).

	N Surplus
Netherlands	321
Belgium	170
Germany	121
France	73
UK	59
Portugal	6
United States	3

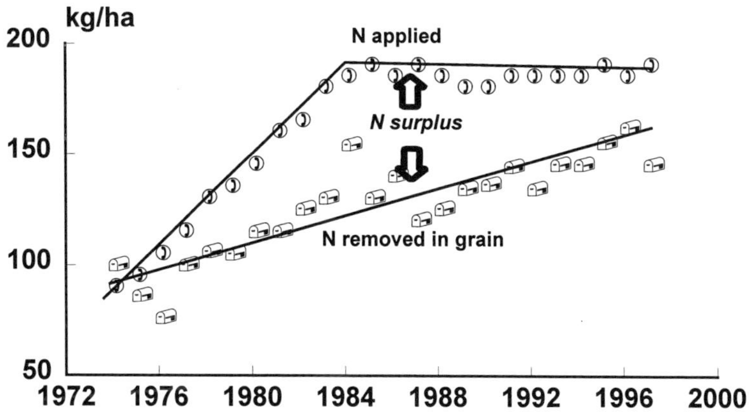

Figure 3. Nitrogen fertilizer applied and removal by winter wheat in England and Wales, 1974-1997.

The position is most clear for arable and horticultural systems, and a set of best management practices for optimum N use efficiency are globally applicable, including:

- Farmers should choose the highest-yielding variety appropriate to maximize the use of available N (bearing in mind quality, e.g., for milling).
- A green cover should be maintained as much as practicable. Use a cover crop if necessary and drill autumn-sown crops early. However, this must be balanced against effective weed, pest, and disease control.
- Fertilizer requirements should be calculated using a fertilizer recommendation system, allowing for soil mineral N, mineralized N from soil organic matter, crop residues, legumes, and manures.
- Nitrogen management strategies should start with a good understanding of precipitation patterns and variability in order to minimize N loss but not be N deficient with the crop.
- As reasonable as possible, synchronize N applications with crop needs. Nitrogen applications should be timed for optimal N use by the crop. Fall N should be applied only to those crops that need it. Unnecessarily early spring applications of fertilizer should be avoided. Applications should be timed to provide N when the crop is growing rapidly. Use of a starter fertilizer and banding for vegetables can greatly increase the efficiency with which the N is used.
- Leaching losses may be reduced by splitting spring fertilizer applications, but yield benefits should not be expected. For sandy soils, timing of N applications with crop need is crucial since leaching potential is high.

- Fertilizers and manures should be applied evenly, and well away from watercourses, with a properly calibrated spreader.
- Appropriate controls to minimize pest, disease, and weed infestation are essential because a diseased crop is less able to use its N.
- If irrigation is required, this should be done carefully, i.e., only to support crop yield and using a scheduling system that accounts for precipitation.
- Irrigation systems that deliver water non-excessively (irrigation rate. infiltration rate) and evenly over the field can be used for spoon-feeding N in the irrigation water (fertigation).

These best management practices have been proven throughout the world. In the UK, limitations on total N application rates and the timing of manure applications were tested in Nitrate Sensitive Areas (NSAs). In December 1998, enforcement was initiated in 68 Nitrate Vulnerable Zones (NVZs) covering 600,000 ha. Results from measurements and modeling studies in NSAs showed a significant reduction (about 20%) in N usage and losses (Dampney *et al.*, 2000). Experiments at the International Maize and Wheat Improvement Centre (CIMMYT), Mexico, showed that changing N application and irrigation schedules allowed inputs to be reduced by almost 30% (250 down to 180 kg ha^{-1}) and leaching losses reduced by 49 to 70 kg ha^{-1}, while yields were maintained. For horticultural crops, research has shown that fertilizer applications to brassica rotations can be reduced by 50% without loss of yield if residual N is taken into account, and using starter or banded fertilizers on vegetable crops can reduce leaching losses by up to 75% (Rahn *et al.*, 1993).

On a larger spatial and temporal scale, a change of rotation or type of farming system can reduce losses. Organic farming can result in smaller leaching losses over a rotation, but careful management during the plowing out of the leguminous phase is required because this releases large amounts of N through mineralization. Some crops may not be able to utilize all the N released, resulting in large losses in that year. A very thorough review of N efficiency in organic agriculture was made by Kristensen (1995). Integrated Farming Systems (IFS) are proving to be both profitable and less polluting, but evidence suggests that the system must be tailored to the local conditions (Goulding *et al.*, 1999).

Some other specific management practices for improving N efficiency deserve special mention. Crop yield and N use efficiency has been improved under some field conditions with nitrification (Prasad and Power, 1995) or urease (Schlegel *et al.*, 1986) inhibitors, but results are inconsistent. Coarse textured soils appear to be best suited for inhibitor use. A review of the nitrification inhibitor DCD (Dicyandiamide) in the U.S. found increased rice (*Oryza sativa* L.) yield under a variety of cultural practices (Wells *et al.*, 1989). Inhibitors have not been extensively adopted in Europe. A recent reviews of inhibitors (McCarty, 1999) did not even address practical issues but only modes of action. Prasad and Power (1995) pointed out that the need for a 270 to 450 kg ha^{-1} increase in yield to cover the costs had prevented many nitrification inhibitors from reaching the farm. At this time the low cost of N fertilizer and the high cost of inhibitors makes it unlikely that they will play a major role in improving N efficiency. However, legislative moves such as an N tax to reduce N losses to water and the atmosphere could change the economics.

Experiments have shown that planting a cover crop [such as rye (*Secale cereale* L.), white mustard, (*Brassica phaecelia* L.), or hairy vetch (*Vicia villosa* Roth)] between harvest and planting a late winter or spring crop is the single most effective way of retaining N, as reviewed in several chapters in Hargrove (1988). However, when the cover crop is killed its N is released back into the soil at a rate that depends on climate and management. This re-mineralized N can be effectively used by the following crop, but can also be leached in subsequent seasons (Harrison & Peel, 1996).

The introduction of buffer strips between agricultural land and water courses or bodies can help prevent the movement of nitrate, phosphate, and pesticides into water courses at some sites (Leeds-Harrison *et al.*, 1999). They have proved to be very effective in some circumstances (e.g., for New Zealand see Downes *et al.*, 1997 and for the U.S. see Dickson and Schaeffer, 1997). However, buffer strips remove nitrate by denitrification; this increases nitrous oxide emissions - swapping one pollutant for another (Goulding *et al.*, 1996). Such measures are, at best, short-term and are better replaced by actions that reduce all off-field N losses. In other words, treat the cause rather than the problem.

Multiple cropping, those systems with an average of more than one crop per year, includes sequential crops, intercrops, or combinations of the two. Multiple crop systems are most effective in improving both N and water use efficiency for climatic regions where precipitation and temperature allow an effective growing season beyond the time needed for monocrop culture (Hook and Gascho, 1988). Crops and crop rotations that are designed to minimize erosion and nitrate leaching, to utilize crops capable of biological fixation of N, and to allow for timely N application (whether with fertilizer, manure, or crop residue management) will generally achieve efficient N use (Kurtz *et al.*, 1984).

3. DETERMINING NITROGEN FERTILIZER REQUIREMENT (NFR)

For decades, the starting point for producers in determining crop N need has been to multiply a target crop yield by the concentration of N in the harvested plant material. This calculation produces a number that is, in essence, an estimate of the amount of N that will be removed from the field (Stanford and Legg, 1984; Meisinger and Randall, 1991). This mass-balance approach excludes the unharvested plant material left in the field since it decomposes over time and releases N for subsequent crops. When N is not a limiting factor for crop growth, the amount of N removed from the field with harvest will, even under ideal conditions, be 30 to 50 % less than the sum of available soil N and fertilizer N (Hauch, 1973; Pierce and Rice, 1988; Kolberg *et al.*, 1997). This lack of crop usage results from a plethora of interacting soil, climate, and management factors that either cause N loss from the crop-soil system (through processes such as denitrification, leaching, and volatilization) or change N into forms unavailable to the crop (such as immobilization).

The crop NFR (i.e., the amount of fertilizer or manure needed so that N does not become limiting, but that inorganic N is not in excess) is usually adjusted for the lack of 100% efficiency. Input recommendations typically include a crop NUE for the soil and fertilizer N of around 50 to 70% (Dahnke and Johnson). In the UK, fertilizer recommendations for arable crops, issued by the Ministry of Agriculture, Fisheries, and Food, are based on measured N use efficiencies of 55-70 %, varying with soil type (MAFF, 2000). Producers generally want a simplified crop-specific equation for estimating the N input requirement. As an example, many corn producers in humid regions of the U.S. currently use a rule of applying about 23 kg N for every Mg of target grain yield. Based on average corn grain N content (16.5 g kg^{-1}, dry weight basis), this rule assumes a N use efficiency of about 60 %. Adjustments also should be included for the contribution of soil N as well as other credits, such as the N available from a preceding leguminous crop, manure, or irrigation water.

While we recognize that plant uptake from each source of N has a unique NUE (Pierce and Rice, 1988), we use a simplified calculation for determining the NFR as follows:

$$NFR = \frac{[(TY)(CNC) - SN - NC]}{NUE} \quad [1]$$

Where: NFR = crop N fertilizer input requirement; TY = target yield (as dry matter); CNC = crop N concentration in the harvested portion of the crop; SN = soil N measured or estimated to be available for the crop; NC= N credits from other potential sources; and NUE = N use efficiency (expressed as a fraction).

As previously indicated, optimal NUE will be achieved when implementing the practices identified earlier under *Tried and True Practices*. On-farm tools and practices that have been developed to help producers obtain reasonably accurate values for target yield (TY), and soil N (SN) are considered further.

3.1. Deriving Target Yield

In Eq. 1, target yield influences NFR more than any other term. Deriving an accurate and realistic (unbiased by false hopes and a desire to keep up with neighboring farmers) estimate of the target yield is challenging, particularly for rain-fed cropland with precipitation varying seasonally as well as annually. A number of approaches for determining target yield have been considered.

3.1.1. Historical Yield

Averaging yields over a number of years can be used, but this method will inevitably result in inadequate N for years when conditions provide better than average yield. A target yield that is based upon only the best years will generally meet crop N needs, but potentially will leave inorganic N in the soil when growing conditions have not been ideal. In dryland agriculture where nitrate-N leaching is minimal, left-over N is not considered problematic, particularly since it can be accounted for with soil sampling and credited toward subsequent crops (Hergert, 1987). In humid areas, such as eastern U.S. and Western Europe, leftover N has a much greater potential for loss from the crop-soil environment and thus a much less chance of being available for subsequent crops.

Target yield is often determined by adding 5 to10 % to the average yield of the most recent 5 to 7 years (Rice and Havlin, 1994). Surveys have demonstrated that a majority of producers over-estimate their target yield when determining N recommendations (Schepers and Mosier, 1991; Goos and Prunty, 1990) because of the low cost to apply ample N fertilizer to insure it will not be limiting, regardless of the type of year. Inflated target yield may also suggest producers do not use actual whole-field averages but rather rely upon yield expectations from the highest producing field areas. Even before the availability of combines with yield monitoring systems farmers intuitively have known that, for a field-average 10 Mg ha^{-1} corn yield, there were areas within that same field that probably produced 12 to 14 Mg ha^{-1}. Nitrogen fertilization at or even only slightly higher than actual field average levels will underestimate NFR for the most productive soils of a field and over-estimate NFR for chronic poor producing soils of a field.

3.1.2. Yield Mapping

Yield variation within fields is a major disadvantage of using a single target yield to represent the entire field. If yield variability could be predicted, it would be a basis for variable application of N. Since the early 1990s yield monitoring and mapping have offered producers a direct method for measuring spatial variations in crop yield (Lark and Stafford, 1996). Yield mapping has shown within-field variation as high as 150 % (Kitchen *et al.*, 1999b). Producers view these maps and intuitively see an opportunity for variable-rate N applications. However, yield maps are confounded by many potential causes of yield variability (Pierce *et al.*, 1997) as well as potential error sources from combine yield sensors (Lamb *et al.*, 1995; Blackmore and Marshall, 1996). Using yield maps to predict crop production for N management without also relying on spatial measurement of soil/landscape properties, as well as other potential and often transient yield-limiting factors (e.g., pest incidence, nutrients, and management variation), is almost certainly futile. Averaging multiple years of yield maps has been suggested as one way of establishing stable yield productivity patterns related to soil properties (Kitchen *et al.*, 1995; Stafford *et al.*, 1996; Colvin *et al.*, 1997). However in some regions, high producing areas of a field during "dry" years can be low producing areas of the same field in "wet" years (Wibawa *et al.*, 1993; Colvin *et al.*, 1997; Sudduth *et al.*, 1997). Averaging yield maps may neutralize the information needed to better understand the interaction between soil/landscape properties and climate for crop production (Sawyer, 1994).

3.1.3. Remote Sensing for Yield

High resolution remote sensing from airborne or satellite systems has also been used with varying success in quantifying within field yield variation (Moran *et al.*, 1997; Shanahan *et al.*, 2000). Yield prediction accuracy is greatly improved when early to mid-season remotely sensed images are used to estimate vegetative growth, such as normalized difference vegetation index (NDVI), and then are combined with agrometerological models. Since images taken late in the growing season express the cumulative seasonal effects of soil, pest, management, and climate, these can be used to predict crop yield maps using simple regression techniques (Moran *et al.*, 1997). Remotely sensed data for yield mapping have advantages over on-the-go combine yield monitoring including higher resolution and with less error associated with data collection (e.g., time lags from harvest point to sensor, combine speed variation, combine vibration). While a certain amount of on-going ground calibration may also be necessary, Pierce and Nowak (1999) have speculated that remotely sensed data for constructing yield maps may someday replace combine yield monitors.

3.1.4. Yield Potential from Soil and Landscape Maps and Measurements

Soil types have been used as a guide for describing field yield variation. Traditional soil surveys usually report the target grain yield of major crops by soil map unit. Soil surveys in the U.S. have not been conducted at a scale precise enough for effective use of site-specific N management (Mausbach *et al.*, 1993). In the UK, recommendation systems are still largely based on target yields, as explained in the *Fertilizer Recommendations* (MAFF, 2000). The procedure links an established requirement for optimum yields of a particular crop to a soil supply index based on soil type and previous cropping. However, the most progressive recommendation systems in the UK use computer models (Dampney *et al.*, 2000) and some scientists are moving away from a yield-based system toward one based on crop canopy management (Gillett *et al.*, 1999).

Slope position and landform characteristics are topographic features that also have been used to explain crop productivity (Hanna *et al.*, 1982; Jones *et al.*, 1989; Khakural *et al.*, 1996; McConkey *et al.*, 1997; McGee *et al.*, 1997; Sudduth *et al.*, 1997; Timlin *et al.*, 1998). Generally, footslope positions out-yield upslope positions unless poor drainage causes ponding. Real-Time Kinematic (RTK) GPS receivers have made possible the automated collection of highly accurate elevation data, thus providing an efficient way of obtaining high-resolution digital elevation models (DEM) of agricultural fields (Clark and Lee, 1998). Field topography plays an important role in the hydrological response of rainfall catchment and has a major impact on water availability to crop production. The increasing availability of DEMs and the advent of computerized terrain analysis tools have made it possible to quantify the topographic attributes of a landscape (Weibel and Heller, 1991; Bell *et al.*, 1995).

Soil productivity indices have also been developed using specific soil properties to characterize the suitability of the root zone for crop growth (Pierce *et al.*, 1983; Scrivner *et al.*, 1985a; Scrivner *et al.*, 1985b). However, the measurements that are required to calculate soil productivity indices on individual fields are expensive, time consuming, and require follow-up laboratory analysis.

Rapid spatial measurement of soil profile apparent soil electrical conductivity (EC_a) has potential for predicting variation in crop production potential as caused by soil differences (Jaynes *et al.*, 1993; Kitchen *et al.*, 1999b; Lund *et al.*, 1999). For example, soil EC_a has been used to estimate topsoil thickness (i.e., depth to first Bt horizon) on claypan soils (Doolittle *et al.*, 1994; Kitchen *et al.*, 1999b). For these soils, crop yield is depressed with decreasing topsoil thickness for average and below-average precipitation years (Gantzer and McCarty, 1987; Thompson *et al.*, 1991; Kitchen *et al.*, 1999b). Predicting target corn yields from topsoil thickness has been estimated by Kitchen *et al.* (1999a) as follows:

$$TY = 6.08 + 0.054(X) \qquad [2]$$

Where TY = target yield (Mg ha^{-1}), but with a maximum TY of 11.5 Mg ha^{-1}, and X = topsoil thickness (i.e., depth to the Bt horizon) in cm. The line shown in Figure 4 represents Eq. 2 along with corn yield data (in kg ha^{-1}) in relation to EC_a-derived topsoil thickness, from two claypan soil fields.

Field variation in claypan soil topsoil thickness as determined by soil EC_a has been used to predict target corn yield (Eq. 2) for variable-rate N applications (Kitchen *et al.*, 1999a). Figure 5 (top) displays actual soil EC_a values obtained on 5 m transects for a 14 ha field. On the same day that EC_a measurements were taken, 19 field points selected to span the field's

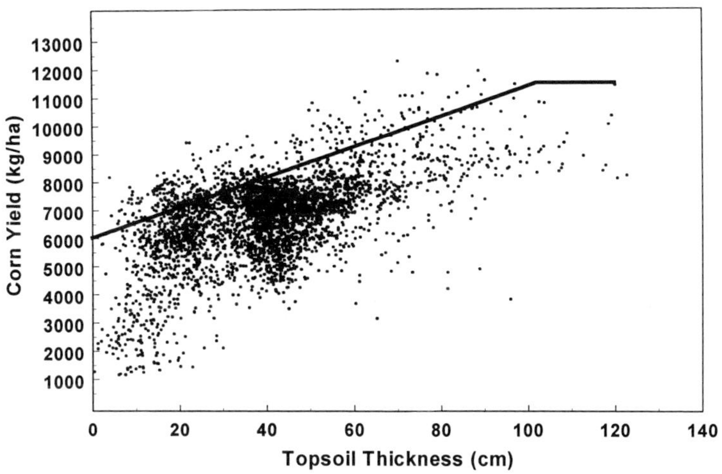

Figure 4. Relationship between topsoil thickness and 1997 corn yield, combined over two fields. The line represents the target yield equation (Eq. 2) used to determine variable-rate N applications

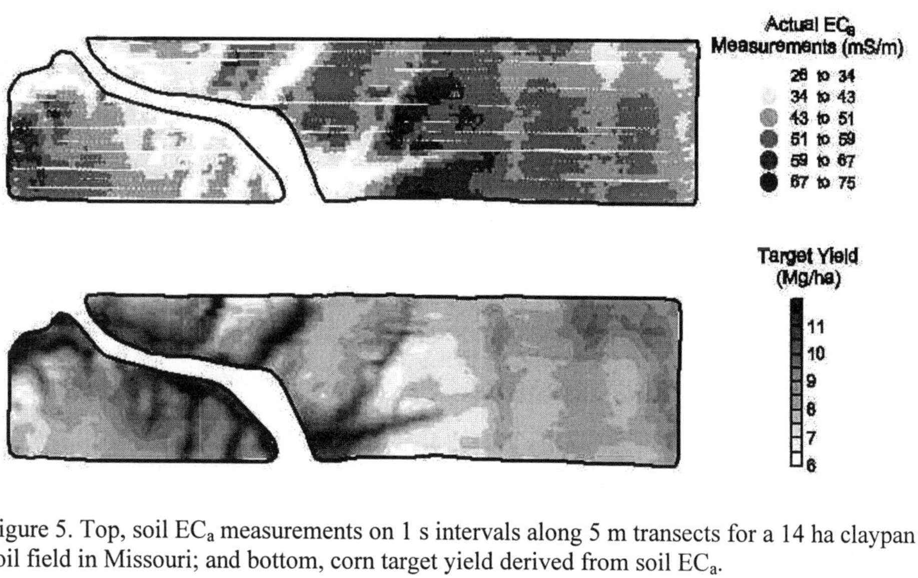

Figure 5. Top, soil EC_a measurements on 1 s intervals along 5 m transects for a 14 ha claypan soil field in Missouri; and bottom, corn target yield derived from soil EC_a.

range of EC_a values were soil sampled with a hand probe and the topsoil thickness determined. A regression equation relating EC_a to topsoil thickness was obtained for the calibration dataset (R^2=0.84). Figure 5 (bottom) is the resultant target yield derived from EC_a-estimated topsoil depth (Eq. 2), and from which a variable-rate N application map was obtained. Variable-rate N application compared to adjacent strips of conventional single-rate N treatments (one yield goal) were equal in corn yield where topsoil thickness was < 38 cm, with variable rate N producing about 0.5 Mg ha^{-1} more where topsoil thickness areas were > 38 cm.

3.2. Assessing Soil Nitrogen

Soil N available for crop uptake and growth fluctuates within and between growing seasons because of many environmental factors (including soil moisture, organic matter quality, temperature, pH, and oxygen). Yet, if producers hope to optimize N inputs they need accurate cost-effective tools for estimating soil N.

3.2.1. Potential Mineralizable Nitrogen

Nitrogen availability tests employing biological assays, where net mineralization is measured after incubation under controlled soil moisture and temperature, have been explained extensively earlier (Stanford and Smith, 1972; Stanford and Epstein, 1974; Keeney, 1982; Stanford, 1982; Meisinger, 1984). Since N mineralization in the field is largely controlled by unpredictable factors, like temperature and soil moisture, correlation with incubation tests can be inconsistent (Fox and Piekielek, 1984). Procedures for in situ measurement of N mineralization, such as enclosing a soil sample in a buried polyethylene bag or tube for incubation under ambient conditions, have been shown to correlate well with season-long mineralization (Eno, 1960; Westermann and Crothers, 1980; Poovarodom *et al.*, 1988). The advantages of these methods include the prevention of nitrate leaching and the control of N mineralization rates at field temperatures. Various methods of chemically or physically extracting that fraction of soil organic matter which will most easily decompose and make N available (Keeney, 1982; Christensen, 1992) are less time consuming than incubation tests. These procedures also vary in their agreement to field measurements of N mineralization because of year-to-year climatic variation (Fox and Piekielek, 1984; Gelderman *et al.*, 1984). While biological and chemical extraction tests are routinely used in research, their application for on-farm decisions has seen limited use.

3.2.2. Inorganic Nitrogen

Inorganic N soil tests–referred to as soil mineral N (SMN) measurements in Europe and parts of North America–assess soil nitrate-N and sometimes ammonium-N from soil samples either taken in the fall (for arid and colder regions) or just before planting or early in the growing season (for humid and warmer regions) and have been widely used for N fertilization decisions (Madgoff *et al.*, 1984; Blackmer *et al.*, 1989; Fox *et al.*, 1989; Magdoff, 1991; Bundy and Meisinger, 1994). In Europe samples for SMN are generally sampled in spring for modifying N recommendations. The UK recommendations (MAFF, 2000) advise farmers to measure SMN rather than use tables of soil N supply, especially in fields where manures have been applied regularly or large crop residues remained.

Soil sampling depth for these tests varies from 30 to 90 cm; sample depth guidelines depend upon a variety of factors, including crop, climate, soil type (Dahnke and Johnson, 1990), and producers' willingness to obtain subsoil samples. Under arid conditions, inorganic N soil tests are used to determine the mass of available N and could be used as the SN parameter in Eq. 1 (Peterson and Voss, 1984; Westfall, 1984). Elsewhere inorganic tests are more often used as indicators of soil N sufficiency. In this way, the test is calibrated with N fertilizer response and used directly for making N recommendations, as opposed to the mass balance approach of Eq. 1. Tests of soil N sufficiency include the pre-plant soil nitrate test (PPNT) and the pre-sidedress soil nitrate test (PSNT). Variations of these two tests are used in humid and semi-humid regions of North America and Europe. Calibrations with the PSNT found that nitrate-N levels > 20 to 25 mg N kg^{-1} typically show little or no response to the application of additional N fertilizer (Blackmer et al; 1989; Fox *et al.*, 1989; Meisinger *et al.*; 1992; Bundy and Andraski, 1995).

Table 2. Critical soil nitrate-N levels and percent of sites where soil tests failed to predict N response, derived from linear response plateau models using all observations.

Previous crop or cropping system	Time of soil sampling	Soil depth	N	Critical soil nitrate-N level	Failed soil test† TYPE A	TYPE B
				ppm	------- % of sites	--------
all observations	PPNT	0	292	15.7	1	35.3
		0-60	292	9.3	6.8	22.6
	PSNT	0	301	16.9	2.3	25.2
		0-60	239	12	4.6	18
Corn (without manure in study year)	PPNT	0	127	19.2	1.6	26.8
		0-60	126	16.1	11.1	14.3
	PSNT	0	125	18.9	3.2	21.6
		0-60	115	14.2	3.5	11.3
Corn (with manure in study year)	PPNT	0	28	11	3.6	42.9
		0-60	28	12.2	3.6	14.3
	PSNT	0	29	16.6	3.5	24.1
		0-60	24	22.4	8.3	16.7
Alfalfa	PPNT	0	27	na	0	92.6
		0-60	27	na	0	77.8
	PSNT	0	28	na	0	39.3
		0-60	26	na	0	38.4

† TYPE A failure = soil test predicted non-N-responsive, but was responsive
TYPE B failure = soil test predicted N-responsive, but was not responsive
Adapted from Bundy *et al.* (1999).

The PPNT and PSNT have been simultaneously evaluated under various management practices at more than 300 sites in ten U.S. cornbelt states (Bundy *et al.*, 1999). They concluded that a more practical way of assessing the economic and environmental consequences of management decisions made with these two tests was based on the *rate of failure* by the tests to predict non-N-responsiveness (Table 2). Two types of failure were identified. TYPE A failure resulted when the soil test predicted a non-N-responsive site, but the site actually responded to N fertilization (an economic loss due to lost yield). TYPE B failure resulted when the soil test predicted a N-responsive site, but the site did not respond to N fertilization (both an economic loss from applying un-needed N and increased risk for environmental loss due to excess N). Total failures (TYPE A and B) of the soil tests using a linear-plateau response model (average critical soil nitrate-N level of 13.5 mg kg^{-1}) were 3 to 8 % lower than the quadratic-plateau response model (average critical soil nitrate-N level of 20.2 mg kg^{-1}), primarily from a reduction in TYPE B failures (over application) and a concurrent modest increase in TYPE A failures (under application). Incidence of TYPE B failure occurred more frequently than TYPE A failure, but was much less with latter soil sampling (PSNT) and deeper soil sampling (0-60 cm sampling depth) (Table 2.) Sampling later and deeper was also especially important in corn cropping systems that included manuring and/or a preceding alfalfa crop.

Since the spatial variation of inorganic N can be high (Cahn *et al.*, 1994; Cambardella *et al.*, 1994; Selles *et al.*, 1999), producers are encouraged to composite a minimum of 15 to 20 cores. For fields with obvious landform variation, subdivision following soil and landscape patterns will improve accuracy in predicting crop NFR and N use efficiency (Dahnke and Johnson, 1990; James and Wells, 1990; Walter and Goesch; 1999).

The successful use of inorganic N soil tests has not been universal. Some soils are too stony to make sampling practicable. Following a crop like potato that is expected to supply significant N to the next crop, the spatial variability of soil test N may not be as important to predicting N supplying capacity of the soil as the spatial variability of potentially mineralizable N remaining in roots and plant residues (Franzen *et al.*, 1999). Calibration efforts under similar soil, climate, and cropping systems help establish the conditions under which the tests are most successful (Bundy *et al.*, 1999). In some situations, grower adoption of SMN tests are enhanced by governmental policy. As an example, in central Nebraska, ground water nitrate contamination in the Platte River aquifer has resulted in the Central Platte Natural Resources District requiring soil nitrate sampling on corn production fields. Use of the soil test has helped producers identify those fields high in residual soil N contributing to groundwater contamination and adjust N inputs accordingly (Schepers *et al.*, 1997). Adoption of N soil tests has been high for crops such as sugar beets where close scrutiny is needed to maintain crop quality (Ulrich et al, 1993).

3.3.3. Spatial Variability of Soil Nitrogen

As previously noted, soil N availability is often highly variable within fields. Schepers and Meisinger (1994) succinctly captured the reason for this variability:

> Nitrogen mineralization is a complex process that involves a vast collection of microorganisms (bacteria, fungi, and actinomyces) acting on a wide array of substrates (crop residues, soil humus, dead microbial tissue, and

> manure) under varying soil environments (temperature, water content, and aeration) to produce a remarkably simple product (nitrate-N) that can be used by plants, lost to the atmosphere as N gases, immobilized, accumulated in soil, or leached from the soil-crop system.

Little doubt is left as to why soil N–in both its organic and inorganic forms–is spatially variable as we consider that each condition and process mentioned varies within fields.

With inexpensive tools (like GPS) available to make spatial soil and plant measurements and from which the spatial structure can be determined (e.g., geostatistics) and maps created, interest in quantifying patterns of within-field availability of soil N has been spurred (Pierce and Nowak, 1999). Variable-rate N application maps derived from rootzone nitrate-N grid soil samples on a field considered uniform resulted in a 60 % increase in area correctly fertilized over fields of fixed-rate applications (Ferguson *et al.*, 1996). For fields with areas of high leaching potential, profile nitrate-N can be highly variable with short-scale (e.g., < 5 m) spatial structure, rendering spatial soil sampling for N management decisions ineffective (Everett and Pierce, 1996). Soil sampling intensity can be reduced and yet provide accurate N availability maps with "targeted" soil sampling, meaning like soil areas are grouped into zones and sampled and analyzed independently. Success with target sampling has been achieved using aerial image/spectral reflectance data (Diker and Bausch, 1999; Franzen *et al.*, 1999) and soil EC_a (Franzen and Kitchen, 1999) to derive sampling zones.

While the soil sampling density required for accurate N application maps varies from field to field, time and expense constraints limit use of spatially-dense sampling for N in most crop production systems (Ferguson *et al.*, 1996). Exceptions are with those high value crops such as potatoes and sugar beets where profit margins permit the additional expense. Alternatively, new technologies and tools may allow for on-the-go insitu measurement of soil N. For example, near infared soil sensing has been effectively used in predicting inorganic N content as long as a calibration set included the same interfering soil constituents as the unknown samples (Ehsani *et al.*, 1999). Further development is needed in sensors that can rapidly measure soil properties associated with estimating soil N.

4. PLANT MEASUREMENTS

Plant measurements for determining crop N status are a sufficiency/deficiency strategy, not a mass balance strategy as shown in Eq. 1. Plant measurements serve as indicators for within season N additions, or if measured at crop maturity to diagnose whether or not conditions provided deficient, sufficient, or excessive N for the crop. Since plants integrate soil, climate, management, and other environmental influences on crop N health, they provide an opportunity for improving NUE over relying only on yield prediction and preplant or early season soil N measurements. However, issues related to plant N measurements need to be considered before including these tools in the N management plan, including: 1) uncertainty of determining full-season N status and fertilizer needs from young crop plants, when an opportunity for N addition still exists; 2) a reported wide-range in sufficiency critical values; 3) varying sufficiency critical values as the crop matures; 4) varying critical values from various plant parts (e.g., leaves vs stems); and 5) the need for maintaining a N sufficiency block or strip for reference that adequately represents N needs of the remaining field (Schröder *et al.*, 2000).

Plant tissue sampling for N management decisions has previously been extensively reviewed (Westerman *et al.*, 1990; Bennett, 1993; Barraclough, 1997) and will not be detailed here. Generally, tissue N tests are highly variable and unstable indicators for within season N decisions (Schröder *et al.*, 2000). Exceptions exist on a crop-by-crop and region-by-region basis, particularly when a specific plant sampling procedure can be identified. Successful examples include petiole sampling for potatoes (Westermann and Kleinkopf, 1985; Williams and Maier, 1990a and 1990b) and sugar beets (Ulrich et al, 1993)], wheat tissue sampling combined with tiller density measurements (Scharf and Alley, 1993), end of growing season corn stalk nitrate test (Binford *et al.*, 1990 and as reviewed by Schröder *et al.*, 2000), pre-harvest plant tissue and post-harvest grain N for spring wheat (Peltonen, 1992), and stem testing for linola (Hocking, 1995).

4.1. Leaf and Canopy Greenness

Since N is a primary constituent of plant chlorophyll pigments, leaf or crop canopy greenness can be used to evaluate crop N health for within-season N input decisions. An obvious advantage of using plant greenness is there is little time delay between measurement and interpretation, such as occurs in soil sampling and analysis. Further, since each plant expresses crop N status for its given location, greenness sensing over the crop canopy provides the best opportunity for quantifying detailed spatial variability of crop N needs.

4.1.1. Spectral Reflectance Sensing

Measurement of crop canopy reflectance, either from ground-based or air borne platforms using image and photographic cameras, can provide a valuable measure of potential N status of the crop. Plant transformation of light energy to chemical energy (photophosphorylation) is most efficiently accomplished in chloroplasts by absorbing red (630-680 nm) and blue (450-520 nm) wavelength light. Green light (520-600 nm) is absorbed much less by plants, producing higher reflectance in this wavelength range. Hence sensing reflectance at these three wavelengths (RGB light) provides a measure of active plant chlorophyll.

By definition, crop reflectance is the ratio of the amount of light leaving the canopy to the amount of incoming light. Digital reflectance sensors (spectral radiometers) and photographic images are commonly calibrated against a standardized reference panel to assess the amount of incoming light. This is needed because radiometers vary in wavelength discrimination and light intensity sensitivity. Film types also vary in sensitivity to different light. Reflectance can also be successfully calculated for crop N status by obtaining a relative reference by comparing reflectance leaving the crop canopy of an area known to be non-limiting in N to reflectance from the test area. This has been accomplished with both spectral radiometer measurements (Chappelle *et al.*, 1992; Blackmer *et al.*, 1996) and photography (Blackmer *et al.*, 1996; Flowers et. al., 2000). Image interpretation is merely qualitative unless referenced with standardized panels under the same light conditions, or non-limiting N reference is obtained. Reflectance measurements are affected by many environmental factors other than N such as canopy architecture (Jackson and Pinter, 1986) and hybrid (Blackmer *et al.*, 1996). Referencing reflectance to a non-limiting N area within the same field can account for many of these factors (Blackmer *et al.*, 1996). Also for ground-based sensing of corn prior to tasseling, a 75° view angle allowed for more plant and less soil reflectance and was more accurate in predicting plant N than reflectance measurements taken from a nadir view (Bausch *et al.*, 1996; Diker and Bausch, 1999).

Green and red light reflectance alone can be a strong indicator of plant N content (Blackmer *et al.*, 1994; Blackmer *et al.*, 1996). From digitized film images RGB wavelength can be separated and intensity counted (0 to 255) for analysis with crop N (Blackmer *et al.*, 1996; Flowers *et al.*, 2000). Brightness of red light was shown to be a better indicator of corn N deficiency than chlorophyll meter readings (Blackmer and Schepers, 1996). Inclusion of other reflectance information related to plant biomass has often been shown to be a better index for assessing crop N health and making management decisions than just using RGB reflectance. Plants absorb much less near infrared (NIR) light (700-1400 nm) than does soil. This difference in absorption between soil and plants provides a contrast that has been the basis for numerous biomass or vegetative indices as reviewed by Myneni *et al.* (1995) and Moran *et al.* (1997). Calculations combining red and/or green light reflectance (a measure of the plant's photosynthetic health) with NIR reflectance (a measure of the plant's structure and capacity to assimilate carbon) have been successfully used in evaluating crop N health and making N fertilizer additions. Stone *et al.* (1996) was able to reduce N fertilizer input and increase NUE for wheat by variably applying N using a plant N spectral index (PNSI) derived from red and NIR reflectance values. Corn canopy NIR and green reflectance were used to develop a N reflectance index that was strongly correlated to chlorophyll meter readings, plant N content (Bausch *et al.*, 1996) and within season soil N (Diker and Bausch, 1999). To date, photographic images have mainly provided qualitative assessment of those fields that are N deficient. Verification of crop N deficiency has been needed since other environmental stresses can produce a similar reflectance signature. However, recent work has shown N recommendations can be obtained from aerial photographs. For example, NIR photographs taken during early spring were accurate in estimating soft red winter wheat tiller density and making correct N fertilizer recommendations (Flowers *et al.*, 2000).

Aerial images of crop fields are appealing to producers because it is low cost, has quick turn around, provides whole field information that is spatially accurate, can be used as a diagnostic tool for assessing many different types of crop stress, and gives producers an immediate visual assessment of conditions. With well-known field landmarks also visible on an image (such as field boundaries, trees, or structures), producers are quickly able to estimate the extent of the crop stress as well as associate stress areas with soil and landform features.

4.1.2. Chlorophyll Meter Sensing

A hand-held chlorophyll meter (Minolta SPAD-502) measures leaf transmittance centered at red (650) and NIR (940 nm) wavelengths and has been shown to be sensitive to N stress in corn (*Zea Mays* L.) (Dwyer *et al.*, 1991; Schepers *et al.*, 1992; Wood *et al.*, 1992; Piekielek *et al.*, 1995), wheat (*Triticum aesitivum* L.) (Follett *et al.*, 1992; Fox *et al.*, 1994; Miller *et al.*, 1999), rice (*Oryza sativa* L.) (Turner and Jund, 1991), and tall fescue (Kantety *et al.*, 1996). The meter has been effectively used as a decision-aid in correcting N deficiencies for irrigated corn (Blackmer and Schepers, 1995; Varvel *et al.*, 1997); but under rain-fed conditions the meter was ineffective (Bullock and Anderson; 1998). Corn growth stage, variety (Sunderman *et al.*, 1997; Varvel *et al.*, 1997; Bullock and Anderson, 1998), and water stress (Schepers *et al.*, 1996) are factors that will influence chlorophyll readings. To minimize the impact of these non-N effects on chlorophyll meter readings, a normalized measurement (referred to as a N sufficiency index) can be calculated by dividing the readings from N deficient plants by readings from N sufficient plants (Piekielek *et al.*, 1994; Varvel *et al.*, 1997). This step is similar to that discussed under *Spectral Reflectance Sensing.*

To operate, the SPAD-502 is clamped onto a single leaf to prevent interference from external light. The meter is limited to sensing transmittance through a very small area of leaf (about 6 mm^2) with each reading. While individual readings can be rapidly obtained, acquiring a representative value for an entire field is time consuming and for fields with spatial variability in
soil N it is difficult to obtain representative measurements (Schepers *et al.*, 1995). For this reason, chlorophyll meter sensing to assess production scale crop N health is not practical for most producers (Diker *et al.,* 1999). The SPAD-502 will continue to aid N research primarily as a diagnostic tool but has limited use in N management decisions for large-scale production agriculture.

Table 3. Allowed N surpluses in the Netherlands, MINAS Nutrient Budgeting Scheme (kg N ha^{-1} yr^{-1}).

Year	Arable	Grassland
1998	175	300
1999	175	300
2000	150	275
2002	(125)	(250)
2005	(110)	(200)
2008	(100)	(180)

Figures in parentheses are not yet agreed upon.

5. NUTRIENT BUDGETS

Nutrient budgets have been compiled around the world, using a variety of scales and methodological approaches (Meisinger and Randall, 1991; Watson and Atkinson, 1999). They measure or estimate the inputs and outputs of nutrients (usually N, P and K) to a field, farm or system, usually at the farm gate. Nutrient budgeting may operate on daily, monthly, or annual time frame. More frequent tracking requires more user input, but also provides the greatest opportunity for synchronizing nutrient inputs with crop needs. Farm gate budgets usually include inputs in feed, fertilizers, manures, composts, and bedding and outputs in saleable produce. They do usually include the necessarily very detailed measurements of losses such as leaching, denitrification, and ammonia volatilization, consider each field separately, or measure transfers between fields. Nor do they provide information on soil processes or biological inputs or outputs of nutrients, which are particularly important for N. By their nature they cannot improve N use efficiency but only highlight problems and raise awareness of the need for better techniques. For many producers and agronomists, however, raising awareness is an essential first step.

To counter their large N surpluses (see Table 1), The Netherlands have introduced a compulsory nutrient budgeting policy. This requires nutrient budgets to be made on all farms with > 2.5 livestock units per hectare and gives allowed surpluses (Table 3). If these values are exceeded, farmers are taxed about 75c (£0.5 or 1 euro) for each kg N above the limit. However, it should be noted that farmers do not have to include atmospheric deposition or fixation by legumes in their calculations of inputs, and some ammonia losses are allowable. Despite these relatively generous regulations, Dutch farmers are not happy with the arrangements and will have great difficulty meeting the requirements because of the focus on livestock production systems.

6. CONSIDERATIONS FOR DEVELOPING NEW ON-FARM TECHNOLOGIES

Some of the diagnostic tools for assessing crop N needs discussed here have been available to producers for several decades. Researchers and extension agronomists have advocated the adoption of such tools, but with limited success. For example, in 1999 knowledgeable representatives from the U.S. were asked what percentage of their state's corn acreage was tested annually using the PPNT, PSNT, early-season chlorophyll meter sensing, and stalk nitrate testing. Each of these diagnostic tests were designed to help producers make better N management decisions. A summary of their responses (Table 4) indicates that adoption has been generally low, but high where adaptive into specific cropping systems. In the humid regions of the northeastern U.S. the PPNT test has been put into practice on about 13 % of that region's corn acreage, but this area represents a very small percentage of corn grown nationally. The PSNT has also seen significant use in the north central region, predominantly on irrigated acres in the western portion of the region (reaching a high of about 60 to70 % of corn acreage in Nebraska). Many may find this level of adoption discouraging until they reflect upon the nature of N in a biologically-complex agricultural production system. One test, one technology, or one practice should not be the goal. Instead the goal should be a myriad of options from which N management can be tailored. A recent review of the potential use of precision agriculture technologies in Northern Europe (Sylvester-Bradley *et al.*, 1999) concluded that they were most likely to be adopted where prior knowledge identified large heterogeneity and predicted treatment zones, but that the main obstacle was the lack of appropriate sensors.

"On farm" implies that producers will be at the center of implementing changes; but "change" also means there will be attractive new choices available to motivate producers. Many N management technologies and practices, though soundly developed and tested, have been left on the shelf by growers. Prerequisites for grower adoption require that new and innovative practices be reliable, incur minimal additional expense (time and equipment), and integrate with ease into current operations. When these cannot be met, external incentives (e.g., regulation, private or government cost sharing programs) may be needed.

Table 4. From a survey about corn grain grown in the U.S., what percentage of the acreage in 1999 used these soil and plant diagnostic tools for N management? (Numbers represent the upper limit when a range was given).

diagnostic test	New England/ Mid-Atlantic region (11 states) representing 4.2 M acres	North Central region (13 states[†]) representing 61.2 M acres
	------------------------ % of acres	------------------------------
pre-plant soil nitrate test	13.3	1.8
pre-sidedress soil nitrate test	≈0	14.0[*]
early-season chlorophyll meter	≈0	< 1
stalk nitrate test	< 1	< 1

[†] includes one Canadian providence
[*] primarily from states with a majority of irrigated acreage (e.g., Kansas, Nebraska)

7. INTERNET TOOLS FOR NITROGEN MANAGEMENT DECISIONS

Accessibility to computer-assisted decision tools has exploded with the Internet, not to the exclusion of N management tools for on-farm decisions. While sophistication–and therefore power–varies with each tool, producers, researchers, and government regulators can find these helpful in considering various "what if" management scenarios that may improve NUE. Below is a listing of some tools obtained through extensive searches on the Internet (verified 11 Aug 2000).

7.1 Interactive Web Tools

Nitrogen Crop Response Model
http://www.qpais.co.uk/nable/nitrogen.htm
Greenwood, D.J., C.R. Rahn, A. Draycott. 1999
Simulates the growth response of 25 crops to applications of N fertilizer. Includes the effects of organic material, climate and leaching. Based on the N ABLE model developed by the authors. Developed primarily for vegetable crops.

Nitrogen Ready Reckoner
http://www.nre.vic.gov.au/cgi-bin/exsysweb.exe?KBNAME=nut04
Thomas, G. 1999.
Calculates the response of pasture to N fertilizer, and cost of the response in cents per megajoule.

Nitrogen Management Calculator
http://www.hollysugar.com/grower/nitrogen6.htm

Imperial Sugar Company. 2000

Calculates applications for sugar beets based on residual nitrate, irrigation water nitrate, inches of irrigation water, soil OM, other N credits, and N fertilizer form.

Forage, Straw, and Crop Nutrient Use Calculators

http://www.agric.gov.ab.ca/calculator/nutrientuse.html

Alberta Agriculture Food and Rural Development. 2000.

Calculates N, P, K, S, and Ca removed with various crops, forages and straw. The calculations are based on Canadian Fertilizer Institute crop nutrient uptake and removal numbers for wheat barley, oats, canola, and peas.

Nitrogen Carryover Calculator

http://www.soils.wisc.edu/wimnext/nitrogen.html

Bland, W. DATE.

Estimates N carryover after corn, based on county, soil type and weather. Designed for Minnesota and Wisconsin farmers, as a decision aid in predicting the need for pre-plant soil nitrate testing.

Terra Nitrogen Calculators

http://www.farmline.com/folcorps/terra/calculators/ForageCrops.asp

Terra Industries. DATE.

Calculates N recommendations for a wide variety of crops and forages considering credits from previous crops or rotations. This is a European site focused on small grains and vegetable crops.

Liquid Animal Waste Calculators

http://cumberla.ces.state.nc.us/fertpage/liquid_waste.html

Bailey, K. 1999.

Several calculators for manure management. Calculates land required for total nutrient load based on any single nutrient component. Includes some fertigation calculations.

7.2 Downloadable Spreadsheets and Applications

Decision Support Systems for the use of Nitrogen in Intensive Pasture Systems

http://www.nitrogen.landfood.unimelb.edu.au/ (Use decision support link)

Eckard, R. 2000.

Decision tool for dairy pasture management practices designed to minimize losses while maintaining productivity.

Nitrogen Decision Aid

http://www.mrsars.usda.gov/morris/products/nitrogen.htm

Olness, A., W. Voorhees, D. Archer. 2000.

This program predicts the amount of N mineralized from corn planting to side-dress, or 5-leaf growth-stage, in Northern climates.

8. CONCLUSIONS

Modern agriculture is increasing in complexity as demands for more food, feed and fiber, at higher quality, while concurrently safe-guarding the environment are requested by the consumer. The economics of food, feed, and fiber production are now embracing the costs of environmental impact. Fine-tuned N management that minimizes off-field losses remains a challenge for farmers and agronomists. Tried and tested old as well as new technologies offer ways of increasing NUE, sometimes by significant amounts. Tools that indicate N in excess of crop needs for the year in question may have little economic appeal to producers, because of inexpensive N, but these same tools used under these conditions will grant the greatest environmental benefit. Opportunity for improvement largely lies with technologies that enable timely, quick, and accurate measurement of the spatial variability of crop yield potential, soil N availability, and within season indication of crop N health. Soil N excess and deficiency can exist on the same field. "Thus, it is the variability in space and time of the processes that regulate the availability of N to plants and the fate of N in soil that make precision N management attractive" (Pierce and Nowak, 1999, pp 32-33). Airborne or satellite remotely sensed images are being aggressively tried. In most cases, the decision rules for transforming images into N management decisions are not well developed or validated yet, but limitations of remotely sensed data are likely to be remedied soon (Moran, 2000).We predict by the year 2010 remotely sensed data will be commonly used in crop N management in the U.S. and European countries.

Environmentally, some of the biggest problems of poor NUE are associated with poor utilization of animal manures, and here progress has been slow. Nutrient heterogeneity within stockpiled manures along with transport logistics are issues that magnify in significance as animal confinement operations become larger and more concentrated. Whenever animal feeding is a component of an agriculture production operation, we strongly encourage whole-farm nutrient budgeting and planning.

One final point. A focus just on increasing NUE can lead to, in some situations, other environmental problems. For example, early sowing to obtain effective crop cover and an increase in N uptake and reduction in N losses can promote the risk of pest and disease carry-over and pesticide use. The overriding need is for technologies that embrace all aspects of farm efficiency to ensure optimum improvement.

9. REFERENCES

Algero, P.A. and L. Thylen. 1999. Variable nitrogen application: Effects on crop yield and quality. p. 709-717. *In* P.C. Robert *et al.* (ed.), Precision Agriculture- Proc. of the 4th Intl. Conf., St. Paul, MN, 19-22 July 1998. ASA, CSSA, and SSSA, Madison, WI.

Barraclough, P.B. 1997. N requirement of winter wheat and diagnosis of deficiency. Aspects of Applied Biology 50: 117-123.

Bausch, W.C., H.R. Duke, and C.J. Iremonger. 1996. Assessment of plant nitrogen in irrigated corn. p. 23-32. *In* P.C. Robert *et al.* (ed.), Precision Agriculture- Proc. of the 3rd Intl. Conf., Minneapolis, MN, 23-26 June 1996. ASA, CSSA, and SSSA, Madison, WI.

Bell, J.C., C.A. Butler, and J.A. Thompson. 1995. Soil-terrain modeling for site-specific agricultural management. p. 210-227. *In* P.C. Robert *et al.* (ed.) Site-specific management for agricultural systems. Proc. of the 2nd Intl. Conf. Minn., MN, 27-30 Mar. 1994. ASA, CSSA, and SSSA, Madison, WI.

Bennett, W.F. (ed.). 1993. Nutrient deficiencies and toxicities in crop plants. APS Press, St. Paul, MN.

Binford, G.D., A.M. Blackmer, N.M. El-Hout. 1990. Tissue test for excess N during corn production. Agron. J. 82:124-129.

Blackmer, A.M., D. Pottker, M.E. Cerrato, and J. Webb. 1989. Correlations between soil nitrate concentrations in late spring and corn yields in Iowa. J. Prod. Agric. 2:103-109.

Blackmer, T.M., and J.S. Schepers. 1995. Use of chlorophyll meter to monitor nitrogen status and schedule fertigation for corn. J. Prod. Agric. 8:56-60.

Blackmer, T.M. and J.S. Schepers. 1996. Aerial Photography to Detect Nitrogen Stress in Corn. J. Plant Physiol. 148:440-444.

Blackmer, T.M., J.S. Schepers and G.E. Varvel. 1994. Light reflectance with other nitrogen stress measurements in corn leaves. Agron. J. 86:934-938.

Blackmer, T.M., J.S. Schepers, G.E. Varvel, and E.A. Walter-Shea. 1996. Nitrogen deficiency detection using reflected shortwave radiation from irrigated corn canopies. Agron. J. 88:1-5.

Blackmer, A.M. and S.E. White. 1996. Remote sensing to identify spatial patterns in optimal rates of nitrogen fertilization. p. 33-41. *In* P.C. Robert *et al.* (ed.), Precision Agriculture- Proc. of the 3rd Intl. Conf., Minneapolis, MN, 23-26 June 1996. ASA, CSSA, and SSSA, Madison, WI.

Blackmer, A.M., and C.J. Marshall. 1996. Yield mapping; errors and algorithms. p. 403-415. *In* P.C. Roberts et al. (ed) Precision agriculture, Proc of the 3rd Intl. Conf. Minn., MN, 23-29 June 1996. ASA, CSSA, and SSSA, Madison, WI.

Bock, B.R. 1984. Efficient use of nitrogen in cropping systems. p. 273-294. *In* R.D. Hauck (ed.) Nitrogen in crop production. ASA, CSSA, and SSSA, Madison, WI.

Braum, S.M., P. Hinds, G.L. Malzer, J. Bell, D. Mulla, and P.C. Robert. 1999. Terrain attributes and soil nitrogen: Spatial effects on corn yield responses to nitrogen fertilization for a northern, glaciated landscape. p. 195-201. *In* P.C. Robert *et al.* (ed.), Precision Agriculture- Proc. of the 4th Intl. Conf., St. Paul, MN, 19-22 July 1998. ASA, CSSA, and SSSA, Madison, WI.

Bruulsema, T.W., G.L. Malzer, P.C. Robert, J.G. Davis, and P.J. Copeland. 1996. Spatial relationships of soil nitrogen with corn yield response to applied nitrogen. p. 505-512. *In* P.C. Robert *et al.* (ed.), Precision Agriculture- Proc. of the 3rd Intl. Conf., Minneapolis, MN, 23-26 June 1996. ASA, CSSA, and SSSA, Madison, WI.

Bundy, L.G., D.T. Walters, and A.E. Olness. 1999. Evaluation of soil nitrate tests for predicting corn nitrogen response in the north central region. North Central Reg. Res. Publ. 342. Univ. of Wisconsin-Madison.

Cahn, M.D., J.W. Hummel, and B.H. Brouer. 1994. Spatial analysis of soil fertility for site-specific crop management. Soil Sci. Soc. Am. J. 58:1240-1248.

Cambardella, C.A., T.B. Moorman, J.M. Novak, T.B. Parkin, D.L. Karlen, R.F. Turco, and A.E. Konopka. 1994. Field-scale variability of soil properties in central Iowa soils. Soil Sci. Soc. Am. J. 58:1501-1511.

Campbell, C.A., Y.W. Jame, and O.O. Akinremi. 1994. Evaluating potential nitrogen mineralization for predicting fertilizer nitrogen requirements of long-term field experiments. p. 81-100. *In* J.L. Havlin *et al.* (ed.) Soil testing: Prospects for improving nutrient recommendations. SSSA Spec. Publ. 40. SSSA, Madison, WI.

Chappelle, E.W., M.S. Kim and J.E. McMurtrey III. 1992. Ratio analysis of reflectance spectra (RARS): An algorithm for the remote estimation of the concentrations of Chlorophyll A, Chlorophyll B and carotenoids in soybean leaves. Remote Sen. Environ. 39:239-247.

Christensen, B.T. 1992. Physical fractionation of soil and organic matter into primary particle size and density separates. Adv. Soil Sci. 20:1-90.

Clark, R.L, and R. Lee. 1998. Development of topographic maps for precision farming with kinematic GPS. Trans. of the ASAE 41(4): 909-916.

Coelho, A.M., J.W. Doran, and J.S. Schepers. 1999. Irrigated corn yield as related to spatial variability of selected soil properties. p. 441-452. *In* P.C. Robert *et al.* (ed.), Precision Agriculture- Proc. of the 4th Intl. Conf., St. Paul, MN, 19-22 July 1998. ASA, CSSA, and SSSA, Madison, WI.

Colburn Jr., J.W. 1999. Soil doctor multi-parameter, real-time soil sensor and concurrent input control system. p. 1011-1021. *In* P.C. Robert *et al.* (ed.), Precision Agriculture- Proc. of the 4th Intl. Conf., St. Paul, MN, 19-22 July 1998. ASA, CSSA, and SSSA, Madison, WI.

Colvin, T.S., D.B. Jaynes, D.L. Karlen, D.A. Laird, J.R. Ambuel. 1997. Yield variability within a central Iowa field. Trans. ASAE. 40(4):883-889.

Dahnke, W.C. and G.V. Johnson. 1990. Testing soils for available nitrogen. p. 127-139. *In* R.L. Westerman (ed.) Soil testing and plant analysis. 3rd Ed. SSSA, Madison, WI.

Dampney, P.M.R., G. Goodlass, M.A. Froment, and J.V. Stafford. 1999. Environmental and production effects from variable rate nitrogen fertilizer to winter wheat in England. p. 697-708. *In* P.C. Robert *et al.* (ed.), Precision Agriculture- Proc. of the 4th Intl. Conf., St. Paul, MN, 19-22 July 1998. ASA, CSSA, and SSSA, Madison, WI.

Dampney, P.M.R., E.I. Lord, and B.J. Chambers. 2000. Development of improved advice for farmers and advisers. Soil Use Manage. 16:162-166.

Davies, D.B. 2000. Tackling nitrate from agriculture. Soil Use Manage. 16:141-174.

Davis, J.G., G.L. Malzer, P.J. Copeland, J.A. Lamb, P.C. Robert, and T.W. Bruulsema. 1996. Using yield variability to characterize spatial crop response to applied N. p. 513-519. *In* P.C. Robert *et al.* (ed.), Precision Agriculture- Proc. of the 3rd Intl. Conf., Minneapolis, MN, 23-26 June 1996. ASA, CSSA, and SSSA, Madison, WI.

Dickson, B.C. and D.J. Schaeffer. 1997. Ecoforestation of riparian forests for non-point source pollution control: Policy and ecological considerations in agroecosystem watersheds. p. 221-227. *In* N.E. Haycock *et al.* (ed.), Buffer zones: Their processes and potential in water protection. Quest Environmental, UK.

Diker, K. and W.C. Bausch. 1999. Mapping in-season soil nitrogen variability assessed through remote sensing. p. 1445-1455. *In* P.C. Robert *et al.* (ed.), Precision Agriculture- Proc. of the 4th Intl. Conf., St. Paul, MN, 19-22 July 1998. ASA, CSSA, and SSSA, Madison, WI.

Diker, K., M. Unlu, and W. C. Bausch. 1999 . Determining spatial variability of plant nitrogen in corn fields using a chlorophyll meter. p. 593-602. *In* J.V. Stafford (ed.)

Precision Agriculture '99- Proc. of the 2nd Eur. Conf. on Precision Agriculture. Odense Congress Centre, Denmark, 11-15 July 1999. SCI, Sheffield, UK.

Doolittle, J.A., K.A. Sudduth, N.R. Kitchen, and S. J. Indorante. 1994. Estimating depths to claypans using electromagnetic induction methods. J. Soil Water Conservation 49:572-575.

Downes, M.T., C. Howard-Williams and L.A. Schipper. 1997. Long and short roads to riparian zone restoration: Nitrate removal efficiency. p. 244-254. *In* N.E. Haycock *et al.* (ed.), Buffer zones: Their processes and potential in water protection. Quest Environmental, UK.

Dwyer, L.M., M. Tollenaar, and L. Houwling. 1991. A nondestructive method to monitor leaf greeness in corn. Can. J. Plant Sci. 71:505-509.

Ehsani, M.R., S.K. Upadhyaya, D. Slaughter, S. Shafi and M. Pelletier. 1999. A NIR technique for rapid determination of soil mineral nitrogen. Precision Agriculture 1: 217-234.

Eno, C.F. 1960. Nitrate production in the field by incubating the soil in polyethylene bags. Soil Sci. Soc. Am. Proc. 24:277-279.

Evans, R.G., S. Han, M.W. Kroeger, and S.M. Schneider. 1996. Precision center pivot irrigation for efficient use of water and nitrogen. p. 75-84. *In* P.C. Robert *et al.* (ed.), Precision Agriculture- Proc. of the 3rd Intl. Conf., Minneapolis, MN, 23-26 June 1996. ASA, CSSA, and SSSA, Madison, WI.

Everett, M.W. and F.J. Pierce. 1996. Variability of corn yield and soil profile nitrates in relation to site-specific N management. p. 43-53. *In* P.C. Robert *et al.* (ed.), Precision Agriculture- Proc. of the 3rd Intl. Conf., Minneapolis, MN, 23-26 June 1996. ASA, CSSA, and SSSA, Madison, WI.

Ferguson, R.B., C.A. Gotway, G.W. Hergert, and T.A. Peterson. 1996. Soil sampling for site-specific nitrogen managment. p. 13-22. *In* P.C. Robert *et al.* (ed.), Precision Agriculture- Proc. of the 3rd Intl. Conf., Minneapolis, MN, 23-26 June 1996. ASA, CSSA, and SSSA, Madison, WI.

Ferguson, R.B., G.W. Hergert, J.S. Schepers, and C.A. Crawford. 1999. Site-specific nitrogen management of irrigated corn. p. 733-743. *In* P.C. Robert *et al.* (ed.), Precision Agriculture- Proc. of the 4th Intl. Conf., St. Paul, MN, 19-22 July 1998. ASA, CSSA, and SSSA, Madison, WI.

Flowers, M., R. Heiniger, and R. Weisz. 2000. Aerial photographic determination of nitrogen application timing and rate recommendations in winter wheat. p ??. *In* ??? , Precision Agriculture- Proc. of the 5th Intl. Conf..????

Follett, R.H., R.F. Follett, and A.D. Halvorson. 1992. Use of a chlorophyll meter to evaluate the nitrogen status of dryland winter wheat. Commun. Soil Sci. Plant Anal. 23:687-697.

Follett, R.F., D.R. Keeney, and R.M. Cruse (ed.). 1991. Managing nitrogen for groundwater quality and farm profitability. SSSA. Madison, WI.

Fox, R.H. and W.P. Piekielek. 1984. Relationships among anaerobically mineralized nitrogen, chemical indexes, and nitrogen availability to corn. Soil Sci. Soc. Am. J. 48:1087-1090.

Fox, R.H., W.P. Piekielek, and K.M. Macneal. 1994. Using a chlorophyll meter to predict nitrogen fertilizer needs of winter wheat. Commun. Soil Sci. Plant Anal. 25:171-181.

Fox, R.H., G.W. Roth, K.V. Iverson, and W.P. Pieckielek. 1989. Soil and tissue nitrate tests compared for predicting soil nitrogen availability to corn. Agron. J. 81:971-974.

Franzen, D.W., L.J. Cihacek, and V.L. Hofman. 1996. Variability of soil nitrate and phosphate under different landscapes. p. 521-529. *In* P.C. Robert *et al.* (ed.), Precision Agriculture- Proc. of the 3rd Intl. Conf., Minneapolis, MN, 23-26 June 1996. ASA, CSSA, and SSSA, Madison, WI.

Franzen, D.W., V.L. Hofman, L.J. Cihacek, and L.J. Swenson. 1999. Soil nutrient relationships with topography as influenced by crop. Precision Agriculture 1: 167-183. Franzen, D.W. and N.R. Kitchen. 1999. Developing management zones to target nitrogen applications. Site-Specific Management Guidelines. SSMG-5, Potash and Phosphate Institute. Norcross, GA.

Franzen, D.W., L. Reitmeier, J.F. Giles, and A.C. Cattanach. 1999. Aerial photography and satellite imagery to detect deep soil nitrogen levels in potato and sugarbeet. p. 281-290. *In* P.C. Robert *et al.* (ed.), Precision Agriculture- Proc. of the 4th Intl. Conf., St. Paul, MN, 19-22 July 1998. ASA, CSSA, and SSSA, Madison, WI.

Gantzer, C.J., and T.R. McCarty. 1987. Predicting corn yields on a claypan soil using a soil productivity index. Trans. ASAE 30:1347-1342.

Gelderman, R.H., W.C. Dahnke, and L. Swenson. 1988. Correlation of several N indices for wheat. Commun. Soil Sci. Plant Anal. 19:755-772.

Gillett, A.G., N.M.J. Crout, D.T. Stokes, R. Sylvester-Bradley, and R.K. Scott. 1999. Simple winter wheat green area index model under UK conditions. J. Agric. Sci. 132:263-271.

Goos, R.J., and L. Prunty. 1990. Yield variability and the yield goal decision. p. 187-189. *In* J.L. Havlin (ed.) Proc. of the Great Plain Soil Fertility Conf., Denver, CO. 6-7 Mar. 1990. Kansas State Univ., Manhattan.

Goulding, K.W.T., N.J. Bailey, N.J. Bradbury, P. Hargreaves, M. Howe, D.V. Murphy, P.R. Poulton, and T.W. Willison. 1998b. Nitrogen deposition and its contribution to nitrogen cycling and associated soil processes. New Phyt.139:49-58.

Goulding, K., V. Jordan, A. Leake, K. Warman, D. Donaghy, F. Reynolds, L. Stockdale, A. Crew, C. Webster, J. Conway, and S. Jarvis. 1999. Interactions between agricultural emissions to the environment: The value of system studies in minimizing all emissions. p. 55-63. *In* S.T.D. Turner *et al.* (ed.) Agriculture and the environment. Challenges and conflicts for the new millennium. Warwick, UK, 14-16 April 1999, ADAS, Wolverhampton.

Goulding, K.W.T., L.S. Matchett, G. Heckrath, C.P. Webster and P.C. Brookes. 1996. Nitrogen and phosphorus flows from agricultural hillslopes. p. 213-227. *In* M.G. Anderson *et al.* (ed.), Advances in hillslope processes. Proc. British Geomorphological Research Group Conf., September 1996, John Wiley & Sons, Chichester.

Goulding, K.W.T., E.A. Stockdale, S. Fortune, and C. Watson. 2000. Nutrient cycling on organic forms. J. Roy. Ag. Soc. Eng., in press.

Haahr, V., R.N. Jorgensen, A. Jensen, and J. Overgaard. 1999. A method for optimal site-specific nitrogen fertilization. p. 709-718. *In* J.V. Stafford (ed.) Precision Agriculture '99- Proc. of the 2nd Eur. Conf. on Precision Agriculture. Odense Congress Centre, Denmark, 11-15 July 1999. SCI, Sheffield, UK.

Hanna, A.Y., P.W. Harlan, and D.T. Lewis. 1982. Soil available water as influenced by landscape position and aspect. Agron. J. 74:999-1004.

Hargrove, W.L. (ed). 1988. Cropping strategies for efficient use of water and nitrogen. ASA Special Publ. 51. ASA, Madison, WI.
Harrison, R. and S. Peel. 1996. Nitrogen uptake by cover crops and its subsequent fate in arable systems. Aspects of Appl. Biol. 47:51-58.
Hauck, R.D. 1973. Nitrogen tracers in nitrogen cycle studies–past use and future needs. J. Environ. Qual. 2:317-327.
Havlin, J.L. and J.S. Jacobsen (ed.). 1994. Soil testing: Prospects for improving nutrient recommendations. SSSA Special Publ. 40. SSSA, Madison, WI.
Hergert, G.W. 1987. Status of residual nitrate-nitrogen soil tests in the United States of America. p. 73-88. *In* J.R. Brown (ed.) Soil testing: Sampling, correlation, calibration, and interpretation. SSSA Special Publ. 21. SSSA, Madison, WI.
Hocking, P.J. 1995. Critical nitrate-nitrogen and total nitrogen concentrations for vegetative growth and seed yield of linola (edible-oil linseed) as affected by plant age. Australian Journal of Experimental Agriculture 35: 239-246.
Hook, J.E. and G.J. Gascho. 1988. Multiple cropping for efficient use of water and nitrogen. p. 7-20. *In* W.L. Hargrove (ed.) Cropping strategies for efficient use of water and nitrogen. ASA Special Publ. 51. ASA, CSSA, and SSSA, Madison, WI.
Howarth, R.W. 1998. An assessment of human influences on fluxes of nitrogen from the terrestrial landscape to the estuaries and continental shelves of the North Atlantic Ocean. Nutr Cycling Agroecosystems 52:213-223.
Jackson, R.D. and P.J. Pinter, Jr. 1986. Spectral response of architecturally different wheat canopies. Remote Sens. Environ. 20:43-56.
James, D.W. and K.L. Wells. 1990. Soil sample collection and handling: Technique based on source and degree of field variability. p. 25-44. *In* R.L. Westerman (ed.) Soil testing and plant analysis. 3rd Ed. SSSA, Madison, WI.
Jaynes, D.B., T.S. Colvin, and J. Ambuel. 1993. Soil type and crop yield determinations from ground conductivity surveys. ASAE paper 933552, Am. Soc. of Agric. Engrs., St. Joseph, MI.
Jones, A.J., L.N. Mielke, C.A. Bartles, and C.A. Miller. 1989. Relationship of landscape position and properties to crop production. J. Soil and Water Cons. 44:328-332.
Kantety, R.V., E. Van Santen, F.M. Woods, and C.W. Woods. 1996. Chlorophyll meter predicts nitrogen status of tall fescue. J. Plant Nutr. 19:881-899.
Keeney, D.R. 1982. Nitrogen availability indices. p. 711-733. *In* A.L. Page *et al.* (ed.) Methods of soil analysis. Part 2, 2nd ed. Agron. Monogr. 9. ASA and SSSA, Madison, WI.
Khakural, B.R., P.C. Robert, and D.J. Mulla. 1996. Relating corn/soybean yield to variability in soil and landscape characteristics. p. 117-128. *In* P.C. Robert *et al.* (ed.) Precision Agriculture- Proc. of the 3rd Intl. Conf. Minn., MN, 23-26 Jun 1996. ASA, CSSA, and SSSA, Madison, WI.
King, B.A., J.C. Stark, I.R. McCann, and D.T. Westermann. 1996. Spatially varied nitrogen application through a center pivot irrigation system. p. 85-94. *In* P.C. Robert *et al.* (ed.), Precision Agriculture- Proc. of the 3rd Intl. Conf., Minneapolis, MN, 23-26 June 1996. ASA, CSSA, and SSSA, Madison, WI.
King, B.A., J.C. Stark, and J.P. Taberna, Jr. 1999. In-season spatial variability of potato petiole nitrogen. p. 55-65. *In* P.C. Robert *et al.* (ed.), Precision Agriculture- Proc. of

the 4th Intl. Conf., St. Paul, MN, 19-22 July 1998. ASA, CSSA, and SSSA, Madison, WI.

Kitchen, N.R., S.A. Sudduth, J.G. Davis, C.W. Fraisse, and S.T. Drummond. 1999a. Precision nitrogen strategies for Midwest claypan soils. p. 46-47. *In* Proc. of the 3rd National Workshop on Constructed Wetlands/BMPs for Nutrient Reduction and Coastal Water Protection., New Orleans, LA, June 9-12, 1999.

Kitchen, N.R., K.A. Suddduth, and S.T. Drummond. 1999b. Soil electrical conductivity as a crop productivity measure for claypan soils. J. of Prod. Agric. 12:607-617.

Kitchen, N.R., K.A. Sudduth, D.F. Hughes, and S.J. Birrell. 1995. Comparison of variable rate to single rate nitrogen fertilizer application: corn production and residual soil NO3-N. p. 427-442. *In* P.C. Robert et al. (ed.) Site-specific management for agricultural systems, Proc. of the 2nd Intl. Conf. Minn., MN, 27-30 Mar. 1994. ASA, CSSA, and SSSA, Madison, WI.

Kolberg, R.L., N.R. Kitchen, D.G. Westfall, and G.A. Peterson. 1997. Cropping intensity and nitrogen management impact of dryland no-till rotations in the semi-arid western Great Plains. J. of Prod. Agric. 10(3):517-522.

Kristensen, L. (ed.) 1995. Nitrogen leaching in ecological agriculture. Biol. Ag. Hort. 11:1-340.

Kurtz, L.T., L.V. Boone, T.R. Peck, and R.G. Hoeft. 1984. Crop rotations for efficient nitrogen use. p. 295-306. *In* R.D. Hauck (ed.) Nitrogen in crop production. ASA, CSSA, and SSSA, Madison, WI.

Lamb, J.A, J.L. Anderson, G.L. Malzer., J.A. Vetch, R.H. Dowdy, D.S. Onken, and K.I. Ault. 1995. Perils of monitoring grain yield on-the-go. *In* P.C. Robert *et al.* (ed.), Site-specific management for agricultural systems- Proc. of the 2rd Intl. Conf., Minneapolis, MN, 27-30 Mar. 1994. ASA, CSSA, and SSSA, Madison, WI.

Lægrid, M., O.C. Bøckman, and O. Kaarstad. 1999. Agriculture, fertilisers and the environment. CAB International, Wallingford, UK.

Lark, R.M. and J.V. Stafford. 1996. Consistency and change in spatial variability of crop yield over successive seasons: methods of data analysis. p. 117-128. *In* P.C. Robert et al. (ed.) Precision agriculture, Proc of the 3rd Intl. Conf. Minn., MN, 23-26 Jun 1996. ASA, CSSA, and SSSA, Madison, WI.

Leclerc, H,. F. Laurent, D. Boisgontier, J.M. Le Bars, and P. Zwaenepoel. 1999. Using field and crop characteristics for site specific nitrogen management on winter wheat crop. p.719-731. *In* P.C. Robert *et al.* (ed.), Precision Agriculture- Proc. of the 4th Intl. Conf., St. Paul, MN, 19-22 July 1998. ASA, CSSA, and SSSA, Madison, WI.

Leeds-Harrison, P.B., J.N. Quinton, M.J. Walker, C.L. Sanders, and T. Harrod. 1999. Grassed buffer strips for the control of nitrate leaching to surface waters in headwater catchments. Ecol. Eng. 12:299-313.

Lefroy, R.D.B. and Y. Konboon. 1998. Studying nutrient flows to assess sustainability and identify areas of nutrient depletion and imbalance: An example for rainfed rice systems in Northeast Thailand. p. 77-93. *In* J.K. Landa *et al.* (ed) Rainfed lowland rice: Advances in nutrient management research. IRRI, Los Banos, Philippines.

Lilienthal, H., E. Haveresch, and D. Garbriel. 2000. Effective cost reduction of nitrogen measures based on the evaluation of hyperspectral data. p. 430-435. *In* Proc. of the 2nd

Intl. Conf. on Geospatial Information in Agriculture and Forestry. Vol.2., Lake Buena Vista, FL, 10-12 January 2000. ERIM International, Inc., Ann Arbor, MI.

Linquist, B., P. Sengxua, A. Whitbread, J. Schiller, and P. Lathvilayvong. 1998. Evaluating nutrient deficiencies and management strategies for lowland rice in Laos PDR. p. 59-73. *In* J.K. Landa *et al.* (ed) Rainfed lowland rice: Advances in nutrient management research. IRRI, Los Banos, Philippines.

Lund, E.D., C.D. Christy, and P.E. Drummond. 1999. Practical applications of soil electrical conductivity mapping. p. 771-779. *In* J.V. Stafford (ed.) Precision Agriculture '99- Proc. of the 2nd Eur. Conf. on Precision Agriculture. Odense Congress Centre, Denmark, 11-15 July 1999. SCI, Sheffield, UK.

Lutticken, R.E. 1999. Implementation of precision fertilizing concepts on practical farms in Western Germany. p. 859-867. *In* P.C. Robert *et al.* (ed.), Precision Agriculture- Proc. of the 4th Intl. Conf., St. Paul, MN, 19-22 July 1998. ASA, CSSA, and SSSA, Madison, WI.

MAFF (UK Ministry of Agriculture, Fisheries and Food).1994. Fertiliser recommendations for agricultural and horticultural crops, RB209, 6th edition. HMSO, London.

MAFF (UK Ministry of Agriculture, Fisheries and Food).2000. Fertiliser recommendations for agricultural and horticultural crops, RB209, 7th edition. HMSO, London.

Magdoff, F. 1991. Understanding the Magdoff pre-sidedress nitrate test for corn. J. Prod. Agric. 4:297-305.

Magdoff, F.R., D. Ross, and J. Amadon. 1984. A soil test for nitrogen availability to corn. Soil Sci. Am. J. 48:1301-1304.

Mausbach, M.J., D.J. Lytle, and L.D. Spivey. 1993. Application of soil survey information to soil specific farming. p. 57-68. *In* P.C. Robert *et al.* (ed.) Soil specific crop management, Proc. of the 1st Intl. Conf. Minn., MN, 14-16 Apr. 1992. ASA, CSSA, and SSSA, Madison, WI.

McCarty, 1999. Modes of action of nitrification inhibitors. Biol. Fert. Soils 29:1-9.

McConkey, B.G., D.J. Ullrich, and F.B. Dyck. 1997. Slope position and subsoiling effects on soil water and spring wheat yield. Can. J. of Soil Sci. 77:83-90.

McGee, E.A., G.A. Peterson, and D.G. Westfall. 1997. Water storage efficiency in no-till dryland cropping systems. J. Soil and Water Cons. 52:131-136.

McMurtrey, J., P. Doraiswamy, J. Schepers, P. Zara, A. Stern, and L. Corp. 2000. Evaluation of hyperspectral algorithms and methods to detect changes in the primary plant pigments for nitrogen management in field corn. p. 395-401. *In* Proc. of the 2nd Intl. Conf. on Geospatial Information in Agriculture and Forestry. Vol.2., Lake Buena Vista, FL, 10-12 January 2000. ERIM International, Inc., Ann Arbor, MI.

Meisinger, J.J. 1984. Evaluating plant-available nitrogen in soil-crop systems. p. 391-416. *In* R.D. Hauck (ed.) Nitrogen in crop production. ASA, CSSA, and SSSA, Madison, Wi.

Meisinger, J.J., V.A. Bandel, J.S. Angle, B.E. O'Keefe, and C.M. Reynolds. 1992a. Presidedress soil nitrate test evaluation in Maryland. Soil Sci. Soc. Am. J. 56:1527-1532.

Meisinger, J.J. and G.W. Randall. 1991. Estimating nitrogen budgets for soil-crop systems. p 85-124. *In* R.F Follett *et al.* (ed.) Managing nitrogen for groundwater quality and farm profitability. SSSA. Madison, WI.

Miller, R.O., S. Pettygrove, R.F. Denison, L.F. Jackson, M. Cahn, R. Plant, and T. Kearney. 1999. Site-specific relationships between flag leaf nitrogen, SPAD meter values and grain protection in irrigated wheat. p. 113-122. *In* P.C. Robert *et al.* (ed.), Precision Agriculture- Proc. of the 4th Intl. Conf., St. Paul, MN, 19-22 July 1998. ASA, CSSA, and SSSA, Madison, WI.

Moore, G. A. and J.P. Tyndale-Briscoe. 1999. Estimation of the importance of spatially variable nitrogen application and soil moisture holding capacity to wheat production. Precision Agriculture 1: 27-38.

Moran, M.S. 2000. Image-based remote sensing for agricultural management- Perspectives of image providers, research scientists and users. p. 23-30. *In* Proc. of the 2nd Intl. Conf. on Geospatial Information in Agriculture and Forestry. Vol.1., Lake Buena Vista, FL, 10-12 January 2000. ERIM International, Inc., Ann Arbor, MI.

Moran, M.S., Y.Inoue, and E.M. Barnes. 1997. Opportunities and limitations for image-based remote sensing in precision crop management. Remote Sens. Environ. 61:319-346.

Moulin, A.P., H.J. Beckie, and D.J. Pennock. 1999. Strategies for variable rate nitrogen fertilization in hummocky terrain. p. 839-846. *In* P.C. Robert *et al.* (ed.), Precision Agriculture- Proc. of the 4th Intl. Conf., St. Paul, MN, 19-22 July 1998. ASA, CSSA, and SSSA, Madison, WI.

Myeni, R.B., S. Maggion, J. Iaquinta, *et al.* 1995. Optical remote sensing of vegetation: Modeling caveats and algorithms. Remote Sens. Environ. 51:169-188.

Nolan, S.C., T.W. Goddard, D.C. Penney and F.M. Green. 1999. Yield response to nitrogen within landscape classes. p.479-485. *In* P.C. Robert *et al.* (ed.), Precision Agriculture- Proc. of the 4th Intl. Conf., St. Paul, MN, 19-22 July 1998. ASA, CSSA, and SSSA, Madison, WI.

Ortega, R.A., D.G. Westfall, W.J. Gangloff, and G.A. Peterson. 1999. Multivariate approach to N and P recommendations in variable rate fertilizer applications. p. 387-396. *In* J.V. Stafford (ed.) Precision Agriculture '99- Proc. of the 2nd Eur. Conf. on Precision Agriculture. Odense Congress Centre, Denmark, 11-15 July 1999. SCI, Sheffield, UK.

Paz, J.O., W.D. Batchelor, T.S. Colvin, S.D. Logsdon, T.C. Kaspar, D.L. Karlen, B.A. Babcock, and G.R. Pautsch. 1999. p. 1279-1289. Model-based technology to determine variable rate nitrogen for corn. *In* P.C. Robert *et al.* (ed.), Precision Agriculture- Proc. of the 4th Intl. Conf., St. Paul, MN, 19-22 July 1998. ASA, CSSA, and SSSA, Madison, WI.

Peikielek, W.P., R.H. Fox, J.D. Toth, K.E. Macneal. 1995. Use of a chlorophyll meter at the early dent stage of corn to evaluate nitrogen sufficiency. Agron. J. 87:403-408.

Peltonen, J. 1992. Plant tissue tests for predicting nitrogen fertilizer requirements in spring wheat. Agricultural Science in Finland 1: 509-517.

Pennock, D.J., F.L. Walley, M.P. Solohub, and G. Hnatowich. 1999. Yield response of wheat and canola to a topographically based variable rate fertilization program in Saskatchewan. p. 797-805. *In* P.C. Robert *et al.* (ed.), Precision Agriculture- Proc. of the 4th Intl. Conf., St. Paul, MN, 19-22 July 1998. ASA, CSSA, and SSSA, Madison, WI.

Peters, M.W., I.T. James, R. Earl, and R.J. Godwin. 1999. Nitrogen management strategies for precision farming. p. 719-728. *In* J.V. Stafford (ed.) Precision Agriculture '99-

Proc. of the 2nd Eur. Conf. on Precision Agriculture. Odense Congress Centre, Denmark, 11-15 July 1999. SCI, Sheffield, UK.

Peterson, G.A., and R.D. Voss. 1984. Management of nitrogen in the west north central states. p. 721-723. *In* R.D. Hauck (ed.) Nitrogen in crop production. ASA, CSSA, and SSSA, Madison, WI.

Pierce, F.J., N.W. Anderson, T.S. Colvin, J.K. Schueller, D.S. Humburg, and N.B. McLauglhlin. 1997. Yield mapping. p 211-243. *In* F.J. Pierce and E.J. Sadler (ed.) The state of site-specific management for agriculture. ASA Misc. Publ., ASA, CSSA, and SSSA, Madison, WI.

Pierce, F.J., W.E. Larson, R.H. Dowdy, and W.A.P. Graham. 1983. Productivity of soils: Assessing long-term changes due to erosion. J. Soil and Water Cons. 38:39-44.

Pierce F.J. and P. Nowak. 1999. Aspects of precision agriculture. Advances In Agron. 67:1-85.

Pierce, F.J., and C.W. Rice. 1988. Crop rotation and its impact on efficiency of water and nitrogen use. p. 21-42. *In* W.L. Hargrove *et al.* (ed.) Cropping strategies for efficient use of water and nitrogen. ASA Special Publ. 51. ASA, CSSA, and SSSA, Madison, WI.

Poovarodom, S., R.L. Tate, and R.A. Bloom. 1998. Nitrogen mineralization rates of the acidic, xeric soils of the New Jersey pinelands: Field rates. Soil Sci. 145:257-263.

Prasad, R. and J.F. Power, 1995. Nitrification inhibitors for agriculture, health, and the environment. Adv. Agron. 54:233-281.

Rahn, C.R., C.D. Paterson, and L.V.V. Vaidyanathan, 1993. Improving the use of nitrogen in brassica rotations. Acta Horticulturae 339: 207-218.

Rauscholb, R.S. and A.G. Hornsby. 1994. Nitrogen management in irrigated agriculture. Oxford University Press, New York, New York.

Rengel, Z. 1998. Nutrient use in crop production. Food Products Press, New York. (Also published as J. Crop Prod. 1, No 2, 1998)

Rice, C.W. and J.L. Havlin. 1994. Integrating mineralizable nitrogen indices into fertilizer nitrogen recommendations. p. 1-13. *In* J.L. Havlin *et al.* (ed.) Soil testing: Prospects for improving nutrient recommendations. SSSA Spec. Publ. 40. SSSA, Madison, WI.

Roloff, G., K.B. MacDonald, and A.R. Couturier. 1999. Potential effects of nitrogen-fertilization scenarios in small watersheds in Southern Ontario. p. 1395-1407. *In* P.C. Robert *et al.* (ed.), Precision Agriculture- Proc. of the 4th Intl. Conf., St. Paul, MN, 19-22 July 1998. ASA, CSSA, and SSSA, Madison, WI.

Romstad, E., Simonsen, J., and Vatn, A. 1997. Controlling mineral emissions in European agriculture. Economics, policies and the environment. CAB International, Wallingford, UK.

Runge, E.C.A. and F.M. Hons. 1999. Precision Agriculture- Development of hierarchy of variables influencing crop yields. p. 143-158. *In* P.C. Robert *et al.* (ed.), Precision Agriculture- Proc. of the 4th Intl. Conf., St. Paul, MN, 19-22 July 1998. ASA, CSSA, and SSSA, Madison, WI.

Sawyer, J.E. 1994. Concepts of variable rate technology with considerations for fertilizer application. J of Prod. Agric. 7:195-201.

Scharf, P. C. and M.M. Alley. 1993. Spring nitrogen on winter wheat. II. A flexible multicomponent rate recommendation system. Agron. J. 85:1186_1192.

Schepers, J.S., T.M. Blackmer, W.W. Wilhelm, and M. Resende. 1996. Transmittance and reflectance measurements of corn leaves from plants with different nitrogen and water supply. J. Plant Physiol. 148:523-529.

Schepers, J.S., D.D. Francis, M.F. Vigil, and F.E. Below. 1992a. Comparisons of corn leaf nitrogen concentration and chlorophyll meter readings. Commun Soil Sci. Plant Anal. 23:2173-2187.

Schepers, J.S. and J.J. Meisinger. 1994. Field indicators of nitrogen mineralization. p. 31-46. *In* J.L. Havlin *et al.* (ed.) Soil testing: Prospects for improving nutrient recommendations. SSSA Spec. Publ. 40. SSSA, Madison, WI.

Schepers, J.S., M.G. Moravek, R. Bishop, and S. Johnson. 1997. Impact of nitrogen and water management on ground water quality. p. 267-278. *In* A. Ahmed (ed.) Groundwater Protection Alternatives in the U.S.A. Amer. Assoc. Civil Eng.

Schepers, J.S. and A.R. Mosier. 1991. Accounting for nitrogen in nonequilibrium soil-crop systems. p. 125-138. *In* R.F. Follett et al. (eds.) Managing nitrogen for groundwater quality and farm profitability. SSA, Madison, WI.

Schepers J.S., G.E. Varvel, and D.G. Watts. 1995. Nitrogen and water management strategies to reduce nitrate leaching under irrigated maize. J. of Contam. Hydrol. 20:227-239.

Schlegel, A.J., D.W. Nelson, and L.E. Sommers. 1986. Field evaluation of urease inhibitors for corn production. Agron. J. 78:1007-1012.

Schröder, J.J., J.J. Neeteson, O. Oenema, and P.C. Struik. 2000. Does the crop or the soil indicate how to save nitrogen in maize production? Reviewing the state of the art. Field Crops Res. 66:151-164.

Scrivner, C.L., B. L. Conkling, and P. G. Koenig. 1985a. Soil productivity indices and soil properties for farm_field sites in Missouri. Missouri Agric. Exp. Stat. Pub. EC0947.

Scrivner, C.L., B. L. Conkling, and P. G. Koenig. 1985b. The effects of soil erosion upon soil productivity in Missouri farm fields. Missouri Agric. Exp. Stat. Pub. EC0950.

Selles, Fernando, C.A. Campbell, B.G. McConkey, D. Messer, and S.A. Brandt. 1999. Spatial distribution of soil nitrogen supplying power: A tool for precision farming. p. 407-415. *In* P.C. Robert *et al.* (ed.), Precision Agriculture- Proc. of the 4th Intl. Conf., St. Paul, MN, 19-22 July 1998. ASA, CSSA, and SSSA, Madison, WI.

Shanahan, J.F., J.S. Schepers, D.D. Francis, G.E. Varvel, W.W. Willhlem, J.M. Tringe, M.S. Schlemmer, and D.J. Major. 2000. Use of remote sensing imagery to estimate corn yield. Agron. J. 93(3).

Shrestha, R.K. and J.K. Ladha. 1998. Nitrogen dynamics in intensive rainfed lowland rice-based cropping systems and strategies to improve efficiency. p. 95-108. *In* J.K. Landa *et al.* (ed) Rainfed lowland rice: Advances in nutrient management research. IRRI, Los Banos, Philippines.

Solohub, M.P., C. Van Kessel, and D.J. Pennock. 1996. The feasibility of variable rate N fertilization in Saskatchewan. p. 65-73. *In* P.C. Robert *et al.* (ed.), Precision Agriculture- Proc. of the 3rd Intl. Conf., Minneapolis, MN, 23-26 June 1996. ASA, CSSA, and SSSA, Madison, WI.

Stafford, J.V., B. Ambler, R.M. Lark, and J. Catt. 1996. Mapping and interpreting the yield variation in cereal crops. Computers and electronics in agriculture. 14:101-119.

Stafford, J.V., R.M. Lark, and H.C. Bolam. 1999. Using yield maps to regionalize fields into potential management units. p. 225-237. *In* P.C. Robert *et al.* (ed.), Precision Agriculture- Proc. of the 4th Intl. Conf., St. Paul, MN, 19-22 July 1998. ASA, CSSA, and SSSA, Madison, WI.

Stanford, G. 1982. Assessment of soil nitrogen availability. p. 651-688. *In* F.J. Stevenson (ed.) Nitrogen in agricultural soils. Agron. Monogr. 23. ASA, CSSA, and SSSA, Madison, WI.

Stanford, G., and E. Epstein. 1974. Nitrogen mineralization water relations in soils. Soil Sci. Soc. Am. Proc. 38:103-106.

Stanford G. and J.O. Legg. 1984. Nitrogen and yield potential. p. 263-272. *In* R.D. Hauck (ed.) Nitrogen in crop production. ASA, CSSA, and SSSA, Madison, WI.

Stanford, G., and S.J. Smith. 1972. Nitrogen mineralization potential in soils. Soil Sci. Soc. Am. Proc. 36:465-472.

Stone, M.L., J.B. Solie, W.R. Raun, R.W. Whitney, S.L. Taylor, and J.D. Ringer. 1996. Use of spectral radiance for correcting in-season fertilizer nitrogen deficiencies in winter wheat. Trans. ASAE. 39(5):1623-1631.

Sudduth, K.A., S.T. Drummond, S.J. Birrell, and N.R. Kitchen. 1997. Spatial modeling of crop yield using soil and topographic data. Vol. 1, p. 439-447. *In* J.V. Stafford (ed.) Precision agriculture '97- Proc. of the 1st Eur. Conf. on Precision Agriculture. Warwick Univ. Conf. Centre, UK, 7-10 Sept. 1997. SCI, London, UK.

Sunderman, H.D., J.S. Pontius, and J.R. Lawless. 1997. Variability in leaf chlorophyll concentration among fully fertilized corn hybrids. Commun. Soil Sci. Plant Anal. 28: 1793-1803.

Sylvester-Bradley, R., Lord, E., Sparkes, D.L., Scott, R.K., Wiltshire, J.J.J., and Orson, J. 1999. An analysis of the potential of precision farming in Northern Europe. Soil Use Manage. 15:1-8.

Thompson, A.L., C.J. Gantzer, and S.H. Anderson. 1991. Topsoil depth, fertility, water management, and weather influences on yield. Soil Sci. Soc. Am. J. 55:1085-1091.

Timlin, D.J., Ya. Pachepsky, V.A. Snyder, and R.B. Bryant. 1998. Spatial and temporal variability of corn grain yield on a hillslope. Soil Sci. Soc. Am. J. 62:764-773.

Turner, F.T. and M.F. Jund. 1991. Chlorophyll meter to predict nitrogen topdress requirement for semidwarf rice. Agron. J. 83:926-928.

Ulrich A., J.T. Moraghan, and E.D. Whitney. 1993. Sugar beet. p. 91-98. *In* W.F. Bennett (ed.), Nutrient deficiencies and toxicities in crop plants. APS Press, St. Paul, MN.

Varvel, G.E., J.S. Schepers, and D.D. Francis. 1997a. Ability of in-season correction of nitrogen deficiency in corn using chlorophyll meters. Soil Sci. Soc. Am. J. 61:1233-1239.

Vitousek, P.M. 1994. Beyond global warming: Ecology and global change. Ecology 75:1861-1876.

Walters, D.T. and J.E. Goesch. 1999. Temporal and spatial variation in soil nitrate acquisition by maize as influenced by nitrate depth distribution. p. 43-54. *In* P.C. Robert *et al.* (ed.), Precision Agriculture- Proc. of the 4th Intl. Conf., St. Paul, MN, 19-22 July 1998. ASA, CSSA, and SSSA, Madison, WI.

Watkins, K.B., Y.C. Lu, and W.Y. Huang. 1999. Economic returns and environmental impacts of variable rate nitrogen fertilizer and water applicants. p. 1667-1679. *In* P.C.

Robert *et al.* (ed.), Precision Agriculture- Proc. of the 4th Intl. Conf., St. Paul, MN, 19-22 July 1998. ASA, CSSA, and SSSA, Madison, WI.

Watson, C. A. and D. Atkinson. 1999. Using nitrogen budgets to indicate nitrogen use efficiency and losses from whole farm systems: A comparison of three methodological approaches. Nutr. Cycling in Agroecosystems 53:259-267.

Weibel, R. and M. Heller. 1991. Digital terrain modeling. p. 269-297. *In* M.F. Goodchild and D.Rhines (eds.) Geographic Information Systems, Principles and Applications. Taylor and Francis, New York.

Wells, B.R., P.K. Bollich, W. Ebelhar, D.S. Mikkelsen, R.J. Norman, D.M. Brandon, R.S. Helms, F.T. Turner, and M.P. Westcott. 1989. Dicyandiamide (DCD) as a nitrification inhibitor for rice culture in the United States. Comm. Soil Sci. Plant Anal. 20:2023-2047.

Welsh, J.P., G. A. Wood, R.J. Godwin, J.C. Taylor, R.Earl, B.S. Blackmore, G. Spoor, G. Thomas, and M. Carver. 1999. Developing strategies for spatially variable nitrogen application. p. 729-738. *In* J.V. Stafford (ed.) Precision Agriculture '99- Proc. of the 2nd Eur. Conf. on Precision Agriculture. Odense Congress Centre, Denmark, 11-15 July 1999. SCI, Sheffield, UK.

Westerman, R.L. 1990. Soil testing and plant analysis. 3rd ed. SSSA Book Ser. 3. SSSA, Madison, WI.

Westermann, D.T. and G.E. Kleinkopf. 1985. Nitrogen requirements of potatoes. Agron. J. 77:616_621.

Westfall, D.G. 1984. Management of nitrogen in the mountain states. p. 751-763. *In* R.D. Hauck (ed.) Nitrogen in crop production. ASA, CSSA, and SSSA, Madison, WI.

White, P., T. Oberthur, A. Dobermann, and R. Chhay. 1998. Simple soil classifications are important tools for improving fertility management of infertile rice soils of the rainfed lowlands. p. 43-58. *In* J.K. Landa *et al.* (ed) Rainfed lowland rice: Advances in nutrient management research. IRRI, Los Banos, Philippines.

Wibawa, W.D., D.L. Dludlu, L.J. Swenson, D.G. Hopkins, and W.C. Dahnke. 1993. Variable fertilizer application based on yield goal and soil map unit. J. Prod. Agric. 6:255-261.

Williams, C. M. J. and N.A. Maier. 1990a. Determination of the nitrogen status of irrigated potato crops. 1. Critical nutrient ranges for nitrate_nitrogen in petioles. J. Plant Nutr. 13: 971_84.

Williams, C. M. J. and N.A. Maier. 1990b. Determination of the nitrogen status of irrigated potato crops. 2. A simple on farm quick test for nitrate_nitrogen in petiole sap. J. Plant Nutr. 13: 985_93.

Willis, P.R., P.G. Carter, and C.J. Johannsen. 1999. Assessing yield parameters by remote sensing techniques. p. 1465-1473. *In* P.C. Robert *et al.* (ed.), Precision Agriculture-Proc. of the 4th Intl. Conf., St. Paul, MN, 19-22 July 1998. ASA, CSSA, and SSSA, Madison, WI.

Wood, C.W., D.W. Reeves, R.R. Duffield, and K.L. Edmisten. 1992. Field chlorophyll measurements for evaluation of corn nitrogen status. J. Plant Nutr. 15:487-500.

Wright, D.K., W.C. Bausch, D.G. Westfall, J.A. Delgado, and J.K. Berry. 2000. Using a multispectral video mapping system for remote sensing of crop nitrogen stress. p.181-182. *In* Proc. of the 2nd Intl. Conf. on Geospatial Information in Agriculture and

Forestry. Vol.2., Lake Buena Vista, FL, 10-12 January 2000. ERIM International, Inc., Ann Arbor, MI.

Yang, C., G.L. Anderson, J.H. King, Jr., and E.K. Chandler. 1999. Comparison of uniform and variable rate fertilization strategies using grid soil sampling, variable rate technology, and yield monitoring. p. 675-686. *In* P.C. Robert *et al.* (ed.), Precision Agriculture- Proc. of the 4th Intl. Conf., St. Paul, MN, 19-22 July 1998. ASA, CSSA, and SSSA, Madison, WI.

Nitrogen in the Environment: Sources, Problems, and Management
R.F. Follett and J.L. Hatfield (Eds)

Chapter 14. Developing Software for Livestock Manure and Nutrient Management in the USA

Philip J. Hess[a], Brad C. Joern[a], and John A. Lory[b]

[a]Agronomy Dept., Purdue University, 1150 Lilly Hall of Life Sciences, West Lafayette, IN, 47907, United States. [b]Agronomy Dept., University of Missouri, 210 Waters Hall, Columbia, MO, 65211, United States.

More and more crop and livestock producers will be required to develop nutrient management plans to demonstrate to regulatory agencies that their operations have sufficient crop acreage, seasonal land availability, manure storage capacity, and application equipment to manage commercial fertilizers, animal manures and other land-applied nutrient resources in an environmentally responsible manner. Computer software has been and will continue to be used to help develop these plans. Previous software addressed some pieces of the management plan puzzle, but were generally limited in scope and failed to address fully the temporal and spatial nature of nutrient management or provide ways to accommodate multiple states and changing regulatory reporting requirements. New nutrient management planning software needs to take these issues into account, utilize national databases and standards, and take advantage of modern software technologies. This chapter discusses the limitations of previous planning software, the opportunities this presents for new software, the data requirements of planning software, challenges and barriers to creating this software, and a glimpse of what this new software might look like.

1. INTRODUCTION

1.1 U.S. livestock and poultry production

Livestock and poultry in the U.S. excrete nearly 1.3 billion tons of fresh manure annually (Table 1). If this manure were placed in railcars, the resulting train would circle the earth more than three and a half times! The phosphate (P_2O_5) and potash (K_2O) excreted by these animals nearly equals annual U.S. P_2O_5 and K_2O commercial fertilizer consumption and the nitrogen (N) excreted exceeds 60% of U.S. commercial fertilizer N consumption. While more than half of the manure produced by cattle in the U.S. is excreted directly onto pasture and range lands, most of the manure generated by pigs and poultry is collected and can be managed as a crop nutrient resource. The "collectable" portion of livestock and poultry manure still constitutes more than 50% of U.S. commercial P_2O_5 and K_2O fertilizer consumption (Table 1).

While the total N excreted in the collectable fraction of livestock and poultry manure is nearly 35% of U.S. commercial fertilizer N consumption, 30-85% of this N may be lost to the atmosphere during manure storage and application depending on the manure management system used by an operation. Additional N losses following application may also occur due to application timing, method and weather. Therefore, the potential N value of collectable manure is likely less than 20% of U.S. commercial fertilizer N consumption.

Because of these N losses, when animal manure is applied to meet crop N requirements, P_2O_5 and K_2O application rates can be 2.5-5 times crop uptake. Consequently, fields that frequently receive livestock manure often have soil test P and K levels that greatly exceed those needed for optimum crop production. In addition, the number of livestock and poultry farms in the U.S. has decreased by approximately two-thirds in the last 30 years, while the number of animals in inventory has increased (USDA/NASS, 1999). This increase in animal density on livestock and poultry farms can make it more difficult to find sufficient land near the operation to utilize manure P and K effectively.

Table 1. Estimated quantities of manure and nutrients produced annually in the United States compared to commercial fertilizer consumption.

	Animals†	Manure‡	N‡	P_2O_5‡	K_2O‡
	millions	millions of tons§	---- thousands of tons§ ----		
Beef cattle	65	773	4,284	2,057	3,049
Milk cattle	18	364	1,825	837	1,212
Hogs	61	94	777	496	407
Broilers	1,103	36	463	282	221
Turkeys	104	13	180	154	77
Layers	419	17	234	181	107
All animals	**1,770**	**1,297**	**7,763**	**4,007**	**5,073**
Collectable manure¶		**651**	**4,143**	**2,371**	**2,736**
Fertilizer consumption#			**11,897**	**4,424**	**5,178**

† Number of animals in inventory from 1997 Census of Agriculture (USDA/NASS, 1999).
‡ Excretion values adapted from Midwest Plan Service (MWPS, 2000).
§ To convert to metric tonnes, multiply values by 0.907.
¶ Assumes 100% of manure from beef cows and dairy replacement animals (heifers and heifer calves) and 15% of manure from pigs, poultry and other cattle is *not* collected.
# Average 1990-1999 U.S. commercial fertilizer consumption (AAPFCO-TFI, 1999).

1.2. Policy drivers

According to the U.S. Environmental Protection Agency (USEPA) 1998 national water quality inventory, nearly 40% of U.S. waters surveyed are too polluted for fishing or swimming. USEPA estimates that agricultural sources are responsible for 60% of this pollution in rivers and streams and 45% of this pollution in lakes (USEPA, 2000a). In response to water quality impairments attributed to agricultural sources, the U.S. Department of Agriculture (USDA) and USEPA published the Unified National Strategy for Animal Feeding Operations (USDA and USEPA, 1999). This document emphasizes the need for livestock and poultry operations to develop comprehensive nutrient management plans

(CNMPs) to minimize the impact of these operations on water quality and public health. The USDA Natural Resources Conservation Service (NRCS) has published guidelines that define the specific components of CNMPs (USDA/NRCS, 2000). If all livestock and poultry operations are eventually required to develop CNMPs, over 300,000 plans will be needed. The USEPA is developing specific regulations to address water quality issues for large concentrated animal feeding operations via a permit nutrient management plan (PNP) (USEPA, 2000b). PNPs may be required for approximately 30,000 animal feeding operations. In addition, nutrient criteria are being developed by USEPA for surface waters based on total N, total P, turbidity, and chlorophyll *a* in the water column (USEPA, 2000c). Total Maximum Daily Loads (TMDLs) in nutrient impaired water bodies may be based on these nutrient criteria, which could significantly increase the total number of nutrient management plans needed. To satisfy both current and future regulatory reporting requirements, nutrient management planning software that supports sophisticated add-in reports is needed since both federal and state-specific reports may be required for the same operation. This will require software that is national in scope.

1.3. What *is* a livestock nutrient management plan?

Nutrient management planning helps farmers develop strategies to use manure and other fertilizers as nutrient sources for crop production in a manner that protects environmental quality. The planning process is documented in a written nutrient management plan for possible review by regulatory personnel, consultants or agency personnel and for use by the farmer as a working plan. The plan includes information about the operation's animal types and numbers, the quantity of manure produced, the manure nutrients available for land application, the storage facilities where animal manure is collected and stored, the crop rotations and fertilizer needs of crops where manure will be applied, the methods used for applying manure, and the timing of planned manure and other fertilizer applications. The plan often accounts for manure produced over a multiple-year period. Local, state, and national regulations can determine the specific information required in a plan.

At its simplest, a plan includes an estimate of the total quantity of manure and nutrients produced by the operation during the plan period and shows how those nutrients will be distributed (when, where and how much) on the available crop land. A more comprehensive plan takes into account temporal, spatial and environmental limitations to show that the operation has sufficient crop acreage, seasonal land availability, manure storage capacity, and application equipment to manage manure and other nutrient resources in a responsible and sustainable manner.

1.4. Strategic and tactical nutrient management planning

Strategic nutrient management planning uses long-term data and industry standards to help the farmer create a blueprint for handling manure and other nutrients. These plans evaluate the sustainability of an operation's approach to nutrient management over multiple years. Strategic plans determine if sufficient land is available for manure application, how often and how much manure typically needs to be land-applied, and the typical fertilizer value of manure nutrients. These plans are used by farmers as long-term planning tools and by regulatory agencies to insure that proper measures are in place to collect and handle manure. A strategic plan identifies the fields that will likely receive manure and the typical rates to be applied to each field.

Tactical planning uses current data to solve immediate nutrient management challenges. For example, tactical planning helps a farmer evaluate the impact of above average rainfall on manure storage capacity and identify the best field to apply manure under these wet conditions.

In some cases, the same data can be used both in strategic and tactical planning, but the data will be collected or determined in different ways. Table 2 highlights some of the differences in data used for strategic versus tactical planning.

Table 2. Differences in data used for strategic and tactical planning.

Strategic planning	Tactical planning
30-year normals for weather	Actual on-farm weather data or data from nearest weather station
Most recent soil tests used for entire plan period	Soil nitrate test levels
Yield goals and typical cropping sequences	Actual yields and previous crops
Book value-based estimates or most recent manure analyses used for entire plan period	Actual manure analyses and measured volumes
Month or season of manure applications	Actual dates of manure applications
Planned application methods	Actual application methods

2. THE NEED FOR COMPUTER SOFTWARE

Most of the arithmetic involved in creating a plan is not difficult and can be performed on a pocket calculator. However, the number of calculations required and the repetitive nature of the calculations when many fields and manure applications are involved can be daunting. Furthermore, the amount of reference material and information about the operation that is needed at hand before the calculations can be done is considerable. Added to this is the need to run various what-if scenarios for making a non-sustainable operation sustainable or keeping an operation sustainable if it expands. Computer software has the potential to reduce the work of generating a plan, permit review of different nutrient management options, and generate reports in a standard format. Use of approved computer software can be a form of quality control, ensuring that the plan incorporates approved recommendations and standards.

2.1. Opportunities for new nutrient management software

A survey of software programs that calculate dairy manure application rates revealed that there were at least a dozen such programs available at the end of 1995 (Thompson et al., 1997). Many of these programs could supply several of the basic components of a livestock manure management plan. Most were available from the authors for a nominal fee and few had unusual hardware or software requirements. However, from the perspective of late 2000, these programs already sound somewhat antiquated. In general, they can be characterized as pre-Windows®, pre-Internet, pre-geographic information system (GIS) programs.

Based on the survey's descriptions, these programs suffered from many of the same limitations. The following are areas where new software can bring novel ideas and approaches to bear on the complex challenge of nutrient management planning.

- **Take advantage of Windows- and Web-based interfaces.** Previous programs used the limited text-only interface supported by operating systems such as Microsoft® MS-DOS® that were in use at the time the programs were designed. Today's Windows- and Web-based software support the more powerful and intuitive graphical interface expected by modern computer users.
- **Fully address temporal issues.** Earlier software focused on annual totals for manure production and manure applied to each field. To meet water quality goals, nutrient management planning must address the timing as well as the amount of manure and other fertilizers applied to fields. Effective planning must also help the farmer anticipate times of the year when there is a danger of exceeding manure storage capacity.
- **Enhanced management of spatial data.** Previous software at most noted the distance fields were from manure storage. In practice, nutrient management planning requires evaluating the proximity of manure management practices to many geographical features, including lakes, streams, wells, neighbors and steeply sloping land. Advances in the availability of GISs provide an opportunity to integrate more sophisticated map-based features into nutrient management software.
- **Process-oriented.** Most earlier programs appeared to base manure N availability and loss estimates on an annual or seasonal loss factor. Prediction of N availability may be improved by process-oriented models that account for the effects of time, weather, application method, and source characteristics on transformations occurring in the field.
- **Regional and national focus.** Previous software was developed by individuals who were focused primarily on the specific needs and conditions of a particular state. Nutrient management efforts are more similar than they are different from state to state. Supporting multiple states is not only cheaper in the long run, but is consistent with the current trend of developing common approaches for generating land application standards for multiple states. To satisfy regulatory reporting requirements, software will be needed that supports sophisticated reports. Both federal and state-specific reports may be required for the same operation. This will require software that is national in scope.
- **Designed to be extended or enhanced by users and third parties.** Previous programs were largely limited to enhancements supported by MS-DOS and the DOS-based programming tools prevalent at the time the programs were developed. Today's users require ways to add value to software by incorporating ideas and innovations of their own. This gives users a personal stake in what otherwise could be perceived as impersonal national software.

- **Utilize existing standards for identifying and exchanging data.** Data exchange between earlier programs generally had to be worked out by the developers themselves on a case-by-case basis. Today's users may need more than one nutrient management tool. If the tools adhere to standards and support common ways of identifying and exchanging data, the amount of data that must be manually entered can be kept to a minimum.
- **Take advantage of the Internet.** Most software today can be updated by downloading revisions or updates from the Internet. This requires a more sophisticated way of packaging and installing the software. The Internet can also be used to provide weather and geographic data relevant to nutrient management planning.
- **Link to a GIS.** Working with a GIS is now a common way of putting a friendlier, map-oriented interface on agricultural software.
- **Provide both "strategic" and "tactical" approaches.** A strategic approach requires that the focus be on planning for the next several years; a tactical approach must address immediate and short-term planning and recordkeeping needs; and a bimodal approach requires a way to switch cleanly between modes.

3. DATA REQUIREMENTS AND ACQUISITION

Easy access to relevant data is the foundation of successful computerized nutrient management planning. Automated data acquisition is one of the primary potential benefits that new software can have on nutrient management planning. Automated data acquisition can reduce much of the drudgery of manual data input, provide more accurate and timely information, and include more of the information needed for developing detailed plans.

Transfer of data between programs is not assured; substantial effort must be applied to standardize data among programs and develop protocols for communication between them. For example, the apparently simple task of importing soil test results from different soil testing laboratories into planning software is only now being resolved.

3.1. External and operation-specific data

Data required by planning software consist of two types, external and operation-specific. External data typically are derived from state and national databases. These data can be modified, but only infrequently. Operation-specific data are unique to the operation and often vary from year to year. Strategic planning relies heavily on external, non-temporal data, while tactical planning demands operation-specific, temporal data. Strategic planning requires generalization, while tactical planning forces better data collection. The more recent the data, the better the tactical plan will be.

Examples of external data include:

- Location data, including state, county and watershed boundaries.
- Soil survey attributes.
- Long-term weather data.
- Fertilizer recommendations.
- Plant nutrient removal rates.

- Typical manure excretion volume and nutrient content.
- Soil test response to nutrient inputs.
- Nutrient availability calculations.
- Digitized orthophoto quadrangles (DOQs).

Examples of operation-specific data include:

- Field layout, including predominant soil type, total area, spreadable area, setbacks, buffers, and distance from manure storage.
- Soil test levels.
- Current weather data.
- Crop rotation and yield goals.
- Manure storage facilities.
- Animal numbers, types and rations.
- Manure volume and analysis.
- Equipment types and specifications.
- Fertilizer and manure applications.
- Planting, residue, tillage and harvest information.

3.2. Standards for identifying data

Two obstacles prevent data from one program from being used in another program. The first is the way in which the information is encoded and stored. For example, are the data in a computer file or a paper report? If the data are in a computer file, what is the file's format? Is it a spreadsheet file, a text file, or some other type of file? Software that imports data must be able to identify and understand the file's format to read the file. Overcoming data format barriers is largely a technical issue and can generally be solved through the development of conversion tools or program support.

The second obstacle is the problem of identifying and describing data. A program that imports data must be able to identify each piece of incoming data. With agricultural data, there are a number of areas where standards for describing data would be useful. In some cases, standards have recently emerged or are being developed. In other cases, no standards yet exist.

3.2.1. Farm field identification

Unlike street addresses for houses and businesses, farm fields are fairly amorphous, dividing and combining as they're sold, rented, farmed and managed from year to year. Producers may have more than one way of identifying fields. For example, they might have one system based on field ownership. This could correspond to the traditional USDA Farm Service Agency (FSA) farm/tract/field system of field identification. They might have another system based on how the field is managed, in which a field is divided into subfields based on soil type, drainage, proximity to water, or other factors. Each subfield might be managed differently for manure even though the entire field is planted and harvested the same. A field could also be identified by the latitude and longitude of points on its boundary, although this is less a method of identification than a method of locating the field and is most useful within the context of a GIS.

3.2.2. Soil test data identification

With soil test data, the problem of identification is twofold. Not only does the field from which the sample was taken need to be identified, but the particular test and units used by the laboratory where the sample was processed must also be known. With a sample's field identification, the field ID must travel with the sample's data so it appears alongside the test results in the report or data file returned by the laboratory.

Programs that import soil test results must also know, for example, which value is pH and which is P. Furthermore, the program will need information about the kind of data being provided. For example, the method used to measure soil test P (e.g. Bray P-1, Mehlich III, Olsen, etc.) and the units (e.g. parts per million or pounds per acre) must be known to properly interpret the incoming data.

Data standards can be used to facilitate transfer of information between programs. These standards can be program specific, in which a program has certain expectations about incoming data. For example, a program may insist that all P data be based on the Mehlich III method and in units of part per million. Soil test data that were derived using a different extraction or in different units must be converted to the expected standard before they can be imported. Alternatively, universal standards can be developed where the full range of possible data types are defined. In this scenario, both data source and receiving program agree to use the predetermined definitions. The program receiving data can use the standard to identify the type of data provided and convert the incoming data to its own required format.

The Ag Electronics Association (AEA), a not-for-profit trade association founded in 1995, has worked ambitiously to create some universal standards. The AEA has developed two specifications pertinent to soil test data. The first is a data dictionary for defining soil test data (AEA, 1997). The second is a transfer specification for exchanging soil test data (AEA, 1998). These specifications are voluntary guidelines; no one is required to use them. An informal survey of soil testing labs conducted by two of this chapter's authors in early 2000 did not find any labs that were familiar with the AEA's work. However, these specifications go a long way toward setting an important standard for soil test data identification and exchange.

3.2.3. Crop identification

In some states, numeric codes have been assigned to crops commonly grown in those states. These codes are used to specify crops on soil testing laboratory submission forms and in generating extension fertilizer recommendations. While these codes make for a state standard of sort, they do not make for a national one since each state's numbering system is different from the next. Without a universal way of identifying crops, a different scheme for linking to national crop databases will have to be worked out for each state.

3.2.4. Soil identification

NRCS has developed the National Soil Information System (NASIS) as a central repository for storing and working with soil survey data. This database sets an important standard for identifying soils and describing soil attributes, but it does not solve all problems. In some cases, where a survey area has been remapped or renumbered, the NASIS database differs greatly from the published soil survey in use today. Any data references to the published soil identifiers in remapped or renumbered counties will be incompatible with NASIS.

Closely linked to NASIS is the NRCS Soil Survey Geographic Database (SSURGO). SSURGO was designed for use by natural resource planners working with GISs. NASIS has the ability to export a subset of its data as a SSURGO data set. A SSURGO data set includes digitized versions of the original soil survey maps and associated soil attribute data. A soil survey area typically consists of a single county. Once digitized, a county's SSURGO data is submitted to NRCS for certification. By December 2000, SSURGO data for about one third of U.S. counties had been certified.

3.2.5. Watershed identification

During the 1970's the U.S. Geological Survey (USGS) mapped the U.S. at the subbasin drainage level, developing an 8-digit hydrologic unit code (HUC) to identify each subbasin. An 8-digit HUC represents roughly 448,000 acres. In the late 1970's, recognizing the inadequacy of 8-digit hydrologic units for local water resource planning, NRCS began mapping subbasins into 11-digit watersheds (40,000 to 250,000 acres) and then mapping these watersheds into 14-digit subwatersheds (3,000 to 40,000 acres). By the early 1990's NRCS had decided to map and digitize the entire U.S. at the 14-digit subwatershed level. The objective was to develop a national GIS watershed database that matches USGS topographical maps. By December 2000, NRCS had certified two states' 14-digit maps, with work in progress in most other states.

3.2.6. Crop planting and harvest units

In the U.S., units for seed, planting rates, and harvest yields vary from crop to crop and even from state to state. For example, depending on the crop, seed can be priced per pound, bushel, hundredweight, 80,000-seed bag, 100,000-seed bag, or 50-pound bag. Planting rates are in seeds per acre, pounds per acre, or bushels per acre, while yields are reported in pounds, bushels, hundredweight, or tons. The USDA Census of Agriculture uses a standard set of units for crop yields, but in some cases the census units differ from what producers commonly use.

3.3. Data acquisition challenges

While the public databases mentioned above are indeed public, obtaining them is not easy. For example, while any SSURGO county data set can be downloaded from the NRCS SSURGO Web site, obtaining data sets for multiple counties this way is slow and tedious. Developing software that can be used in more than one state will require a special arrangement with the responsible agency to facilitate the acquisition of necessary data. Furthermore, since most of these national databases are works in progress, with sizeable gaps in the data, the software will also need to be able to make do when the data for a county or state are not available.

In some cases, vital state-specific data may not be part of a national database. For example, in Wisconsin, each soil in the state has been assigned a subsoil group (Kelling et al., 1998). This subsoil group is used in generating extension fertilizer recommendations, but is currently not one of the NASIS soil attributes. Attributes like this will need to be merged with NASIS data on a state-by-state basis to be used by planning software.

Acquiring GIS data can be particularly tricky. Not only is there the challenge of locating and obtaining the right data, but working with and distributing very large image files can be difficult. In addition, there are often numerous file and data formats to choose from. Finally, in the case of digitized aerial photographs, the age of the photograph can be important. If the photograph is too old, changes to farm structures may not be present.

4. OTHER CHALLENGES

4.1. Inadequate research to support process-oriented models.

Process-oriented models must be based on and validated by directly measured field data. At the present time, there is a lack of data for some of the basic processes involving manure applications. For example, while the general concepts of many nutrient transformations are well understood, actual field verification of nutrient transformations and losses have not been widely measured within the context of developing or testing transformation models.

4.2. Limited funding.

In the survey of programs cited earlier, all of the programs appear to have been developed by non-profit organizations, in most cases land grant universities. Because programs like these are often closely associated with university research and government agencies, personnel within those organizations traditionally have developed the programs. This is not likely to change. The relatively small market for nutrient planning software means that this kind of software will continue to be developed by public organizations. As a result, these programs will be constrained by funding; their existence and longevity will be directly related to success in grant writing and the length of those grants.

4.3. Finding qualified computer personnel.

In the past, students wrote much of the agricultural software produced by universities. However, students rarely possess much relevant programming experience and usually move on after two or three years. This sometimes makes software development success hard to achieve. Today's users expect software that is easy to use, reliable and frequently updated. As with a lot of professional activities, the writing of production code should probably be left to qualified professionals.

Even though today's programming tools and environments are much more advanced than just a few years ago, learning to use the new tools can be difficult. In the current job market, finding programmers with the required experience and skills may not be easy.

4.4. Satisfying 50 states.

Often one of the first questions a software developer is asked upon presenting a new program to a diverse group is "How can I use your program for my needs?" With agricultural software, if little thought was given to this possibility in the software's design, support for additional states or regions will have to be retrofitted to the program. This can sometimes conflict with the original design or goals of the program and delay its completion. However, designing a program from the beginning to meet the potential needs of multiple states is challenging. Often each state likes to think of itself as unique from all others, even though its farmers produce many of the same crops under generally the same growing conditions as neighboring states. Overcoming this challenge is less a technical problem than a political one.

If planning software must account for state and local regulations, developers will need links with each regulatory agency to stay apprised of changes in regulatory and reporting requirements.

4.5. Technological limitations.

While most areas of the U.S. can now connect to the Internet at relatively low cost, much of rural America is limited to slow modem speeds and, barring some technological breakthrough, will continue to be limited to slow modems for the foreseeable future. This not only restricts the distribution of software and data via the Internet, but more importantly limits the possibilities for on-line data entry, retrieval, and reporting.

5. WHAT SHOULD THIS NEW SOFTWARE LOOK LIKE?

Software that is national in scope, avoids the limitations of previous programs, satisfies nutrient management planning's data requirements, and overcomes the hurdles described above will not be easy or cheap to develop. What follows is a discussion of the minimum set of features that this software should possess.

5.1. Deals with temporal issues visually

Manure can be added to storage facilities on a regular basis or only during certain periods of the year. Manure can also be removed from storage for field application at potentially any time of the year. A useful way of dealing with these numerous manure "transactions" is a calendar that shows the status of all storage facilities and fields at monthly intervals. A calendar is also a good way of showing projected month-to-month fluctuations in manure accumulation.

Closely connected with the use of a calendar to display manure storage and application status is the concept of "live" data handling. With live data handling, each new or modified transaction forces all "downstream" cells of the calendar to be updated automatically. For example, removing manure from a storage facility not only reduces the amount of manure displayed for the facility, but also updates its projected status for the remaining months of the calendar. Applying manure to a field has a similar effect as well, increasing the number of the field's acres that have been manured and reducing the field's nutrient requirements and manure application priority in subsequent years.

5.2. Deals with spatial issues visually

Understanding the spatial relationships between a livestock operation's geographic features is key to accounting for them properly in planning software. An intuitive way of showing these relationships is to depict them on a multi-layered map. This is best done with a GIS that allows various spatially referenced data layers to be overlaid on each other as needed by the user. These layers can consist of digitized aerial photographs, soil survey, watershed and political boundaries, roads, water bodies, user-drawn farm and field boundaries, and setbacks and buffers. One recently developed GIS application, the Spatial Nutrient Management Planner (SNMP), automatically determines field size, setbacks, spreadable acres, distance from manure storage, soil type, watershed and county for each field in an operation and stores this data in a file for use in other programs (Lory et al., 2000).

5.3. Resolves field identification problem

Planning software must be able to support more than one system of field identification. Obviously it has to support the producer's own system of naming fields, while also supporting a way of identifying portions of a larger field for management purposes. One way of handling this is a three-level hierarchy of farm, field, and subfield ID's. In addition, the software needs to identify FSA farm/tract/field ID's for government program purposes, although it doesn't have to use these in its primary system of field identification. Knowing the field's 14-digit watershed and the spatial coordinates of its boundaries would also be useful. A GIS can determine these last two items.

5.4. Maximizes use of publicly available databases

NRCS and other federal agencies have put a tremendous amount of work into constructing a number of national databases. Wherever possible, planning software should defer to standards set by these agencies rather than trying to develop new ones. For the most part these databases are more than adequately documented, although the amount of information present in these databases can be daunting. At the same time, planning software needs to use local data such as state-specific soil attributes as well.

5.5. Generates state-specific fertilizer recommendations for multiple states

Traditionally, most programs that generate crop fertilizer recommendations have embedded both the crop-specific logic and the recommendation data in the program code. Even those programs that stored a portion of the data in an external file embedded the logic of how to interpret the file's structure and data in the program code. While this approach may be adequate for individual states, it poses several problems when trying to support multiple states: (i) changes to recommendations usually require program changes, (ii) nearly everything about each state's recommendations must be known in advance, and (iii) knowledge of recommendation decisions and changes tends to become lost over time if the program is the only source for the recommendations.

A more general approach is to move much of the recommendation implementation out of the program itself and into an external file in the form of expressions that can be evaluated at runtime (Hess, 1998). The recommendation expressions can be written in any suitable scripting language and loaded and evaluated as needed during program execution. Writing the recommendations as external expressions means that a great deal of conditional logic can be moved out of the program into an external form that is easier to access, document, modify, and maintain. This approach also allows outside parties to verify the resulting recommendations. In addition, since the recommendation expressions can be as general as the scripting language they're written in, a well-designed implementation will be able to handle a wide range of future demands.

5.6. Estimates N availability based on N transformations

Using a single factor for estimating N loss ignores temporal, soil, management and climate variables. With only a single factor, manure applied in September will have the same N loss estimate as manure applied in December. A process-oriented approach models the N transformations that take place between the time the manure is applied and the time that the crop needs the manure's N. This is the approach taken in an N transformation model developed for Ohio, Indiana and Michigan (Johnson, 1999). Purdue University's Manure Management Planner (MMP) program implements this model, modifying it to take into account other factors and generalizing it for use in additional states (Joern and Hess, 2000). The steps in MMP's algorithm are as follows.

1. Determine manure's ammonium N. (Based on estimated or measured analysis.)
2. Determine manure's organic N. (Based on estimated or measured analysis. With poultry manure, treat portion of organic N as ammonium N to account for readily mineralizable uric acid.)
3. Determine amount of organic N that can be mineralized in first year. (Based on estimates for various manure types.)
4. Determine amount of ammonium N lost due to volatilization during application. (Based on temperature at application.)
5. Determine amount of ammonium N lost due to volatilization during first month following application. (Based on application method, high temperature following application, application rate, and amount of water in manure.)
6. Calculate amount of ammonium N remaining at end of application month. (Subtract total volatilized N (steps 4 and 5) from step 1's total ammonium N.)
7. Determine amount of organic N converted to ammonium N during application month. (Based on month's temperature. Reduce organic N by this amount and increase ammonium N by this amount.)
8. Determine amount of ammonium N converted to nitrate N (nitrification) during application month. (Based on month's temperature. Reduce ammonium N by this amount and increase nitrate N by this amount.)
9. Determine amount of nitrate N lost due to denitrification during application month. (Based on month's temperature, soil type, and number of days soil is continuously saturated with water. Reduce nitrate N by this amount.)
10. Determine amount of nitrate N lost due to leaching during application month. (Based on month's temperature and time since application. Reduce nitrate N by this amount.)
11. Repeat steps 7-10 for each month until N is needed by crop.
12. Sum the remaining ammonium N and nitrate N. This is the N available to crop.

For strategic planning, estimated manure analyses and long-term temperature and rainfall data must be used since actual data will not yet be known. For tactical planning, actual manure analyses and weather data can be used. The transformation factors used in each step of the algorithm can be adjusted for each state as necessary.

Figure 1 shows the results of running MMP's N transformation model with various hypothetical manure applications. Note the importance of application timing to total N loss.

Figure 1. Percent N remaining after applying 6,000 gallons/acre liquid pig manure (15,000 gallons/acre lagoon effluent for irrigation) at various months of application in northern Indiana using MMP's N transformation model (Bailey and Joern, 2000).

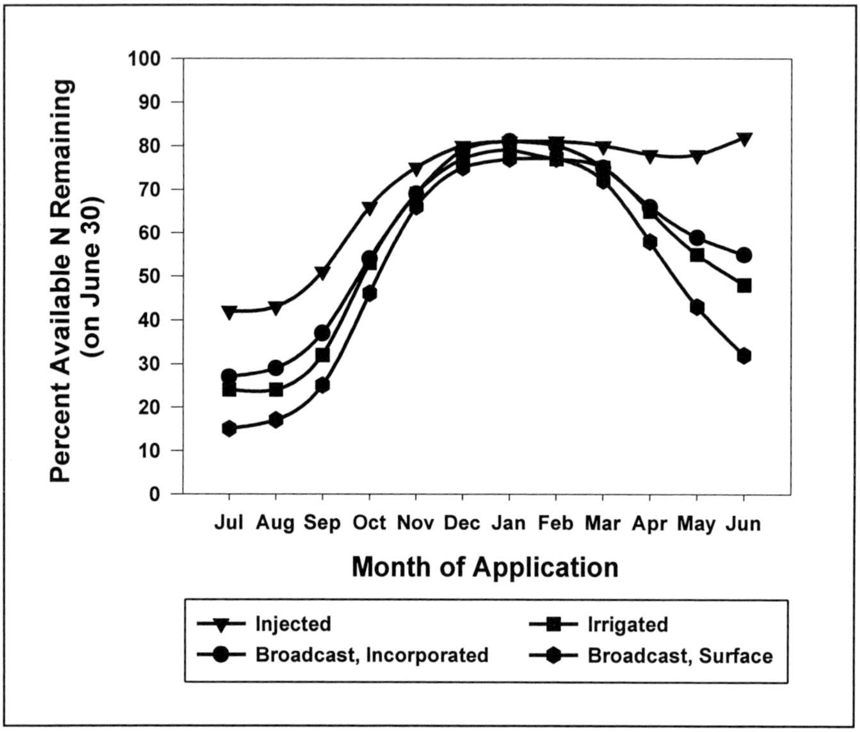

5.7. Open architecture

The term "open" can have several meanings when applied to computer software, but in the context of agricultural software it usually means the ability to import and export data in popular file formats. This definition needs to be expanded to include other types of access, including program-to-program links and facilities for enhancing and customizing the software by users.

5.7.1. Imports and exports data in multiple formats

Traditionally, programs have linked to each other by importing and exporting data via external files. By supporting various standard file formats, data can be extracted from one program's usually proprietary format and used in another program. By working through a neutral format, neither program needs to know very much about the other program. Unfortunately, importing and exporting is not the same as merging data or exchanging data dynamically. Neither do these capabilities address the problem of data identification, only the physical exchange of data. Efforts like those of the AEA can help by establishing a standard way of identifying and exchanging soil test data via a Transfer Support Layer (TSL) file created for each data source. However, creating a TSL is not a trivial task and will probably require software developers to work closely with soil testing laboratories to create each TSL. A similar effort to establish standard ways of identifying crops would be helpful. For example, a four-level hierarchy for identifying crops by genus, species, use, and cultivar could be one approach.

5.7.2. Supports modern computer technologies

A smoother way of linking programs is to utilize newer software technologies. One of these is Microsoft's Component Object Model (COM), which is built into all of its Windows operating systems beginning with Windows 95. COM supports a mechanism called Automation, which permits one program to operate another program remotely without user activity. The program being manipulated is the Automation "server" and the program doing the manipulation is the Automation "client." Programs can be both a client and a server. Automation offers a number of advantages over previous technologies such as Dynamic Link Libraries (DLL). To start, the language used to develop an Automation client or server is not relevant, as the COM specification is language independent. This permits developers greater latitude in their choice of development tool. In addition, an Automation server can be largely self-documenting.

Automation also permits more than just data interchange. Any capability of a program can be made available to other programs. This allows complex systems to be built in a modular fashion and allows programs to take advantage of the unique computational abilities of other programs. For example, the previously cited GIS application, SNMP, can utilize Automation to exchange non-spatial data and fertilizer recommendations with Purdue's MMP software (Joern et al., 2000). Publicly funded agricultural software should support Automation both as client and server to make its data and algorithms as widely available as possible.

5.7.3. Add-in facility

For planning software to be widely adopted, it needs a way to be customized and enhanced by third parties. One way of doing this is to provide a facility for adding components developed by others. Most modern software has this capability. For example, word processors support document templates, browsers support helper applications for viewing foreign file formats, and so on. With planning software, specialized reports are necessary. Typically, these reports use the program's data but are developed with commercial database and reporting software. If the planning software has a way of adding in these reports, they will appear to the user as an integrated part of the program.

5.7.4. Permits state-specific customization

In addition to specialized reports, planning software will need to be customized for use in other states with minimal program changes and without forking into multiple versions of the program. One way to help with this problem is to place all program data that could vary from state to state into external files. This not only provides a common way of customizing the program, allowing state-specific files to "drive" the program, but also places data where others can view and verify the values.

5.7.5. Technical documentation

To realize the goals outlined above, planning software will need documentation not only for the user (a given), but also for other developers. To make it as publicly available as possible, this technical documentation should be posted on a Web site and updated regularly as the program progresses.

6. FUTURE DIRECTIONS

Developing software that models natural processes is an ever-changing landscape. Assumptions made today about hardware, software, methodology, data, or users may not be true tomorrow or may become irrelevant in light of technological developments. However, it is clear that software will continue to be asked to do more and more. Software will need to be more open, more flexible and customizable, and easier to use. Several areas where these needs could be addressed are described below.

6.1. Linking private and public software

Privately developed software is available for practically every commercial aspect of agriculture, including software that manages and analyzes site-specific data and documents farm activities such as pesticide applications, equipment depreciation, tax form filing, and bookkeeping. In addition, a great deal of publicly developed software that incorporates research done by universities and government agencies is also available, including software that estimates soil loss, schedules irrigation, creates conservation plans, and helps agency personnel keep track of farmer clients. Much of the information entered about an operation or activity in one of these programs could potentially be used in other programs. As more and more aspects of agriculture become measurable, the amount of data generated by each operation will increase. Exchanging data between various programs needs to be dynamic and simple to do, if not completely transparent to the user. Furthermore, to avoid duplication of programming effort, programs need to be able to utilize each other's computational abilities.

6.2. Utilizing the Internet to distribute software and data

A great deal of software and data are already distributed via the Internet, either by manual downloads or, in some cases, automatic updates. More and more software will undoubtedly gain the ability to update itself automatically. How much software will eventually be run over the Internet as Java™ or other programs remains to be seen. Faced with the age-old problems of software compatibility and distribution, many developers embraced the Internet as a way of avoiding these problems, only to discover that Web-based software presented its own set of new problems involving bandwidth, security, privacy, and data ownership.

Another area where new possibilities exist is for software to assist the user in locating the increasing amounts of publicly available on-line data. Rather than requiring that each user hunt for necessary data more or less manually, software that knows where to look for the data could simplify the search process.

6.3. Emergence of GIS as a true end-user management tool

Until recently, GIS was limited by its considerable hardware and data requirements. Many of these limitations have eased or will disappear as the digitizing and mapping efforts now underway are completed and the resulting data are made available in more compact and selective forms. The world of commercial GIS software is also changing, with more modern and standard programming tools becoming available. These changes will allow GIS to become a ubiquitous tool for visually managing many types of agricultural data and supplant the traditional row-by-column way of presenting agricultural data to the user.

The marriage of GIS and the Internet appears to be a natural match. Few users require more than a tiny fraction of publicly available GIS data. Yet packaging data to be used by a large group requires that most of this GIS data be included. Why not provide GIS data on an as-needed basis via the Internet? For example, a farmer or consultant could locate a farm on a Web-based map server. The server then packages all relevant data for downloading. To keep the amount of download data to a minimum, bulky layers such as digitized aerial photographs can be clipped to the farm boundaries.

Another approach would be a Web-based GIS that supports many of the nutrient management tasks performed by today's standalone software. With this approach, all GIS data would remain on the server.

6.4. Modeling of nutrient loss potential on a site-specific basis

As the impacts of land use on nutrient and sediment loss potential become more quantifiable, new software that can estimate the effects of management on these losses will be needed. For example, research is currently underway to assess the risk of P loss from agricultural soils on a site-specific basis. When surface water nutrient criteria are developed for various water body types, TMDLs in nutrient- and sediment-impaired watersheds will likely follow. Software that can model N, P, and sediment loss potential will be needed in direct response to these guidelines or regulations.

7. SUMMARY

Nutrient management planning software that accurately models management impacts on natural processes and the environment can help crop and livestock producers document the sustainability of their operations in an increasingly regulated environment. However, a number of challenges must be addressed for this software to be effective both locally and nationally. These challenges include the problems of data acquisition, standardization and exchange. The software must also accommodate the spatial and temporal nature of manure and nutrient management, as well as be flexible enough to satisfy emerging regulatory reporting requirements. Finally, and most importantly, the software must provide useful, practical tools such as farm and field maps, planning calendars and rate calculators to help producers better manage their operations.

REFERENCES

AAPFCO-TFI. 1999. Commercial Fertilizers 1999. A joint publication of the Association of American Plant Food Control Officials and The Fertilizer Institute. Washington, DC.

Ag Electronics Association. 1997. A*E*A soil fertility data dictionary specification. Public review draft. AEA, Chicago, IL.

Ag Electronics Association. 1998. A*E*A transfer support layer specification. Version 1.0 draft. AEA, Chicago, IL.

Bailey, A., and B. Joern. 2000. Nutrient management in Indiana. *In* Agronomy abstracts. ASA, Madison, WI.

Hess, P.J. 1998. A general approach to programming multi-state fertilizer recommendations. Manure management planner program technical documentation. Purdue University, West Lafayette, IN.

Joern, B.C., and P.J. Hess. 2000. Manure management planner program. Purdue University, West Lafayette, IN.

Joern, B.C., P.J. Hess, and J.A. Lory. 2000. Software programs for nutrient management planning and recordkeeping. *In* Agronomy abstracts. p. 40. ASA, Madison, WI.

Johnson, J. 1999. Ohio nitrogen transformation and loss program. The Ohio State University, Columbus, OH.

Kelling, K.A., L.G. Bundy, S.M. Combs, and J.B. Peters. 1998. Soil test recommendations for field, vegetable, and fruit crops. University of Wisconsin, Madison, WI.

Lory, J.A., C. Barnett, and C.X. Fulcher. 2000. The spatial nutrient management planner (SNMP). *In* Agronomy abstracts. p. 419. ASA, Madison, WI.

MWPS. 2000. Manure characteristics. Manure management system series. MWPS-18 Section 1. MidWest Plan Service, Ames, IA.

Thompson, R.B., D. Morse, K.A. Kelling, and L.E. Lanyon. 1997. Computer programs that calculate manure application rates. J. Prod. Agric. 10:58-69.

USDA/NASS. 1999. 1997 census of agriculture. United States summary and state data. Volume I, part 51. March 1999. U.S. Department of Agriculture, National Agricultural Statistics Service. Washington, DC.

USDA/NRCS. 2000. Comprehensive nutrient management planning technical guidance. December 1, 2000. U.S. Department of Agriculture, Natural Resources Conservation Service. Washington, DC.

USDA and USEPA. 1999. Unified national strategy for animal feeding operations. March 9, 1999. U.S. Department of Agriculture and U.S. Environmental Protection Agency. Washington, DC.

USEPA(a). 2000. Water quality conditions in the United States: a profile from the 1998 national water quality inventory report to congress. EPA 841-F-00-006. June 2000. U.S. Environmental Protection Agency, Office of Water. Washington, DC.

USEPA(b). 2000. Proposed regulations to address water pollution from concentrated animal feeding operations. EPA 833-F-00-016. December 2000. U.S. Environmental Protection Agency, Office of Water. Washington, DC.

USEPA(c). 2000. Nutrient criteria technical guidance manual. Lakes and reservoirs. First edition. EPA 822-B00-001. April 2000. U.S. Environmental Protection Agency, Office of Water. Washington, DC.

Nitrogen in the Environment: Sources, Problems, and Management
R.F. Follett and J.L. Hatfield (Eds)
Published by Elsevier Science B.V.

Chapter 15. Field Techniques for Modeling Nitrogen Management

Marvin J. Shaffer[a] and Jorge A. Delgado[b]

[a]USDA-ARS, Great Plains System Research, P.O. Box E, Fort Collins, CO 80522

[b]USDA-ARS, Soil Plant Nutrient Research, P.O. Box E, Fort Collins, CO 80522

Nitrogen use efficiency (NUE) in production agriculture often is too low resulting in losses of excess N to ground water as NO_3^--N, to gaseous emissions of NH_3 and N_2O, and to N losses in surface runoff and erosion. Best management practices (BMPs) are needed to improve efficiency levels while maintaining proper nutrition for crops. Field studies designed to investigate potential BMPs are both time consuming and costly, and cannot cover all scenarios. Application of simulation models with N cycling components in conjunction with associated field investigations offers methodology that can help identify BMPs that show promise in increasing NUE, but at reduced cost and time expended. Credible BMP studies employing simulation tools need to proceed along a well-defined path involving model selection, model adaptation and calibration, sensitivity analyses, data requirements and availability, model application, and model results interpretation and limitations. Early and continuing interaction with local producers and field research programs are essential parts of these BMP modeling studies. Examples from irrigated agriculture in Colorado and rainfed agriculture elsewhere illustrate cases where models have been successfully used to identify potential BMPs to improve NUE and reduce leaching of NO_3^--N.

1. INTRODUCTION

Nitrogen is the most important nutrient used in agricultural systems and contributes greatly to the economical viability, sustainability, and improvement of cropping systems across the world. It is important to have an adequate supply of this element in the rooting zone of cropping systems to maintain and increase yields needed to supply the nutritional demands of over six billion people and the continuing population growth. This element has been key to sustained increases in agricultural production, however, NUE is usually reported to be lower than 50% (Newbould, 1989). Worldwide NUE for cereal production is reported at approximately 33% which is equivalent to losses of billions of dollars (Raun and Johnson, 1999). Those NUE's lower than 50% can contribute not only to economic losses across all continents, but when N is transported off site it can potentially have negative impacts on important natural resources (Smith et al., 1990; Milburn et al., 1990; Follett et al., 1991; McCracken et al., 1994; Owens and Edwards, 1994). Drinking water with NO_3^--N concentrations above 10 ppm has been established to be unsafe by the United States Environmental Protection Agency (USEPA, 1989). Those most susceptible to high NO_3^--N concentrations are infants under 3 months of age that can be affected by blue baby syndrome (clinical methemoglobinemia) (Follett and Walker, 1989). It is imperative to continue the development, evaluation, and implementation of new management practices that increase N recovery and reduce potential losses to the environment. Excess NH_4^+-N and NO_3^--N in soils have been linked with N_2O

greenhouse gas emissions (Mosier et al., 1991; Duxbury et al., 1993). Recently, N cycle models such as NLEAP (Shaffer et al., 1991) and DAYCENT (Parton et al., 1998) have been extended to simulate emissions of N_2O from soils (Xu et al., 1998; Del Grosso et al., 2001). Oxygen hypoxia problems in the Gulf of Mexico have been attributed, in part, to nonpoint NO_3^--N from agriculture (Antweiler et al., 1996), and NUE plays a role in this regard.

Researchers are constantly working to develop and improve Best Management Practices (BMPs) that increase NUE. Due to the variability of geographical areas, cropping systems, management scenarios, and weather, it is impossible to conduct field plot or whole-farm studies that cover every possible scenario. Computer simulation and decision support (DSS) models for soil-crop systems that emphasize the N cycle, especially when coupled with economics and geographical information systems (GIS), are viable alternatives that can contribute to evaluating different combinations of management scenarios and how they impact the recovery of N by a cropping system for a given set of conditions. These models represent a complex series of algorithms and databases that can interact with different conditions and serve as mechanistic tools to evaluate different scenarios and their effects on NUE, and the sustainability of a system.

Farmers, consultants, and the developers of public policy need these kind of efficient tools to help them identify, prioritize, and learn about how management practices will affect economic returns and regional environmental quality. The coupling of computer models with GIS techniques can help develop public policy that promotes the improvement of economic, environmental, and social-well being of a specific region. System models are thus important tools in the evaluation of how new practices will affect the sustainability and economical viability of agricultural systems. Models can be used for studies on the effects of management on N dynamics over long term periods. For example, they can simulate the effect of a set of management scenarios and cropping systems such as the incorporation of crop residue versus removal of straw on soil and water quality over a 25 to 50 year period, or longer. Computer algorithms allow the use of large databases that interact with the parameters and management scenarios to identify best alternatives. These simulation analyses can be used to develop and implement the best management policies that can contribute to maximized economical returns, and improvements in NUE and environmental conservation.

2. APPLYING MODELS TO FIELD SITUATIONS

Application of models in field studies where conditions are variable and a wide range of potential management scenarios exist can be challenging to agricultural managers and others who need answers quickly, but also need to be sure that the answers are credible and meet project needs. Users of models are quickly faced with a number of issues such as selection of models and databases, collection of model input data in the field, configuring the model for the study, developing management scenarios for the model, installation and operation of the model, model calibration and validation locally, and interpretation of the results. Effective and efficient handling of all these areas is necessary if successful modeling study results are to be achieved.

2.1. Model Selection

Various agricultural systems models are available worldwide with the ability to simulate carbon/nitrogen cycling in soil-crop systems. Selection of an appropriate model for a given region and application is not a trivial task and requires knowledge of model capabilities and limitations, and the problem and location to be addressed.

Some examples of available models that can be used to simulate C/N dynamics are Crop Estimation through Resource and Environmental Synthesis, CERES (Ritchie et al., 1985); Erosion /Productivity Impact Calculator, EPIC (Williams et al., 1983); Nitrogen Tillage Residue Management Model , NTRM (Shaffer and Larson, 1987); LEACHM (Wagenet and Hutson, 1989); Root Zone Water Quality Model, RZWQM (Ahuja, et al., 2000); Nitrate Leaching and Economic Analysis Package, NLEAP (Shaffer et al., 1991); Great Plains Framework for Agricultural Resource Management, GPFARM (Ascough et al., 1998); the University of Minnesota NCSOIL model (Molina et al., 1983); GLEAMS (Knisel, 1993); CENTURY carbon model (Parton and Rasmussen, 1994); the Danish Nitrogen simulation system, DAISY (Hansen et al., 1991); the German model, HERMES (Kersebaum, 1989); the Rothamstead N turnover model, SUNDIAL (Bradbury et al., 1993); the German UFZ model, CANDY (Franko, 1996); the Canadian model, *ecosys* (Grant, 1997); Introductory Carbon Balance Model, ICBM (Andren and Katterer, 1997); the Swedish model, SOILN (Eckersten et al., 1998); and the Dutch model, ANIMO (Groenendijk and Kroes, 1997). Many of these models have Internet Web-sites that contain model descriptions and in some cases the latest versions of the models and their associated databases. A summary of the Web-site addresses is given in Table 1.

C/N models have been applied to a range of environmental and management problems such as NO_3^- -N leaching, greenhouse gas emissions, carbon sequestration, and soil fertility management to name a few. Detailed descriptions of typical applications involving C/N models can be found in Shaffer et al. (2001b). The amount of detail contained in these models is highly variable and ranges from highly detailed research models to more user-oriented screening tools. Comprehensive reviews and comparisons of these models are presented by Ma and Shaffer (2001) for U.S. models, McGechan et al. (2001), for models in Europe, and Grant (2001) for the Canadian model *ecosys*. The potential user needs to review and judge the model capabilities versus project requirements and select the tool that best fits these needs. The best model for a given application usually lies somewhere in the middle near the maximum usability shown in Fig. 1. Selecting a model that either is too simple or too detailed for a given application, or that is inappropriate has caused many problems with model application studies in the past and needs to be avoided. Potential model users especially need to look at model capabilities, applicability, reliability, ease of use, data needs, and supplied databases relative to the needs and requirements of their project. For example, if a project contains a specific cropping system, but a model cannot handle this scenario, then that particular model probably cannot be used. Also, if some models do not contain soil and climate databases for the area of interest, then additional work will be needed to develop these resources, and this could play a role in final model selection. If a model was developed and tested in a region with considerably different conditions than the proposed project, then extra effort probably will be needed to configure and calibrate the model for the local area.

Table 1.
Internet Web sites for C/N Cycling Models.

Model	Web site address.
ANIMO	http://dino.wiz.uni-kassel.de/model_db/mdb/animo3_3.html
CENTURY	http://www.nrel.colostate.edu /PROGRAMS/MODELING/CENTURY/CENTURY.html
CERES	http://www.icasanet.org/. http://nowlin.css.msu.edu/.
DAISY	http://www.dina.kvl.dk/~abraham/daisy/
EPIC	http://www.brc.tamus.edu/epic/ http://arsserv0.tamu.edu/
GLEAMS	http://arsserv0.tamu.edu/ http://sacs.cpes.peachnet.edu/sewrl/
GPFARM	http://www.gpsr.colostate.edu/gpsr/products/gpfarm.htm
HERMES	http://www.gsf.de/ufis/ufis/modell47/modell_EN.html
LEACHM	http://www.es.flinders.edu.au/LeachmDownload/LEACHWEB.HTM
NCSOIL	http://soils.umn.edu/research/ncswap-ncsoil/
NLEAP	http://www.nleap.usda.gov http://www.gpsr.colostate.edu/gpsr/products/nleap/nleap.htm ftp://ftp.nrcs.usda.gov/centers/itc/applications/wqmodels/
NTRM	http://www.wiz.uni-kassel.de/model_db/mdb/ntrm.html http://www.wcc.nrcs.usda.gov/water/quality/common/h2oqual.html
RZWQM	http://www.gpsr.colostate.edu/gpsr/products/rzwqm.htm
SOILN	http://www.mv.slu.se/bgf/ http://amov.ce.kth.se/coup.htm

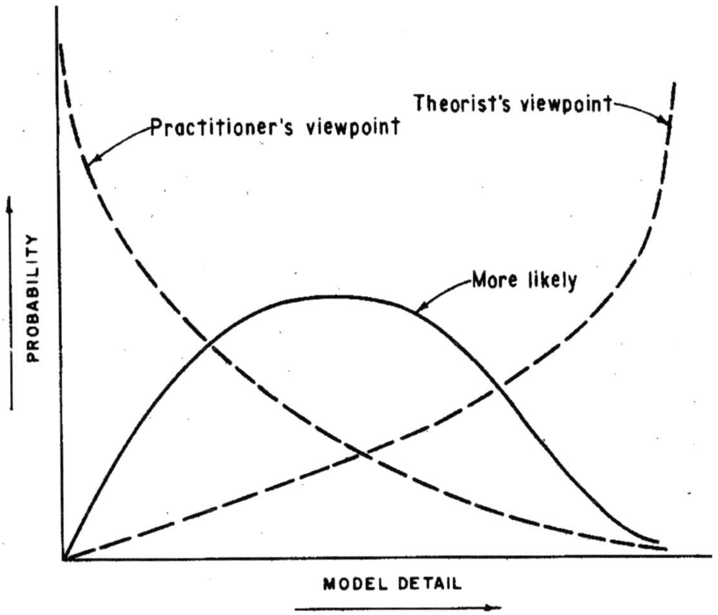

Fig. 1. Selecting the Best Model for a Field Project.

2.2. Model Adaptation and Calibration

Once a model has been selected, work must be done to configure and calibrate it for the local regional area and cropping systems. This includes the general layout of the model application, the databases, and the model parameters. The general layout includes items such as the scope of the model options and submodels to use, linkage considerations to other models (e.g. – economics packages and GIS), and the types of output variables needed. Databases may need to be customized or extended for local conditions. For example, regional soil and climate databases may not adequately represent local conditions for specific farms. Often, model parameters will need to be determined or refined locally. This may include crop parameters, process rate coefficients, and other functional coefficients. For example, yield and N uptake are NLEAP model functions that can be affected by several parameters. NLEAP uses algorithms that are driven by the expected yield and the N uptake index, to simulate the N sink (uptake). Yields can be affected regionally by evapotranspiration, precipitation, temperature (degree days) and other parameters. Additionally, varieties may change from region to region. There are varieties that have a higher NUE and a lower N uptake index, so the amount of N needed to produce a unit of yield will be lower. Rooting depth parameters can also change with varieties.

The model calibration/validation process should first define the management practices to be evaluated. The effect of soil type needs to be taken into consideration as well as the selection of crops that are grown or that are anticipated to be the dominant crops in the region. For example, the NLEAP "region.idx" model parameter file often needs to be fine-tuned to the local area with additional crops and parameters such as the N uptake indexes. For nutrient management studies involving NUE and NO_3^--N leaching, calibration of soil residual NO_3^--N should be done by comparing simulated residual soil NO_3^--N values with observed NO_3^--N values for the root zone and below the root zone. Observed and simulated root zone soil water content should be compared and tested in a similar manner.

2.3. Field Setup for Model Calibration

Residual NO_3^--N, % soil organic matter (SOM), soil water content, and crop N uptake in commercial fields should be monitored using selected field plots. A good working plot configuration is at least four 20.9 m^2 plots established for replication and size considerations. The plot borders should be identified with field transponders installed at the corners so the same plots can be re-sampled. Transponders will facilitate the location of the corners within a one-inch variability. New technologies such as real time kinemantic GPS can also be used to identify the border plot with accuracies at the sub-centimeter level (Zuydam, 1999). Plot data collected under commercial operations that are monitored more intensively with yield monitors or clipping should be used for the calibration/validation process, and whole field simulations with farmer yields should be used for technology transfer of information (Delgado et al., 2000). Farmer yield data from the entire field could be used (truck loads), or if the field is divided by areas, yield monitors or truck loads from the respectively areas should be used as inputs into the model.

Aboveground and belowground plant samples from different crops such as small grain, and vegetables are collected by harvesting 0.4 m^2 in each plot. All above- and belowground plant compartments need to sampled. For example, aboveground vines need to be collected for potatoes, and tubers harvested. Main roots need to be picked from the plot, especially those for grains including a significant sink such as the crowns. The mean root depth also needs to be measured for all crops. Plant samples need to be collected prior to farmers harvesting of their fields. Plant samples need to be dried at 55 °C, ground and analyzed for total C and N content with procedures such as automated combustion using a Carlo Erba automated C/N analyzer©[1]. For the NLEAP model, total N uptake by all compartments needs to be added up and divided by total yield, to calculate a mean N uptake index. Water content of the harvested portion needs to be accounted for by collecting a clean fresh weight as soon as the samples have been collected, and determining the water content of the sample.

One or two soil cores should be taken for the initial and final soil samples collected in each plot. In the case of whole fields, at least 20 cores need to be taken and composited for the initial and final soil samples. If the field is subdivided into areas, each area should then be sampled with up to 20

1 Names are necessary to report factually on available data; however USDA neither guarantees nor warrants the standard of the product, and the use of the name by the USDA implies no approval of the product to the exclusion of others that may be suitable

cores. Soils are sampled in 0.3 m or more frequent intervals down to 1.5 m depending on model needs. Other chemical and physical variables such as the % coarse fragments by weight and by volume, % SOM, pH, CEC, and soil water content are also measured for the initial samples. Soil samples need to be collected from each core and should be kept in cool sealed bags to measure the initial % water content. After harvesting, the same procedure is used for soil samples collected to measure residual soil NO_3^--N and soil water content.

The soil samples collected from each 0.3 m (or other) depth increment should be placed immediately into coolers and transported to the laboratory where they are air dried, and sieved through a 2 mm sieve. The percentage weight of the coarse fragments is used to calculate the percentage coarse fragments by volume (Delgado et al., 1998*a*). Bulk densities need to be determined or estimated from texture as described by USDA-SCS (1988). Soil samples need to be extracted with 2*N* KCl, and the NO_3^--N and NH_4^+-N contents determined colorimetrically by automated flow injection analysis. Records of the irrigation, N fertilizer application, planting, harvesting, cultivation, and other agricultural management practices must be collected to be used as inputs into the model as part of calibration. All N inputs such as amount and type of N fertilizer, amount of N in the irrigation water, crop residue mass and its N content, and the initial soil inorganic N content and any other N input required by the system needs to be accounted. Center-pivot irrigation sprinklers need to be calibrated for accuracy and irrigation water samples need to be collected at least three times during the growing season and analyzed for NO_3^--N. Climatic data need to be collected at the site or from the nearest weather station. Rain and/or snow amounts need to be measured locally during the growing season at all sites.

The model can then be used to simulate the effects of crop management on residual soil NO_3^--N in the profile and of the available soil water in the root zone. The simulated residual NO_3^--N for the root zone, bottom of the root zone to the bottom of the soil profile desired, and for the whole soil depth (e.g. 1.5 m) can then be compared to observed values. Correlations between predicted and observed available soil water and between predicted and observed residual soil NO_3^--N can be conducted. For these analyses the intercept (b_0) and slope (b_1) of the regression line can be tested statistically for differences from 0 and 1, respectively.

After the collection of the basic input data, and the conclusion of the calibration and validation process on field plots, the model then can be tested for technology transfer on whole field scenarios (Delgado, et al., 2000). The simulation of whole field scenarios will give an idea of the level of accuracy and variability explained by the simulations (r^2) between measured and simulated values across the whole field. The user needs to include in the analysis and interpretation, the potential changes in chemical and physical characteristics across the field due to variability in soil type. This will help to account for variability in the measured residual soil NO_3^--N (variability of x, observed NO_3^--N) versus variability due to model simulations (variability of y, predicted NO_3^--N) (Delgado, 1999 & 2001).

Until we have a new generation of simulation models capable of conducting spatial simulation across fields and farms that include changes in soil surface characteristics as well as changes in the subsoil (*changes in x, y and z*), for a series of layers, we will need to use simulation models that simplify the system.

Data collection needed to take the model through the calibration/validation and technology transfer effort should allow completion of the technology transfer process and determination of the effect of

BMPs on NUE and transport of NO_3^--N in the soil profile. In general, this can require several years and should encompass two or more crop rotation cycles. If the user wants to refine and fine tune the model one step further for more advanced and long-term simulations of N transformations and how these changes may affect NO_3^--N leaching, then longer term studies are needed that can evaluate changes in soil organic matter and N cycling. In other words, we may need to fine tune and calibrate the simulations of the N pools of the model on longer term scenarios (> 10 years) if users want to extend the simulation of these N pools, otherwise the basic assumptions with calibration/validation and technology transfer processes (6 – 10 years) have been proven sufficient to simulate the effect of BMPs on NO_3^--N dynamics.

Additional crop, soil, and weather inputs are important and need to be accounted for when comparisons are made of BMPs across a region or across years. Model algorithms have the advantage that they can be used for evaluation of BMPs based on a case by case scenario. An example of why weather data are important is that simulation of BMPs evaluated for a region will be affected by local rain and or evapotranspiration scenarios, since there may be significant variability of local thunderstorms and elevation. This variability needs to be factored in by regions. In the case of soil inputs, soil texture could be the same, a loamy sand, but the coarse fragments can significantly vary across the region and can impact the simulations. An example of a crop parameter will be the use of different varieties that can have different rooting depths and N uptake indices.

2.4. Sensitivity Analyses

The user can set up a sensitivity analysis to evaluate which model parameters are more important for a region. A series of simulations can be conducted by changing one parameter at a time by a factor of 0.25, 0.50, 1.50, and 1.75. This can give an idea of the relative impact and importance of the input data as well as the impact of the variability of an input parameter. In other words, how are the conclusions affected due to the variability of the input parameters (e.g. results of soil organic matter content from two different laboratories). Sensitivity analysis is important in determining the relative importance of each input, but the analysis needs to be conducted considering long-term scenarios since it could also be confounded by specific initial conditions and events. For example, the interpretation of the sensitivity could be confounded by the initial data used for the respective growing season conditions such as a higher residual soil nitrate content (150 kg NO_3^--N ha^{-1}) versus a lower initial content (5 Kg NO_3^--N ha^{-1}).

2.5. Types of Field Analysis

Application of models to field conditions involves a number of options that must be considered before proceeding with the study. The introduction of GIS technology now allows spatial simulations and mapping to be done across fields, farms, and regions. This capability requires that geo-referenced databases be available for roads, towns, legal boundaries, soils, climate, and management across the study area. Also, the simulation model of interest should be linked with these databases and be capable of providing output back to the GIS software. These linkages can be accomplished using hand techniques, but considerably more time and effort will be required than with an established C/N cycling-GIS interface. Some progress in this area has been made recently using the Internet to link C/N model and GIS servers with GIS databases (Shaffer et al., 2001a). When the required databases and models are available, a GIS analysis of NO_3^--N leaching and NUE for an entire farm or region can provide considerably more information and insight than single site analyses. The user must decide whether the additional expense and effort required to conduct a GIS study are justified.

In general, most studies involving projected simulations of NO_3^--N leaching and NUE need to be run for multiple years to dynamic steady-state conditions. This allows the effects of the initial conditions be reduced and long-term trends to become more visible. For example, the effects of an alternate management scenario need to be evaluated for at least 6 to 10 years and through at least two crop rotation cycles to allow re-establishment of dynamic steady-state. Shorter-term studies are usually reserved for preliminary model testing and for cases where the period to steady-state is of interest. For example, Delgado (1998) conducted short-term simulations of a lettuce, winter cover crop - potato rotation for a period of 2 years. The study reported that not only do the winter cover crops scavenge the NO_3^--N that leached below the rooting systems of the lettuce, but they also reduce the NO_3^--N leaching during the potato growing season. We can evaluate the impact and benefits not only during the current growing season, but those that are observed during the growing season of the following crop (Delgado, 1998).

In addition to GIS applications, C/N cycling models for soils can be linked with applications such as ground water models and economics programs. These tools will require specific types of data from a C/N cycling model such as daily or monthly water and NO_3^--N leached and management details, usually in the form of a text file or database. If these types of extended applications are going to be used, linkages with potential C/N cycling models should be investigated during the model selection process.

The NUE for each crop-system needs to be evaluated. Each system NUE can be calculated as follows: NUE_{sys} = ((total N uptake by crop / total N available in the soil profile, e.g. 0 - 1.5 m) * 100). Total N available includes initial NO_3^--N in the soil profile, added fertilizer, added fertilizer in irrigation, background N in water, and simulated N cycling from soil and crop residue mineralized N (Delgado, 1998 & 2001; Delgado et al., 2001). Another analysis for well irrigated systems is the net NO_3^--N recovery from underground irrigation water. This net recovery from underground water will represent the potential for mining NO_3^--N by the system (Delgado 2001 and Delgado et. al., 2001). A large negative number will represent a system with a high potential to contribute NO_3^--N to the underground water system since we do not know if all the NO_3^--N leached from the system will eventually reach the underground water (e.g. some may be lost by denitrification, or may be recovered by a scavenger crop). A high positive number will represent a system that is serving as a scavenger crop for the NO_3^--N added as irrigation water. A positive net recovery simulates a mining process for NO_3^--N from underground water. We could calculate then, this potential for mining NO_3^--N for the root zone or for a similar soil depth. For a rotation that includes shallow and deeper-rooted crops such as lettuce-winter wheat, a simulation on a similar soil depth is important since deeper rooted systems can serve as a scavenger and recover residual soil NO_3^--N from below the rooting systems of shallower rooted crops such as lettuce and potato (Delgado 1998 & 2001; Delgado et al., 1998b & 2001). This NO_3^--N mining potential is calculated as follows: a) NO_3^--N mining for the root zone equals NO_3^--N in the groundwater added as irrigation water to the field minus NO_3^--N leached from the root zone; and b) NO_3^--N mining for the soil profile equals NO_3^--N in the groundwater added as irrigation water to the field minus NO_3^--N leached from a similar soil profile for the rotation.

2.6. Data Requirements and Availability

In general, plot studies for model calibration and validation have shown that local inputs of field data have improved model predictability of residual NO_3^--N, NUE, and NO_3^--N leaching. However, the large areas involved with whole-farms and regions have made data collection at field-plot intensities unfeasible from the standpoints of cost and logistics of time and personnel. The question now arises as to how much use can be made of large-scale soil and climate databases without undesirable losses in model accuracy? An even more fundamental question is how accurate do simulation results need to be on a whole-farm or regional scale to allow management decisions at these scales? Issues of scaling-up models developed and tested at smaller (field plot) scales arise here as well. Definitive answers to these questions are beyond the scope of this paper, but we do know the accuracy levels being achieved by C/N cycling models for a wide range of field plot studies around the World (Shaffer et al., 2001b).

To maintain this accuracy, input data needs to be provided at the same resolution and detail. Fortunately, recent advances in remote sensing and GPS technology have improved the chances of developing higher resolution soil and climate datasets over large areas. Much faster computers coupled with GIS and database technology are now making detailed simulations of regional areas more feasible. Also, geo-referenced climate and soils GIS databases are becoming available at a national scale in the United States (http://www.wcc.nrcs.usda.gov/water/climate/prism/prism.html; http://www.ftw.nrcs.usda.gov/ssurgo_ftp3.html). Although these GIS databases do not yet provide data resolution at the field plot scale, they are a significant step in the right direction.

2.7. Model Interpretation and Limitations

Model limitations need to be accounted for and be understood to ensure that the interpretation of the simulated outputs is correct and the conclusions are representative of the dynamics of the natural systems. If the model is going to be used for prediction, the user will have to consider the effects of unknown conditions that may affect yields. For example, the effect of diseases, micronutrient deficiencies, salinity, acidity, weeds, and other effects that may affect yield may not be simulated by some models. The user will need to account for such effects when the expected yield is entered or simulated, or to realize that the simulations are being conducted for average expected yields under the respective BMP simulated scenario.

Using most models for prediction of scenarios under BMPs where pests are controlled and yields are maximized will be more accurate than conducting simulations under weed infestation or with disease problems, since the model or user generally will have to predict the reduction in yield due to such an adverse condition. Additionally, the model may not account for the N uptake by the weeds. Usually the model is being used under BMPs that supply the necessary fertilizer inputs and control weeds, diseases and pest problems. If other adverse conditions need to be simulated, N partition and uptake by weeds will need to be accounted for, as well as the effects on reduction of yields.

Additionally, crop N deficiency is not well simulated in some models since they do not reduce the rate of growth at the inflection point where N deficit appears. The user should understand that this type model is better applied under best maximum conditions, but that it could also be applied to N deficiency conditions. For example, if leaching events are higher than expected due to extraordinary rain events, that may impact yields, the user will need to interpret the effect of this N loss and the potential reduction on yield, and N uptake. The NO_3^--N leaching under such N deficiencies may be underestimated, since the model will keep assuming the same rate of uptake for the crop. This first simulation could then be followed by a second simulation with a lower expected yield to account for the lower yield and expected lower N uptake, however, by shifting the rate of uptake to a lower curve, the N uptake to the point of high precipitation may then be underestimated, and the NO_3^--N leaching may be overestimated.

Some models may underestimate the expected N uptake for those crops and or varieties that can exhibit exuberant or succulent N uptake. For example, NLEAP uses the expected yield and a mean N uptake index based in total units of N uptake per unit of yield. The N uptake indices were developed under Best Management Practices under commercial operations. Under extremely high N applications, succulent N uptake may be higher than under BMPs and may be underestimated.

There are climate limitations when some models are applied over a region. If using GIS for simulations of BMPs over a region, it is important to consider for example, that the model may not simulate the changes in temperature with changes in altitude. These changes in temperature can affect the simulation of N dynamics due to mineralization of soil organic matter and crop residue, which decreases with higher altitudes and lower temperatures. For such a condition, the region will have to be divided by ecological or climatic variability and the simulations conducted within each division will better simulate these dramatic changes in temperature, or precipitation and evapotranspiration. Regional simulations should also consider that single point simulations will not account for differences in yields due to soil type. This can be achieved by dividing the simulations by soil type where the respective expected yields by soil type can be entered as an individual input. Similarly, if specific fields are simulated, it is important then to collect local precipitation at the site during the period that the simulation is conducted to fine tune and better account for variability in local precipitation.

Other potential considerations that need to be accounted for is the time interval. For example, NLEAP conducts it simulations using a daily or event-based time interval, so rapid infiltration events, denitrification, or NH_3 volatilization effects, at intervals smaller than a day can create high N losses and may be underestimated by the model. Spatial variability across the field such as the occurrence of gravel bars, salinity at the lower spots and significant differences in soil type will not be simulated by the model. These spatially variable fields can be divided into soil type regions and or topographic regions within a field, and NLEAP would more accurately simulate these separate conditions (Delgado, 1999 & 2001; Delgado et al., 2000 & 2001). Major differences in soil layers within the profile will not be simulated since the NLEAP model uses the soil's physical and chemical characteristics across two (and more recently three) soil layers. The new model 1.20 (Delgado et al., 1998*b*; Shaffer et al., 1998) can enter the rooting depth and the soil profile depth desired, and simulate up to 0.03 m accuracy. This may help by using a more uniform soil depth within a 1.5 m profile. For example using a 0.9 m profile for simulations.

It is important that the user understands how the model inputs data and how to use the correct input variables. The user also needs to understand how to calculate the predicted or simulated results and have a basic knowledge of N dynamics and effects of management practices.

Results should be summarized and presented in graphs, tables, and text. The simulation of the transport of NO_3^--N in the soil profile is very important and as well as the simulation of soil water content. If the simulated residual soil NO_3^--N in the root zone and below the root zone is in agreement with the observed residual soil NO_3^--N, and the simulated soil water content in the root zone is also in agreement with the observed, then model assumptions are sufficient to simulate the effects of BMPs on the dynamics and transport of NO_3^--N. These graphs of predicted soil water content, and residual soil NO_3^--N in the root zone and below the root zone versus the observed need to be presented. The evaluation of cropping sequences as well as different soil types, crops and varieties is also important and should be part of the presentation of the data as well as the effects on NUE and on underground irrigation water NO_3^--N mining potential (Delgado, 2001; Delgado et al., 2001).

NLEAP uses local databases to simulate N dynamics and has a leaching index that is descriptive on how sensitive each specific area is. This index for NO_3^--N leaching can help identify areas that are susceptible and vulnerable to groundwater contamination. It is important that we use the correct inputs that consider the variability of weather, soil type, yields, evapotranspiration, etc.

3. EXAMPLES FROM IRIGATED AGRICULTURE

The capability of NLEAP to simulate NO_3^--N dynamics in the South Platte region of Northeastern Colorado and the San Luis Valley (SLV) of South Central Colorado has been extensively studied. The land management of the South Plate alluvial aquifer in Northeastern CO is mainly dominated with irrigated agriculture. Both center pivot sprinkler and furrow irrigation are used for corn, potatoes, onions, sugar beets, beans, alfalfa hay, and a number of other specialty crops. The uplands of the South Platte alluvial aquifers are dominated by dryland agriculture and grazing lands. The are numerous confined animal operations (CAFOs) in this region of Northeastern Colorado. The manure from these CAFOs is recycled into the adjacent cropland areas after harvest. The results of these simulations for these two regions of Colorado have been published extensively in the literature (Delgado, 1998, 1999 & 2001; Delgado et al., 1998a & b, 1999a & b, & 2001; Hall et al., 2000; Shaffer et al., 1995; Wyllie et al., 1994).

The SLV is an important agricultural base for the State of Colorado with 90% of the potato, 77% of the spring wheat, 81% of the barley, 32% of the oat and 12% of the hay produced in the state of Colorado during 1996 (CDA and USDA, 1997). During 1996 Colorado was the fifth highest producer of potato in the USA (USDA, 1997), thus the SLV region represents an important potato producer of the Nation. Other vegetable crops such as lettuce, carrot and spinach represent an important and viable production base of the valley with about 7,000 acres. Irrigated agriculture for this region is of most importance since it impacts the economics of most of the residents of the valley (Eddy-Miller, 1993). The SLV, with a mean elevation of 2,348 m and a mean precipitation of 180.3 mm, is a high altitude intermountain dessert valley that extends 105 miles long and 20 to 50 miles wide (Edelmann and Buckles, 1984; Hearne and Dewey, 1988).

Austin (1993) reported that irrigation started in the SLV with the earlier Spanish settlers who established the first irrigation system in Colorado to divert water from the Rio Grande. Initially irrigation was limited prior to 1880, but between 1880 and 1890 an intensive network of canals was constructed increasing the area of furrow irrigation (Hearne and Dewey, 1988). Underground water resources became a more important source with the introduction of the high capacity pumps during the 1950's, (Hearne and Dewey, 1988). The efficiency of water use increased significantly during the 1970's, with the introduction of sprinkler irrigation systems that contributed to an increased irrigated area under these systems, since they increased from 262 wells in 1973 to over 2000 by 1996. Each center pivot irrigation system covers on average 54.7 ha. Furrow irrigation is also still used extensively across this region.

Although there are a variety of soil types across the SLV, the soil texture of this region is dominated by the sandy textured soils or soils over a coarse textured substratum (USDA-SCS, 1973). Nitrate contamination of local wells in excess of EPA standards have been extensively documented in the literature by USGS (Emery et al., 1973, Edelman and Buckles, 1984). The USDA established the San Luis Valley Water Quality (SLVWQDP) to evaluate the effect of BMPs for this region. The SLVWQD, USDA-NRCS and USDA-ARS worked in cooperation to use the NLEAP model to evaluate the effect of BMPs across different cropping systems. Commercial operations were monitored extensively over the whole valley on over 25 farms, with over 400 different simulations. A list of BMPs for this region was published by Ristau (1999).

NLEAP model simulations for this region show that inclusion of early planted winter cover crops, after lettuce harvest on a lettuce - potato rotation, significantly increases NUE and decreases NO_3^--N leached during the potato growing season (Figs. 2 and 3, Data adapted from Delgado, 1998). Delgado's (1999) sequential simulation shows how important it is to evaluate the crop rotation on a similar soil depth for all the crops and to consider the previous year management practices that can affect NO_3^--N leaching from the system. He used the new version of the NLEAP model 1.20 that allowed the simulation of multiple crops with different rooting depths (Shaffer et al., 1998; Delgado 1998 & 2001; Delgado et al., 2000). The winter cover crops (WCC) planted immediately following lettuce harvest, have enough degree days to develop a deep rooting system that can scavenge large amounts of NO_3^--N from the soil profile (Delgado et al., 1999*a*). Early planted WCC reduce the amount of NO_3^--N potentially available to leach, and also reduce the NO_3^--N leaching during the potato growing season (Fig. 2) and contributed to conserve soil and water quality (Delgado et al., 1999*a*). Delgado (2001) reported a significant correlation between rooting depth and NUE, NO_3^--N leaching and with the capacity to recover NO_3^--N from underground water sources for small grains and winter cover crops. When well water is used for irrigation, the winter cover crops and small grains act as filters scavenging the NO_3^--N and reducing the NO_3^--N losses from the system.

The NLEAP model was capable of simulating different cropping systems from the SLV (Delgado et al., 1999; Delgado, 2001). Fig. 4 presents a correlation for seven irrigated crops grown in South Central Colorado and two irrigated crops from Northeastern Colorado for the observed vs. predicted residual soil NO_3^--N. NLEAP was capable of simulating the effects of management practices on the soil N dynamics for corn and sugarbeets grown in Northeastern Colorado. The residual soil NO_3^--N for the whole soil profile (0 - 0.9 m) was lower for the small grain than for the shallower rooted crops that were grown in the San Luis Valley (Fig. 5, Data adapted from Delgado, 2001). The model simulated the transport of NO_3^--N and NO_3^--N leaching below the rooting zone of shallower and deeper rooted crops (Delgado et al., 2000). Delgado (2001) reported that BMPs can potentially contribute to the savings of millions of dollars by increasing NUE in this region of South Central Colorado and decreasing NO_3^--N leaching into underground water. They can potentially remove NO_3^--N from irrigation water that is applied to the field, if deeper rooted crops are rotated with the shallower rooted crops and if recommended BMPs for N fertilization and irrigation are implemented.

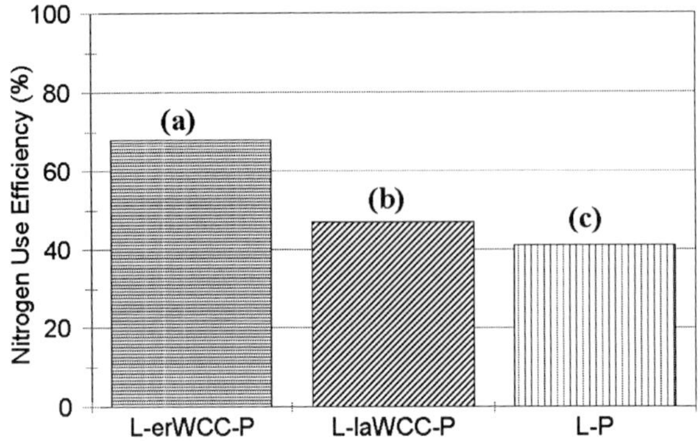

Fig. 2. Mean Nitrogen Use Efficiency (NUE) for lettuce - early winter cover crop - potato (L-erWCC-P); for L- late winter cover crop - P (L-laWCC-P); and for a L-P rotations. Letters indicate significant differences at LSD P < 0.05 (Data adapted from Delgado, 1998).

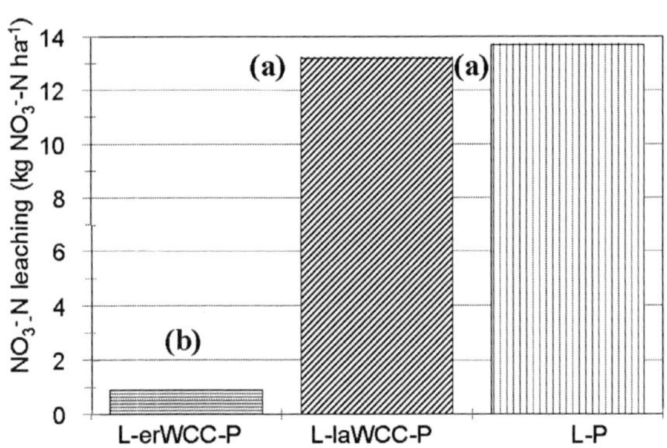

Fig. 3. Mean Nitrate leaching for lettuce - early winter cover crop - potato (L-erWCC-P); for L- late winter cover crop - P (L-laWCC-P); and for a L-P rotations. Letters indicate significant differences at LSD P < 0.05 (Data adapted from Delgado, 1998).

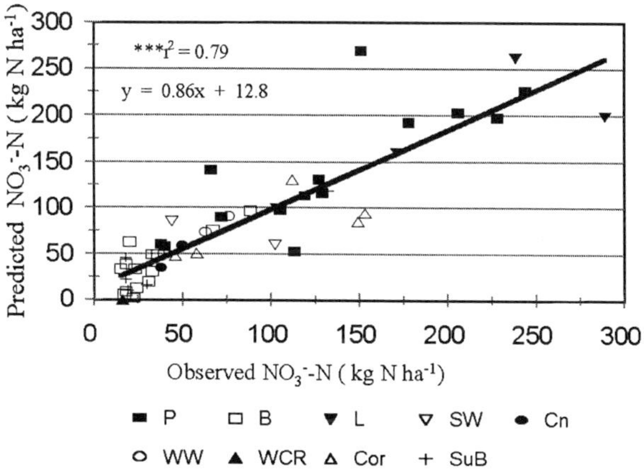

Fig. 4. Observed and NLEAP simulated residual soil nitrate (NO_3^--N) in the soil profile. Observed and simulated data for potato (P), barley (B), lettuce (L), spring wheat (SW), canola (Cn), winter wheat (WW), and winter cover rye (WCR) grown in Southcentral Colorado and of corn (Cor) and sugarbeets (SuB) grown in Northeastern Colorado (*** = r^2 significant at $P < 0.001$).

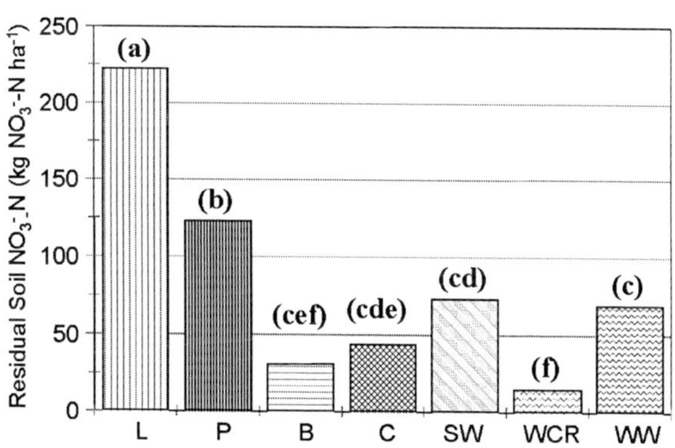

Fig. 5. Observed residual soil nitrate (NO_3^--N) for lettuce (L); potato (P); barley (B), canola (Cn) spring wheat (SW) winter cover rye (WCR) and winter wheat (WW) for a 0 - 0.9 soil depth. Letters indicate significant differences at LSD $P < 0.05$ (Data adapted from Delgado, 2001).

4. EXAMPLES FROM RAINFED AGRICULTURE

The NLEAP model has been applied to rainfed agriculture throughout the U.S. and in foreign countries. Walthall et al. (1996) used the NLEAP model to investigate NO_3^--N leaching from fertilizers used in cotton production on the Macon Ridge in Louisiana. Results helped to establish a linkage between NO_3^--N concentrations in the shallow groundwater and leaching from the crop root zone in terms of lag times, annual rainfall distribution, and NO_3^--N available for leaching.

Kaap et al. (1995) used NLEAP to develop strategies for municipal wellhead protection in Central Wisconsin sands. The study involved both rainfed and irrigated areas and found that no simulated NO_3^--N leached to groundwater under alfalfa stands, with moderate amounts (30-50 kg/ha) leached under rainfed corn and irrigated snap bean, and large amounts (61 – 130 kg/ha) leached under irrigated corn. The modeling study helped to establish that the use of proposed BMPs alone failed to meet the 10 mg/l NO_3^--N groundwater quality standard. Other scenarios were proposed that could help to meet this standard including retirement of agricultural lands to forest or grassland, changing the agricultural crop rotations to more hay, and converting the agricultural lands to residential.

Stoichev et al. (2001) working in Bulgaria compared NLEAP simulated NO_3^--N leaching for sunflower-winter wheat and corn-sunflower-winter wheat rotations with simulated NO_3^--N leaching from irrigated home vegetable gardens. The modeling results helped to establish that the majority of the NO_3^--N leaching in the local village was from the gardens rather than from agricultural fields as initially assumed by the villagers. As a result, remedial measures were recommended to the villagers involving reduced N input to their gardens.

5. DISCUSSION AND CONCLUSIONS

Field applications of models for nitrogen management are challenging from the standpoints of selecting an appropriate model from the long list of available tools and then applying the model to field situations that often are removed from conditions and locations where the model was developed and tested. The potential user must be prepared to collect a reasonable amount of field data to calibrate the model for the soils, management, and climate conditions in the study area. This will involve one or more seasons or years of field work to collect the required crop yield, soil nitrogen, and soil water data that are needed. Once this is accomplished, calibration of the model should be done using a systematic approach based on a prior sensitivity analyses run on the tool. The more sensitive parameters for the study region should be adjusted to improve agreement with field data. Some additional testing should also be done with data not used in the calibration to help develop a reasonable amount of validation experience with the model.

Once the calibration procedure has been completed, the model can be applied to test alternative scenarios involving nitrogen management. Potential scenarios should be developed in cooperation with local producer, commodity, and action agency groups. Early buy-in of these organizations is essential for later adoption of the best management practices that are identified. In general, simulated scenarios should be run for a number of years until dynamic steady-state is achieved in terms of residual nitrate in the soil profile. This provides for a better test of long-term management impacts on the system and minimizes the effects of the initial conditions, which may be uncertain across the region or farm. In some cases, shorter term studies may be needed to test, for example, methods of mitigating problems with existing NO_3^--N accumulation in the soil profile. Comparisons among simulations of various management scenarios should be done taking into account the uncertainty in the results obtained from the calibration and validation studies. For most nitrogen studies in the field, this means that small differences for simulated residual soil nitrates and nitrate leached will not be statistically meaningful for comparisons of some management scenarios. Larger potential differences should be targeted when selecting management scenarios to be tested, especially if producers are expected to demonstrate positive benefits from the adoption of BMPs. Indeed, helping to identify scenarios with substantial potential benefits is one of the better uses for modeling in the N management area.

Examples, where field modeling studies have identified significant potential differences in nitrate leaching potentials for alternate nutrient management include a fertilizer and manure management study reported by Hall et al., 2000 where long-term manuring at high rates was shown to be leaching excessive nitrates from the root zone; the leaching study of Kaap et al. (1995) in Wisconsin; the Bulgarian NO_3^--N study in the village of Parvomaitsi reported by Stoichev, et al. (2001); and the leaching simulation work done in the San Luis Valley of Colorado by Delgado (2001). These studies have demonstrated how the application of a C/N model such as NLEAP can make a difference in recommended N management scenarios. Basically, these authors applied the procedures outlined in this chapter to implement and complete successful nitrogen modeling studies under field conditions.

Models for N dynamics are tools that can be used to help identify and improve BMPs and to transfer research results to producers, consultants, and extension personnel. Successful field applications of these tools need to proceed along a well defined path as outlined in this chapter. This begins with model selection and proceeds through field data collection; initial BMP selection; model adaptation, calibration, and testing phases; model application; and results presentation and evaluation phases.

REFERENCES

Ahuja, L.R., K.W. Rojas, J.D. Hanson, M.J. Shaffer and L. Ma 2000. Root Zone Water Quality Model. Water Resources Publications, LLC, Highlands Ranch, CO. 372 pp.

Andren, O. and T. katterer. 1997. ICBM: The introductory carbon balance model for exploration of soil carbon balances. Ecological applications 7:1226-1236.

Antweiler, R.C., D.A. Goolsby, and H.E. Taylor. 1996. Nutrients in the Mississippi River, In Contaminants in the Mississippi River, Meade, R.H., ed.: U.S. Geological Survey Circular 1133, pp. 73-85.

Ascough II, J.C., G.S. McMaster, M.J. Shaffer. J.D. Hanson, and L.R. Ahuja. 1998. Economic and environmental strategic planning for the whole farm and ranch: The GPFARM decision support system. Proc. First Interagency Hydrologic Modeling Conference, Las Vegas, Nevada. April 19-23.

Austin, B. 1993. Report to the Commissioner of Agriculture, Colorado Department of Agriculture, Groundwater monitoring activities San Luis Valley Unconfined Aquifer. Colorado Department of Public Health and Environment. Denver, CO.

Bradbury, N.J., A.P. Whitmore, P.B.S. Hart, and D.S. Jenkinson. 1993. Modelling the fate of nitrogen in crop and soil in the years following application of ^{15}N-labeled fertilizer to winter wheat. Journal of Agricultural Science. Cambridge. 121:363-???.

CDA and USDA. 1997. (Colorado Department of Agriculture and United States Department of Agriculture) 1997. Colorado Agricultural Statistics. Colorado Agricultural Statistics Service. Lakewood, CO

Delgado, J.A. 1998. Sequential NLEAP simulations to examine effect of early and late planted winter cover crops on nitrogen dynamics. J. Soil and Water Conserv. 53:241-244.

Delgado, J.A. 1999. NLEAP simulation of soil type effects on residual soil NO_3-N in the San Luis Valley and potential use for precision agriculture. In P.C. Robert, R.H. Rust, and W.E. Larson (eds.) Proc. 4th Int. Conf. on Precision Agriculture. 1367-1378. ASA Madison, WI.

Delgado, J. 2001. Use of simulations for evaluation of best management practices on irrigated cropping systems. Chapter 10. In M.J. Shaffer et al. (eds.) Modeling Carbon and Nitrogen Dynamics for Soil Management. CRC Press. Boca Raton, FL.

Delgado, J.A., R.F. Follett, and M.J. Shaffer. 2000. Simulation of NO_3^--N dynamics for cropping systems with different rooting depths. J. Soil Sci Soc Am. 64:1050-1054.

Delgado, J.A., R.F. Follett, J.L. Sharkoff, M.K. Brodahl, and M.J. Shaffer. 1998*a*. NLEAP facts about nitrogen management. J. Soil Water Conservation 53:332-338.

Delgado, J., R. Riggenbach, R. Sparks, M. Dillon, L. Kawanabe, and R. Ristau. 1999*b*. Poster presented at the 16th Southern Rocky Mountain Potato/Grain Conference, Center Colorado. http://www.gpsr.colostate.edu/ SPNR /NLEAP/nleap3.htm

Delgado, J.A., R.J. Ristau, M.A. Dillon, H.R. Duke, A Stuebe, R.F. Follett, M.J. Shaffer, R.R. Riggenbach, R.T. Sparks, A. Thompson, L.M. Kawanabe, A. Kunugi, and K. Thompson. 2001 Use of Innovative Tools to Increase Nitrogen Use Efficiency and Protect Environmental Quality in Crop Rotations. Commun. Soil Sci. Plant Anal. *32:(7&8)(Invited Special Issue, Accepted, In Print)*

Delgado, J.A., M.J. Shaffer and M.K. Brodahl. 1998*b*. New NLEAP for shallow and deep rooted crop rotations. J. Soil and Water Conserv. 53:338-340.

Delgado, J.A., R.T. Sparks, R.F. Follett, J.L. Sharkoff, and R.R. Riggenbach. 1999*a*. Use of winter cover crops to conserve soil and water quality in the San Luis Valley of South Central Colorado. pp 125-142.R. Lal (ed.) Soil quality and soil erosion. CRC Press Boca Raton Fl.

Del Grosso, W.J. Parton, A.R. Moser, M.D. Hartman, J. Brenner, D.S. Ojima, and D.S. Schimel. 2001. Simulated interaction of soil carbon dynamics and nitrogen trace gas fluxes using the DAYCENT model.. Chapter 8. In M.J. Shaffer et al. (eds.) Modeling Carbon and Nitrogen Dynamics for Soil Management. CRC Press. Boca Raton, FL.

Duxbury, J.M., L.A. Harper, and A.R. Mosier. 1993. Contributions of agroecosystems to global climate change. In Harper, L.A., A.R. Mosier, J.M. Duxbury, and D.E. Rolston (eds.) Agricultural Ecosystems Effects on Trace Gases and Global Climate Change. ASA Special Pub. No 55. ASA Madison WI pp 1-18.

Eckersten, H., P-E. Hansson, and H. Johnsson. 1998. *SOILN model user's manual, version 9.2*, Division of Hydrotechnics, Communications 98:6, Swedish University of Agricultural Sciences, Uppsala, 113pp.

Eddy-Miller, C.A. 1993. Evaluation of shallow ground water wells as a method of monitoring nitrate leaching in the San Luis Valley. M.S. Thesis. Department of Agricultural and Chemical Engineering. Colorado State University.

Edelmann, P. and D.R. Buckles. 1984. Quality of ground water in agricultural areas of the San Luis Valley, south-central Colorado. Water-Resources Investigations Report 83-4281. U.S. Geological Survey.

Emery, P.A., R.J. Snipes, J.M. Dumeyer, J.M. Klein. 1973. Water in the San Luis Valley, south-central Colorado: Colorado Water Conservation Board Water Resources Circular 18, 26 p.

Follett, R.F., D.R. Keeney, and R.M. Cruse, (eds.). 1991. Managing nitrogen for groundwater quality and farm profitability. SSSA, Madison WI.

Follett, R.F. and D.J. Walker. 1989. Groundwater quality concerns about nitrogen. In Nitrogen Management and Groundwater Protection. R.F. Follett ed. Elsevier Sci. Pub. Pp 1-22.

Franko, U, 1996. Modelling approaches of soil organic matter turnover within the CANDY system. *In* "Evaluation of soil organic matter models using excisting, long-term datasets" (D. S. Powlson, P. Smith and J. U. Smith, eds.), Vol. NATO ASI Series I, pp. 247-254. Springer-Verlag, Heidelberg.

Grant, R.F. 1997. Changes in soil organic matter under different tillage and rotation: mathematical modelling in ecosys. Soil Sci. Soc. Am. J. 61:1159-1174.

Grant, R.F. 2001. A review of Canadian ecosystem model – ecosys. Chapter 6. In M.J. Shaffer et al. (eds.) Modeling Carbon and Nitrogen Dynamics for Soil Management. CRC Press. Boca Raton, FL.

Groenendijk, P. and J.G. Kroes. 1997. Modelling the nitrogen and phosphorus leaching to groundwater and surface water; ANIMO 3.5. Report 144, DLO Winand Staring Centre, Wageningen, Netherlands.

Hall, M.D., M.J. Shaffer, R.M. Waskom, and J.A. Delgado. 2000. Regional Nitrate Leaching Variability: What Makes a Difference in Northeastern Colorado. J. Water Resources Bulletin *(Accepted, In Print).*

Hansen, S., Jensen, H. E., Nielsen, N. E., and H. Svendsen. 1991. Simulation of nitrogen dynamics and biomass production in winter wheat using the Danish simulation model DAISY. Fertil. Res. 27. Pp. 245-259.

Hearne, G.A. and J.D. Dewey. 1988. Hydrologic analysis of the Rio Grande Basin north of Embudo, New Mexico, Colorado and New Mexico. Water-Resources Investigations Report 86-4113. U.S. Geological Survey.

Kaap, J.D., W. Ebert, G. Kraft, and M.K. Brodahl. 1995. Using NLEAP to develop land use strategies for municipal wellhead protection in the central Wisconsin sands. Proceedings of the Animal Waste and the Land-Water Interface Conference. Arkansas Water Resources Center. Fayetteville, AR.

Kersebaum, K.C. 1989. Die Simulation der Stickstoffdynamik von Ackerböden, Ph. D. thesis, University Hannover. 141p.

Knisel, W. G. (ed). 1993. GLEAMS, Groudwater Loading Effects of Agricultural Management Systems. Version 2.10. UGA-CPES-BAED, publication No. 5. 259pp.

Ma, L. and M.J. Shaffer. 2001. A review of carbon and nitrogen processes in nine U.S. soil nitrogen dynamics models. Chapter 4. In M.J. Shaffer et al. (eds.) Modeling Carbon and Nitrogen Dynamics for Soil Management. CRC Press. Boca Raton, FL.

McCracken, D.V., M.S. Smith, J.H. Grove, C.T. MacKown, and R.L. Blevins. 1994. Nitrate leaching as influenced by cover cropping and nitrogen source. Soil Sci. Soc. Am. J.58:1476-1483.

McGechan, M.B. and L. Wu. 2001. A review of carbon and nitrogen processes in European soil nitrogen dynamics models. Chapter 5. In M.J. Shaffer et al. (eds.) Modeling Carbon and Nitrogen Dynamics for Soil Management. CRC Press. Boca Raton, FL.

Milburn, P., J. E. Richards, C. Gartley, T. Pollock, H. O'Neill, and H. Bailey. 1990. Nitrate leaching from systematically tilled potato fields in New Brunswick, Canada. J. Environ. Qual. 19:448-454.

Molina, J.A.E., C.E. Clapp, M.J. Shaffer, F.W. Chichester, and W.E. Larson. 1983. NCSOIL, a model of nitrogen and carbon transformations in soil: Description, calibration, and behavior. Soil Sci. Soc. Am. J. 47:85-91.

Mosier, A., D. Schimel, D. Valentine, K. Bronson and W. Parton. 1991. Methane and nitrous oxide fluxes in native, fertilized and cultivated grasslands. Nature. 350:330-332.

Newbould, P. 1989. The use of nitrogen fertilizer in agriculture. Where do we go practically and ecologically? Plant Soil. 115:297-311.

Owens, L. B., and W. M. Edwards. 1994. Groundwater nitrate levels under fertilized grass and grass legume pastures. J. Environ. Qual. 23:752-758.

Parton, W.J., M. Hartman, D.S. Ojima and D.S. Schimel, 1998, DAYCENT: Its land surface submodel: description and testing, Global Planetary Change, 19: 35-48.

Parton, W. J., and P. E. Rasmussen, 1994, Long-term effects of crop management in wheat/fallow: II. CENTURY model simulations, Soil Science Society of America Journal, 58, 530-536.

Raun, W.R. and G.V. Johnson. 1999. Improving nitrogen use efficiency for cereal production.Agron. J. 91:357-363.

Ristau R.J. (ed.) 1999. Best Management Practices for Nutrient and Irrigation Management in the San Luis Valley. Colorado State University Cooperative Extension, Fort Collins, CO

Ritchie, J.T., D.C. Godwin, and S. Otter-Nacke. 1985. CERES Wheat. A simulation model of wheat growth and development. Texas A & M Univ. Press. College Station.

Shaffer, M.J., Brodahl, M.K. and. J.A. Delgado. 1998. NLEAP verion 1.20. USDA-ARS-GPRS Fort Collins, CO.

Shaffer, M.J., A.D. Halvorson, and F.J. Pierce. 1991. Nitrate leaching and economic analysis package (NLEAP): Model description and application. *In* R.F. Follett, D.R. Keeney and R.M. Cruse (Eds.), Managing Nitrogen for Groundwater Quality and Farm Profitability, Chapter 13. Soil Science Society America, Madison, WI. pp 285-322.

Shaffer, M.J., and W.E. Larson. 1987. NTRM, a soil-crop simulation model for nitrogen, tillage, and crop residue management. USDA-ARS Conserv. Res. Rep. 34-1.

Shaffer, M. J., K. Lasnik, X. Ou, and R. Flynn. 2001a. NLEAP Internet tools for estimating NO_3-N leaching and N_2O emissions. In: M. J. Shaffer, L. Ma, and S. Hansen (eds.) Modeling Carbon and Nitrogen Dynamics for Soil Management. CRC Press.

Shaffer M.J., L. Ma, and S. Hansen (eds.) 2001b. Modeling Carbon and Nitrogen Dynamics for Soil Management. CRC Press. Boca Raton, FL.

Shaffer, M.J., B.K. Wylie, and M.D. Hall. 1995. Identification and mitigation of nitrate leaching hot spots using NLEAP-GIS technology. J. Contaminant Hydrology 20:253-263.

Shaffer, M.J.and C. Xu. 1998. Simulation modeling in precision agriculture: Soil fertility P55 - 60. *In* Proc. Great Plains Soil Fertility Conference. Denver, Colorado. March 3-4, 1998, A.J. Schlegel (ed.) KSU Manhattan KS.

Smith S.J., J.S. Schepers, and L.K. Porter. 1990. Assessing and managing agricultural nitrogen losses to the environment. Pp 1-43. In Advances in Soil Science. B.A. (Stewart ed.). Springer-Verlag. New York USA.

Stoichev, D., M. Kercheva, and D. Stoicheva. 2001. NLEAP water quality applications in Bulgaria. Chapter 9. In M.J. Shaffer et al. (eds.) Modeling Carbon and Nitrogen Dynamics for Soil Management. CRC Press. Boca Raton, FL.

USDA-SCS (United States Department of Agriculture-Soil Conservation Service). 1973. Soil Survey of Alamosa Area, Colorado. USDA-SCS, Washington, DC.

USDA-SCS (United States Department of Agriculture-Soil Conservation Service). 1988. National Agronomy Manual. 2nd Edition, Washington, DC.

USDA (United States Department of Agriculture). 1997. Agricultural Statistics. U.S. Government Printing Office Washington D.C.

USEPA, (U.S. Environmental Protection Agency). 1989. Federal Register. 54 FR 22062, 22 May. USEPA, Washington, DC.

Wagenet, R.J. and J.L. Hutson. 1989. LEACHM: Leaching Estimation and Chemistry model- a process based model of water and solute movement, transformations, plant uptake and chemical reactions in the unsaturated zone. Continuum Vol. 2. Water Resour. Inst., Cornell Univ., Ithaca, NY.

Walthall, P. M., W. D. Brady, and R. L. Hutchinson, 1996. Cotton Production on the Macon Ridge. Louisiana Agriculture, Spring, 1996, pp. 5-9.

Williams, J.R.,P.T. Dyke, and C.A> Jones. 1983. EPIC- A model for assessing the effects of erosion on soil productivity. P. 555-572. *In* W.K. Lauenrroth et al., (ed.) Proc. of the Third Int. Conf. on State of the art in Ecol. modeling, Colorado State Univ., May 24-28, 1982. Elsevier Scientific, New York.

Wylie, B.K., M.J. Shaffer, M.K. Brodahl, D. Dubois, and D.G. Wagner. 1994. Predicting spatial distributions of nitrate leaching in northeastern Colorado. J. Soil and Water Conserv. 49:288-293.

Xu, C., M.J. Shaffer and M. Al-kaisi. 1998. Simulating the impact of management practices on nitrous oxide emissions. Soil Sci. Soc. Am. J. 62:736-742.

Zuydam, R.P., 1999. Centimeter-precision guidance of agricultural implements in the open field by means of real time kinematic DGPS. In P.C. Robert, R.,H. Rust, and W.E. Larson (eds.) Proc. 4th Int. Conf. on Precision Agriculture. 1367-1378.. ASA Madison, WI.

Nitrogen in the Environment: Sources, Problems, and Management
R.F. Follett and J.L. Hatfield (Eds)

Chapter 16. Simulated effects of land use, soil texture, and precipitation on N gas emissions using DAYCENT

S.J. Del Grosso,[a] W.J. Parton,[a] A.R. Mosier,[a,b] M.D. Hartman,[a] C.A. Keough,[a] G.A. Peterson,[c] D.S. Ojima,[a] and D.S. Schimel[d]

[a] Natural Resource Ecology Laboratory, Colorado State University, Fort Collins.

[b] Agriculture Research Service, U.S. Department of Agriculture, Fort Collins, CO.

[c] Soil and Crop Sciences Department, Colorado State University, Fort Collins.

[d] Max Planck Institut für Biogeochemie, Jena, Germany.

We describe the N gas (N_2O, NO_x, N_2) flux submodel used in the DAYCENT ecosystem model and demonstrate the ability of DAYCENT to simulate the low N gas emissions observed from native soils, the intermediate emissions associated with dryland agriculture, and the high emissions observed for irrigated agricultural soils. Although DAYCENT did not always capture the observed daily variability in N gas emissions, it did simulate the observed seasonal patterns of N gas emissions within systems and differences in mean emissions among systems reasonably well. DAYCENT was used to compare N gas emissions from soils for native range grass, winter wheat/fallow conventional till and no till, winter wheat/corn/fallow no till, irrigated corn and irrigated silage cropping. NO_x made up the majority of N gas fluxes in all cases followed by N_2O (2-21% of N gas fluxes) and N_2 (0.1-11% of N gas fluxes). Yearly N_2O emissions ranged from 0.06-0.1 kgN ha^{-1} yr^{-1} for native range grass, 0.25-0.67 kgN ha^{-1} yr^{-1} for winter wheat/fallow and silage cropping, and 1-1.5 kgN ha^{-1} yr^{-1} for the irrigated corn. NO_x emissions declined while N_2O and N_2 emissions increased as soil texture became finer. Soil water inputs, tillage, timing of crop/fallow periods and fertilizer application interact to control N gas emissions so generalizations regarding land use are difficult to make. Switching to no till without changing the winter wheat/fallow cropping schedule resulted in higher N_2O emissions because the increased soil water content induced by no till supported higher denitrification rates. However, the soil water savings associated with no till also allows a reduction in the fallow period and the 3-year winter wheat/corn/fallow rotations had lower N_2O and NO_x emissions than the 2-year winter wheat/fallow systems considered.

1. INTRODUCTION

Flows of nitrogen (N) between the atmosphere, soil, and biota strongly influence the carbon (C) cycle and atmospheric chemistry. Net primary productivity (NPP) is limited by N availability in most terrestrial ecosystems (Vitousek and Howarth, 1991; Levin, 1989) and N transformations in soils are a major source of N gas emissions to the atmosphere (Davidson and Kingerlee, 1997; Kroeze et al., 1999). Nitrous oxide (N_2O) is a long-lived, important greenhouse gas (Prather et al., 1995). Nitric oxide (NO) and its oxidized counter part, NO_2, (together referred to as NO_x) are major ozone regulators and limit the overall oxidizing capacity of the troposphere (Williams et al., 1992). Human activity has profoundly altered fixation rates of atmospheric N_2 and emission rates of N_2O and NO_x to the atmosphere. The amount of reactive N induced into the biosphere from fertilizer production, N-fixation in crops and fossil fuel combustion exceeds the N fixed annually in natural systems (Smil, 1999; Vitousek et al., 1997; Galloway et al., 1995). Anthropogenic activities, mainly fossil fuel burning and agriculture, are major sources of atmospheric NO_x (Prather et al., 1995) while biogenic processes are the major source of N_2O (Kroeze et al., 1999).

The atmospheric concentration of N_2O has been well documented for current and historical time periods (Prather et al., 1995). In contrast, the amount of N in soils and the biota, as well as the flows of N that contribute to the observed changes in atmospheric N_2O, cannot be measured directly on the global scale. Ecosystem models are necessary to scale up results of plot sized experiments and calculate the contributions of natural and managed systems to global N budgets. Simple empirical models correlate N gas flux with N additions (IPCC, 1997) or with soil water content (Davidson and Verchot, 2000). At the opposite extreme, highly mechanistic models explicitly simulate the biological, physical, and chemical processes involved in N transformations and flows (Grant and Pattey, 1999). Simple models tend to be over-generalized and cannot represent the heterogeneity of real world systems while mechanistic models require detailed parameterization and intensive computation. DAYCENT is an ecosystem model of intermediate complexity, some processes are represented mechanistically but the model requires a relatively small number of site specific parameters. In this chapter we begin with a brief overview of the DAYCENT model and describe the N gas submodel of DAYCENT in detail. Then we present comparisons of simulated and observed values of N gas emissions to demonstrate the validity of DAYCENT. Lastly, we use DAYCENT to compare annual N gas (NO_x, N_2O and N_2) emissions associated with different land uses, soil textures, and weather data.

2. DAYCENT MODEL DESCRIPTION

DAYCENT (Kelly et al., 2000; Parton et al., 1998; Del Grosso et al., 2001) is the daily time step version of the CENTURY model. CENTURY (Parton et al., 1994) operates on a monthly time step and was developed to simulate changes in soil organic matter (SOM), plant productivity, nutrient availability and other ecosystem parameters in response to changes in land management and climate. However, finer time scale resolution is required to simulate N gas emissions from soils because the processes that result in N gas fluxes respond nonlinearly to important controls such as soil water content. DAYCENT simulates exchanges of carbon, nutrients, and trace gases among the atmosphere, soil, and plants as well as events and management practices such as fire, grazing, cultivation, and organic matter or fertilizer additions. To run DAYCENT for a particular site, soil texture, current and historical land use, and daily maximum/minimum temperature and precipitation data are required. Soil water content, temperature, mineral N concentration, trace gas flux, and SOM decomposition are simulated on a daily time step while plant growth is updated weekly.

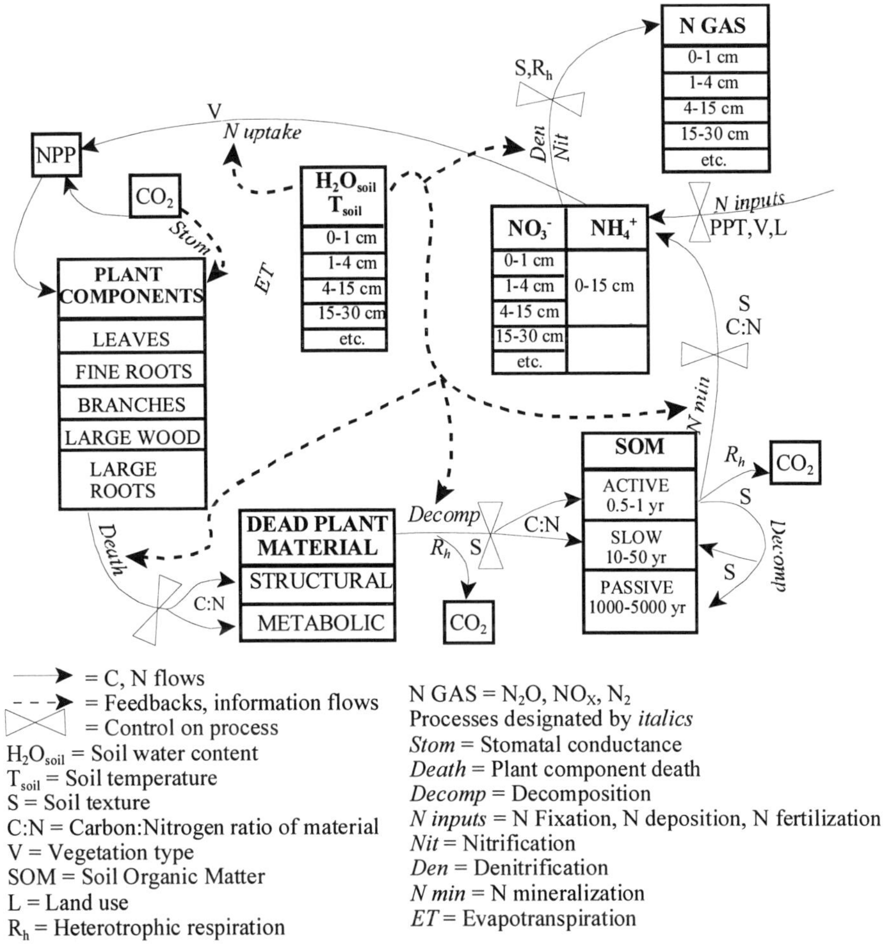

Figure 1. Conceptual diagram of the DAYCENT ecosystem model, modified form Del Grosso et al. (2001).

DAYCENT (Figure 1) includes submodels for plant productivity, decomposition of dead plant material and SOM, soil water and temperature dynamics, and N gas fluxes. Flows of C and nutrients are controlled by the amount of C in the various pools, the N concentrations of the pools, abiotic temperature/soil water factors, and soil physical properties related to texture. Net Primary Productivity (NPP) is a function of nutrient availability, soil water and temperature, shading, and vegetation type (Metherell et al., 1993). NPP is divided among leafy, woody, and root compartments based on plant type. The root to shoot ratio of NPP allocation is a function of soil water content and mineral N availability. The death rate of plant compartments is controlled by soil water, temperature, season, and plant specific senescence parameters. Structural detritus has a higher C:N ratio and is more difficult to decompose than metabolic detritus. SOM is divided into three pools based on decomposition rates (Parton et al., 1993; 1994). Decomposed detrital material that has a low C:N ratio flows to the active SOM pool, which includes microbial biomass and the highly labile by-products of decomposition that turnover in approximately 1 year or less. The products of detrital decomposition that have a wider C:N ratio flow to the slow SOM pool, which includes the relatively resistant (10-50 year turnover rate) by-products of decomposition. The passive SOM pool consists of humus that is extremely resistant to further decomposition. As soils become finer textured a lower portion of SOM is respired as CO_2 and more SOM is retained in stable form due to physical and chemical protection. Decomposition of SOM and external nutrient additions supply the nutrient pool, which is available for plant growth and microbial processes that result in trace gas fluxes. Ammonium (NH_4^+) is modeled for the top 15 cm while nitrate (NO_3^-) is distributed throughout the soil profile. Nutrients and SOM are concentrated near the soil surface and decrease exponentially with depth. The land surface submodel of DAYCENT simulates water flow through the plant canopy, litter and soil profile, as well as soil temperature throughout the profile (Parton et al., 1998; Eitzinger et al., in press). DAYCENT has been shown to reliably model soil water content, N mineralization, and NPP for a shortgrass steppe in Colorado (Kelly et al., 2000). The SOM and N cycling submodels used in DAYCENT have been validated for various systems including grasslands and forests (Kelly et al., 1997), as well as agricultural soils in Sweden (Paustian et al., 1992) and Oregon (Parton and Rasmussen, 1994).

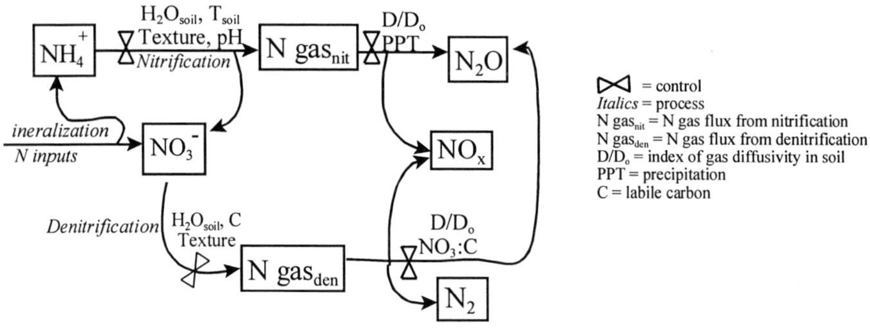

Figure 2. The nitrogen gas flux submodel of DAYCENT, modified from Del Grosso et. al. (2001).

The N gas submodel of DAYCENT (Figure 2) simulates soil N_2O and NO_x gas emissions from nitrification and denitrification as well as N_2 emissions from denitrification. Nitrifying microbes oxidize NH_4^+ to NO_3^-, with some N_2O and NO_x released during the intermediate steps. N gas flux from nitrification is assumed to be a function of soil NH_4^+ concentration, water content, temperature, and pH (Parton et al., 1996; Paron et al., in press). Nitrification rates increase linearly with soil NH_4^+ concentration. Nitrification is limited by moisture stress on biological activity when soil water-filled pore space (WFPS = % relative saturation) is too low and by O_2 availability when WFPS is too high. Nitrification increases exponentially with temperature and stabilizes when soil temperature exceeds the site specific average high temperature for the warmest month of the year. Nitrification is not limited when pH is greater than neutral but decreases exponentially as soils become acidic.

Denitrification is an anaerobic process in which heterotrophic microbes reduce NO_3^- to NO_x, N_2O, and N_2. Denitrification is a function of soil NO_3^- (e^- acceptor) concentration, labile C (e^- donor) availability, WFPS, and soil physical properties related to texture that influence gas diffusivity (Parton et al., 1996; Del Grosso et al., 2000a). Denitrification increases exponentially with increasing soil NO_3^- concentration when NO_3^- concentration is low (< 50 ppm) and approximately linearly at higher NO_3^- levels. Denitrification increases approximately linearly with soil heterotrophic respiration, a proxy for labile C availability. No denitrification is assumed to occur until WFPS values exceed 50-60%, then denitrification increases exponentially until WFPS reaches 70-80% and it stabilizes as soil water content approaches saturation. The model calculates $N_2 + N_2O$ emissions from denitrification by assuming that the process is controlled by the input (NO_3^-, respiration, WFPS) that is most limiting. N_2O emissions are calculated from $N_2 + N_2O$ gas emissions and an N_2:N_2O ratio function. The ratio of N_2:N_2O gases emitted due to denitrification is assumed to increase as the ratio of e^- acceptor (NO_3^-) to e^- donor (labile C) decreases and as soil gas diffusivity and O_2 availability decrease. N_2O can act as an alternative e^- acceptor and be reduced to N_2 when labile C is in excess compared to NO_3^-. D/D_o, a relative index of gas diffusivity in soils, is calculated as a function of WFPS and soil physical properties (bulk density and field capacity) that influence gas diffusion rates using equations presented by Potter et al. (1996). As D/D_o decreases, the residence time of N_2O in soil increases, thus increasing the probability that N_2O will be further reduced to N_2 before diffusing from the soil.

NO_x emissions from soil are a function of total N_2O emissions, a NO_x/N_2O ratio equation, and a precipitation initiated pulse multiplier (Parton et al., submitted). Simulated N_2O gas emissions from nitrification and denitrification are summed to obtain total N_2O flux. The NO_x/N_2O ratio is high (maximum of ~20) when D/D_o is high and decreases to a minimum of ~1 as D/D_o decreases. This is based on the following observations. The majority of NO_x emissions from soils are from nitrification because NO_x is highly reactive under the reducing conditions that facilitate denitrification (Conrad, 1996). Total N gas flux is due primarily to nitrification when soils are well aerated (high D/D_o) and mainly to denitrification under anaerobic conditions (Davidson, 1993; Linn and Doran, 1984). Thus, the model assumes that NO_x becomes a larger proportion of total N gas emissions as soil gas diffusivity increases. The modeled total N_2O emission rate is multiplied by the ratio function to obtain a base NO_x emission rate. The base NO_x emission rate may be modified by a pulse multiplier. Large pulses of NO_x are often initiated when precipitation falls on soils that were previously dry (Smart et al., 1999; Martin et al., 1998; Hutchinson et al., 1993). The pulses are thought to be related to substrate accumulation and activation of water stressed bacteria upon wetting (Davidson, 1992). To account for these pulses the model incorporates the pulse multiplier submodel described by Yienger and Levy (1995). The magnitude of the multiplier is proportional to the amount of precipitation and the number of days since the latest precipitation event, with a maximum multiplier of 15.

On a daily time step simulated values of soil NH_4^+, NO_3^-, heterotrophic CO_2 respiration, water content, temperature and site specific values for soil texture and physical properties are used to calculate N_2O emissions from nitrification and denitrification and N_2 emissions from denitrification. Total N_2O emissions, a NO_x/N_2O ratio function, and a pulse multiplier are used to calculate NO_x emissions. N balance is verified on a daily basis and calculated potential N gas emission rates are revised downwards if there is not enough NO_3^- and NH_4^+ available to supply the potential N gas emissions for a particular time step.

3. DAYCENT MODEL VALIDATION

We first summarize results of previous tests of the DAYCENT model for various natural and managed systems then present results of tests with an updated version of the model on soils used for winter wheat/fallow cropping. Frolking et al. (1998) compared simulated and observed values of soil water content, mineral N, and N_2O emissions for soils in Colorado, Scotland, and Germany. The Colorado site is a dry shortgrass steppe (annual ppt. ~360 mm), the Scotland site is a fairly moist ryegrass pasture (annual ppt. ~850 mm), and the German sites are perennially cropped (annual ppt. ~830 mm). DAYCENT correctly simulated the observed low N_2O fluxes for the shortgrass steppe, the moderate N_2O emissions for the Scotland site and the high N_2O emissions observed for the cropped soils. DAYCENT also simulated the observed daily variability in N_2O emission rates and soil water content reasonably well for the different sites. Parton et al. (in press) tested DAYCENT simulations of soil water content, soil temperature, and N_2O and NO_x gas emissions from rangeland soils of varying texture and fertility levels. DAYCENT simulated soil temperature, WFPS, and NO_x emissions generally well on daily and seasonal bases, although winter season WFPS values were not well represented. DAYCENT did not accurately represent the observed daily variability in N_2O emissions. However, DAYCENT simulated the observed seasonality of N_2O emissions fairly well, although winter season N_2O emissions tended to be underestimated. Del Grosso et al. (2001) tested the ability of DAYCENT to simulate soil water, temperature, NH_4^+, NO_3^-, and N_2O emissions for irrigated, fertilized barley (*Hordeum vulgare*) and corn (*Zea mays*) crops. Similar to the rangeland soils, soil temperature and water were generally simulated rather well, but winter season WFPS values were not accurately represented by the model. DAYCENT correctly simulated high values of NH_4^+ and NO_3^- after fertilization in spring and decreasing values during the growing season. Simulated values of N_2O emissions compared favorably with the data for the corn crop, but N_2O emissions were overestimated for the barley crop.

The latest version of DAYCENT was tested with N_2O gas flux data collected from winter wheat/fallow fields during 1992-1995 (Mosier et al., 1997). The fields are on private land in northeastern Colorado and had been used for winter wheat/fallow cropping for over 60 years. We tested DAYCENT with data from two fields, one (WE) had been planted with winter wheat in September of even years and the other (WW) had been planted with winter wheat in September of odd years. Both fields are classified as sandy loams and were fertilized with sewage sludge during the fallow season starting in 1990. Herbicides were used to control weeds and conventional tillage cultivation was used. Figure 3 shows a time series of simulated and observed N_2O fluxes for each field. The day to day variability in N_2O flux is very high, but DAYCENT correctly simulated the observed patterns of N_2O flux fairly well. Although some individual points are greatly over or under estimated, the model correctly simulated many of the observed high N_2O fluxes associated with snow melt events during the spring as well as some of the high N_2O fluxes observed during the summer. The model often mistimes the large N_2O flux events by a few days but this may be at least partially due to differences between the precipitation data used to drive the model and the precipitation that

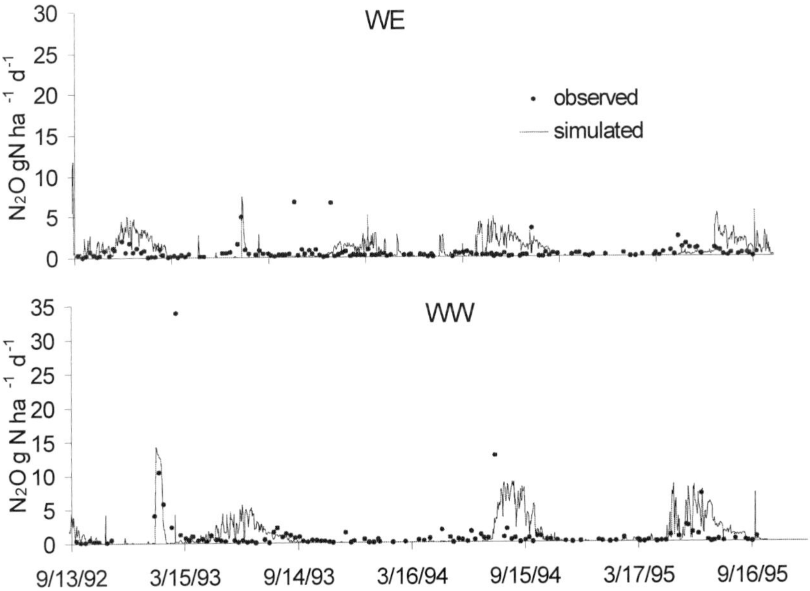

Figure 3. Comparison of observed and simulated daily N_2O gas emission rates from two fields used for winter wheat/fallow rotations (Mosier et al., 1997).

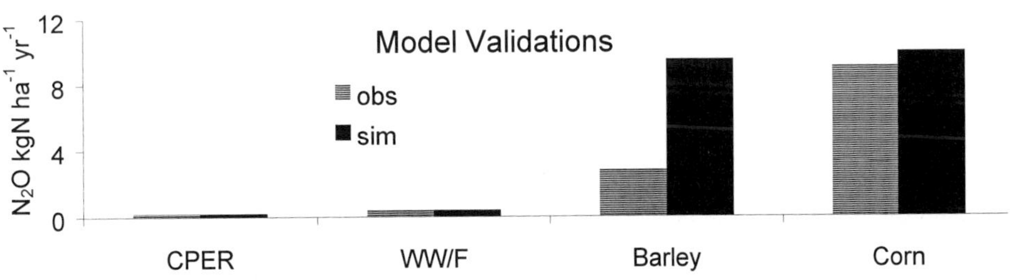

Figure 4. Comparison of observed and simulated average annual N_2O emissions from the Central Plains Experimental Range (Mosier et al., 1996), winter wheat/fallow rotations (Mosier et al., 1997), and irrigated barley and corn cropping (Mosier et al., 1986).

fell on the fields. The model was driven with weather data from the Central Plains Experimental Range (CPER) which is ~20 km from the winter wheat/fallow fields. Comparisons of average simulated and observed annual N_2O emissions for the winter wheat/fallow fields and soils under alternative land uses that were tested show that the model does well in most cases (Figure 4). The barley field was fertilized at the same level as the corn field and the spring the barley was planted was relatively wet (Mosier et al., 1986) but the simulated high N_2O emissions due to denitrification episodes were not observed. A possible explanation is that growth patterns of the barley affect N availability for microbial processes that result in N_2O gas emissions.

Results of tests with many soils show that the DAYCENT model does not always reliably simulate the observed daily variability in N_2O fluxes but does accurately simulate differences in N_2O fluxes between different sites and among seasons for a given site. Difficulties in modeling soil water content in winter and spring are responsible for some of the errors in simulated N_2O emission rates. WFPS is an important driver of the processes that control N_2O emission rates but heterogeneity in snow drifting and snow melting make it difficult to simulate soil WFPS during winter and spring. This variability of important model drivers on smaller scales than are resolved by the model contributes to the observed model error. The ability of DAYCENT to reliably simulate NO_x emissions has not been extensively tested but Parton et al. (in press) showed that the model represented the observed monthly patterns of NO_x flux and captured the observed differences in average NO_x flux from rangeland soils of varying texture and fertility levels. The ability of DAYCENT to simulate N_2 emissions has not been extensively validated because little field data for N_2 emissions exist. However, the denitrification submodel reliably modeled (r^2=0.47) daily N_2+N_2O emissions from agricultural soils in Pakistan (Del Grosso et al., 2000a).

4. DAYCENT MODEL APPLICATION

The effects of land use, soil texture, and precipitation on N gas emissions from soils in the U.S. Great Plains were compared using the DAYCENT model. The U.S. Great Plains is a semi-arid region with annual precipitation ranging from 300-600 mm. The winter wheat/fallow cropping system has been used extensively in the Great Plains because soil water stored during the fallow season leads to higher water availability during the cropped season. However, this system has led to a 50% decrease in soil organic matter (SOM) after 30-50 years (Peterson et al., 1998). Alternatives to winter wheat/fallow cropping have been proposed which store C in SOM depleted soil and compensate for CO_2 gas emissions (Bruce et al., 1999; Lal et al., 1998). However, the greenhouse effects of N_2O emissions must also be included in calculations of net greenhouse gas exchanges between the atmosphere and the soil (Del Grosso et al., 2000b; Robertson et al., 2000). NO_x budgets are important because NO_x affects the oxidizing capacity of the troposphere. From an N balance perspective it is desirable to account for total N gas (N_2O, NO_x and N_2) emissions from soils. We performed DAYCENT simulations and compiled annual N_2O, NO_x and N_2 emissions from soils of different texture, moisture, and cropping scenarios. We used weather data from the Central Plains Experimental Range (CPER) in northern Colorado (ann. ppt. ~350 mm, ann. temp. ~9.5 ^{o}C) to simulate N gas emissions from sandy loam, loam, and clay loam soils used for conventional winter wheat/fallow, no till winter wheat/fallow, irrigated corn, irrigated silage, and native range grass. Weather data from Sterling, Colorado (ann. ppt. ~400 mm, ann. temp. ~9.7 ^{o}C) were used to compare N gas emissions from a loam soil under winter wheat/fallow conventional tillage, winter wheat/fallow no till, and winter wheat/corn/fallow rotations under no till. Land use alternatives are summarized in Table 1.

Table 1. Land management practices simulated by DAYCENT

crop/vegetation	N addition ($gN\ m^{-2}\ yr^{-1}$)	CPER weather month(s) of N addition	irrigation	tillage
wwf-ct	2.5	9,4	no	conventional
wwf-nt	2.5	9,4	no	no till
corn	10	5,7	yes	conventional
sil	6.25^m	12	yes	conventional
rg	0		no	none
		Sterling weather		
wwf-ct	2.5	9,4	no	conventional
wwf-nt	2.5	9,4	no	no till
wwcf-nt	3.3	$9^w,4^w,5^c$	no	no till

wwf-ct = winter wheat/fallow rotations, conventional till
wwf-nt = winter wheat/fallow rotations, no till
corn = perennial corn cropping, conventional till
sil = perennial silage cropping, conventional till
rg = native range grass, grazed from April-November
wwcf-nt = winter wheat/corn/fallow rotations, no till
^{m} = N from manure, all others from ammonium nitrate fertilizer
^{w} = fertilizer application for winter wheat years
^{c} = fertilizer application for corn year

Winter wheat/fallow rotations were simulated by planting in September, harvesting the following July, and leaving the field fallow from August of the harvest year through August of the following year. Corn and silage were planted in May and harvested in September. To simulate winter wheat/corn/fallow rotations, winter wheat was planted in September of the 1st year and harvested the following July. The field was left fallow until the following year when dryland corn was planted in May and harvested in October. The field was left fallow after the corn harvest until winter wheat was planted again the following September.

Before simulating the different land use scenarios, SOM and mineral N (NH_4^+, NO_3^-) pools were initialized for the different soils by running the model for 75 years of winter wheat/fallow cropping under conventional tillage using the respective weather files and soil parameters. The various land use scenarios were simulated for 50 years and average annual N_2O, NO_x and N_2 emission rates were calculated.

To illustrate the inter-annual variability in N gas emissions, Figure 5 shows 15-year time series of NO_x and N_2O emissions for the three management alternatives along with precipitation data from the Sterling weather file. The winter wheat/fallow rotations had high N gas emissions during the fallow year and low emissions during the planted year. The no till winter wheat/fallow option had the highest NO_x and N_2O emissions during most years but some years show emissions from the other two scenarios that equal or exceed those from the no till winter wheat/fallow. The 3-year winter wheat/corn/fallow rotation is less variable because fallow periods during the summer months are reduced. In addition to the timing of crop/fallow seasons, variability is driven by interactions involving the timing of planting, harvesting, tillage, N addition and climate patterns. 1990 had very high N gas emissions from the winter wheat/fallow no till because it was fallow during the summer of that year and precipitation was high. The emissions for the winter wheat/corn/fallow alternative were low

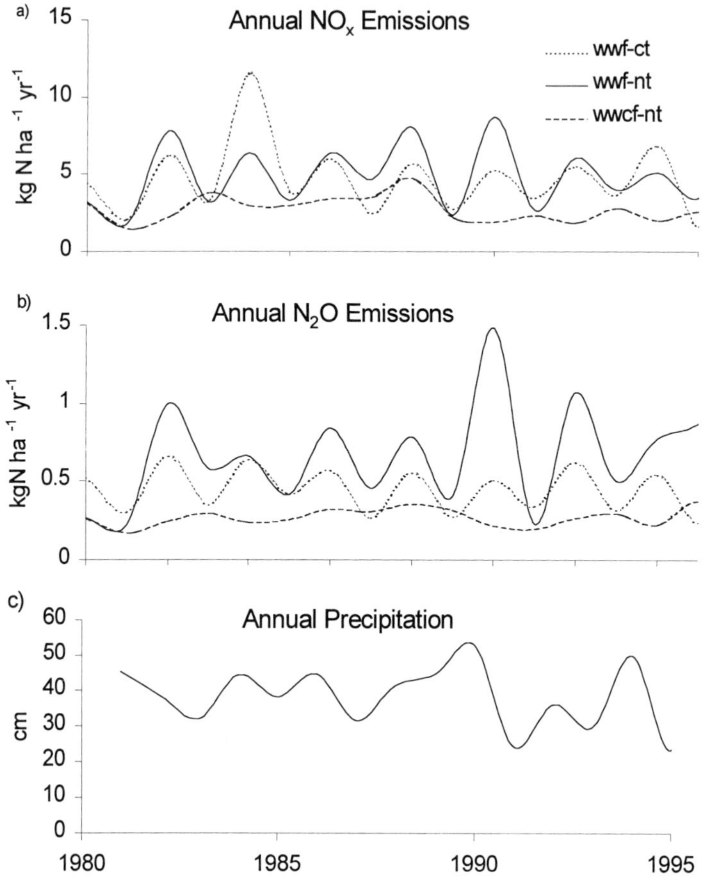

Figure 5. Simulated annual NO_x (a) and N_2O (b) gas emissions and precipitation data (c) from Sterling, CO.

in 1990 because corn was growing during the summer of that year so less N was available for nitrification and denitrification. To isolate the effects of land use, soil texture, and precipitation, we averaged N gas emissions from 50-year simulations of each scenario.

Average annual N gas emission rates for the sandy loam, loam, and clay loam soils driven with the CPER weather data were highest for the irrigated corn and lowest for the native range grass (Figure 6a). NO_x emissions were strongly affected by texture with decreasing sand content associated with decreasing NO_x emissions (Figure 6b). N_2O emissions were highest for the corn, lowest for the range grass, and increased as sand content decreased (Figure 6c). N_2 emissions were highest for the corn and tended to increase as sand content

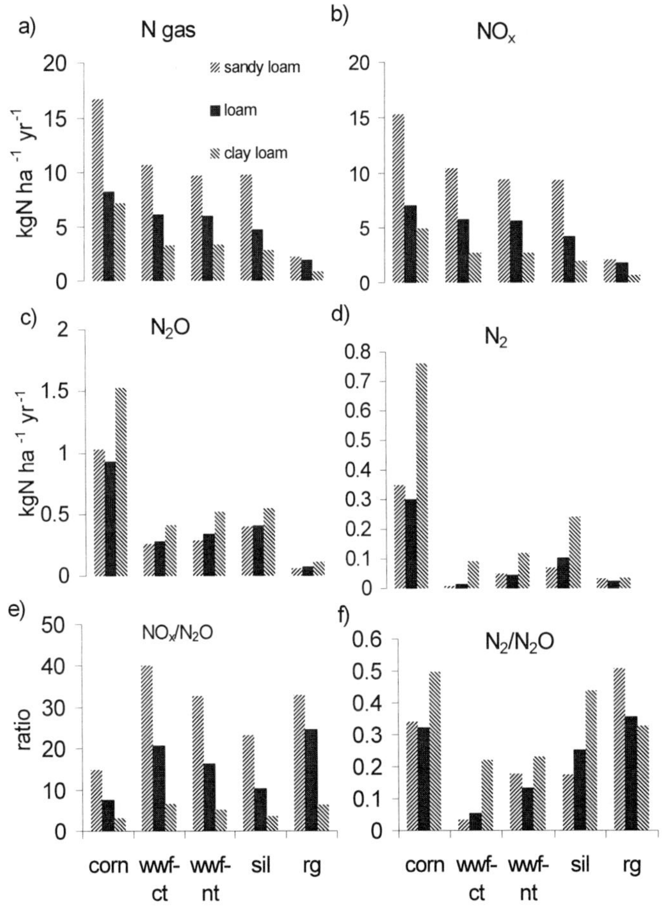

Figure 6. Simulated average annual N GAS (a), NO_x (b), N_2O (c), and N2 (d) emissions and NO_x/N_2O (e) and N_2/N_2O (f) ratios for soils driven with weather data from the CPER.

decreased (Figure 6d). The majority of total N gas emissions from soil were in the form of NO_x. N_2O emissions were responsible for 2 to 21% of the total N gas fluxes. N_2 emissions represented a low proportion of the total N gas emissions in most cases (0.1 - 11%) and were highest for the corn grown in clay loam soil (Figure 6d). We must point out that these results are based on calculations of soil N gas emissions and that ecosystem losses of N due to NO_x emissions are likely to be much lower because 50% or more of soil NO_x emissions may be absorbed by the plant canopy (Yienger and Levy, 1995).

Figures 6e and 6f show the ratios of NO_x/N_2O and N_2/N_2O gases emitted for the various scenarios considered. Average NO_x/N_2O ratios were always greater than one and decreased as soil texture became finer. Finer soil texture is associated with lower gas diffusivity and higher water retention which favor the anaerobic conditions that facilitate reduction of NO_x to N_2O. The ratio of N_2/N_2O gases emitted was always less than one for the simulations performed. It should be noted that the ratio of N_2/N_2O gases emitted from denitrification is usually greater than one but figures 6e and 6f include N_2O fluxes from nitrification and denitrification. The ratio of N_2/N_2O gases tended to increase as clay content increased because the anoxic conditions that favor the reduction of N_2O to N_2 are more prevalent in finer textured soils.

The corn had the highest N gas emissions because it was fertilized and irrigated. Fertilization supplies labile N and irrigation leads to higher soil water contents so denitrification rates increase. The silage was irrigated to the same extent as the corn but had lower emissions because it was amended with manure instead of ammonium nitrate fertilizer and total N applied to the silage was less than the corn (Table 1). N associated with manure addition is mineralized relatively slowly so less is available for nitrification and denitrification than with fertilizer addition. In all cases except for the irrigated corn grown in the clay loam soil, nitrification was responsible for the majority of total N_2O fluxes. The majority of total fluxes being in the form of NO_x and the small N_2 emissions also resulted from nitrification being a more important agent of N gas loss than denitrification for most of the scenarios considered. The relatively high N_2 loss associated with the irrigated corn grown in clay loam results from denitrification being responsible for a the majority of the total N gas flux under this scenario. NO_x, N_2O, and N_2 emissions were not sensitive to no till versus conventional till for winter wheat/fallow cropping simulated with the CPER weather data.

The loam soil driven with weather data from Sterling, Colorado had the highest N gas, NO_x and N_2O emissions for the winter wheat/fallow no till treatment and lowest N gas, NO_x and N_2O emissions for the winter wheat/corn/fallow no till rotations (Figures 7a-7c). N_2 emissions were highest for the winter wheat/fallow no till rotations and roughly equivalent for the other two options (Figure 7d). N_2 emissions contributed less then 3% of the total N gas emissions for each scenario considered with the Sterling weather, indicating that a higher proportion of N gas flux was due to nitrification compared to denitrification. The higher N_2O and N_2 gas emissions for the no till winter wheat/fallow compared to the conventional till winter wheat/fallow is a result of higher denitrification rates in the no till treatment. No till conserves soil moisture which favors the anaerobic conditions that facilitate denitrification. The NO_x/N_2O ratio was lowest for the no till winter wheat/fallow (Figure 7e) while the N_2/N_2O ratio was highest for this option (Figure 7f). Both of these observations suggest that denitrification, while responsible for a minority of total N gas emissions for these simulations, is relatively more important for the no till winter wheat/fallow system. This is a result of higher water levels associated with no till leading to anaerobic conditions that favor the reduction of NO_x to N_2O and of N_2O to N_2.

More N fertilizer was added with the winter wheat/corn/fallow rotations but total N gas emissions were less (Figure 7a). At least two factors besides N additions regulate N gas emissions. First, there is a decrease in the number of fallow months, particularly during the summer, in the 3-year (wwcf-nt) compared to the 2-year rotations (wwf-ct). When the soil is

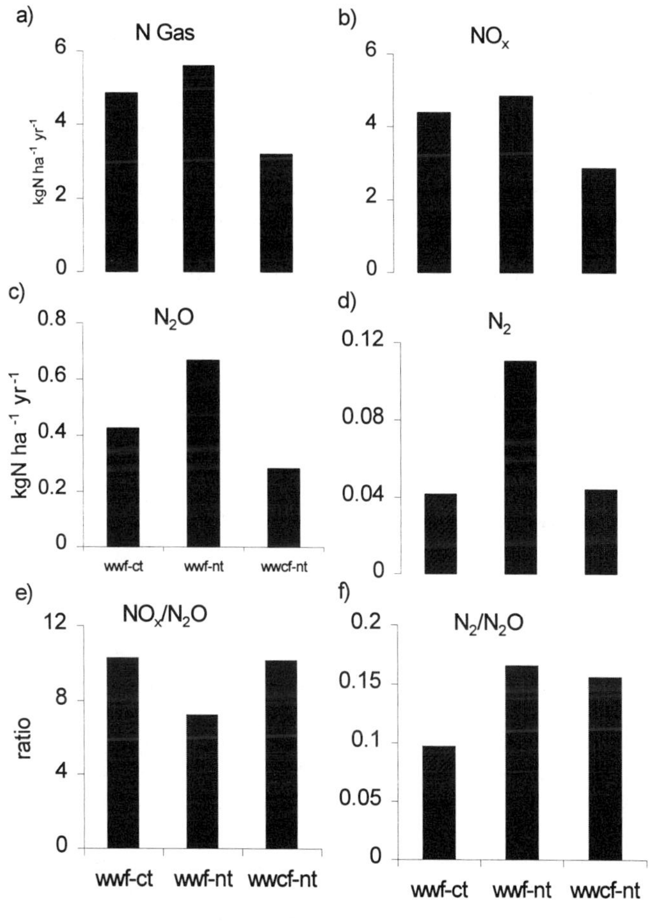

Figure 7. Simulated average annual N gas (a), NO_x (b), N_2O (c), and N_2 (d) emissions and NO_x/N_2O (e) and N_2/N_2O (f) ratios for a loam soil driven with weather data from Sterling, CO.

fallow there is no plant uptake of N so more N is available for microbial processes (nitrification and denitrification) that result in N gas emissions. Second, the 3-year rotation leads to an increase in SOM while SOM levels are stable with the conventional winter wheat/fallow system. When SOM is increasing, N availability for processes that result in N gas loss is likely to decrease. The major benefit of the no till system is that it conserves soil water so the fallow period can be reduced without risk of soil water levels dropping too low to support successful cropping. Independent of other factors, no till may induce higher N gas emissions because higher soil water levels in these systems, which tend to be water limited, can facilitate denitrification. The decrease in N gas emissions from reducing the fallow period and from increasing SOM more than compensates for the increase associated with no till, therefore the 3-year no till rotation has lower emissions than the 2-year conventional till rotation.

Comparison of the winter wheat/fallow conventional till and no till options for simulations of the loam soil driven by the CPER weather versus the same options driven by the Sterling weather (Figure 8) shows that N_2O and N_2 emissions were higher for the wetter (Sterling) simulations and NO_x emissions were higher for the CPER simulations. Both of these result from higher soil water levels leading to higher denitrification rates and lower nitrification rates in the Sterling soil compared to the drier CPER soil. The winter wheat/fallow rotations were sensitive to tillage type with the Sterling weather but not with the CPER weather. This is a result of interactions with climate. Average annual precipitation is somewhat higher in Sterling (~40 vs. ~35 cm). The higher precipitation coupled with the effects of no till on soil moisture

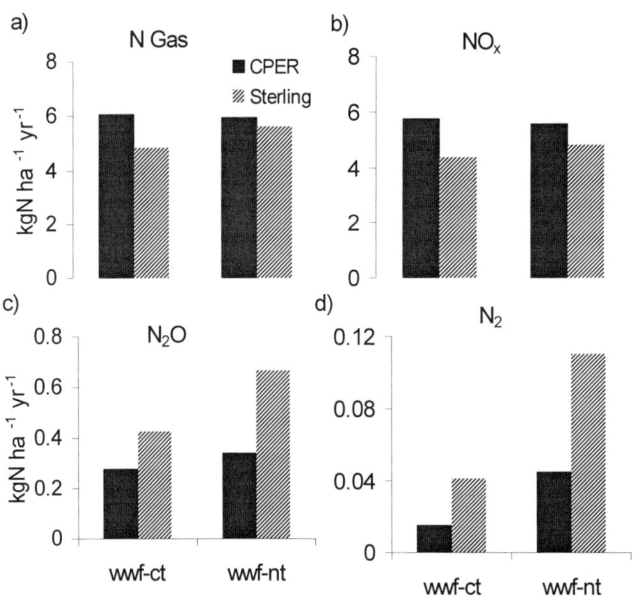

Figure 8. Comparison of simulated average annual N gas (a), NOx (b), N_2O (c), and N_2 (d) emissions for a loam soil driven with weather data from Sterling, CO and the CPER.

induce denitrification episodes. The effects of a moderate increase in precipitation or of no till cultivation in isolation had little effect on N gas fluxes from the winter wheat/fallow system. But the interaction of these two factors led to increased N_2O emissions because the response of denitrification to soil water content is highly non-linear (Del Grosso et al., 2000a).

5. SUMMARY AND CONCLUSIONS

We have described the DAYCENT ecosystem model and shown that annual N_2O emissions can be reliably simulated for some native and managed systems. The model was used to explore how N gas emissions respond to changes in land management, soil texture, and precipitation. Initial values for soil organic C and mineral N levels of the various pools simulated by DAYCENT were obtained by assuming 75 years of winter wheat/fallow cropping. The winter wheat/fallow system has been a common practice in the arid Great Plains but it results in the depletion of SOM. As expected, total N gas and N_2O emissions were higher from fertilized dryland winter wheat/fallow and winter wheat/corn/fallow rotations than native range grass. The highest total N gas and N_2O emissions were from the irrigated perennially cropped corn which had higher levels of N inputs than the dryland cropping. The silage option had N gas emissions similar to the winter wheat/fallow scenarios even though the silage had higher levels of N addition and was irrigated. This can be explained by two factors. First, the silage was amended with manure instead of N fertilizer so a smaller proportion of the applied N was available for nitrification and denitrification. Second, conversion to the silage option resulted in increasing SOM while SOM levels were stable with the winter wheat/fallow systems.

Soil texture had significant effects on total N gas emissions and the relative proportions of the N gas species emitted. NO_x and total N gas emissions tended to decrease as soils became finer textured while N_2O and N_2 emissions tended to increase (Figure 6). This is related to the effects of soil texture on gas diffusivity. As soil texture becomes finer water retention tends to increase and gas diffusivity tends to decrease. These conditions contribute to soil anoxia and increase the probability that N oxides produced from nitrification and denitrification will be reduced to N_2O or N_2 before emission from the soil. Simulated denitrification rates are insensitive to soil texture for loam and coarser textured soils but increased substantially in clay loam and finer textured soils. Total N gas losses decreased as soil texture became finer because NO_x emissions decreased and NO_x emissions formed the majority of N gas fluxes. From a greenhouse gas perspective, fine textured soils are expected to emit more N_2O, but from an N balance perspective, fine textured soils are expected to show smaller total N gas losses from the system.

Simulations show that N gas fluxes from soils respond nonlinearly to controls and that interactions among controls are important. Using the CPER weather, tillage type for winter wheat/fallow rotations had little effect on N gas emissions. But with the wetter Sterling data, the no till winter wheat fallow had somewhat higher N_2O and slightly higher NO_x emissions. It is also necessary to consider the interactions between the C and N cycles. Soils that are gaining C will show lower N gas emissions under a given land use than soils that are C saturated. The simulated clay loam soil showed lower total N gas losses because finer textured soils that are C depleted have a higher potential to store C than coarser textured soils.

We conclude that mitigation of N gas losses from soils is strongly dependent on land management but that generalizations based solely on soil N and water inputs are likely to be limited because soil texture, SOM levels, and the timing of cropping/fallow seasons are also important.

ACKNOWLEDGMENTS

The research for this paper was supported by the National Aeronautics and Space Administration, the National Science Foundation, the Department of Energy and the Environmental Protection Agency through the following grants; NASA-EOS NAGW 2662, NSF-LTER BSR9011659, DOE NIGEC LWT62-123-06516 and EPA Regional Assessment R824939-01-0.

REFERENCES

Bruce, J.B., M. Frome, E. Haites, H. Janzen,R. Lal and K. Paustian. 1999. Carbon sequestration in soils. J. Soil and Water Conserv. 54:382-389.

Conrad, R. 1996. Soil microorganisms as controllers of atmospheric trace gases (H_2, CO, CH_4, OCS, N_2O, and NO). Microbiol. Rev. 60:609-640.

Davidson, E.A. 1992. Sources of nitric oxide and nitrous oxide following the wetting of dry soil. Soil Sci. Soc. Am. J. 56:95-102.

Davidson, E.A. 1993. Soil water content and the ratio of nitrous oxide to nitric oxide emitted from soil. p. 369-386. *In* R.S. Oremland (ed.) Biogeochemistry of Global Change: Radiatively Active Trace Gases. Chapman & Hall, New York.

Davidson, E.A., and W. Kingerlee. 1997. A global inventory of nitric oxide emissions from soils. Nut. Cyc. Agroecosys. 48:37-50.

Davidson, E.A., and L. Verchot. Testing the hole in the pipe model of nitric and nitrous oxide emissions from soils using the TRAGNET database. Global Biogeochem. Cycles. 14:1035-1043.

Del Grosso, S.J., W.J. Parton, A.R. Mosier, D.S. Ojima, A.E. Kulmala and S. Phongpan. 2000a. General model for N_2O and N_2 gas emissions from soils due to denitrification. Global Biogeochem. Cycles. 14:1045-1060.

Del Grosso, S.J., W.J. Parton, A.R. Mosier, D.S. Ojima, and M.D. Hartman. 2000b. Interaction of soil carbon sequestration and N_2O flux with different land use practices. p. 303-311. *In* J. van Ham et al. (ed.) Non-CO_2 Greenhouse Gases: Scientific Understanding, Control and Implementation. Kluwer Academic Publishers, The Netherlands.

Del Grosso, S.J., W.J. Parton, A.R. Mosier, M.D. Hartman, J. Brenner, D.S. Ojima, and D.S. Schimel. 2001. Simulated interaction of carbon dynamics and nitrogen trace gas fluxes using the DAYCENT model. p. 303-332. *In* M. Schaffer et al. (ed.). Modeling Carbon and Nitrogen Dynamics for Soil Management. CRC Press LLC, Boca Raton, Fl.

Eitzinger, J., W.J. Parton, and M.D. Hartman, improvement and validation of a daily soil temperature submodel for freezing/thawing periods. Soil Sci. (In press.)

Frolking, S.E., A.R. Mosier, D.S. Ojima, C. Li., W.J. Parton, C.S. Potter, E. Priesack, R. Stenger, C. Haberbosch, P. Dörsch, H. Flessa and K.A. Smith. 1998. Comparison of N_2O emissions from soils at three temperate agricultural sites: simulations of year round measurements by four models. Nutr. Cyc. Agroecosys. 52:77-105.

Galloway, J.N., W.H. Schlesinger, H. Levy II, A. Michaels, and J.L. Schnoor. 1995. Nitrogen fixation: Anthropogenic enhancement-environmental response. Global Biogeochem. Cycles. 9:235-252.

Grant, R.F., and E. Pattey. 1999. Mathematical modeling of nitrous oxide emissions from an agricultural field during spring thaw. Global Boigeochem. Cycles. 13:679-694.

Kelly, R.H., W.J. Parton, G.L. Crocker, P.R. Grace, J. Klir, M. Korschens, P.R. Poulton and D.D. Richter. 1997. Simulating trends in soil organic carbon using the century model. Geoderma 81:77-90.

Kelly, R.H., W.J. Parton, M.D. Hartman, L.K. Stretch, D.S. Ojima and D.S. Schimel. 2000. Intra and interannual variability of ecosystem processes in shortgrass steppe. J. Geophys. Res. 105:20,093-20,100.

Kroeze C, A. Mosier, and L. Bouwman. 1999. Closing the global N_2O budget: A retrospective analysis 1500-1994. Global Biogeochem. Cycles. 13:1-8.

Hutchinson, G.L., G.P. Livingston and E.A. Brams. 1993. Nitric and nitrous oxide evolution from managed subtropical grassland. p. 290-316. *In* R.S. Oremland (ed.) Biogeochemistry of Global Change: Radiatively Active Trace Gases. Chapman & Hall, New York.

IPCC (Intergovernmental Panel on Climate Change). 1997. Greenhouse gas emissions from agricultural soils. *In* J.T. Houghton et al. (ed.) Greenhouse Gas Inventory Reference Manual; Revised 1996 IPCC Guidelines for National Greenhouse Gas Inventories. Volume 3, Section 4.5. Agriculture IPCC/OECD/IEA. UK Meteorological Office, Bracknell, UK.

Kelly, R.H., W.J. Parton, G.L. Crocker, P.R. Grace, J. Klir, M. Korschens, P.R. Poulton and D.D. Richter. 1997. Simulating trends in soil organic carbon using the century model. Geoderma 81:77-90.

Kelly, R.H., W.J. Parton, M.D. Hartman, L.K. Stretch, D.S. Ojima and D.S. Schimel. 2000. Intra and interannual variability of ecosystem processes in shortgrass steppe. J. Geophys. Res. 105:20,093-20,100.

Kroeze C, A. Mosier, and L. Bouwman. 1999. Closing the global N_2O budget: A retrospective analysis 1500-1994. Global Biogeochem. Cyc. 13:1-8.

Hutchinson, G.L., G.P. Livingston and E.A. Brams. 1993. Nitric and nitrous oxide evolution from managed subtropical grassland. p. 290-316. *In* R.S. Oremland (ed.) Biogeochemistry of Global Change: Radiatively Active Trace Gases. Chapman & Hall, New York.

Lal, R., J.M. Kimble, R.F. Follett and C.V. Cole. 1998. The potential of U.S. cropland to sequester carbon and mitigate the greenhouse effect. Ann Arbor Press, Chelsea, MI.

Levin, S.A. 1989. Challenges in the development of a theory of community and ecosystem structure and function. p. 245-255. *In* J. Roughgarden et al. (ed.) Perspectives in Ecological Theory. Princeton University Press, Princeton, New Jersey.

Linn, D.M., and J.W. Doran. 1984. Effect of water-filled pore space on carbon dioxide and nitrous oxide production in tilled and nontilled soils. Soil Sci. Soc. Am. J. 48:1267-1272.

Martin, R.E., M.C. Scholes, A.R. Mosier, D.S. Ojima, E.A. Holland, and W.J. Parton. 1998. Controls on annual emissions of nitric oxide from soils of the Colorado shortgrass steppe. Global Biogeochem. Cycles. 12:81-91.

Metherell, A.K., L.A. Harding, C.V. Cole, and W.J. Parton. 1993. CENTURY soil organic matter model environment, Technical documentation, Agroecosystem version 4.0. Great Plains System Research Unit Technical Report No. 4. USDA-ARS, Fort Collins, Colorado.

Mosier, A.R., W.D. Guenzi, and E.E. Schweizer. 1986. Soil losses of dinitrogen and nitrous oxide from irrigated crops in northeastern Colorado. Soil Sci. Soc. Am. J. 50:344-348.

Mosier, A.R., W.J. Parton, D.W. Valentine, D.S. Ojima, D.S. Schimel, and J.A. Delgado. 1996. CH_4 and N_2O fluxes in the Colorado shortgrass steppe: 1. Impact of landscape and nitrogen addition. Global Biogeochem. Cycles. 10:387-399.

Mosier, A.R., W.J. Parton, D.W. Valentine, D.S. Ojima, D.S. Schimel, and O. Hienemeyer. 1997. CH_4 and N_2O fluxes in the Colorado shortgrass steppe: 2. Long-term impact of land use change. Global Biogeochem. Cycles. 11:29-42.

Parton, W. J., and P. E. Rasmussen. 1994. Long-term effects of crop management in wheat/fallow: II. CENTURY model simulations. Soil Sci. Soc. Am. J. 58:530-536.

Parton, W.J., J.M.O. Scurlock, D.S. Ojima, T.G. Gilmanov, R.J. Scholes, D.S. Schimel, T. Kirchner, J.C. Menaut, T. Seastedt. E. Garcia Moya, Apinan Kamnalrut, and J.L. Kinyamario. 1993. Observations and modeling of biomass and soil organic matter dynamics for the grassland biome worldwide. Global Biogeochem. Cycles. 7:785-809.

Parton, W.J., D.S. Ojima, C.V. Cole and D.S. Schimel. 1994. A general model for soil organic matter dynamics: Sensitivity to litter chemistry, texture and management. p 147-167. *In* Quantitative Modeling of Soil Forming Processes. SSSA, Spec. Pub. 39. Madison, WI..

Parton, W.J., A.R. Mosier, D.S. Ojima, D.W. Valentine, D.S. Schimel, K. Weier, and K.E. Kulmala. 1996. Generalized model for N_2 and N_2O production from nitrification and denitrification. Global Biogeochem. Cycles. 10:401-412.

Parton, W.J., M. Hartman, D.S. Ojima and D.S. Schimel. 1998. DAYCENT: Its land surface submodel: description and testing. Global Planetary Change. 19:35-48.

Parton, W.J., E.A. Holland, S.J. Del Grosso, M.D. Hartman, R.E. Martin, A. R. Mosier, D.S. Ojima and D.S. Schimel. Generalized Model for NO_x and N_2O Emissions from Soils. J. Geophys. Res. (In press.)

Paustian, K., W.J. Parton and J. Persson. 1992. Modeling soil organic matter in organic amended and nitrogen-fertilized long term plots. Soil Sci. Soc. Am. J. 56:476-488.

Peterson, G.A., A.D. Halvorson, J.L. Havlin, O.R. Jones, D.J. Lyon, and D.L. Tanaka. 1998. Reduced tillage and increasing cropping intensity in the Great Plains conserves soil C. Soil Till. Res. 47:207-218.

Potter, C.S., E.A. Davidson, and L.V. Verchot. 1996. Estimation of global biogeochemical controls and seasonality in soil methane consumption. Chemosphere. 32:2219-2245.

Prather, M.J., R.Derwent, D. Ehhalt, P. Fraser, E. Sanhueza, and X. Zhou. 1995. Other trace gases and atmospheric chemistry. p. 73-126. *In* J.T. Houghton et al. (ed.) Climate Change 1994. Cambridge University Press, Cambridge, UK.

Robertson, G.P., E.A. Paul, and R.R. Harwood. 2000. Greenhouse gases in intensive agriculture: Contributions of individual gases to the radiative forcing of the atmosphere. Science. 289:1922-1925.

Smart, D.R., J.M. Stark, and V. Diego. 1999. Resource limitation to nitric oxide emissions from a sagebrush-steppe ecosystem. Biogeochem. 47:63-86.

Smil, V., 1999. Nitrogen in crop production: An account of global flows. Global Biogeochem. Cycles. 13:647-662.

Vitousek, P.M. and R.W. Howarth. 1991. Nitrogen limitation on land and in the sea: How can it occur? Biogeochem. 13:87-115.

Vitousek, P.M., J. Aber, E.H. Howarth, G.E. Likens, P.A. Matson, D.W. Schindler, W.H. Schlesinger, D.G. Tilman. 1997. Human alteration of the global nitrogen cycle: causes and consequences. Issues in Ecology. 1:1-15.

Williams, E.J., G.L. Hutchinson, and F. Fesenfeld. 1992. NO_x and N_2O emissions from soil. Global Biogeochem. Cycles. 6:351-388.

Yienger, J.J. and H. Levy. 1995. Empirical model of global soil biogenic NO_X emissions. J. Geophys. Res. 100:11447-11464.

Nitrogen in the Environment: Sources, Problems, and Management
R.F. Follett and J.L. Hatfield (Eds)
Published by Elsevier Science B.V.

Chapter 17. Remediation of Drinking Water for Rural Populations

W. J. Hunter

Soil-Plant Nutrient Research Unit, USDA-ARS, P. O. Box E, Fort Collins, CO 80521 USA

Nitrate is the most common nitrogen contaminant in raw water supplies. In rural areas agricultural activities that involve the use of fertilizers and animal manures are major sources of nitrate contamination. Several processes are currently available that can effectively remove nitrate from raw water. Systems that are suitable for small rural communities include reverse osmosis, ion exchange, and electrodialysis. However, all of these systems move nitrate from raw water and concentrate it in a reject water or brine. Disposal of the reject water can be a major expense as well as an environmental issue. Several emerging systems are under development that convert nitrate to harmless nitrogen gas. These include biological denitrification systems and catalytic systems. The ability of these systems to convert the nitrate to nitrogen gas is a major advantage

1. INTRODUCTION

Ammonia and nitrite can cause water quality problems but it is nitrate that is most often associated with the contamination of drinking water. Nitrate in rural drinking water supplies is a common and growing world-wide water-quality problem. Natural sources of nitrate can contaminate groundwater sources (Edmunds and Gaye, 1997) but nitrate-N concentrations that exceed 2 - 3 mg L^{-1} usually indicate that the source is anthropogenic (Foster et al., 1982; Kross et al., 1993; Mueller et al., 1995). Contamination can result from a number of commercial activities (Table 1) but it is the contribution by fertilizers that has been especially important in rural areas. Since 1950 the use of nitrogen fertilizers has increased sharply in most countries due to the expansion of intensive crop production, and contamination of groundwater supplies continues to increase as nitrates derived from animal wastes or fertilizers deposited years ago migrate slowly downward through the overlying soils to the aquifer (Gormly and Spalding, 1979; Hiscock et al., 1991; Spalding and Exner, 1993; Green and Shelef, 1994; Hamilton and Helsel, 1995), or migrates slowly within the aquifer (Hallberg, 1989). In Germany water samples containing 4.5 to 11.3 mg L^{-1} nitrate-N have shown a steady increase through most of the century, rising from 2% of samples in 1915 to 21% of samples in 1975 and to 23 % of samples in 1989 (Piotrowski and Kraemer, 1998). In England the number of groundwater sites with nitrate levels that exceeded the European drinking water standard increased three fold between 1970 and 1990, and in sections of Denmark and the Netherlands nitrate in groundwater samples has increased at an annual rate of 0.04 to 0.29 μg nitrate-N L^{-1} (Green and Shelef, 1994).

In Europe the greatest problems occur in the northwest (Schrama, 1998) in Belgium, Denmark, Germany, and the Netherlands. Models indicate that soil concentrations of nitrate are

Table 1

Releases of nitrate and nitrite to land and water in 1991 through 1993 by commercial activities (USEPA, 1999).

Major industry	Nitrate and nitrite
	Metric tonnes
N-fertilizers	22,766
Industrial inorganics	15,326
Metal ores	2,615
Industrial	2,309

sufficiently high that groundwaters in major parts of these countries are likely to exceed the European drinking water standard, while in France, Italy, and England problems are likely to be more localized (O'Tool, 1998). In rural areas of the United States, where groundwater is the main source of drinking water, it was estimated in 1993 that 2.4% of rural domestic wells exceed the US drinking water standard for nitrate (Benjamin and Belluck, 1994). The problem however is much greater than the national figure suggests because nitrate contamination problems tend to be localized. Drinking water problems, though clearly not limited to that portion of the country, are of greatest concern in the Great Plains (Spalding and Exner, 1993; Nolan et al., 1998). Hamilton and Helsel (1995) surveyed five regions in the US and found that in central and western Connecticut, 12%; in south-central Kansas, 17%; in Long Island NY, 27%; in the Delmarva Peninsula of Delaware and Maryland, 33%; and in the high plains of Nebraska, 46% of well water samples collected were above the US standard for nitrate in drinking water. In Iowa 18% and in Kansas 28% of private drinking water wells exceeded the drinking water standard (Kross et al., 1993). A northeastern Colorado survey found that 70% of sampled wells exceeded the US drinking water standard (Schuff, 1992; Wylie et al., 1994; Wylie et al., 1995). Nitrate contamination of drinking water supplies coupled with the difficulty of removing nitrate from water has forced a number of rural communities to abandon their wells and seek cleaner sources of drinking water (Schuff, 1992; Spalding and Exner, 1993). Nitrate in water causes the abandonment of more drinking water sources than does contamination by toxic chemicals (O'Tool, 1998).

Ingestion of nitrate is a concern because of the effects that nitrite can have on human health (National Academy of Sciences, 1978; Train, 1979; Kross et al., 1993). Nitrate is not very toxic, but its presence in drinking water is a health concern because nitrate can be transformed to nitrite following ingestion. In unweaned infants this transformation of nitrate to nitrite occurs more readily than in older children or adults, and nitrate ingestion can cause methemoglobinemia or blue-baby syndrome, a condition where nitrite binds to hemoglobin. Its presence interferes with the binding of oxygen to hemoglobin and reduces the amount of oxygen that can be transported by the blood (Train, 1979). Methemoglobinemia is intensified by intestinal infections and most cases have occurred with rural water supplies of bad bacteriological quality where nitrate-N concentrations exceeded ~22 mg L^{-1}. The condition is rare in properly disinfected water systems with nitrate-N concentrations below ~22 mg L^{-1} (Croll, 1994). In adults, it has been suggested that nitrosamines may form following nitrate ingestion (National Academy of Sciences, 1978) and animal studies have shown that nitrosamines are potent carcinogens. Theoretically, nitrite might react with amines in the intestine to form nitrosamines, though intestinal conditions do not favor the reaction (Croll, 1994). A number of studies have investigated the relationship between nitrate in drinking water and stomach cancer but results have been conflicting with some studies finding a correlation and others failing to find any correlation (Croll, 1994). Nitrate in water may indicate other water quality problems. If the nitrate is coming from human or animal wastes then microbial contamination may also be a concern.

In the United States the USEPA (1973) recommends that water used for human consumption contain no more than 10 mg L^{-1} nitrate-N and in Canada (Liem et al., 1996) the recommended guideline is 45 mg L^{-1} as nitrate (~10 mg L^{-1} nitrate-N). In Europe the guideline is that drinking water should never exceed 50 mg L^{-1} as nitrate (11.3 mg L^{-1} nitrate-N). The World Health Organization recommends 50 mg L^{-1} nitrate as the maximum long-term exposure though, under its guidelines, short term exposure to amounts in excess of 50 mg L^{-1} as nitrate are acceptable (Croll, 1994). The National Academy of Sciences-National Research Council (Bruning-Fan and Kaneene, 1993) recommends that drinking water supplied to farm animals contain no more than 100 mg L^{-1} nitrate-N.

2. CURRENT PROCESSES

At the present time methods for removing nitrate from drinking water include reverse osmosis, electrodialysis, ion exchange, and distillation. Carbon adsorption filters, mechanical filters of various types, and standard water softeners do not remove nitrate-nitrogen.

2.1 Reverse osmosis

This process, as the name implies, is the reverse of osmosis. With reverse osmosis water pressure is used to force water through a thin-film composite or cellulose triacetate membrane (Harries et al., 1991; Kunz, 1997). In the process water moves from the more concentrated solute side of the membrane to the less concentrated solute side of the membrane. The pressure used to drive the process must be sufficient to overcome the osmotic pressure; thus, the higher the concentration of salts in the supply and reject waters, the higher the water pressure must be to operate the system. Higher water pressure also improves the efficiency of the system. Under pressure, molecules of water dissolve into the membrane and pass through the membrane to the permeate side by the process of diffusion. Dissolved ions, such as salts, that are charged are likely to be rejected by the membrane. The larger the particle and the larger the charge the less likely that the molecule will pass through the membrane. Contaminants such as nitrate, nitrite, ammonia and other salts cannot dissolve into the membrane and remain on the concentrate side of the membrane. Uncharged molecules, such as organic contaminants are more likely to pass through the membrane. Thus reverse osmosis produces a permeate water with very low inorganic mineral content, and a brackish reject concentrate with high inorganic mineral content. Reverse osmosis works well with nitrate and nitrite. About 96 to 98% of monovalent and 98 to 99% of divalent ions are rejected by the system (Harries et al., 1991). With small home systems about

80 to 90% of nitrate is rejected (Kamrin et al., 1991; Olson et al., 1994). Better rejection of contaminants is achieved at higher pressures (Kamrin et al., 1991). Calcium can clog a reverse-osmosis membrane and systems should not be used with water that contains calcium at levels that exceed ~180 mg L^{-1} (Harries et al., 1991; Kunz, 1997). Pretreatment of the raw water with nanofiltration can reduce the amount of calcium present in the raw water (Bohdziewicz et al., 1999). Disposal of the reject water can be a problem with reverse osmosis systems. The reject water or brine from systems used to cleanse nitrate from drinking water would contain the rejected nitrate and other rejected salts. With community systems, concentration and disposal of these rejected salts can represent as much as 60% of the cost of operation of systems such as reverse osmosis that use physicochemical processes to separate nitrates from groundwater (Green and Shelef, 1994).

Reverse osmosis can be used for community or home water systems. In terms of water usage home systems are inefficient; 1 to 9 liters of water will be rejected by the system for each liter of clean water produced (Kamrin et al., 1991; Kunz, 1997). With larger community systems about 0.33 L of water are rejected for each liter produced (Harries et al., 1991). Water pressure has a important impact on water use efficiency (Kamrin et al., 1991). Systems, like those that might be used by a small rural community, may require 1,400 kPa of water pressure when used with a supply water containing dissolved solids at a concentration of 1,000 mg L^{-1} and water pressures of up to 10,000 kPa are required for water with a dissolved solids concentration of 35,000 mg L^{-1} (National Research Council, 1997). Residential systems that produce only 10 to 15 liters of water a day for drinking and cooking may operate with as little as 280 kPa of water pressure (Kunz, 1997).

Reverse osmosis units used in homes may fit under a sink or may be installed on top of a counter. Such systems produce from 8 to 40 liters of water per day that can be used for drinking or cooking (Kamrin et al., 1991; Kunz, 1997). The units usually consist of i) a sediment filter, ii) a reverse-osmosis membrane, iii) a small storage tank, and iv) an activated carbon filter. The activated carbon filter used with home systems would remove organic contaminants that might have passed through the reverse osmosis membrane. The type of membrane that is used in the system, thin-film composite or cellulose triacetate, influences water use efficiency. Thin-film composite membranes are more efficient than cellulose triacetate membranes but are degraded by chlorine (Kamrin et al., 1991; Kunz, 1997). A prefilter can be used to protect the thin-film composite membrane from chlorine.

2.2 Electrodialysis

This is a water treatment process that produces demineralized water from waters with high salt content. The process is suitable for small communities. For this process an electric current is used to force ions through a pair of semipermable membranes, removing them from the contaminated supply source. The system employs two types of flat sheet membranes that are arranged in an alternating pattern; one membrane is permeable to cations and the other is permeable to anions. Contaminated water is cleansed of anions and cations as it flows between the two membranes (Figure 1). Feed waters supplied to electrodialysis units should have a turbidity that is less than 2.0 nephelometric turbidity units. In addition hydrogen sulfide and manganese levels should be less than 0.3 mg L^{-1} each and free chlorine levels should be less than 0.5 mg L^{-1} (Conlon, 1990). With electrodialysis systems about 70 to 85% of the water that is supplied to the system is available for use as low nitrate water (Harries et al., 1991). The other

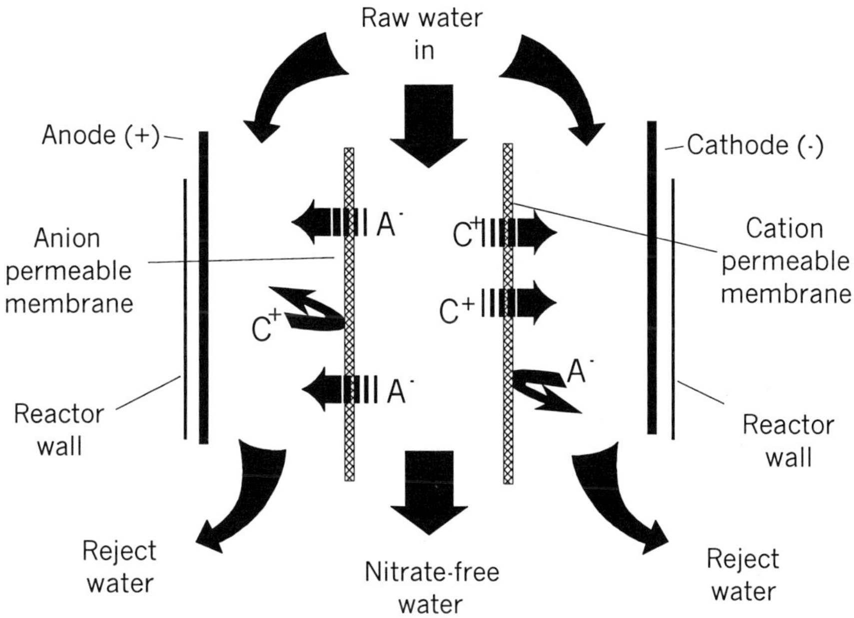

Figure 1. Electrodialysis membrane arrangement. As contaminated supply water flows throug the center of the cell, between the two selective membranes, anions (A^-) such as nitrate and nitrite move through the anion permeable membrane toward the anode while cations (C^+) such ammonia move through the cation permeable membrane toward the cathode. Water with greatl reduced ionic content exits from the center cell. Contaminated reject waters concentrate in the left and right cells.

15 to 30% of the water will contain high levels of nitrate and other ions that present disposal problems similar to those noted for reverse osmosis. Unlike water from reverse osmosis systems, water cleansed by an electrodialysis system does not pass through a membrane. Thus microorganisms and suspended particles will not be removed during the electrodialysis step and another means of filtration must be provided to remove these water contaminants.

A similar process, electrodialysis reversal, periodically reverses the polarity of the electrodes reversing the movement of the ions. During the polarity reversal an automatic valving arrangement reverses the water flow in order to prevent the mixing of cleansed and contaminated waters. The reversal reduces the buildup of deposits on the membranes and prolongs membrane life. Accumulation of deposits that can foul the membranes are a problem with electrodialysis systems (Osmonics, 1992; National Research Council, 1997).

2.3 Ion exchange

Ion exchange can be used to supply nitrate-free water to households, animal operations, or communities. For this process charged beads are used to remove ionic contaminants from flowing water. There are two basic types of beads. Anion exchange beads are made from resins that have positive charges and cation exchange beads from resins that have a negative charge. Nitrate and nitrite, which have a negative ionic charge, will bind to the positively charged sites on the anion exchange beads. Thus water flowing through a bed composed of anion exchange beads would be cleansed of nitrate, nitrite and other negatively charged ions. Ammonia, which carries a positive ionic charge, will not bind to an anion exchange bead but will bind to the negatively charged sites on a cationic exchange beads. Thus a cation exchange bed would be required to remove ammonia and other positively charge ions from water. Water that is nearly free of both anions and cations can be produced by flowing water sequentially through both types of exchange beds or by flowing water through a mixed bed reactor, a reactor that contains both anionic and cationic exchange beads.

Ion exchange reactors stop removing ions and must be regenerated when most of the charged sites on the beads become occupied. With anion exchange beads, hydroxide ions from a sodium hydroxide solution or chloride ions from a sodium chloride solution are used to displace the bound anions. With cation exchange beads, regeneration involves the use of protons from an acid solution or sodium or potassium ions. The waste solution produced during regeneration will contain used regenerate solution and high concentrations of the ions that were removed from the water. It is more difficult to dispose of ion exchange brine than to dispose of reverse osmosis or electrodialysis reject water because of the counterions that are added during the ion exchange regeneration process (Cevaal et al., 1995). Disposal of the waste solution produced during regeneration can be a difficult and costly process (Croll, 1994; Green and Shelef, 1994; National Research Council, 1997).

During operation, water use efficiency with ion exchange reactors is almost 100%. Ion exchange units have no reject water; all of the water that enters an ion exchange reactor becomes treated water. The only losses that occur are during the regeneration process and the volume used here would represent 0.7 to 2% of the volume of treated water (Green and Shelef, 1994).

2.4 Distillation

Distillation removes a wide range of contaminants from drinking water. The process cleanses raw water of contaminants by heating the water until it turns to steam. The steam is then condensed back to water in a condensation coil and purified water is collected in a separate vessel. Nonvolatile contaminants that were present in the raw water will remain in the boiler vessel and are periodically flushed into the septic or sewer system. The condensed water that collects in the second vessel is cleansed of nonvolatile contaminants. The process is an effective method for removing inorganic salts such as nitrate and nitrite from water, but this process may not remove some volatile organic and inorganic compounds.

Maintenance of a distillation system involves periodic cleaning of the boiler side of the unit to remove contaminants that build up over time. The amount of energy required by the unit and the small volume of water produced limits distillation to point-of-use applications such as home use or use in some commercial markets. Units may be mounted on the wall or placed on the countertop. Distillation removes beneficial minerals from the water and water produced by distillation may have a flat taste. The costs of operating a home distillation system may be higher than those with some other forms of home treatment systems (Kamrin et al., 1990).

2.5 Abandonment and blending

Abandonment of an existing contaminated drinking water supply is not a form of remediation but is an approach that is often used by rural households and communities to obtain drinking water that meets the EPA's guidelines for nitrate. With small water systems, such as those that are often used in rural areas, abandonment may be less expensive than attempting to remediate a contaminated water source (Nugent et al., 1993; National Research Council, 1997). Abandonment would include importing water from a distant source as well as the drilling of a new well or wells into a less contaminated aquifer. A new well may be drilled into a deeper or adjacent aquifer that is less contaminated, or the well may be placed in the same aquifer but distant from the source of contamination (i.e. a livestock pen or septic tank). Water from the less contaminated well may then be used directly or mixed with contaminated water to produce a blended-water that is acceptable for drinking. The use of bottled water purchased from a store or from a bottling company may also be considered a form of abandonment. This approach might be considered when the primary concern is for a short period of time such as water for infant food and drinking.

3. EMERGING TECHNOLOGIES

Several technologies are being investigated or proposed as methods for removing nitrate from drinking water. These emerging technologies include biological and abiotic denitrification, as well as several catalytic approaches.

3.1 Biological denitrification

Biological denitrification has been used to remove nitrate from wastewaters for many decades and in recent years the idea of using this process to remove nitrate from drinking water has gained ground, especially in Europe. Drinking water can be denitrified in above ground bioreactors, or *in situ* while it is still below ground. *In situ* treatment may be the most economical (Streile et al., 1991). Biological denitrification is a microbial respiratory process where facultative and anaerobic microorganisms use nitrate, rather than oxygen, as an electron acceptor for respiration. Many soil bacteria are able to carry out this process and are likely to do so in soils and waters where nitrate and a suitable electron donor (usually a carbon substrate) are present but oxygen is limiting. Naturally occurring microbial denitrifiers are ubiquitous in soil and water. Facultative microorganisms can use either nitrate or oxygen as an electron acceptor for respiration, though if oxygen is available these microorganisms will use the oxygen first. It is advantageous to use the oxygen first because oxygen respiration yields about 20% more energy than nitrate respiration. Generally, when oxygen is present respiratory denitrification for energy generation is inhibited (Carter et al., 1995). The steps involved in respiratory denitrification are:

$$NO_3^- \rightarrow NO_2^- \rightarrow NO \rightarrow N_2O \rightarrow N_2 \quad (1)$$

Not all denitrifiers are able to reduce nitrate all the way to nitrogen gas, and nitrite, nitric oxide or nitrous oxide may accumulate when pure cultures of these microorganisms are incubated under denitrifying conditions (Hiscock et al., 1991). In nature a consortium of microorganisms would often be involved in the process and nitrogen gas would normally be the principal end product. However, environmental conditions such as nutrients, pH, or electron donor availability

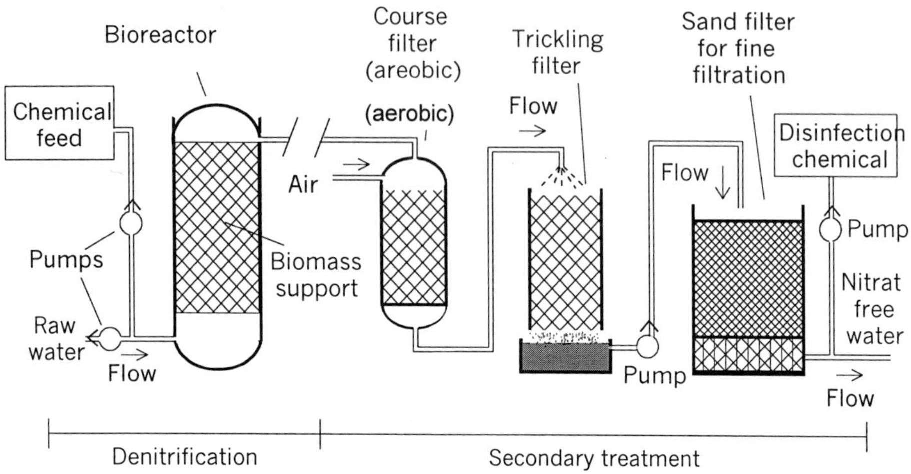

Figure 2. Flow diagram of a hypothetical biological denitrification reactor showing the major system components. Systems may have more than one bioreactor and components of the secondary treatment process may be combined into a single unit.

may influence the reduction process and may cause intermediates to accumulate. While a number of factors affect the denitrification process it is usually the availability of an electron donor that limits the rate at which denitrification proceeds, and the addition of a carbon source or other electron donor to contaminated water will often stimulate denitrification (Myrold and Tiedje, 1985; and others).

3.2 *Ex situ* biological reactors

Above ground bioreactors using biological denitrification may be suitable for community water systems in rural areas. A system (Figure 2) would consist of one or more large tanks partially filled with a physical support for the denitrifying biomass. These tanks are where the denitrification process takes place and they are fed the raw water, carbon substrate, and the nutrients needed to drive the process. The effluent water from the tank(s) would contain much less nitrate than the influent water but would contain high numbers of bacteria, suspended solids, organic matter content, and turbidity. These waters would also be low in dissolved oxygen (Dahab and Sirigina, 1994; Hunter and Follett, 1997). These are water quality problems that must be corrected before the water can be consumed. Secondary treatment to improve the quality of the denitrified water would involve filtration, aeration and disinfection to achieve the desired water quality (Roennefahrt, 1986; Dahab and Sirigina, 1994; Green and Shelef, 1994; Hunter and Follett, 1997; Silverstein and Carlson, 1999).

In the US a commercial-sized above ground heterotrophic denitrification reactor was constructed in the town of Wiggins in rural eastern Colorado, USA. The system consisted of two upflow bioreactors, a single roughing filter that served to both filter and aerate the denitrified water, and a slow sand filter. Influent water was pumped into the first bioreactor at a rate of 38 L min^{-1} and contained ~20 mg L^{-1} nitrate-N mixed with a high-fructose corn syrup (52% fructose and 48% glucose) and phosphate. The bioreactors were 2.7 meters high, 0.9 meters in diameter and contained a buoyant (specific gravity = 0.96) and highly porous (94%) cylindrical polypropylene support material. Periodic air scour was used to remove excess microbial biomass. Effluent from the denitrification bioreactors flowed to a roughing filter that was 2.1 meters high, 0.8 meters in diameter and contained the same polypropylene support as the bioreactors but was operated as a downward flow aerobic reactor. A slow sand filter follows the roughing filter. The system was operated for about seven months as a demonstration project and yielded an oxygenated (3.8 mg L^{-1}) product water with acceptable turbidity (0.4 NTU), dissolved organic carbon (3.1 mg L^{-1}), and nitrate-N (4.3 mg L^{-1}) content (Silverstein and Carlson, 1999).

In Europe a number of pilot and demonstration systems have been constructed. The first above ground denitrification reactor was installed in France in 1983 with later installations in Germany and Italy (Green and Shelef, 1994). Systems have been both fixed bed and fluidized bed reactors and have employed a number of different biomass supports. A partial listing of systems is presented in Table 2. Carbon compounds such as ethanol, methanol, acetate, hydrogen, sulfur, and natural gas have all been used or proposed as substrates for microbial denitrification processes (Green and Shelef, 1994; Rajapakse and Scutt, 1999; Houbron et al., 1999) with phosphate normally added as a nutritional supplement. Problems with the systems include a product water that is low in dissolved oxygen and high in bacteria and bacterial products. Also, if the system is operated with too much carbon substrate then residual substrate may be present in the finished water, but if too little substrate is supplied then nitrite may be present. Secondary treatment and disinfection can oxygenate and remove bacteria from the finished water. Careful monitoring of the amount of nitrate entering the system and metering of the amount of carbon substrate added is required to prevent the presence of carbon substrate and nitrite in the finished water.

Mansell and Schroeder (1999) conducted a series of studies with a membrane reactor that should produce a product water that is cleaner than that produced by other primary denitrification reactors. In their bench scale reactor a polytetrafluoroethylene membrane with a pore size of 0.02 microns and porosity of 50% was used to separate the reactor into two sections or flow channels (Figure 3). A suspended culture of denitrifying bacteria, from a culture vessel, was pumped by one side of the membrane, and the raw water to be denitrified was pumped by the other side of the membrane. The pore size of the membrane did not allow the bacteria to enter the raw water but nitrate in the raw water was able to flow through the membrane and was converted to nitrogen gas by the bacterial suspension. Methanol, the carbon source, and phosphate were supplied to the culture vessel. The raw water supplied to the reactor contained 20 or 30 mg L^{-1} nitrate-N and the reactor was able to reduce the nitrate in the water by 41 to 72%. The system should produce water with fewer bacteria than other primary biological denitrification reactors.

Laboratory and pilot scale studies suggest that autotrophic denitrification using sulfur, rather than an organic carbon substrate, as an electron donor might also be used to remove nitrate from pumped groundwater or from surface water. Pumped water would be treated in denitrification reactors (Kruithof et al., 1988; Lampe and Zhang, 1997; Flere and Zhang, 1999) but contaminated surface water ponds might be treated by adding sulfur and limestone directly to the pond (Lampe and Zhang, 1997). The limestone serves as a buffer. One reported advantage of this system is the low amount of biomass produced (Lampe and Zhang, 1997; Flere and Zhang, 1999). The accumulation of microbial biomass can block the flow channels in a denitrification

Table 2
Examples of configurations and electron donors used in denitrification reactors.

Biomass support	Electron donor	Nitrate removed	Scale	Reference
Buoyant polypropylene	Corn syrup	0.3 kg $N/m^3/d$	2.28 m^3/h	Silverstein and Carlson, 1999
Rotating bed	Acetic acid	1.8 to 9.8 kg $N/m^2/d$	Pilot	Mohseni-Bandpi et al., 1999
Sand fluidized bed	Ethanol/acetic acid	4.6 kg $N/m^3/d$	Pilot	Green and Shelef, 1994
Biolite fixed bed	Ethanol	1 kg $N/m^3/d$	50 m^3/h	Green and Shelef, 1994
Clay fixed bed	Ethanol	1.2 kg $N/m^3/d$	400 m^3/h	Rogalla et al., 1990
Polystyrene fixed bed	Methanol/acetic acid/ Ethanol	1.4 kg $N/m^3/d$	800 m^3/h	Roennefahrt, 1986
Sand fluidized bed	Methanol	5.4 kg $N/m^3/d$	Pilot	Green and Shelef, 1994
Sand moving bed	Methanol	2.7 kg $N/m^3/d$	11 m^3/h	Green and Shelef, 1994
Sand fluidized bed	Methanol	3.5 kg $N/m^3/d$	250 m^3/h	MacDonald, 1990
Fixed bed	Sulfur	0.2 to 0.4 kg $N/m^3/d$	Laboratory	Flere and Zhang, 1999
Fixed bed	Thiosulfate	1.5 kg $N/m^3/d$	Pilot	Trouve and Chazal, 1999
Fixed bed	Hydrogen	0.25 kg $N/m^3/d$	100 m^3/h	Green and Shelef, 1994

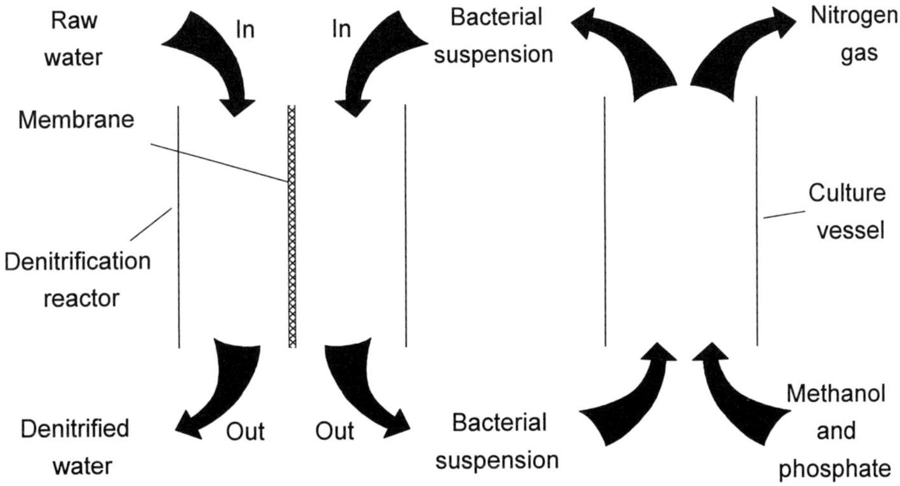

Figure 3. Schematic of a membrane denitrification reactor. Raw water is cleansed of nitrate as the water flows by one side of a porous membrane and a suspended culture of denitrifying bacteria flows by the other side of the membrane (Mansell and Schroeder, 1999).

reactor decreasing its ability to remove nitrate.

Above ground denitrification reactors, because of their high installation costs and complexity of operation, are not suitable for home use though such units might be used by small rural communities with trained operators. Costs, however, might still be an issue. Green and Shelef (1994) compared biological denitrification with ion exchange and concluded that the two had similar costs of operation but that the biological denitrification unit was 2 to 3 times more expensive to install and more complex to operate. Tannehill et al. (1997) estimated the potential cost of removing nitrate from groundwater in six small communities in rural Nebraska and concluded that ion exchange would be the most cost-effective method for all six communities. Above ground denitrification was estimated to be slightly more expensive than ion exchange at the present time though it was suggested that either of these two processes could become the best treatment method in the near future. Reverse osmosis, largely because of the cost of disposal of the waste brine produced, was predicted to be the most expensive option for all six communities.

3.3 *In situ* biological denitrification

Several research and demonstration projects have looked at different approaches and schemes to remove nitrate from contaminated groundwater before it is pumped to the surface. It is envisioned that some approaches will offer an inexpensive method for removing nitrate from contaminated groundwater (Streile et al., 1991; Green and Shelef, 1994). Costs are reduced because less equipment and less oversight are needed for these processes. *In situ* treatment of groundwater, which uses a portion of the aquifer as a denitrification reactor to remove nitrate, can also provide a portion of the secondary treatment. Secondary treatment processes, which include the removal of organic residues, particulate filtration, oxygenation, and disinfection (Dahab and Sirigina, 1994; Hunter and Follett, 1997), can take place in the aquifer (Green and Shelef, 1994).

Denitrification is a natural process that takes place in soils, surface waters and groundwaters. Microorganisms capable of removing nitrate from water by denitrification are naturally present in soil and water. However, in soils that are below the root zone, the activity of denitrifying microorganisms is often severely restricted because of the absence of an appropriate electron donor. Most *in situ* treatment processes involve injecting an electron donor, usually a soluble carbon source, into the contaminated aquifer.

One approach simply involves the use of a single recharge well for the injection of a carbon substrate and a single pumping well to extract the denitrified water. This approach was used to remove nitrate from a gravel aquifer in the Netherlands. Groundwater, containing 18.1 mg L^{-1} nitrate-N, was pumped from the ground at a rate of 33 m^3 h^{-1}, mixed with methanol (49 mg L^{-1}) and injected back into the aquifer at a rate of 20 m^3 h^{-1} for 22 days. During this time the nitrate-N content of the water was reduced by 30%, although both an accumulation of nitrite and a decrease in the hydraulic conductivity between the two wells was observed (Hiscock et al., 1991). Another approach involves the use of small-diameter injection wells arranged in a circle or daisy-pattern around a large-diameter uptake well. The carbon source, often ethanol, is diluted with water from the uptake well and the mixture injected into the aquifer to provide an underground denitrification zone within the aquifer. A major portion of the water flowing to the uptake well would flow through this area and nitrate in the water would be removed by microbial activity as the water passes through this zone. In a study in France about 70% (Hamon and Fustec, 1991) and in a study in the US about 16% of the nitrate was removed (McMahon et al., 1998) by this process, although aquifer plugging was a major problem in both studies. In contrast, Janda et al. (1988) had no problems with aquifer plugging in a full-scale study in a sand and gravel aquifer. Denatured ethanol was used as the carbon source to remove ~ 40% of the nitrate from water containing 22.6 mg L^{-1} nitrate-N during this 141 day study. A modification of the daisy-pattern, the "Nitredox" method developed by Braester and Martinell (1988), uses a second set of wells arranged in a concentric circle within the outer ring of injection wells. The outer wells are used for the injection of the carbon substrate to establish a denitrification zone and the wells in the inner circle are used to inject aerated water to establish a zone for secondary water treatment and to oxygenate the water. A "Nitredox" system consisting of a pumping well, eight oxidation wells at a radius of 10 m, and 16 injection wells at a radius of 18 m was used to reduce groundwater nitrate-N in a gravel aquifer in Austria from 22.6 to 5.7 mg L^{-1} using methanol as the substrate. No problems with aquifer plugging were reported.

A simple approach for *in situ* denitrification involves the use of insoluble substrates to form denitrification walls or barriers. The barriers are placed between the source of nitrate contamination and the point of uptake (Figure 4). The walls are constructed by digging a trench and backfilling it with a mixture of substrate and fine gravel. Thus the substrate is added when the wall is constructed and the need for much of the equipment associated with above ground denitrification reactors is eliminated. Nitrate is removed via denitrification when contaminated water flows through the wall. Sawdust is an inexpensive substrate that has been shown to work well in denitrification walls (Robertson and Cherry, 1995; Schipper and Vojvodic-Vukovic,

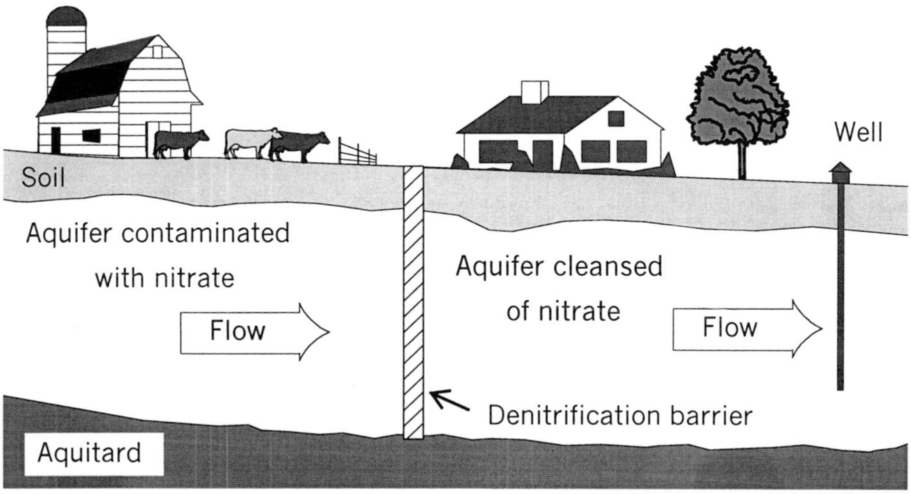

Figure 4. Hypothetical use of a denitrification barrier to remove nitrate from groundwater contaminated by farm animals.

1998; Robertson and Anderson, 1999). Denitrification walls should have a functional life of many years. Robertson and Cherry (1995) used a sawdust denitrification wall that contained 2% carbon to remove nitrate from a sewage leach field and estimated that it would last the 20 year design life of the leach field under the *in situ* conditions at the site. Blowes et al. (1994) estimated that a reactor that contained 5% carbon (as cellulose) might not require additional substrate for several decades. A number of organic and inorganic substrates could be used in denitrification walls and the best choice might depend on what is locally available. Crop residue, newspaper, cotton, sulfur, and innocuous oils have all been shown to support denitrification and might be used in denitrification walls (Blowes et al., 1994; Volokita et al., 1996a; Volokita et al., 1996b; Hunter et al., 1997; Lampe and Zhang, 1997; Flere and Zhang, 1999).

A major limitation and expense associated with denitrification walls is the digging and backfilling of the trench. The technology exists to dig deep trenches, but the problem is the cost associated with the trenches. Shallow trenches can be dug quickly and at relatively low cost but deeper trenches are much more difficult to dig and considerably more costly (Gavaskar et al., 1998). Vegetable oils differ from the other substrates that have been proposed as electron donors for denitrification walls in that they are liquids that are immiscible with water. Laboratory and pilot scale studies have shown that vegetable oils are a good substrate for denitrification and that they can be used to form stationary denitrification walls that are effective at removing nitrate from flowing water (Hunter et al., 1997; Hunter, 1999). Denitrifying walls containing vegetable oil might be formed by filling trenches with sand coated with oil, as can be done with solid substrates, or, it is hypothesized, vegetable oil emulsions might be injected into an aquifer to create a denitrifying wall without the expense associated with digging a trench. *In situ* systems, because of their simplicity, may prove useful in rural areas once the technology is fully developed.

3.4 Abiotic *in situ* denitrification

Abiotic processes might also be used *in situ* to remove nitrate from groundwater. *In situ* iron walls, composed of zero-valent iron mixed with sand or gravel, have been used to remove trichloroethene and other chlorinated organic compounds from groundwater. Zero-valent iron is a strong reducing agent that can displace the chloride from chlorinated organic compounds via a mechanism that is not completely understood. For example, when groundwater contaminated with trichloroethene flows through an iron wall, the iron becomes oxidized and the trichloroethene becomes dechlorinated to yield ethene and chloride as the primary products (Gavaskar et al., 1998). Zero-valent iron can also serve as an election donor to reduce nitrate (Szabo and Bartha, 1952; Young et al., 1964; Huang et al., 1998; and others). The mechanism may be due to the direct reduction of nitrate by Fe^0, or due to its indirect reduction by hydrogen derived from a proton (Huang et al., 1998; Chew and Zhang, 1999). For the reaction to proceed at a significant rate, the pH must be low (Huang et al., 1998) or hydrogen must be supplied (Siantar et al., 1996). Huang et al. (1998) report that pH is a critical factor in the reduction of nitrate by zero-valent iron and that significant reduction does not occur at pHs higher than ~5. In addition, in their studies, ammonia in nearly stoichiometric amounts was the final end-product.

The production of ammonia and requirement for a low pH would be major drawbacks in the use of the iron wall technology for drinking water remediation. However, an enhancement of the process that involves the coupling of the zero-valent iron reaction with electrokinetics may overcome these drawbacks. Electrokinetics is an electrical process where two electrodes are placed in the ground and a low-intensity direct current applied. The applied current causes the migration of ionic species in the soil (EPA, 1995) and the generation of H^+ ions at the anode. Chew and Zhang (1999) conducted laboratory scale studies coupling electrokinetics with an iron wall located next to the anode. The system removed 93 to 96% of the nitrate from an artificial groundwater contained in a soil/sand column. Nitrogen gases (46-50%) and ammonia (~45%) were the principal end-products. They suggest that the reaction to nitrogen gas may proceed according to the following equation:

$$5Fe^0 + 2NO_3^- + 12\,H^+ \rightarrow 5Fe_2^+ + N_2 + 6H_2O \qquad (2)$$

Considerable refinement of this process would be needed before it could be used to remediate drinking water.

3.5 Catalytic systems

The two most developed catalytic approaches are one which uses a bimetallic catalyst with hydrogen gas as the reducing agent, and one that uses immobilized enzymes with reducing power supplied by an electric current. A 5% rhodium on carbon catalyst (Reddy and Lin, 2000) has also been evaluated as a means of removing nitrate from water as have photocatalysts (Mori et al., 1999). These systems might be well suited for small point-of-use units such as home systems, as well as large scale operations. Advantages that these systems offer are the ability to produce water that is free of nitrate but that is not devoid of other beneficial minerals, and the ability to remove nitrate from the environment by converting the nitrate to nitrogen gas. Catalytic systems would not produce a nitrate-laden wastewater. This is a major advantage that these systems have over conventional water treatment processes such as reverse osmosis, electrodialysis, and ion exchange that simply separate nitrate from the water stream and produce a waste that can be difficult to discard in an economically and environmentally acceptable manner. Also, catalytic systems, once fully developed, may be easier to operate and maintain than biological denitrification systems that depend on a living consortium of microorganisms to reduce the nitrate.

Bimetallic catalysts use supported palladium catalysts with copper or tin serving as the catalytic promoter and hydrogen as the source of electrons to reduce nitrate to nitrogen gas. The reduction of nitrate (NO_3^-) to nitrogen gas (N_2) involves its stepwise reduction with nitrite (NO_2^-), nitric oxide (NO), and nitrous oxide (N_2O) forming as intermediate products (Wärnå, et al., 1994). Control problems exist with systems based on bimetallic catalysts in that they may take the reduction reaction too far and produce ammonium, or may fail to completely reduce the nitrate to nitrogen gas and produce nitrite instead (Hörold et al., 1993). Both of these products are more toxic than nitrate. The rhodium catalyst resembles the bimetallic catalysts in that it also requires hydrogen, an electric current, and follows the same reaction path. In studies with this catalyst only nitrate and nitrite were monitored. Nitrate was observed to disappear with time and nitrite was not detected as a reduction product (Reddy and Lin, 2000).

UV light, with or without a catalyst, can be used to reduce nitrate. Silver or platinum in combination with titanium oxide (Ohtani et al., 1988; Kudo et al., 1987), zinc sulfate (Ranjit et al., 1994), and hollandite (Mori, 1999) have been used as photocatalysts to convert nitrate to ammonia in the presence of methanol or propanol. Gonzalez and Braun (1996) observed that a mixture of nitrate and methanol would react under UV light to yield nitrate, ammonia and carbon dioxide. Unfortunately, the final nitrogen product of these systems is ammonia and not nitrogen gas.

Biological catalysts use enzymes immobilized on a matrix that can be packed into small reactors or columns to catalyze the reduction of nitrate to nitrogen gas (Holton, 1996). The steps and intermediates involved are the same as those given above for respiratory denitrification. Three enzymes are involved in the reduction of nitrate to nitrogen gas. These are nitrate reductase, nitrite reductase and nitrous oxide reductase. Nitrate reductase reduces nitrate to nitrite, nitrite reductase reduces nitrite to nitrous oxide, and nitrous oxide reductase reduces nitrous oxide to nitrogen gas. The enzymes are bound to a support matrix and are placed into a reactor. The first reactor contains support matrix with bound nitrate reductase and the second part of the reactor contains support matrix with bound nitrite reductase and nitrous oxide reductase (Figure 5). The two reactors are connected in sequence and as contaminated water flows through these reactors nitrate in the water is reduced to nitrite by the first reactor and then to nitrogen gas by the second reactor. An electric current provides reducing energy for the process (Mellor et al., 1992; Holton, 1996). The system, currently in the development stage, is being promoted as a system for home and farm use (Campbell and Campbell, 2000). Problems that must be overcome before the system can be marketed include improving the stability of the enzymes and reducing the cost of production (Holton, 1996).

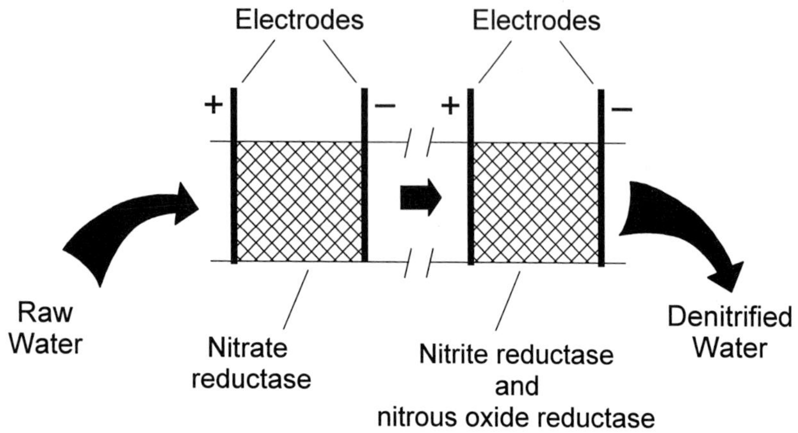

Figure 5. Schematic of a denitrification reactor that utilized enzymes to reduce nitrate to nitrogen gas (Campbell and Campbell, 2000)

4. CONCLUSIONS

Reverse osmosis, ion exchange, and distillation can be used in home systems to remove nitrate from raw water. Reverse osmosis and distillation are small point-of-use systems that provide water for cooking and drinking. With these systems the small volume of reject water generated would be flushed to the septic system and would not normally present a disposal problem. A more important concern might be the volume of water used by some of the systems. Some emerging systems might prove suitable for home or farm use in the future. Biobarriers might be used to protect a well from a contaminated aquifer or to protect an aquifer used for drinking water from a source of nitrate pollution. Above ground denitrification reactors might be used to provide nitrate-free water to rural communities in the not too distant future.

REFERENCES

Benjamin, S., and D. Belluck. 1994. State groundwater regulation: Guide to laws, standards, and risk assessment. Bureau of National Affairs, Inc., Washington, DC.

Blowes, D. W., W. D. Robertson, C. L. Ptacek, and C. Merkley. 1994. Removal of agricultural nitrate from tile-drainage effluent water using in-line bioreactors. J. Contaminant Hydrology 15:207-221.

Bohdziewicz, J., M. Bodzek, and E. Wasik. 1999. The application of reverse osmosis and nanofiltration to the removal of nitrates from groundwater. Desalination 121:139-147.

Braester, C., and R. Martinell. 1988. The Vyredox and Nitredox method of *in-situ* treatment of groundwater. Wat. Sci. Tech. 20:149-163.

Bruning-Fan, C. S., and J. B. Kaneene. 1993. The effects of nitrate, nitrite, and n-nitroso compounds on animal health. Vet. Hum. Toxicol. 35:237-253.

Campbell, W. H., and E. R. Campbell. 2000. Biotechnology people can use! Nitrate Elimination Co., Inc. (NECi). Available at http://nitrate.com/ (verified 5 Feb. 2000)

Carter, J. P. , Y. H. Hsaio, S. Spiro, and D. J. Richardson. 1995. Soil and sediment bacteria capable of aerobic nitrate respiration. Appl. Env. Microbiol. 61: 2852-2858.

Cevaal, J. N., W. B. Suratt, and J. E. Burke. 1995. Nitrate removal and water quality improvements with reverse osmosis for Brighton, Colorado. Desalination 103:101-111.

Chew, F. C., and T. C. Zhang. 1999. Abiotic degradation of nitrates using zero-valent iron and electrokinetic processes. Environ. Eng. Sci. 16:389-401.

Conlon, W. J. 1990. Membrane processes. p. 709-746 *In* F. W. Pontius (ed.) Water quality and treatment. McGraw - Hill, New York.

Croll, B. T. 1994. Nitrate - best agricultural practice for water - the UK experience. p. 2-24 *In* Proceedings of the Fertiliser Society #359. Greenhill House, Peterborough.

Dahab, M. F., and S. Sirigina. 1994. Nitrate removal from water supplies using biodenitrification and GAC-sand filter systems. Wat. Sci. Tech. 30:133-139.

Edmunds, W. M., and C. B. Gaye. 1997. Naturally high nitrate concentrations in groundwaters from the Sahel. J. Env. Qual. 26:1231-1239.

Flere, J. M., and T. C. Zhang. 1999. Nitrate removal with sulfur-limestone autotrophic denitrification processes. J. Env. Eng. 125:721-729.

Foster, S. S. D., A. C. Cripps, and A. Smith Carington. 1982. Nitrate leaching to groundwater. Phil. Trans. Rol. Soc. London 296:477-489.

Gavaskar, A. R., N. Gupta, B. M. Sass, R. J. Janosy, and D. O'Sullivan. 1998. Permeable barriers for groundwater remediation: Design, construction, and monitoring. Battelle Press, Columbus, OH.

Gonzalez, M.C., and A. M. Braun. 1996. Vacuum UV photolysis of aqueous solutions of nitrate. Effect of organic matter II. Methanol. J. Photochem. Photobiol. A: Chem. 95: 67-72.

Gormly, J. R., and R. F. Spalding. 1979. Sources and concentrations of nitrate-nitrogen in ground water of the central Platte region, Nebraska. Ground Water 17:291-301.

Green M., and G. Shelef. 1994. Treatment of nitrate contaminated groundwater for drinking purposes. p. 587-606 *In* U. Zoller (ed.), Groundwater Contamination and Control, Marcel Dekker Publishers, New York.

Hallberg, G. R. 1989. Nitrate in ground water in the United States. p. 35-74 *In* R. F. Follett (ed.), Nitrogen Management and Ground-Water Protection. Elsevier Science Publishing Co., New York.

Hamilton, P. A., and D. R. Helsel. 1995. Effects of agriculture on ground-water quality in five regions of the United States. Ground Wat. 33:217-226.

Hamon, M., and E. Fustec, 1991. Laboratory and field study of an *in situ* groundwater denitrification reactor. Res. J. WPCF 63:942-949.

Harries, T. C., D. Elyanow, D. N. Heshka, and K. L. Fischer. 1991. Desalination of brackish groundwater for a prairie community using electrodialysis reversal. Desalination 84:109-121.

Hiscock, K. M., J. W. Lloyd, and D. N. Lerner. 1991. Review of natural and artificial denitrification of groundwater. Wat. Res. 25:1099-1111.

Holton, C. 1996. Nitrate elimination. Environ. Health Perspect. 104:36-38.

Hörold, S., K. D. Vorlop, T. Tacke, and M. Sell. 1993. Development of catalysis for a selective nitrate and nitrite removal from drinking water. Catalysis Today 17:21- 30.

Houbron, E. and M. Torrijos, and B. Capdeville. 1999. An alternative use of biogas applied at the water denitrification. Wat. Sci. Tech. 40:115-122.

Huang, C. P., H. W. Wang, and P. C. Chiu. 1998. Nitrate reduction by metallic iron. Wat. Res. 32:2257-2264.

Hunter, W. J. 1999. Use of oil in a pilot scale denitrifying biobarrier: p. 47-51 *In* A. Leeson and B. C. Alleman (eds) Bioremediation of metals and inorganic compounds. Battelle Press. Columbus, OH.

Hunter, W. J., and R. F. Follett. 1997. Removing nitrate from groundwater using innocuous oils: Water quality studies. p. 415 - 420 *In* B. C. Alleman and A. Leeson (eds) *In Situ* and On-Site Bioremediation: Volume 3. Battelle Press, Columbus, OH.

Hunter, W. J., R. F. Follett, and J. W. Cary. 1997. Use of vegetable oil to stimulate denitrification and remove nitrate from flowing water. Trans. ASAE 40:345-353.

Janda, V., J. Rudovsky, J. Wanner, and K. Marha. 1988. *In situ* denitrification of drinking water. Wat. Sci. Tech. 20:215-219.

Kamrin, M., N. Hayden, B. Christian, D. Bennack, and F. D'ltri. 1990. A guide to home water treatment. Michigan State University Extension. Extension Bulletin WQ-21. Available at http://www.gem.msu.edu/pubs/msue/wq21p4.html (verified 6 Feb. 2000.)

Kamrin, M., N. Hayden, B. Christian, D. Bennack, and F. D'ltri. 1991. Reverse osmosis for home treatment of drinking water. Wat. Qual. 14:1-4.

Kross, B. C., G. R. Hallberg, D. R. Bruner, K. Cherryholmes, and J. K. Johnson. 1993. The nitrate contamination of private well water in Iowa. Am. J. Pub. Health 83:270-272.

Kruithof, J. C., C. A. van Bennekom, H. A. L. Dierx, W. A. M. Hijnen, J. A. M. van Paassen, and J. C. Schippers. 1988. Nitrate removal from ground water sulphur/limestone filtration. Wat. Supply 6:207-217.

Kudo, A., K. Domen, K. Maruyama, and T. Onishi. 1987. Photocatalytic reduction on NO_3^- to form NH_3 over Pt-TiO_2. Chem. Lett. 6:1019-1022.

Kunz, R. 1997. Reverse Osmosis. Available at http://www.problemsolved.com/Papers/water/Reverse_ Osmosis.htm (verified 5 Jan. 2000)

Lampe, D. G., and T. C. Zhang. 1997. Sulfur-based autotrophic denitrification for remediation of nitrate-contaminated drinking water. p. 423 - 428 *In* B. C. Alleman and A. Leeson (eds) *In Situ* and On-Site Bioremediation: Volume 3. Battelle Press, Columbus, OH.

Liem, L. E., S. J. Stanley, and D. W. Smith. 1996. Northern river basins study project report No. 88: A review of literature on the removal of inorganic contaminants from drinking water. Northern River Basins Study, Edmonton, Alberta.

Mansell, B. O. and E. D. Schroeder. 1999. Biological denitrification in a continuous flow membrane reactor. Wat. Res. 23:1845-1850.

Mellor, R. B., J. Ronnenberg, W. H. Campbell, and S. Diekmann. 1992. Reduction of nitrate and nitrite in water by immobilized enzymes. Nature 355:717-719.

MacDonald, D. V. 1990. Denitrification by fluidized biofilm reactor. Wat. Sci. Tech. 22:451-461.

McMahon, P. B., K. F. Dennehy, and B. W. Bruce. 1998. In-situ bioremediation of Nitrate-contaminated ground water--A pilot test, Julesburg, Colorado, 1996-97. USGS, Water-Resources Investigators Report 98-4046.

Mohseni-Bandpi, A., D. J. Elliott, and A. Momeny-Mazdeh. 1999. Denitrification of groundwater using acetic acid as a carbon source. Wat. Sci. Tech. 40:53-59.

Mori, T., J. Suzuki, K. Fujimoto, M. Watanabe, and Y. Hasegawa. 1999. Reductive decomposition of nitrate ion to nitrogen in water on a unique hollandite photocatalyst. Appl. Catalysis 23:283-289.

Mueller, D. K., P. A. Hamilton, D. R. Helsel, K. J. Hitt, and B. C. Ruddy. 1995. Nutrients in ground water and surface water of the United States--An analysis of data through 1992. U.S. Geological Survey Water Resources Investigations Report 95-4031.

Myrold, D. D., and J. M. Tiedje. 1985. Establishment of denitrification capacity in soil: effects of carbon, nitrate and moisture. Soil Biol. Biochem. 17:819-822.

National Academy of Sciences. 1978. Nitrates: An environmental assessment. Washington, D.C.

National Research Council. 1997. Safe water from every tap: Improving water service to small communities. National Academy Press, Washington, DC.

Nolan, B. T., C. B. Ruddy, K. J. Hitt, and D. R. Helsel. 1998. A national look at nitrate contamination of ground water. Wat. Cond. Purification 39:76-79.

Nugent, M., M. A. Kamrin, L. Wolfson, and F. M. D'Itri. 1993. Nitrate -- a drinking water concern. Michigan State University Extension Bulletin WQ-19. Available at http://www.gem.msu.edu/pubs/msue/wq19p1.html (verified 5 Feb. 2000)

Ohtani, B., M. Kakimoto, and H. Miyadzu. 1988. Effect of surface-adsorbed 2-propanol on the photocatalytic reduction of silver and/or nitrate ions in acidic TiO2 suspension. J. Physical Chem. 92:5773 - 5776.

Olson, W. W., B. L. Yust, S. M. Danes, and K. Chung. 1994. Household management of drinking water: use of reverse osmosis systems. J. Home Eco. 86:13-19.

Osmonics. 1992. Pure water handbook. Osmonics®, Inc. Minnetonka, Minnesota. Available at http://www.osmonics.com/ products/Page 716.htm (verified 5 Feb. 2000)

O'Tool, L. J. 1998. Policy framework - United States. p. 307 - 339 *In* G. J. I. Schrama, (ed) Drinking water supply and agricultural pollution. Kluwer Academic Publishers, the Netherlands.

Piotrowski, R., and R. A. Kraemer. 1998. Policy framework - Germany. p. 93 - 122 *In* G. J. I. Schrama, (ed) Drinking water supply and agricultural pollution. Kluwer Academic Publishers, the Netherlands.

Rajapakse, J. P., and J. E. Scutt. 1999. Denitrification with natural gas and various new growth media. Wat. Res. 33:3723-3734.

Ranjit, K. T., R. Krishnamoorthy, and B. Viswanathan. 1994. Photocatalytic reduction of nitrite and nitrate on ZnS. J. Photochem. Photobiol. A: Chem. 81:55-58.

Reddy, K. J. and J. Lin. 2000. Nitrate removal from groundwater using catalytic reduction. Wat. Res. 34:995-1001.

Robertson, W. D., and M. R. Anderson. 1999. Nitrogen removal from landfill leachate using an infiltration bed coupled with a denitrification barrier. Ground Wat. Monit. Remed.

19:73-80.

Robertson, W.D., and J. A. Cherry. 1995. *In situ* denitrification of septic-system nitrate using reactive porous media barriers: field trials. Ground Wat. 33:99-111.

Roenefahrt, K. W. 1986. Nitrate elimination with heterotropic aquatic microorganisms in fixed bed reactors with buoyant carriers. Aqua 5:283-285.

Rogalla, F., P. Ravarini, G. De Larminat, and J. Couttelle. 1990. Large- scale biological nitrate and ammonia removal. J. Wat. Env. Manag. 4:319-329.

Schipper, L., and M. Vojvodic-Vukovic. 1998. Nitrate removal from groundwater using a denitrification wall amended with sawdust: Field trial. J. Env. Qual. 27: 664-668.

Schrama, G. J. I. 1998. Introduction. p. 3 - 18 *In* G. J. I. Schrama (ed) Drinking water supply and agricultural pollution. Kluwer Academic Publishers, the Netherlands.

Schuff, S. 1992. Nitrates can leach, but they can't hide. Colorado Rancher and Farmer 46(14):6-12.

Siantar, D. P., C. G. Schreier, C. S. Chou, and M. Reinhard. 1996. Treatment of 1,2-dibromo-3-chloropropane and nitrate-contaminated water with zero-valent iron or hydrogen-palladium catalysts. Wat. Res. 30:2315-2322.

Silverstein, J., and G. Carlson. 1999. Biological denitrification of drinking water for rural communities: Demonstration of a novel treatment process in the town of Wiggins, Colorado. University of Colorado, Boulder.

Spalding, R. F., and M. E. Exner. 1993. "Occurrence of nitrate in groundwater--a review. J. Env. Qual. 22:392-402.

Streile, G. P., J. W. Cary, and J. K. Fredrickson. 1991. Innocuous oil and small thermal gradients for *in situ* remediation. p. 32-37 *In* Proceedings of the National Research and Development Conference on the Control of Hazardous Materials. Anaheim, CA., Feb. 20-22, 1991.

Szabo, Z. G. and L. G. Bartha. 1952. The alkalimetric determination of nitrate by means of a copper-catalyzed reduction. Anal. Chem. Acta 6:416-419.

Tannehill, C. C., M. F. Dahab, W. E. Woldt, and I. Bogardi. 1997. Evaluation of nitrate treatment methods under uncertainty. J. Env. Sys. 25:421-444.

Train, R. E. 1979. Quality criteria for water. Castle House Publications, London.

Truove, C., and P. M. Chazal. 1999. Autotrophic denitrification by Thiobacillus denitrificans with thiosulphate as sole energy source. Pilot scale experiments at low temperature. Env. Tech. 20:387-395.

USEPA. 1973. Nitrogenous compounds in the environment. U.S. Government Printing Office, Washington, D.C. EPA-SAB-73-001.

USEPA. 1995. *In situ* remediation technology status report: Electrokinetics. U.S. Government Printing Office, Washington, D.C. EPA-542-K-94-007.

USEPA. 1999. Nitrate and nitrite releases to water and land: 1991 to 1993. Available at http://www.epa.gov/OGWDW/dwh/c-ioc/nitrates .html (verified 6 Feb. 2000)

Volokita, M., A. Abeliovich, and M. I. M. Soares. 1996a. Denitrification of groundwater using cotton as energy source. Wat. Sci. Tech. 34:379-385.

Volokita, M., S. Belkin, A. Abeliovich, and M. I. M. Soares. 1996b. Biological denitrification of drinking water using newspaper. Wat. Res. 30:965-971.

Wärnå, J., I. Turunen, T. Salmi, and T. Maunula. 1994. Kinetics of nitrate reduction in Monolith reactor. Chem. Eng. Sci. 49:5763-5773.

Wylie, B. K., M. J. Shaffer, M. K. Brodahl, D. Dubois, and D. G. Wagner. 1994. Predicting spatial distributions of nitrate leaching in northeastern Colorado. J. Soil Wat. Cons. 49:288-293.

Wylie, B. K., M. J. Shaffer, and M. D. Hall. 1995. Regional assessment of NLEAP NO3-N leaching indices." Wat. Res. Bull. 31:399-408.

Young, G. K., H. R. Bungay, L. M. Brown, and W. A. Parsons. 1964. Chemical reduction of nitrate in water. J. Wat. Poll. Control Fed. 36:395-398.

Nitrogen in the Environment: Sources, Problems, and Management
R.F. Follett and J.L. Hatfield (Eds)

Chapter 18. Remediation at the Water Treatment Plant

L.D. McMullen, P.E., Ph.D.
CEO and General Manager
Des Moines Water Works, Des Moines, Iowa

Communities with elevated levels of nitrate in their source water are faced with the dilemma of figuring out how to optimally lower the nitrate level in the drinking water provided to its systems in the most cost effective manner. Most systems approach the selection process by first studying alternate sources of water; then, physical/chemical systems; and lastly, biological systems.

SELECTION ALTERNATIVES FOR NITRATE REDUCTION

Alternate Sources of Water - For most cities, establishing alternate supplies of source water is the most cost-effective alternative. Generally, nitrates are a problem in surface waters or shallow ground waters. If a community drills a new, deeper well, generally, the deeper well would produce low nitrate water that could be blended with the higher nitrate water to reduce the nitrate below ten milligrams per liter (mg/l). However, there is a drawback in using deeper wells for blending. Usually, the deeper the ground water, the more mineralized it becomes which may then result in a negative impact on homeowners who have home water treatment devices or commercial establishments that have special water quality needs. In spite of the potential negativity of dissolved minerals, obtaining source water from a deeper well is usually very cost effective.

A second alternate source of water would be for the community to connect to a regional water supply, such as a rural water system, and utilize that water for its citizens. The regional connection could be utilized in one of two ways. First, they could abandon their existing treatment facility and buy the water totally from a regional supplier, which would meet drinking water standards. Or, secondly, if the regional water supply is low enough in nitrate, it could be utilized as a blending source with the existing facility's supply to bring the nitrate down below the 10 mg/l standard.

Physical/chemical processes - Generally, the physical/chemical processes fall into three categories – distillation, membrane technology, and ion exchange.

Distillation has been around for a long, long time and is very effective in removing nitrate, however, it is also very expensive. After the water has passed through the distillation process, it becomes very aggressive when attempting to pass it through any kind of piping system. Aggressive water is corrosive to most metallic piping systems leading to pitting and leaks. As a result, distillation, if used, is only blended back with part of the undistilled water stream in proportionate amount to lower the nitrate level below 10 mg/l. Currently, because of its high cost, there are mostly just home-based, or point-of-use, distillation units in service and no municipal distillation processes for nitrate removal.

Distillers cost from \$200 (U.S.) to \$1500 (U.S.) for home use models. Counter top distillers will range from \$200 (U.S.) to \$500 (U.S.) and automatic models from \$600 (U.S.) to \$1500 (U.S.). In addition to the purchase cost, there are yearly operation costs. These include electricity, chemical cleaners, and possibly replacement of activated carbon filters. Yearly operation costs depend on how often the distiller is used. An average family of four may spend approximately \$40 (U.S.) per month in electrical cost for distilled water. [a]

Membrane technology - This methodology utilizes selective membranes for the removal of nitrate and is a very rapidly advancing technology. What seems to be the most promising membrane technology is reverse osmosis (R.O.), where membranes are used not only to remove nitrate, but other byproducts as well such as pesticides or other organic materials that might be present in the water. R.O. treatment produces very high quality water with only a few volatile organics still present. However, it also requires a relatively high-energy demand because of needing water pressure that is above normal distribution pressure to drive the water to be treated through the semi-permeable membrane. Additionally, R.O. has a significant reject stream. This reject stream is produced during the cleaning and flushing of the membranes. Reject stream volume can reach 30% of the total volume treated. This can be a problem in areas where water resources are limited. Several communities have done evaluation studies on R.O. for nitrate removal and several small communities have looked at it as a way to solve multiple water quality issues.

Reverse osmosis systems cost approximately \$0.55 (U.S.) liter $^{-1}$ to install. The primary operating cost component of R.O. is energy. Therefore, its cost effectiveness is contingent upon the local community electric rate. R.O. technology trend is towards lower pressure membranes, which still have the high reject rate for nitrates. As these new membranes are developed, it could result in a significant reduction in the energy costs associated with reverse osmosis. The operating cost of R.O. equipment is approximately \$0.025 (U.S.) liter. $^{-1}$ [b]

Ion exchange - It is by far the most common physical/chemical technique used in the United States for removal of nitrate. In most cases, a strong anion resin with chloride exchange is utilized. The systems work very effectively and are relatively inexpensive to operate due to the regenerant being sodium chloride or solar salt. Nitrate removal is extremely well suited for ion exchange (I. E.) unless there are high concentrations of sulfate in the source water, which is normally not the case with surface waters or shallow ground water. I.E. units can process large quantities of water with a small amount of regenerant that has to be disposed of. In some locales, regenerant disposal may become expensive because of its high salt content. Sanitary sewer systems may not accept high waste brine and direct discharges to streams may not be allowed. In such cases, an evaporative system may be needed for the disposal of the regenerant brine. Recycling of brine through the units, can reduce the volume of brine, however, the chemical make up of the source water is extremely important in determining the appropriateness of recycle.

---

[a] North Dakota State University Extension Service.

[b] Bill Bellamy, Ch_2M Hill consulting engineers.

Utilities that have installed ion exchange have discovered other advantages with the use of strong anion resins. One such advantage is the removal of large chain organic compounds, which substantially reduce total organic carbon in the finished water. One precaution with anion resins is making sure that any iron or manganese present in the source water is removed to prevent resin fouling. Resin is the most expensive component of the treatment technique.

Ion exchange costs approximately \$0.10 (U.S.) liter $^{-1}$ installed.[b] Operating costs can run upward of \$0.10 (U.S.) 1000 liter $^{-1}$.

Biological Processes - Biological treatment is extremely effective in removal of nitrate in an anaerobic system. The oxygen receptors prefer nitrate, which is then converted to nitrogen gas and released to atmosphere. If designed correctly, the process will remove almost 100% of the nitrate. However, it may also end up generating some taste and odor problems associated with the anaerobic systems. This is especially true if there are significant amounts of sulfate in the source water which can be reduced to hydrogen sulfide and give the water a rotten egg odor. Additionally, with any biological system, there will be certain metabolic byproducts that could also generate taste and odors and/or complicating chlorination byproducts once the water is disinfected, if chlorine is used. Furthermore, as with any biological system, its success is dependent on keeping the anaerobic organisms in adequate numbers to be effective. Sometimes different chemicals, such as, super-saturated oxygen, contained in the source water may upset the biological system and fail to reduce the nitrates to the level where they should be. Capital cost for biological systems are approximately \$0.26 (U.S.) liter $^{-1}$, with operating cost estimated at \$0.08 (U.S.) 1000 liter $^{-1}$. [b]

With all treatment systems, whether biological or physical/chemical, the normal scheme is to treat only a side stream of the source water and then to blend that back with the rest of the water to the level where it will maintain nitrate below 10 mg/l.

THE WORLD'S LARGEST NITRATE FACILITY

In the late 1970's and early 1980's, the nitrate concentration in the source water for Des Moines Water Works had risen to a point where it no longer could be blended with shallow ground water and still meet the nitrate drinking water standards. On two occasions, once each in 1984 and 1991, Des Moines Water Works had to provide public notification of violating the drinking water standards to its customer base of approximately 350,000 people.

In between these years, Des Moines Water Works initiated a study to determine exactly what type of treatment would be appropriate for the utility. The result of that study found that ion exchange using a strong anion resin was the desired treatment process. That conclusion came from an extensive study by the utility's laboratory staff who looked at numerous resins and different regenerants to maximize the removal. It was also determined to locate the nitrate facility in the treatment stream that was downstream of both the utility's lime softening process as well as the sand filters so that a side stream could be routed through ion exchange tanks to adequately treat water once it was blended back into the clearwell of the treatment plant. See schematic Figure 1 of DMWW's treatment process. It was also intended

Figure 1
Water Treatment Process

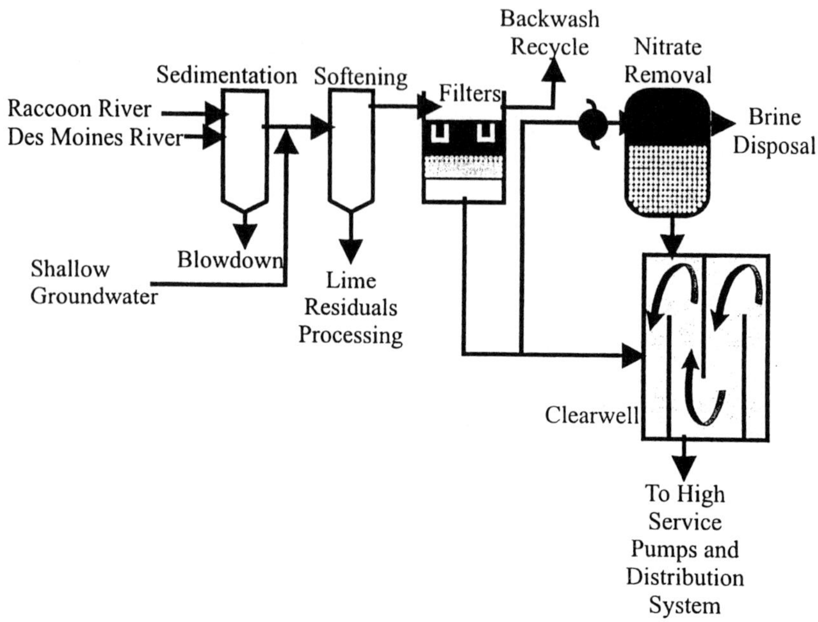

that the facility would be remotely monitored and operated. The utility's laboratory staff studied the chromatographic effect of anion exchange on the other anions and due to the fact the process was after a lime softening operation, found that pH was a good indicator of potential breakthrough for nitrate.

The utility then retained a national consulting firm to prepare a preliminary design report incorporating the findings of the DMWW's laboratory studies. The report concluded a 37.8 Ml day $^{-1}$ (10 million gallon day $^{-1}$) anion exchange treatment process would substantially reduce the probability of the utility exceeding the MCL of 10 mg/l to one day every seven years. The study also confirmed that sodium chloride (NaCl) was the preferred regenerant and the structure housing the facility could be located on top of the clearwell to save space and piping and to facilitate blending. The result is the world's largest nitrate removal facility with a capacity of 37.8 Ml day $^{-1}$ and consisting of eight resin vessels, expandable to ten. Spent brine is diluted with raw source water and discharged to the Raccoon River. See Figure 2 showing the positioning of the nitrate facility.

Figure 2

Figure 3

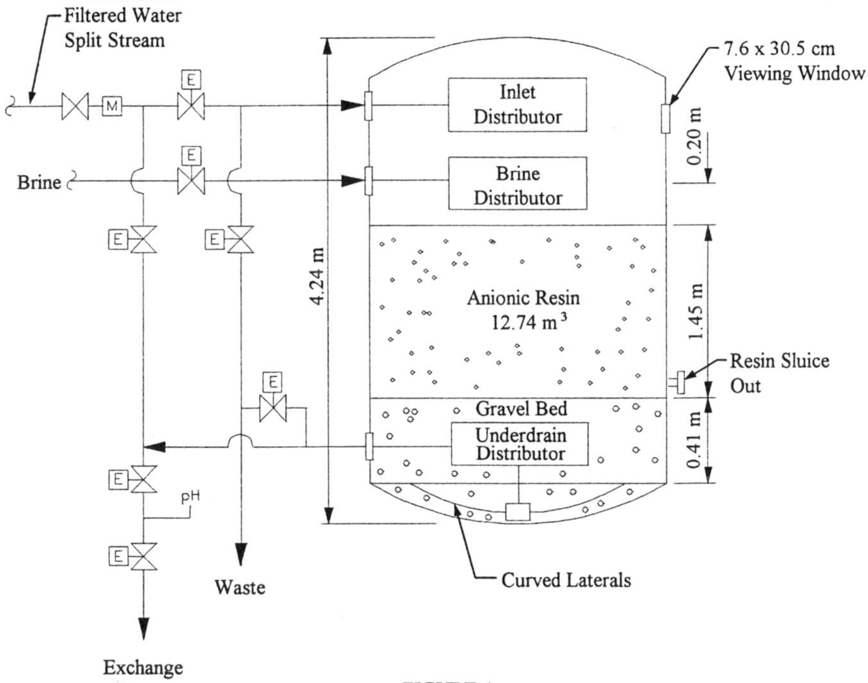

FIGURE 4
ION EXCHANGE VESSEL

The clearwell is a buried concrete structure and the nitrate building is located on top of the southeast ¼. Figure 3 shows the ion exchange vessels and connecting piping header. A detailed view of an ion exchange vessel and header piping is shown in Figure 4.

The facility went into operation in the spring of 1992 and has been used off and on primarily during the spring when nitrate in the source water reaches its peak level. In 1999, Des Moines Water Works set an all time record with 109 days that the facility was run to maintain the drinking water standard below 10 mg/l. During that same time period, an all time record of nitrate concentration in the Raccoon River, the primary source water for Des Moines Water Works, was set at 15.9 mg/l. The facility costs roughly \$3,000 (U.S.) $^{-1}$day to operate, which includes solar salt costs and the energy costs for the booster pumps to pass filtered, pre-chlorinated water through the resin. The cost to construct this facility in 1992 was approximately \$4 million (U.S.).

FUTURE TREATMENT

The future industry trends are to continue with the same physical/chemical processes. However, a new dimension is underway at least in midwestern states of the USA to control nitrogen at the source. The utilities have found that it may be less expensive in the long run to provide education as well as apply existing incentive-based government support programs in an optimal way to reduce the amount of nitrate in the source water. One such USA program is the Conservation Reserve Enhancement Program (CREP). The Conservation Reserve Program under the U.S. Department of Agriculture supports the construction of buffer strips, which are non-tilled grassy areas isolating agricultural cropland from drainage ways. The United States Environmental Protection Agency's (EPA) Clean Water Act created the Section 319 Nonpoint Source Management program to support a wide variety of technical and financial assistance to states, territories, and Indian Tribes. All of these programs are available to water utilities to improve source water. The success of these types of programs will totally determine whether additional treatment techniques or other approaches may be needed into the future.

Section 5:

Economic and Policy Issues

Nitrogen in the Environment: Sources, Problems, and Management
R.F. Follett and J.L. Hatfield (Eds)

Chapter 19. Nitrogen Management by Producers: A Multiple Scale Perspective

Pete J. Nowak[a], Laurence B. Cutforth[b], Perry E. Cabot[b] and Bruce M. Kahn[b]

[a] University of Wisconsin, Department of Rural Sociology, 346D Agriculture Hall, 1450 Linden Drive, Madison, WI 53706

[b] Institute of Environmental Studies, Land Resources Program, University of Wisconsin

Abstract

We discuss social factors of nitrogen management using the farmer as a focal point as a means to inform future research. We focus on two critical questions associated with researching nitrogen management from the perspective of agricultural producers. Our first question is how does the concept of scale help direct a study of the social factors in farmers' nitrogen management decisions? And secondly, how does one analyze these salient social dimensions of nitrogen management? Social factors that influence farmers arise at multiple and overlapping scales in an agro-environmental hierarchy. Management, and therefore mis-management, is a cross-scale phenomena, the resultant social process of many different actors and institutions, including but not limited to farmers. The issue of studying linkages between the social and biophysical factors affecting nitrogen management points to the need for a common area of analysis. We know there is significant variation in the nutrient management behaviors of farmers and that this leads to a disproportionate influence on ecosystem properties. Examining the role of disproportionality will involve linking social to biophysical phenomena at different spatial and temporal scales. We present examples of spatial analysis that uses both social and biophysical data to describe disproportionality at various spatial scales. We conclude by encouraging further research on nitrogen processes that involve both the social and biophysical sciences using common analytical techniques.

1. INTRODUCTION

Nitrogen management in modern commercial agriculture can best be described as the sequential actions taken within a set of farm-system constraints in order to achieve an ecological state of "nitrogen saturation" throughout crop production. That is, the objective is to supply the system with sufficient nitrogen so that water, genetic resources, non-nitrogen nutrients, or other factors become the limiting restrictions on further plant growth. The immobilization of nitrogen in the subsurface by soil biota, its capacity for mineralization and nitrification, and its response to temperature changes has generated a prevailing management pattern of saturation. For the farmer this means applying nitrogen in large amounts to avoid anticipated losses. A diverse set of cultural, economic, institutional and technological conditions "drive" this kind of management, resulting in a very effective, but inefficient crop production system. Indeed, it is effective in the sense that the farmer achieves higher yields; it is inefficient, however, considering the amount of inputs required to compensate for anticipated losses per unit of output during this production process.

If there was uniformity in the behavioral patterns and rationale for pursuing this "saturation" management style, as well as uniformity in the biophysical vulnerability to these management styles, then our task as scientists would be easy. Unfortunately (or fortunately in terms of providing continuing intellectual challenge), this is not the case. There is as much diversity and variation in the social dimensions of farmers as there are in the biophysical resources they manage. Suggesting an approach to understanding the interaction between the two is the objective for this chapter.

It is these "drivers" of nitrogen mis-management, and in particular those that may be characterized as social in nature, that will receive our attention. We acknowledge that many of these social drivers co-exist and interact with more traditional economic processes, structures and constraints. A credible examination of the economic factors associated with nitrogen management is beyond the expertise of the authors and the logistical limitations of this chapter, but there are other sources that discuss them in detail (Shankar et al., 2000; Hopkins, 1996; Bosch et al., 1995).

In this chapter we discuss the social factors of nitrogen management with the farmer as the focal point as a means to inform future research. We focus on two critical questions associated with researching nitrogen management from the perspective of agricultural producers. Social factors have been linked to the farmer's psychological state (Lynne et al, 1988), field management decisions (Hayman and Alston, 1999), how farm firms are managed (Meyer et al., 1997), national policies and programs (Brouwer, 1998), and even the alteration of global cycles (Vitousek, 1997). In short, social factors arise at multiple and overlapping scales in an agro-environmental hierarchy.

For clarity, we define scale as "the spatial, temporal, quantitative or analytical dimensions used by scientists to measure and study objects and processes (Gibson et al.,. 2000, p. 19)." The social factors influencing farmers' also have a scale effect. Consequently, our first question is how does the concept of scale help direct a study of the social factors in farmers' nitrogen management decisions?

The first question relates to the spatial and temporal dimensions that underlie farming systems in general, and nitrogen management in particular. Past social research on decision-making has tested a long list of decision making variables such as age, education, attitudes, and perceptions of risk with limited success (Lockeretz, 1990). Our contention is that a scale-based analysis provides a framework that will limit which social factors need to be deemed important for future analysis. Thus, our second question is how does one analyze these salient social dimensions of nitrogen management? Recent innovations in both techniques and analysis methods offer a number of potential research strategies for social scientists or those engaging in interdisciplinary research. We suggest several approaches and techniques that may be used in future research.

The astute reader will quickly recognize that the authors are very presumptuous for raising these particular questions within a book chapter. Careers, curricula, not to mention numerous books have touched on each of these questions. Nonetheless, we believe that modern developments, such as legislative approaches to nutrient management require a more focused analysis of the farmer's situation. We are not seeking a consensus among researchers in this chapter; rather, we are advocating an analytical framework we believe is capable of developing meaningful and objective insights. The salient point we wish to make is that nitrogen is often mis-managed, and that our understanding of the ecological consequences of these behaviors is increasing each year. Debates over regulatory approaches are characterized by varying levels of belief in technical fixes, all occurring in a context of increasing knowledge of the ecological implications of doing nothing. The nexus point in this complex and dynamic interaction between science, policy, agriculture and the environment is the farmer. Management, and therefore mis-management, is a cross-scale phenomena, the resultant social process of many different actors and institutions, including but not limited to farmers (Beegle, 2000). However, the agriculture producer is a "lynch pin"or "key stone" around whom the analysis of problem causes and solutions must be built. If viable solutions are to be developed, we must understand the farmer fairly and objectively in relationship to all scales of agricultural production and environmental management. This chapter suggests an analytical perspective and associated methods to address that objective.

Our premise is that research on the management of nitrogen in agro-ecosystems cannot unduly focus on generalized best management practices or the consequences resulting from the lack of these techniques. Research on the need for and the design of best management practices within biophysical, technological, or policy settings needs to consider the context of their application, namely the farmers and their farms. While all want to find that true best management practice(s) for nitrogen, it is quite conceivable that this well-intentioned objective may prove difficult, if not impossible to develop without including the social context of farming in the original analytical framework. Any proposed "solution" from the technological or policy arena needs to be grounded in a comprehensive understanding of the farmer's situation. That is, we need a full understanding of the tools and technologies they currently have to work with; how cultural, legal and market structures shape management decisions and technology adoption; and the extent to which biophysical settings interrelate with the consequences of these production decisions and actions.

It is no longer sufficient to focus on understanding the systematic processes or consequences of nitrogen management or mis-management. What is needed now are valid explanations of why it occurs beyond the glib generalizations that have dominated much of the discussion to this point (e.g., risk adversity, profit maximization, traditional attitudes, etc.). Such observations about farmer behavior and farm characteristics are often perceived as being anecdotal to most quantitative data common to the biophysical study of nitrogen in its soil, aqueous, or atmospheric environment. On this note, we propose that nitrogen be evaluated relative to another environment - the social or farm environment. Just as nitrogen behavior is conceptualized in diagrams, for example, how nitrifying bacteria are affected by such biophysical factors as temperature and moisture at various biophysical scales, we propose a similar understanding of nitrogen behavior on farms, with the farmer as an interdependent organism operating not only at biophysical scales, but at various institutional scales as well. As social scientists working in the interdisciplinary setting of nitrogen management research, we feel the two questions imbued within this chapter are starting points in this process.

2. SCALE

We posit that the nature and strength of human interactions with agro-ecosystems is influenced by the observed level of organization. Most environmental variables are considered to be scale dependent, but this same analytical framework has rarely been applied in the social sciences. Why is this important? We suggest that part of the answer to that question lies with the concept of disproportionality. This is the notion that certain places, times or behaviors are more important in explaining the outcome of a system than other similar places, times or behaviors. There is a growing body of research (Heathwaite and Johnes, 1996; Heatwole et al., 1987; Gburek and Sharpley, 1998) that demonstrate that a large proportion of agricultural nutrients lost to surface water comes from a small proportion of the landscape during a very limited time period. What has been lacking, however, is the behavioral dimension of this concept.

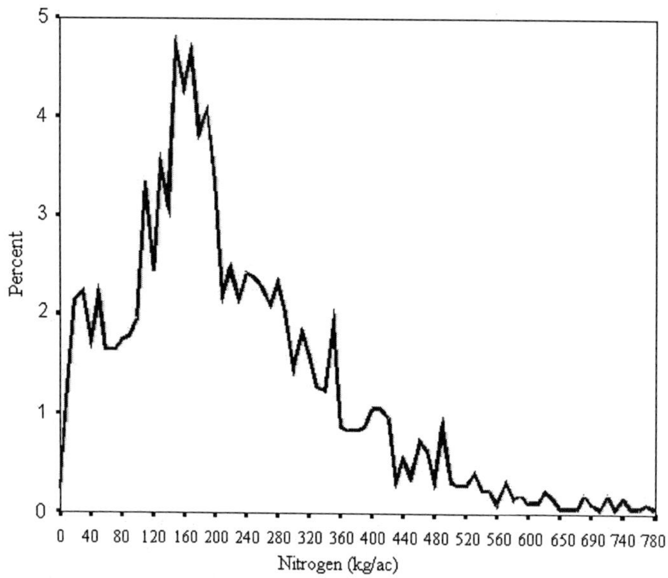

Figure 1: Total N applied during corn production adapted from Shepard (2000)

We know there is significant variation in the nutrient management behaviors of farmers. Figure 1 illustrates the total nitrogen applied in the production of corn by a sample of 2,189 Wisconsin farmers. This includes both commercial nitrogen and that which can be credited from animal manures. This distribution of behaviors, while informative, does not address the question of disproportionality. All one can say is that the probability of a disproportionate contribution to environmental degradation increases as one moves above the measures of central tendency. What is lacking is the biophysical context of the behavior. As Brasier et al.,. (2001) point out, the impact is not so much determined by the nature of these behaviors, but on when and where they occur across the landscape. The concept of disproportionality — the notion that a small percentage of farmer behaviors contribute a disproportionate amount of nutrient mis-management — is a scale-dependent argument. For example, the emergence of a nutrient problem typically occurs at a landscape scale whereas the perturbation occurs at the sub-level scale. Therefore, the concept of scale, by definition, brings in the crucial aspects of the nature, location and timing of farmers' nitrogen management.

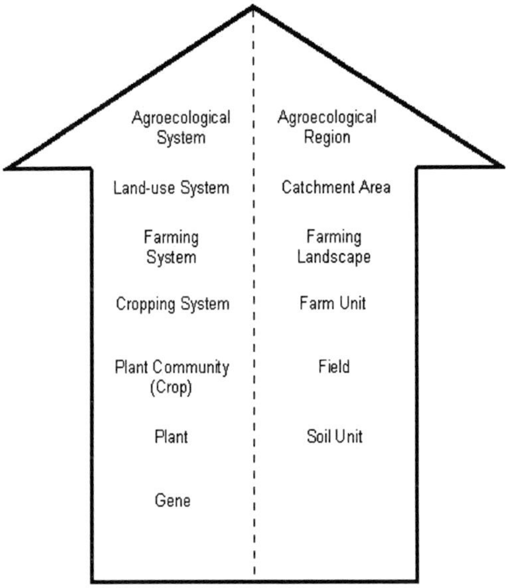

Figure 2: Hierarchy of levels and spatial scales in agriculture (Bouma et al, 1995)

Scale is defined and understood through the lens of the sampling frame used to collect data (Atkinson and Tate, 2000). In agriculture this is traditionally portrayed as a hierarchy of agro-ecosystems where the different levels of scale are defined by spatial extent in a nested hierarchy (see Figure 2). The area of evaluation increases in this hierarchy from the plot all the way to the region. Scale also has a temporal dimension. Climate patterns, pest cycles or even crop rotations can be defined in time as well as space. Change in these patterns occur more slowly at higher levels in the scale hierarchy (Francis and Olson, 1995). While this typical portrayal of scale in agriculture is accurate, it needs to be more inclusive relative to the human dimension. Just as ecological systems are hierarchically organized with interactions across scales (Allen and Starr, 1982), so are human processes and their interactions with agro-ecosystems. One approach to this issue is to examine some general levels of farm management, and the decisions that farmers make at these levels. Farmers make decisions at strategic, tactical and operational levels within the context of the agricultural production process (Bouma, 1997; Beegle and Lanyon, 1994). Strategic decisions apply to the structure and direction of the farm, involving the choice of crops within the context of agricultural markets, managerial capability and tradition. Tactical decisions deal with specific choices regarding the structural variables of the farmer's strategy, such as crop varieties, brands of machinery, and use of external analytical or diagnostic services. Finally, operational decisions relate to very scale-specific decisions such as the amount and timing of inputs for the selected crop varieties, when and where farm equipment will be employed, and the relative amount of human capital allocated to the various production processes. This classification scheme shows an explicit hierarchy; strategic decisions constrain the options available at the tactical level, which in turn constrain the options available at the operational level.

Acknowledgment of this hierarchy reduces both the number and the nature of options available to the producer relative to nitrogen management. As seen in Figure 3, nitrogen management on the farm typically occurs at an operational level of decision-making, but is often pre-determined choice by decisions at the strategic and tactical levels. Other options for nitrogen management may be suggested, but the viability of these options is dependent on their congruence with the other levels of the hierarchy.

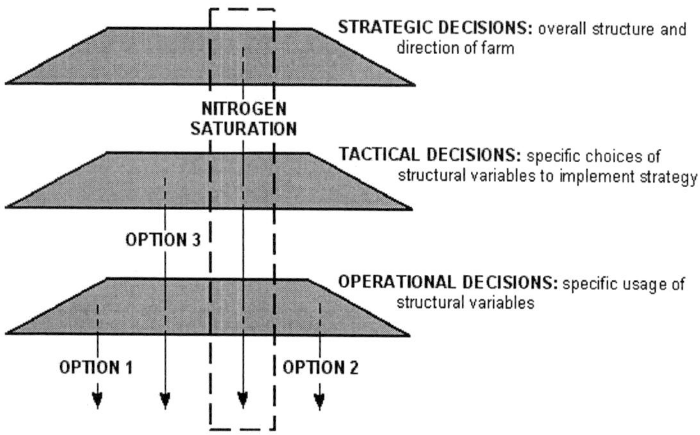

Figure 3: Hierarchy of Farmer Decision Making

The point we wish to make is that farmers have little strategic or tactical control over nitrogen management because of the constrained nature of technologies, tools and recommendations available to them. Such a lack of control reduces the farmer's degrees of freedom in choosing specific options that may be more efficient, profitable or environmentally friendly. This is the case because many of our traditional as well as modern management and production techniques were designed to nullify or mask the influences of ecological variation, such as those factors that influence nitrogen processes within an agro-ecosystems setting. For example, nitrogen is applied on fields by machines designed to provide supposedly uniform application regardless of variance in nutrient holding capacity or historic nutrient levels. Seeds were bred and developed to give a uniform response across a wide range of growing conditions with little consideration given to nitrogen efficiency. Tillage tools are designed to provide uniform results across a wide range of soil conditions with little thought given to how these patterns of soil disturbance may influence nitrogen mobility. In summary, many of the tools and recommendations of modern commercial agriculture were designed to ignore or override rather than actually manage or respond to the natural variation that influences nitrogen processes in crop production.

These strategic and tactical constraints restrict the flexibility of the farmer to make operational decisions oriented towards achieving sufficient nitrogen for crop production. Creating nitrogen management options at the operational level without creating supporting options at the tactical or strategic levels is unproductive and should be resisted (Beegle and Lanyon, 1994). For this reason, research and outreach programs designed to generate and promote nitrogen best management practices must also address the tactical and strategic context of those practices.

The main theme in this discussion is that decisions within a farm system are also hierarchical in nature where options and decisions at higher levels can constrain or restrict the choices available at lower levels. We believe this is related to the concept of disproportionality by recognizing that the nature and number of options available to individual farmers also vary in time and space. There are a number of different products, equipment, analytical and diagnostic services available to assist farmers in the management of nitrogen. These products and services are typically provided through private sector vendors such as crop consultants, input supply dealerships, dealer agronomists and management firms. Public sector support is found in USDA programs and organizations such as the Extension Service and land grant university research and teaching programs. The critical point we make is that these infrastructure products and services are not uniformly distributed in time and space. Some farmers have greater access to a greater variety of these products and services than other farmers. For example, the level of competition among agrichemical dealers in Wisconsin is illustrated in Figure 4. Some producers in certain parts of the state have 10-12 dealers vying for their business while in other parts it is close to a monopolistic relation. The nature and number of options available for strategic, tactical or operational decisions by farmers is directly related to this level of infrastructure support.

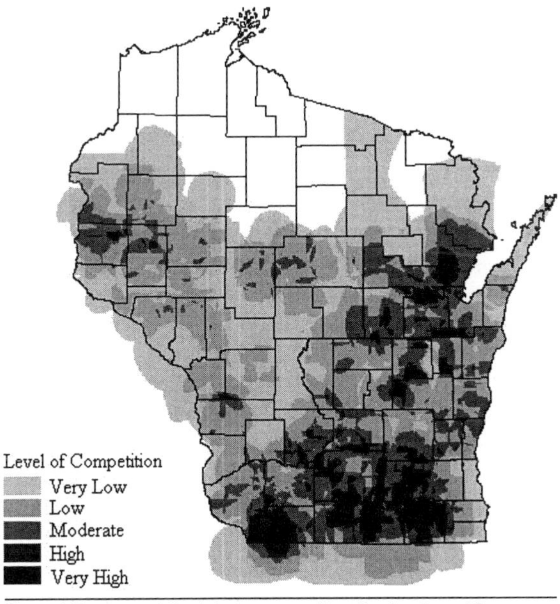

Figure 4: Income weighted dealer competitions levels adapted from Wolf and Nowak (1999)

We have argued that scale is an important consideration when examining nitrogen management by farmers. Decisions by farmers can be ordered in an hierarchical fashion where the slower moving, higher-order decisions can constrain or limit the more rapid, lower-order decisions. Moreover, it was argued that the institutional and infrastructure context of these decisions can also influence the outcome since these also vary in space, and perhaps time. Discussion and analysis of both the institutional (i.e., policy, regulations, programs) and infrastructure (i.e., research, products, services) could be chapters in themselves. This was not our intent. The assumption underlying this chapter is that the concept of disproportionality needs to be the focal point of research on nitrogen management. Examining disproportionality will involve linking social to biophysical phenomena at different spatial and temporal scales. At issue is how this can be accomplished. That is, how do we integrate social and biophysical data to investigate the concept of disproportionality?

3. ANALYTICAL APPROACHES

Without much to guide them, many well-meaning research groups unfortunately perceive social research as a dubious add-on to their scientific investigations. On the other hand, all are aware that agro-ecosystems are increasingly being scrutinized and managed by humans for better yields and environmental quality. From this dichotomy, an urgent need has arisen for practical research approaches that link farmer actions to the same locations as the biophysical processes being managed in both an anticipatory and responsive fashion (Lynne et al., 1988).

What then should be the role of the social sciences in separating nitrogen research into component parts or constituent elements? The quandary underlying this question is that most would agree sociology has something to offer to nitrogen research, but there is little consensus on what this something is, or the direction it should take. Few would disagree that economics as a science has little more to offer nitrogen management research than its most rudimentary techniques (e.g. cost-benefit accounting of the latest technological fix being promoted as a best management practice). The same sentiment is more difficult to argue on behalf of sociology. Tongue-in-cheek as it may sound, many social factors are all too often clumped unceremoniously into a category termed something like “management” and later alluded to at group meetings, but never fully assimilated into the overall goal of farm systems research. We do not wish to further this perception by advocating that nitrogen researchers somehow “graft” unwanted social studies onto their future experimental designs. There are certainly many projects unsuited for the inclusion of social factors, and it just as well that they are left out if studying them adds little to the scientific ideas under examination. However, an analysis of disproportionality, that situation where a minority of inappropriate behaviors in vulnerable biophysical settings causes the majority of degradation, requires an integrated approach.

The issue of studying linkages between the social and biophysical factors affecting nitrogen management points to the need for a common area of analysis. The social sciences are now moving in the direction of “spatially integrated social science,” brought on by the advent of spatial analysis and geographic information systems (Goodchild, 2000; Anselin, 1999). These techniques are now frequently applied to a variety of social science applications ranging from anthropology (Aldenderfer and Maschner, 1996), to criminology (Weisburd and McEwen, 1998), to real estate analysis (Can, 1998). Exploring the applications of spatial data analysis in a rural sociological context will provide further insights and methods that are particularly relevant to nitrogen management.

Measuring social factors for use in a spatial analysis require a departure from the traditional techniques used in rural sociology, such as random sampling (Nowak and Korsching,

1998). Precisely because the data of interest must be georeferenced, the social scientist must use methodology that can meaningfully examine the influence of social factors in the biophysical landscape or the spatial location of management practices. Although the biophysical context is heterogeneous, it is most certainly not random. Furthermore, social factors responding to these biophysical attributes, and influencing nitrogen management practices, will also not likely be random either. For these reasons, the sampling of social must approach that of a proportionate, or stratified, spatial sampling regime. Upon establishing the baseline survey at a chosen scale, more extensive sampling should be directed to locations where patterns are detected. It is this underlying search for spatial patterns of farmer behavior, and the explanation for these patterns, that are guiding current spatially integrated research in rural sociology.

A wide assortment of studies exist where the authors took a spatial approach to the examination of patterns and trends of rural social behavior at a variety of scales. In the general nomenclature of spatial statistics, the data from which these studies were comprised originates from one of the three classes of spatial data: 1) point, 2) area or lattice, and 3) spatially continuous.

3.1 Point Data

The central feature of point data analysis is the presence or absence of an observation at a particular location. The measurement involved is binary in nature; an observation either occurs at a given location or it does not. Analysis methods abound for spatial point data analysis, often focusing on the location of behavior that occur in clusters or patterns under two general groupings (Everitt, 1974). The first group of methods analyze the spatial or geographic distance between observations, with distance used as a criterion to locate clusters, "hot spots," or dispersion. Refined nearest neighbor analysis is one technique under this grouping commonly used to analyze point data based on distances (Diggle and Chetwynd, 1991). A second group of techniques examines the density of observations focusing on regions with a higher density of incidence. An example of this technique is kernel density estimation which estimates the intensity at each point in a set of observations using a weighted function from the surrounding points. Other social disciplines, criminology in particular, make use of point pattern analysis to map the occurrence of criminal activities within a geographic region in order to discern underlying patterns of incidence and make changes in law enforcement strategies. Despite obvious potential application, such as targeting programs for nitrogen management assistance, the field of rural sociology has hitherto left point pattern analysis techniques to other disciplines. We believe there is potential in this technique relative to identifying both the social and biophysical dimensions of disproportionality relative to nitrogen.

Farmers across a region may have access to different technology and information because of their location. For example, innovations have been observed to spread in clusters, influenced by interpersonal communication and agricultural infrastructure such as market vendors or processing plants (Brown et al.,, 1976). A more recent study by Wolf and Nowak (1999) used the ideas behind point data analysis to show spatial patterns of aggregation between input dealers and volume of corn production. The authors assert that dealer competition in Wisconsin is not randomly distributed; rather, it is heavily clustered in regions with relatively highly productive and well-buffered soils. This pattern is further compounded by the spatial distribution of groundwater pollution risk due to agrichemical misuse. They use this analysis to make the statement that the extent of technology transfer and resource allocation is inversely related to areas of high environmental vulnerability. We suggest that a similar analysis of the institutional and infrastructure context in areas with nitrogen-related problems could be very informative.

3.2 Area or Lattice Data

Lattice data are associated with spatial regions, defined by an area measurement and are regulary or irregularly spaced. A critical feature of lattice data analysis is that of a neighborhood structure and neighbor relationships. These relationships can be weighted, based on common boundaries or defined proximities. Andicott et al., (1987) identify two methods for measuring neighborhoods: 1) developing criteria for boundaries of neighborhoods, and 2) measuring the distribution of activity or influence among sampling units. As they note, these criteria will be arbitrary, but must be clearly identified. If the unit of observation varies field by field, the neighborhood boundaries should also follow fields. In the case of an information network, farmers may be expected to interact with input suppliers and others at the farm level. Spatial statistics provides quantitative tools for evaluating these relationships that were not available 10 years ago because of limited computing power.

A hierarchy of ethical concerns and obligations by Noss (1992) shows that our consideration of others extends from self to global levels. Our decisions about agricultural diversity and other environmental interactions occur across these levels of concern (Francis and Olson, 1995). By starting at the individual level of analysis, the social interactions within and commitments to a farm family, a community, landscape, or region could provide more detailed information about the nitrogen management. This type of analysis would be especially fruitful where there are human health concerns associated with nitrogen management.

3.3 Spatially Continuous Data

Brasier et al., (2001) provide an example of integrating social and biophysical data in a spatial analysis with their study of resource degradation in the fictitiously named "Little Mackenzie River watershed." Identity and geographic location of the watershed is withheld to protect the confidentiality of the farmers who provided spatially-specific information. This an agricultural drainage basin of approximately 1076 km^2 (415 mi^2) located in the Cornbelt of the Midwest. The southern and western portion of the watershed has a rolling topography of loess soils, with slopes of 5% to 18%. The northern and eastern portion of the watershed has relatively flat glacial till soils with predominantly 0% to 2% slopes. Land use in the Little Mackenzie River watershed is predominantly agricultural. There are approximately 1500 farms in the watershed, averaging about 142 hectares (356 acres) in size. The majority of farmers in the watershed have mixed corn/soybean and livestock operations. Corn yields average approximately 353 bushels per hectare (141 per acre) while soybeans average about 113 bushels per hectare (45 per acre). Most of the livestock raised in the watershed are hogs and beef cattle, with the proportion of hogs increasing in recent years.

The primary environmental problems associated with the Little Mackenzie River watershed are sedimentation, phosphorus and nitrate contamination of the surface water. As is true in many Midwestern watersheds, a major component of the surface water quality management program is the control of nonpoint source pollution from agricultural areas.

The larger drainage basin that includes the Little Mackenzie River watershed also provides drinking water to hundreds of thousands of urban residents. Average nitrate levels in the Little Mackenzie River are 10-15 mg/l, which means they are at or exceeding Maximum Contaminant Levels (MCLs) as defined by the Environmental Protection Agency (10 mg/l).

This watershed eventually drains into the Mississippi River, so it has contributed historically to high nitrate loading and hypoxic conditions in the Gulf of Mexico.

Because of high nutrient loads coming from the watershed, public and private conservation organizations are working together to study their causes and suggest solutions. Generally, these solutions are based on generalized offers of assistance. That is, any farmer who participates voluntarily in the programs is eligible for technical, financial, and/or informational assistance. The problem with this type of generalized approach is that farmers who are most likely to volunteer are those who typically have the least need. Conversely, the farmers who are least likely to participate – socially and environmentally marginal farmers – are probably contributing disproportionately to the environmental problem. Because the coincidence of social and environmental marginality exacerbates existing environmental problems, policies need to identify and target these "doubly" marginal farmers and offer them the appropriate assistance.

The authors based their spatial analysis upon a composite index integrating the social and biophysical dimensions of resource degradation from farmlands. The first measure represented a parcel's vulnerability due to biophysical variables, based on the Universal Soil Loss Equation (Wischmeier and Smith, 1978). The second measure was based on a measure of social characteristics where all operators of land adjacent to a primary, secondary or tertiary stream were sent a mail survey. The survey contained an aerial photo of a specific parcel bordering a stream, and questions were referenced to this specific area.

Indexing methods such as developed in this study provide a method of differentiating between farmers whose lands have greater vulnerability to environmental problems but are likely to address the problem from those farmers whose lands are vulnerable but are not likely to address their problems. Furthermore, farmers and the tracts of land they manage can be spatially identified as clusters of highly vulnerable regions that are not involved in social networks and/or do not have working knowledge of the best nitrogen management practices for their location. These farmers are the most likely contributors to the nitrogen mis-management problem. In other words, the indexing method allows a statistical and spatial distinguishing between those farmers who can and are addressing their vulnerable landscapes and those who cannot and are not. We believe that this issue of spatially identifying disproportional contributions lies at the core of addressing environmental degradation from agricultural land use.

A traditional social analysis would measure the relation between the vulnerability values and various measures of the farm firm, farm, and perhaps, psychological attributes of the farm decision maker. The intent would be to find some statistical relation between measures of vulnerability and the concepts that have meaning in the discipline (i.e., scale of operations, legal organization, gender roles or perhaps tenure of the parcel). This traditional approach, however, has been unfruitful over the last 25 years (Lockeretz 1990). In this chapter, we acknowledge that these attributes have value, and that the distribution of these values may be skewed, but also to assert that these attributes have definable locations. That is, traditional analysis has pursued the aspatial analysis of the relations between these attributes. We believe that spatial location, particularly regarding the spatial location of the outliers on our measures of vulnerability, may be as informative, if not more informative, as to the environmental consequences of farmer management.

Spatial distributions with values at certain locations showing relationship with values at other locations are spatially autocorrelated. Spatial distribution could indicate patterns of underlying process. Incidents exposed to the impact of similar process tend to follow a similar locating pattern. Hence, study on spatial cluster could reveal information about the underlying geographical process that generates the spatial pattern, which can further aid the comprehension of underlying geographical process and its relationship with the phenomenon under investigation. Quantifying these spatial patterns is important from a social science perspective for two reasons. First, from a policy standpoint, it is important to estimate the distance at which the patterns becomes dissimilar. Knowing this distance gives us a better understanding of cluster sizes and at what distances the patterns become insignificant. The size of these clusters may determine the appropriate allocation of program resources. Secondly, from a conceptual perspective, we want to estimate the size of these clusters so that we can attempt to understand how and why spatial location is related to variation among farm firms.

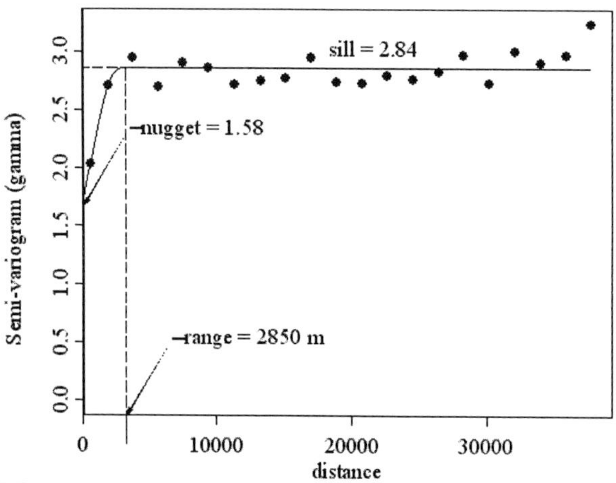

Figure 5: Semi-variogram for vulnerability

A simple semivariogram analysis was used with the aforementioned measures of biophysical and social vulnerability in the Little Mackenzie River watershed. When calculating a semivariogram one calculates the variances between all possible points in the study area, and then averages those variances for given distances between them. These average variances are plotted against the distance, or lag interval. This approach provides an understanding of the extent to which a value at one spatial location is related to values at other locations (Palmer, 1988). Statistically, this means that the correlation between close observations should be high, and this correlation is likely to decrease with distance. The semivariogram was calculated quantifying the size of vulnerable zones based on both the social and biophysical components. As can be seen in Figure 5, the range of 2850 m (9350 ft) suggests that at distances closer than 2850 meters the combined values of vulnerability are similar enough to posit correlation. Using the average farm size of 142 hectares, the range in Figure 5 suggests clusters exist of approximately 2550 ha (6330 ac), or 18 average farms, which are both socially and biophysically vulnerable.

As can be seen in Figure 6, definite "neighborhoods" exist, particularly in the northwestern and southeastern regions of the watershed. There are "clusters" with similar mean values of the combined vulnerability measure. These represent locations where both of these conditions – high biophysical vulnerability and high social vulnerability – occur on the same landscape. The darker areas (high values) indicate clusters where farmers are both environmentally vulnerable (farming soil that is vulnerable to erosion) and socially vulnerable (not integrated into assistance networks and lacking knowledge to address erosion).

The purpose of this extended example from prior work was used to illustrate two themes: 1) techniques are available that will allow a rigorous integration of social and biophysical data, and 2) meaningful results for policy implementation, design of appropriate remedial practices and specifying future research can be derived from the results. Investigating and addressing disproportionality relative to nitrogen will require such an integrated, analytical approach.

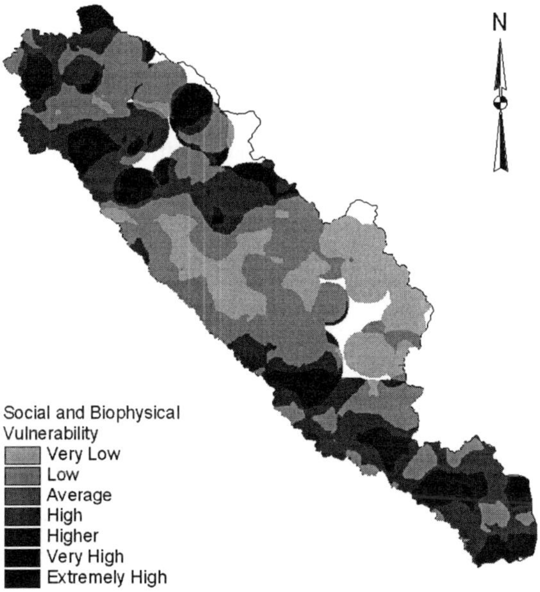

Figure 6: Environmental Vulnerability distributed over watershed adapted from Brasier et al (2001)

4. CONCLUSION

As Tobler stated in his first law of geography, "everything is related to everything else, but near things are more related than distant things" (Tobler, 1979). We use this quote to emphasize that farmers live not only in biophysical space but also in social space. The resulting consequences of this "co-location" emerges from interactions between farmers and the biophysical resources they manipulate. Many of the issues associated with agricultural nitrogen provide a case in point of this relationship. Unfortunately, much of the research on this issue has not been designed around this simple truism.

We point out that the salient social and biophysical processes are not random in space or time. Inappropriate behaviors at vulnerable locations or times create many of the "problems" the policy and research agendas are oriented toward today. This situation of disproportionality should provide a warning that there is not a universal policy or technical solution to these problems. On the other hand, searching for a means to provide an almost infinite set of unique, site-specific solutions is also not feasible. Instead, we have argued for the more classic and scientific approach. One must first be able to describe and measure pattern and process before seeking explanation and prediction. This chapter attempted to contribute toward that objective by describing how concepts such as scale and context apply to the management of nitrogen by farmers. Moreover, we suggested and illustrated several novel — at least to the social sciences — approaches to measuring these concepts in space. Much remains to be accomplished.

We say this because social and biophysical data occur along a continuum of differing scales, and are often incompatible because one was measured at a finer or coarser scale than the other. Further complicating this issue, the data sets may be separately measured either in geographic space or institutional space, without any recognition of the linkage between the two. Other problems associated with the ecological fallacy (i.e., generalizing from the whole to the part) occur too frequently in our explanations of farmer behavior and in attributing causality to observed patterns. Verbal descriptions or narratives of social concepts may be valued within a discipline, but clearly the "bridge" to interdisciplinary efforts must be built with a common metric. Most all would agree that addressing nitrogen management in modern agriculture is going to require an interdisciplinary approach. Thus, while much remains to be accomplished, and there are significant disciplinary obstacles, we conclude that the ideas suggested in this chapter warrant further effort.

Acknowledgements
The Northern temperate Lakes Long Term Ecological Research program provided support allowing this work to be completed.

5. REFERENCES

Aldenderfer, M., and H. Maschner. 1996. Anthropology, space, and geographic information systems. Oxford University Press, New York.

Allen, T.F.H., and T.B. Starr. 1982. Hierarchy: Perspectives for ecological complexity. University of Chicago Press, Chicago.

Andicott, J.F., J.M. Aho, M.F. Antolin, D.K. Padilla, J.S. Richardson, and D.A. Soluk. 1987. Ecological neighborhoods: scaling environmental patterns. Oikos. 49: 340-346.

Anselin, L. 1999. The future of spatial analysis in the social sciences. Geographic Info. Sciences. 5(2): 67-79.

Atkinson, P.M., and N.J. Tate. 2000. Spatial scale problems and geostatistical solutions: a review. Professional Geographer. 52(4): 607-623.

Beegle, D.B., O.T. Carton, and J.S. Bailey. 2000. Nutrient management planning: justification, theory, practice. J. Environ. Qual. 29: 72-79.

Beegle, D.B., and L.E. Lanyon. 1994. Understanding the nutrient management process. J. Soil Water Conserv. 49: 23-30.

Bosch, D.J., Z.L. Cook, and K.O. Fuglie. 1995. Voluntary versus mandatory agricultural policies to protect water quality: adoption of nitrogen testing in Nebraska. Rev. Agric. Econ. 17 (1):13-24.

Bouma, J. 1997. Precision agriculture: introduction to the spatial and temporal variability of environmental quality. p. 5-13. In J.V. Lake, G.R. Bock and J.A. Goode (ed.). Precision agriculture: Spatial and temporal variability of environmental quality. Ciba Foundation Symposium 210. John Wiley and Sons, Chichester, UK.

Bouma, J., and M.R. Hoosbeek. 1995. The contributions and importance of soil scientists in interdisciplinary studies dealing with land. SSSA Special Publication 45. Madison, WI.

Brasier, K., P. Nowak, B.M. Kahn, and P.E. Cabot. 2001. Bringing space into agriculture-environment relations. (Manuscript in review, Rural Sociology).

Brouwer, F. 1998. Nitrogen balances at farm level as a tool to monitor effects of agri-environmental policy. Nutrient Cycling in Agroecosystems. 52:303-308.

Brown, L.A., E.J. Malecki, and A.N. Spector. 1976. Adopter categories in a spatial context: alternative explanations for an empirical regularity. Rural Sociology. 41(1): 99-118.

Can, A. (ed.) 1998. Geographic information systems in housing and mortgage finance. J. of Housing Research 9.

Canter, L.W. 1997. Nitrates in groundwater. CRC. Lewis Publishers, Boca Raton.

Diggle, P., and A.G. Chetwynd. 1991. Second-order analysis of spatial clustering. Biometrics 47:1155-1163.

Everitt, B. 1974. Cluster Analysis. Heinemann Education Books, London.

Francis, C.A., and R.K. Olson. 1995. Diversity from micro to global: overview and conclusions. In R.K. Olson, C.A. Francis, and S. Kafka (ed.). Exploring the role of diversity in sustainable agriculture. ASA, Madison, WI.

Gibson, C.C., E. Ostrom and T.K. Ahn. 2000. The concept of scale and the human dimensions of global change: a survey. Ecological Economics. 32:217-239.

Goodchild, M., L. Anselin, R. Appelbaum, and B. Harthorn. 2000. Towards spatially integrated social science. International Regional Science Review. 23: 139-159.

Gburek, W.J., and A.N. Sharpley. 1998. Hydrologic controls on phosphorus loss from upland agricultural watersheds. J. Environ. Qual. 27:267-277.

Hayman, P.T., and C.L. Alston. 1999. A survey of farmer practices and attitudes to nitrogen management in the northern New South Wales grains belt. Aust. J. Experimental Agric. 39:51-63.

Heathwaite, A.L., and P.J. Johnes. 1996. The contribution of nitrogen species and phosphorus fractions to stream water quality in agricultural catchments. Hydrologic Processes. 10:971-983.

Heatwole, C.D., A.B. Bottcher, and L.B. Baldwin. 1987. Modeling cost-effectiveness of agricultural nonpoint abatement in two Florida basins. Water Resourc. Bull. 23:127-131.

Hopkins, J. 1996. Impacts of nitrogen control policies on crop and livestock farms at two Ohio farm sites. Rev. Agric. Econ. 18(3):311-324.

Kaluzny, S. P., S. C. Vega, T. P. Cardoso, and A. A. Shelly. 1998. S+ Spatial Stats: Users manual for windows and unix. Springer-Verlag, New York.

Lockeretz, W. 1990. What have we learned about who conserves soil? J. Soil Water Conserv. 45:517-523.

Lowrance, R., P.F. Hendrix, and E.P. Odum. 1986. A hierarchical approach to sustainable agriculture. Amer. J. of Altern. Agric. 1(4): 169-173.

Lynne, G.D., J.S. Shonkwiler, and L.R. Rola. 1988. Attitudes and farmer conservation behavior. Amer. J. of Agr. Econ. 70:12-19.

Meyer, D.M. I. Garnett, and J.C. Guthrie. 1997. A survey of dairy manure management practices in California. J. Dairy Sci. 80:1841-1845.

Noss, R.F. 1992. Issues of scale in conservation biology. In P.L. Fiedler and S.K. Jain (Eds.) Conservation biology: the theory and practice of nature conservation, reservation, and management. Chapman and Hall, New York.

Nowak P., and P. Korsching. 1983. Social and institutional factors affecting the adoption and maintenance of agricultural BMPs. p. 349_373. In F. Schaller and G. Bailey (eds.) Agricultural management and water quality. Iowa State University Press, Ames, IA.

Nowak, P. and P. Korsching. 1998. The human dimension of soil and water conservation: a historical and methodological perspective. In Pierce and Frye (ed). Advances in soil and water conservation. Sleeping Bear Press, Boulder, CO: 159-184.

Shankar, B., E.A. DeVuyst, D.C. White, J.B. Braden, and R.H. Hornbaker. 2000. Nitrate abatement practices, farm profits, and lake water quality: a central Illinois case study. J. Soil Water Conserv. 55 (3):296-302.

Shepard, R. 2000. Nitrogen and phosphorus management on Wisconsin farms: Lessons learned for agricultural water quality programs. J. Soil Water Conserv. 55 (1) :63-68.

Tobler, W. 1979. Cellular geography. In Gale, S. and Olson G. (Eds.) Philosophy in Geography. Reidel, Dordrecht: 379_386.

Vitousek, P.M., Aber, J.D., Howarth, R.W., Likens, G.E., Matson, P.A., Schindler, D.W., Schlesinger, W.H. and D.G. Tilman. 1997. Human alteration of the global nitrogen cycle: Sources and consequences. Ecol. Applic. 7(3):737-750.

Weisburd, D., and T. McEwen (eds.) 1998. Crime mapping and crime prevention. Criminal Justice Press, New York.

Wischmeier, W.H., and Smith, D.D. 1978. Predicting rainfall erosion losses: A guide to conservation planning . Washington, D.C.: U.S. Department of Agriculture, Agriculture Handbook No. 537.

Welsh, R. 1997. Vertical coordination, producer response, and the locus of control over agricultural production decisions. Rural Sociology. 62 (4): 491-507.

Wolf, S.A., and P. Nowak. 1999. Institutional failure in agro-environmental management. Research in Social Problems and Public Policy. 7:293-310.

Nitrogen in the Environment: Sources, Problems, and Management
R.F. Follett and J.L. Hatfield (Eds)
Published by Elsevier Science B.V.

Chapter 20. New Policy Directions

Andrew P. Manale

US Environmental Protection Agency, Office of Policy, Economics, and Innovation, Washington, D.C.

Advances in linking the science of ecology with economics and the development of agri-environmental modeling systems coupled with new information technology suggest new public policy approaches that reward agricultural producers for providing ecological services. An incentive-based ecosystem approach that identifies and quantifies an array of environmental services that can be provided by agricultural land, and then facilitates the development of markets in these services can protect environmental quality while improving farm income. Sustainability can be achieved whereby economic needs of society are integrated into environmental protection. Before presenting new agricultural policy approaches for improving environmental quality and, in particular, managing agricultural nitrogen, the nature of the relationship between agriculture and the environment and its implications for policy are discussed .

1. THE POLICY DILEMMA

By design, agriculture alters the natural system of the soil and the associated landscape for the production of crops and animals of benefit to humans (Daily, 1997; Daily and Ehrlich, 1992). The diversion of available energy from maintenance of the ecosystem[1] can, over time, affect the quality or fitness of agricultural ecosystems, as reflected in their resistance and resilience[2] to such physical and biological insults as disease and droughts. These decline well below those of native ecosystems (see Whitford, 1999) with some environmental impacts requiring years to manifest themselves (National Research Council, 1993, p. 3.). Some effects occur slowly with long-term effects that do not become apparent for many years. According to Ludwig [Ludwig et al, 1993], "[t]he delay between predicting and detecting irreversible and

---

[1]According to C. S. Holling, "In short, the success in controlling an ecological variable that normally fluctuated led to more spatially homogenized ecosystems over landscape scales. It led to systems more likely to flip into a persistent degraded state, triggered by disturbances that previously could be absorbed. This is the definition of loss of resilience...So this is the puzzle: The very success in managing a target variable for sustained production of food and fiber apparently leads inevitably to an ultimate pathology of less resilient and more vulnerable ecosystems, more rigid and unresponsive management agencies and more dependent societies" [as quoted in Gunderson, 1995].

[2]Resistance is the capacity of the system to continue to function without change through a disturbance; resilience the ability to recover functional integrity following a disturbance (Herrick, 2000).

deleterious ecosystem-level changes with certainty often delays the receptivity to acknowledging environmental problems and seeking solutions. This uncertainty compounded over the period of delay; the longer the period, the larger the gamble."

Soils, particularly in arid regions or organic soils, can quickly exhibit the signs of degradation, such as reduced water-holding, nutrient-holding, and infiltration capacity; changes to more robust soils may take decades or even generations before the loss in resistance and resilience manifests itself (Kemper, 1997). As quality declines, producers may rely more on chemical inputs, such as chemical fertilizers. As soils lose their tolerance to droughts, producers irrigate more; as soils lose resistance to disease, farmers use more pesticide. These compensatory actions to maintain or improve yields can lead to further declines in environmental quality, such as nutrient contamination of water resources (National Research Council, 1993, p.107, 189-193, and 198-199).

Society, in general, accepts this loss where it perceives the benefits of agricultural production outweighing the environmental costs (Phipps, et al, 1986). However, an ever-growing economy and the steady decline in the relative value of agricultural production changes this ratio. John Antle (2000, 1006.) states that "(t)he growing demand for environmental quality implies that society will increasingly value production processes that provide environmental services--what are becoming known as ecosystem services." As new scientific insight reveals serious long-term environmental effects (see, for example, USGS, 2000 and Crouzet, 2000), the public demands more from agricultural producers to mitigate these impacts (NRCS, 1997). In response, both the United States and the European Union, have, for example, recently enacted new regulations that put increased pressure on state and regional political jurisdictions to resolve the problems (CRS, 2000; EU, 2000).

1.1 Factors Contributing to Agriculture's Role in Pollution

Soils and their constituents, such as nitrogen, phosphorous and carbon are inextricably connected to water in ecological processes. A change in the quality or quantity of the one can, over time, adversely affect the other. It is the latter, however, that commands public attention for environmental protection since water, both above and below ground, is a common pool resource (CPR) shared by many users [Ostrom et al, 1999].

Exploitation of CPRs commonly leads to a policy dilemma called "tragedy of the commons" in which the pursuit of self-interest diminishes the quality and usefulness of the resource for other users (Hardin, 1968). The result is a form of market failure called negative externality whereby short-term interests produce outcomes that are not in anyone's long-term interest (Dynar, 1998). CPR dilemmas are solved by restricting access and creating incentives (usually by assigning individual rights to or shares of the resource) for users to invest in the resource instead of overexploiting it. This usually means privatization of the resource (Ostrom et al, 1999). Privatization of water resources is rarely politically or technically feasible.

Agricultural soils, on the other hand, are generally privately owned and managed. Nevertheless, their interaction with water in the larger ecological dynamic translates the consequence of an insult of one component of the landscape into adverse impacts on the quality of the whole. Individual property ownership and institutions for facilitating communication and negotiation among individual property owners are not likely to lead to rules to solve this social dilemma in these watersheds where there is a high proportion of users who behave in a narrow, self-interested way; where the perceived social benefits accrue to beneficiaries external to watershed; or the costs of doing so are perceived by the users to be high, particularly where the resource is large and complex (Dasgupta, 2000).

Soils and the larger landscape potentially provide multiple services to society, in addition to the service associated with a private benefit. The former, which include flood water retention, drought mitigation, wildlife habitat, water quality, sequestration of carbon and minimizing releases other greenhouse gases thereby helping to mitigate global warming, and even aesthetic open space for the benefit of urban dwellers (Tourbier, 1994) are rarely explicitly monetized (Daily et al, 2000; Daily, 2000; and Haapala, 2000). Though managing the land through its use in agricultural production may reflect its private economic value, the practices may not sufficiently capture its environmental resource value (Dasgupta, 2000).

Farmers have been known to address pollution problems for which resolution can be shown to be in their self-interest, such as the protection of ground water which they use for drinking water. However, there can be a temporal lag between the management practices that cause the insult to the resource and the effect that diminishes the common use of the resource. A relatively common example is nitrate contamination of groundwater supplies resulting from excessive fertilizer use or land application of manure. Producers may not become aware of the commonality of interest in protecting the resource until the damage is manifest, which may not occur for many years. Alternatively, the farmer may be aware of the problem, but does not adopt the necessary practice to resolve the problem because the structure of the agricultural economy does not allow him or her to pass on the additional cost.

The spatial scale of agricultural production complicates the finding of a policy solution to the environmental problems. Production occurs across watershed and political boundaries. Site-specific factors, such as soil type and quality, weather, and proximity to water resources, and history of hydrologic changes, affect the amount and type of pollution associated with the agricultural technology used to produce the agricultural product (Wu and Babcock, 1999).

Superimposed upon the effect of individual farming decisions on the environment are public policies and collective actions that exacerbate or amplify the magnitude of the impacts. In the United States and in most developed countries, government programs have historically affected how farmers farm or steward the land exacerbating agriculture's impact on the environment (NRC, 1993). Agricultural programs that induce a bias towards intensive farming practices that boost yields, expand production onto marginal lands, and concentrate production on a small number of crops can undermine efforts to encourage adoption of conservation practices.[3] Despite major changes in farm policy in the United States and Europe since 1989, the linkages between farm program support and production decisions remain (Crouzet, 2000).[4] Income-support programs in the United States under the 1996 farm bill, such as the commodity loan programs, influence producer decisions regarding the use of marginal lands, the intensity of land use, tillage practice, monocultural cropping practices and habitat protection (Miranowski and Cochran, 1993).

Wide fluctuations in the prices of commodities in domestic and international markets can cause farmers to put marginal land into production, fragile lands that degrade and hence erode

[3]The National Research Council concluded: " [F]ederal policies "significantly influence farmers' choices of agricultural practices. As a whole, federal policies work against environmentally benign practices (NRC, 1989, p.6.).

[4]The 1996 farm bill and subsequent amendments to it and a new crop insurance bill, which expand protection to more crops and livestock, continue these linkages in the United States.

readily (MacGregor and McRae, 2000). An example is the late 1970s in the United States when agriculture boomed with "fence row to fence row" production leading to a significant expansion in cropland and concomitant rates of soil erosion (NRC, 1993, p. 152). Shifting the management of risk from how one crops and manages the land to financial instruments, or "risk management instruments," such as crop or disaster insurance or loan-deficiency payment programs, can encourage producers to shorten the time-scale of management decisions regarding resource use, putting short-term rational individual economic behavior gain into conflict with long-term public resource protection (Skees, 1999).

Recent trends regarding the structure of the industry also have implications for ability of public policy to solve problems (Rickson et al, 1997; Boehlje, 1995; Welsh, 1996). Over time, crop and animal livestock production have diverged with the more and more specialization, a greater percentage of production conducted by fewer farms, and fewer farms producing both whereby crops grown serve as feed for animals and the manure produced by the animals serving as fertilizer for the crops (Crouzet, 2000). The consequence is that livestock production is concentrating within small geographic areas with nitrogen and phosphorous coming into a watershed in grain inputs that are not balanced by removal through food transported out of the area, denitrification, or taken up by plants (Vanderholm, 1994). The excess winds up in ground and surface water supplies or overconcentrating in soils that disturbs soil ecology (Hatfield and Stewart, 1998).

With more and more production occurring under contracts to firms that process and convert raw agricultural produce into value-added products (so-called integrators), key production decisions are no longer necessarily made by owners of the land who may have self-interest in its long-term sustainable management (Schertz and Daft, 1996). The magnitude of production within a watershed or geographic area, where the predominance of production is under contract, has shifted to the integrator who may make decisions with little relationship to resource constraints (Zilberman, 2000; Manale and Narrod, 1994). The solution may require redefinition of property rights to reflect who is better able to pass on the costs of adopting environmental control measures to consumers.

No less important a factor affecting agriculture as a source of pollution and loadings of nitrogen into the environment is the long-term consequence of collective actions and government policies. These have changed landscape hydrology and hence the ecological conditions of soils and their potential to provide a vector of services of social value, such as flood water retention, carbon sequestration, and ground water recharge (Wiener et al, 1998; Guardian, 2000). When Europeans first settled the land that would become the conterminous United States, for example, there were 89.4 million ha (221 million acres) of wetlands. With the introduction of technology facilitating the construction of tile drains, nearly all of the prairie wetlands--prairie potholes--in Iowa, southern Minnesota, and the Red River Valley had been drained. By the mid-1980s, the number of wetlands plummeted to about 42.7 million ha (103 million acres). Many states, such as Ohio, California, Missouri, Illinois, and Indiana, lost more than 70 percent of functioning wetland systems and Iowa 95 percent. In other areas, channelization of rivers contributed to changes in soil ecosystems. Increased transport of sediment and fertilizer nutrients and loss of riparian vegetation often followed channelization.

Altering the rate of overland flow of water affects rates of denitrification or movement of sediment to water bodies (Crumpton and Baker, 1993). "In Iowa, shallow subsurface drainage is "short-circuited" or routed by drain tiles to surface water supplies such as the Des Moines and Raccoon Rivers. This can result in water supplies for cities such as Des Moines exceeding the drinking water standard. Although better N management can help reduce leaching losses, it has not been shown that management alone can reduce nitrate-N concentrations in shallow subsurface drainage to levels below 10 mg/l, unless N applications are eliminated or at least reduced to well below economic optimum rates." Drainage has also caused these soils to become a source of greenhouse gas emissions, rather than a sink, thus contributing to climate change (Armentano, 1980). Because these topsails tend to be deep, degradation can cause nutrients to be released over a long period of time contributing to loadings to surface waters (USGS, 2000)].

Conversion of wetlands and altering the water-holding capacity of agricultural soils add to the risk of floods and droughts. According to Perry and Vanderklein (1996, p. 175), "[while water quality management frequently focuses on small or midscale effects, significant lag-term effects can, and probably do, result from all of the channel modifications, land-use changes, and accumulated connecting impacts." The draining of over 60 percent of the prairie potholes of the northern prairie of the Upper Mississippi River Basin, for example, has resulted in the loss of the potential to store some 38 ha-cm (15 acre-inches) of runoff and more per year (Wiche et al, 1990). During pre-settlement times, forested riparian wetlands adjacent to the Mississippi could store up to 60 days of river discharge. With removal of wetlands through channelization, leveeing, and draining, remaining wetlands have storage capacity of less than 12 days discharge (DeLaney, 1995). Restoring the ability of the soils to retain water can achieve roughly a 36% reduction in runoff (Dyke, 1997).

1.2 United States' and European Union Experience in Addressing the Problem

Traditional approaches have heavily relied on voluntary approaches, such as education technical assistance, directed at users of the land. Education alone has not been effective in promoting adoption of practices that are not profitable (Camboni and Napier, 1994; Feather and Cooper, 1995) except where self interest is at stake (Napier and Brown, 1993; Nowak, 1987). Even where technical and cost-share assistance have been provided, voluntary projects, such as United States Department of Agriculture water quality projects that were implemented on a watershed basis--such as the Model Implementation Program of the 1970s and Rural Clean Water Program of the 1990s--, failed to achieve water quality goals (Shortle and Abler, 1999). Labeling requirements and registration of inputs have been of limited effectiveness because of site-specific conditions.

In the United States, federal programs to protect water resources from agricultural sources have relied on state and local initiatives. Clean Water Act [Federal Water Pollution Control Act] section 208 as well as its 1987 amendments have called for development and implementation of area wide water quality management programs to address point and nonpoint source pollution and for states to develop plans for reducing nonpoint pollution and adoption appropriate land management controls. Congress amended the Clean Water Act (CWA) in 1987 to establish the section 319 Nonpoint Source Management Program because it recognized the need for greater federal leadership to help focus State and local nonpoint source efforts. Under section 319, State, Territories, and Indian Tribes receive grant money which support a wide variety of activities including technical assistance, financial assistance, education, training, technology transfer, demonstration projects, and monitoring to assess the success of specific nonpoint source implementation projects (USEPA, 2001). This gentle federal nudging of states to address what has been perceived as inherently a state or local problem has clearly not been enough [see USEPA, 1998].

Similarly in the EU, traditional command and control and voluntary efforts have met with limited success [European Environment Agency, 2000]. Most of the reduction in nitrogen loading into the environment is likely the consequence of reforms of the EU's agricultural subsidy program which has shifted payments away from support for production to provision for environmental goods. The incentive to use more fertilizer to increase production and hence financial return has thereby been reduced. Implementation of regulatory programs in member states has been unsatisfactory, with only five countries with plans detailing how they are implementing the Nitrate Directive [European Union, 1991] the main EU-wide policy vehicle for reducing nitrogen loadings into the environment from agriculture. Even where programs are in place, regulatory targets for reducing agricultural loadings have been missed [Crouzet et al., 2000]. The explanation has generally been that the measures adopted were accompanied by inadequate control and enforcement, inadequate education, and inadequate use or lack of financial instruments.

Incentive-based approaches have been suggested in recent years (Segerson, 1988). However, agricultural pollutants follow indirect and diffuse routes from agricultural land to air and water resources from a large numbers of agricultural sources. The standard economic prescriptions for negative environmental externalities involving emission-based policy instruments, such as emissions standards or taxes, tradable discharge permits which require metering individual pollution sources would be impractical for nonpoint source pollution [Shorter and Abler, 1999]. Taxes are not necessarily more efficient because of uncertainty regarding farm production decisions and loadings (Weitzman, 1974). Trading between point and non-point sources, which has been urged by some economists, suffers from the difficulty in establishing equivalence between a quantifiable point source emission and a highly variable non-point source emission of a pollutant or pollutants, let alone the high cost of monitoring and verification of reductions.

More recently, federal agricultural farm income support payments in the U.S. have been coupled to implementation of specific practices, such as conservation tillage or protection of cropped wetlands. Though these programs, particularly Conservation Compliance, have succeeded in significantly reducing soil erosion on highly erosive land, they have had far more limited success with reducing nitrogen loadings to the environment [U.S. Geological Survey, 1999].

Also, "green payments" have been paid farmers to adopt of practices that reduce impacts on water resources. As Horan, Shortle, and Abler (1999, p. 1214-1215) have argued, "...major program design issues have yet to be addressed, including how to define environmental performance goals, trade-offs or complementaries between farm income and environmental protection, what types of payments to use (including subsides or contracts), which producers to target, which inputs to target, whether to make payments uniform or partially site-specific, how to control entry and exit, and how to reconcile green payments with the URA (Uruguay Round Agreement)...Under URA..environmental subsidies are limited to producers' participation costs." These programs, where they have been funded by the federal government in the U.S., have been conservatively funded with budgets sufficient only to fund the most closely targeted problems, that is the demand for funding has far exceeded the amounts available (See USDA, 2000). Where these have been implemented broadly, the funds have been spread too thinly to be effective.

Federal green payment programs often suffer from reliance upon uncertain government funding. According to Napier (1999 p. 191), "[o]nce the subsidies used to encourage conservation were withdrawn, or the market changed to the point that it was less profitable to farm with conservation systems, land owner-operators removed conservation structures and reintroduced erosive farming systems that were perceived to be more profitable. The result is a considerable loss of public investment in soil and water conservation...repeated numerous times throughout the U.S. and are likely to continue unless there is significant modification of institutional structure for implementing conservation programs." Because addressing the root causes of environmental problems associated with agricultural may entail restoring the quality of soils--a process that may take years to accomplish, all of which can be lost in a single season--a successful green payment program requires an institutionally secure and enduring funding mechanism.

A more fundamental problem of past efforts has been the narrow, single-objective focus of policy on ameliorating a particular environmental impact of agricultural production at a time--soil erosion, nutrient, runoff, ground-water contamination, for example--rather than at the root causes or at encouraging practices that address multiple impacts (see, for example, Ribaudo, 2000). Policies have assumed the replaceability of the capital stock, despite the amount of time related to the human lifespan necessary to replace or restore them (Phipps and Crosson, p.4). However, instead of an inert chemical matrix in a steady-state relationship with its environment and with agricultural chemical inputs, soils are highly complex dynamic biological systems interconnected into a larger system. "Unfortunately the world doesn't operate on a linear model. Thus, it can be argued that some of the blame for today's increasingly vexing water problems stems from the application of linear thinking to the problems of a cyclical world (Hall 1998, p33).

A more accurate model incorporates a feedback mechanism pertaining to the capacity of the capital stock. According to Hall, "[t]he difference, between the two models is not one of parts, but of interrelationships." Newer models that simulate how the capital stock--soils--provide a variety of ecological services--allow economic evaluation of options that address the causes of the problem--such as the loss of ecological functions of the soil or hydrologic modifications of the greater landscape that prevent farm-level solutions--or even economic policies that have encourage the geographic division of crop and livestock agriculture. Important policy options that may have provided greater economic and environmental benefit simply have or could not be sufficiently considered.[5]

2. AN ECOSYSTEM APPROACH TO AGROENVIRONMENTAL PROBLEMS

A more holistic, ecosystem approach (or integrated resource management, as it is sometimes called) treats the causes of environmental degradation from agricultural activities rather than the symptoms, recognizing the interconnectiveness of problems and the complexity and uncertainty of interactions (Bellamy et al, 1999). Ecological services that have diminished must be restored through the adoption of agricultural systems or sets of practices that either restore or mimic their function. Hence, managing the system for its ecosystem services--that is

[5]"Past attempts have focused on resource that is adversely impacted, such as water and then identifying either the performance standard or the technology that achieves the desired amelioration. The consequence is that the practices or measures adopted will be single-objective and may fail to achieve multi-objective benefits that may accrue from practices that not only achieve the single objective but secondarily also the other benefits (Batie and Ervin, 1997)."

restoring "wholeness"-- means managing the capital stock to provide an array of ecological services, including crop production, to maximize the social good. One thereby minimizes the set of negative externalities associated with the commercial activity.

Management of soils generally follows the value of the land for private benefit. This use of the land, though declining, may continue indefinitely with the availability of modern technologies, such as fertilizers, hybrid seeds, and pesticides. It is the value of the land that derives from management of the land for other ecological functions that declines more precipitously and, in the short-term of social experience, irreversibly for the larger public. Individual property rights, as currently defined, do not suffice to maintain or improve the ecological functions that benefit the larger group or provide too little of these services.

Ecosystems and how they operate and interact are poorly understood. Any economic or policy model that attempts to simulate their functions over time and space to allow for predictions of cost and effects is, from the outset, flawed. As Gunderson (Gunderson et al 1995, p.14) has argued: "In principle, therefore, evolving managed ecosystems and the societies with which they are linked involve unknowability and unpredictability. Therefore sustained development is also inherently unknowable and unpredictable. The essential point is that evolving systems require policies and actions that not only satisfy social objectives, but also achieve continually modified understanding of the evolving conditions and provide flexibility for adapting to surprises. This is the heart of active regional experimentation by management at the scale appropriate to the question--adaptive environmental and resource management. Otherwise the pathologies of exploitive development are inevitable--increasingly brittle ecosystems, rigid management, and dependent societies leading to crises."

A strategy for addressing the environmental problems of agricultural production that incorporates ecological concepts needs "flexible, diverse and redundant regulation; monitoring that leads to corrective responses; and experimental probing of the continually changing reality of the external world" (from Gunderson et al, p.26) The policies that it develops acknowledges the possibility, if not the probability, of failure and hence seeks to move the system toward sustainability--defined in terms no more detailed than the science suggests--by steps that can be reversed as science suggests that the path taken may fail.

It takes a long-term perspective and broad spatial scale focus--that of the landscape, watershed, region, or basin (Burt, 1999, p.260). Many ecological services--such as flood water retention or wildlife habitat--require land-extensive management practices that will likely transcend individual property boundaries. The proper intervention in agriculture may not necessarily be directed at the individual farm, but rather at restoring an ecological service at a broader geographic scale than the individual farm. Thus, the spatial scope should be expanded to encompass a broad range of technical and institutional options to resolve the problem. According to Loucks (1998, p. 38), "[a] more operational view of a watershed, therefore could be the "problem-shed"--a region approach to the issues and problems being studied. If the definition is accepted, a watershed is not necessarily bounded by the hydrological, physical, or even political boundaries. Rather, a watershed is defined by the locations of its stakeholders who have an interest in one or more particular watershed management problems and who have the ability to address and solve them." It provides for a process through which conflicting interests can arrive at consensus for action at the scale appropriate to the problem (Bellamy, 1999; see also Brown and MacLeod, 1996).

The human element must be taken into account as questions of economic viability of communities are raised. Stakeholders and interested parties need to be involved in making the decisions regarding the relative importance of objectives and tradeoffs because solutions to land-extensive environmental problems may require extended periods of time (Burt, 1999). Local commitment provides for greater program continuity while regional, state, or even national involvement allow options to address concerns that extend beyond contiguous political boundaries.

A structured approach facilitates stakeholder involvement. "Stakeholder values are key to the structured decision process because they identify what matters to participants and, in turn, highlight the consequences that require most careful attention and the tradeoffs that matter most." A structured decision approach to public involvement is essential in the following steps: framing the decision, defining key objectives, establishing alternatives, identifying consequences, and clarifying tradeoffs (Gregory, 2000).

The example of soil carbon is illustrative. The amount or concentration of carbon in the soil plays a critical role in soil quality, which in turn affects agricultural productivity. Soil quality affects water retention, which in turns helps reduce flood-water runoff in watersheds. Tilling the soil oxidizes the carbon in the soil, reducing its concentration as less is returned to the soil until a new steady state is reached at a lower concentrations than in its pre-agricultural state. In traditional agriculture, farmers who raised livestock along with crops would apply the manure to soils, returning thereby at least part of the carbon and nutrients to the system. Trends toward specialization of agriculture has witnessed the division of livestock and crop agriculture, such that not only is the farmer not involved in both activities, but the activities can be located a great geographic distances from each other. The output of the former no longer serves as the input of the latter in a holistic whole.

From a public policy perspective, restoring the cycle may not necessarily entail reintegrating the two at the level of a specific farm. The key is to expand the scope of the problem definition to encompass new solutions that may be feasible only at the level of watersheds or larger. Farms may still specialize, but what has been a waste from livestock agriculture is turned into an input for restoring soils in crop production for farms located within the larger geographic area. Reintegration is effected by public or private institutions that provide quality control and otherwise reduce the transaction cost of turning a waste into a valued commodity. Alternatively, restoring the quality of soils may entail adoption of agricultural practices that put carbon and associated nutrients back into the soil, such as through no till and precision agricultural techniques. The byproduct of animal production may serve as an input into electrical energy production that in turn serves to reduce the cost of energy used in agriculture. The economic goal is to convert "waste" into a commodity that has value, thereby reducing the incentives to dispose of in a common pool resource.

2.1 New Science Advances and Policy Analytical Tools Allow for New Approaches

Recent advances in the science of environmental pollution and modeling tools, particularly for nitrogen from agricultural activities allow for more sophisticated interventions to address agricultural non-point source pollution (see, for example, Lakshminarayan and Babcock, 1995; Saleh et al, 2000; Babcock et al, 1997). The new models cover the scale at which agriculture operates and reflect its site-specific nature and interdependence among technologies and landscape modifications (see, for example, Virginia Tech, 2001). Because of constantly changing temporal impacts, the ability to visualize the problem greatly helps in developing policy solutions. According to Perry and Vanderklein (1996, p. 131), the "ability to predict ecological phenomena depends on the relationship between spatial and temporal patterns....as spatial scale increases, the relevant timescale must also increase because at larger scales processses operate at slower rates and have larger time lags and indirect effects are more important...if [they] do not expand temporal scale with spatial scale, predictions are "pseudo-predictions."

Many of the modeling systems and their databases have been developed in a modular manner to allow for substitution among models with different strengths with inputs and outputs connected to geographic information systems that facilitate analyses at various scales. An evaluation of the impacts of practices at the level of 8-digit watershed, for example, can, by making reasonable assumptions regarding what is occurring at larger geographic scales, be extrapolated to the river basins. Hence, analysts and stakeholders can visualize the relationship between how the management of the landscape affects nitrogen or other loadings to water bodies at the top of watersheds or then trace these consequences to ecological impacts hundreds of miles away, such as the Gulf of Mexico. Different types of databases, relating to the physical and political and economic worlds, have been harmonized through geographic information systems (GIS) polygons (see, for example Bauer, 1996). GIS in turn can be used with scanning technology to create high resolution land-use data sets (USGS, 1996). The result is that biophysical models have been dynamically linked with economic models to provide estimates of private and social costs of policies that encourage or constrain the adoption of a practice or sets of practices at the farm, watershed, regional, or national scales. Because of the spatial integration of databases through a GIS, even watershed level impacts, such as flood or drought mitigation, can be included in assessments.

By simulating cause and effect at various spatial scales and the costs associated with the interventions at these scales, policy makers can estimate how prices and policy constraints affect technological choices and in turn the flow of ecological services. This is particularly crucial since the impact of the technology on the resource depends upon many site-specific variables that can vary widely within and across watersheds. Simulating policy impacts at various spatial and temporal scales greatly expand the kinds of policy options that can be evaluated.

Furthermore, since modeling ecological processes is an emerging science, integrated models permit evaluation of "what-if" scenarios to identify and test critical assumptions that may have important influence on policy decisions–to address the inevitable problem of having to make decisions based upon limited scientific or economic knowledge. They also help identify key research and, in particular, monitoring data that need to conducted or collected in order to develop better policies.

The new tools accommodate the site-specific nature of agricultural production with natural resource damages and also provide for a feedback mechanism between the production process and the value of the capital stock (CARD, 1997; Babcock and Paustian, 2001). Because the results of these models can be aggregated up to the national level, the impacts of national policy can be evaluated at a local level and conversely local strategies for managing the resource can be evaluated in the context of national policy. These more sophisticated watershed and basin level models can simulate the massive changes to the ecosystem that have been implemented to enable the agricultural activity. Furthermore, the linkages at various geographic scales show how various assumptions regarding adoption of practices affect water quality and quantity within the watershed[6] and downstream within the larger basin.

For the purpose of modeling the economic and environmental impacts of policies, soils need no longer to be assumed to be in a steady state condition with regard to carbon and the loss (or gain) of associated components, such as nitrogen and phosphorus. Nor is reversibility of the most obvious impact on the soil assumed--that soils can regenerate the carbon that they have lost within a reasonable period of time--with the conclusion that the impact on crop productivity is so small as to be inconsequential. The magnitude of practices over time can be evaluated spatially for the purpose of more targeted policies to protect the productive capacity of the resource. Because the models are beginning to incorporate indicators of soil quality, such as tilth, which affect its regenerative ability and reflect its biological nature (rather than only chemical components), policy makers need no longer discount the important linkages with the quantity and quality of common property resources, such as water or air.

Other new procedures allow for valuation of services provided society by the ecological functions of the greater landscape (see, for example, King, 2000; and Wainger, 2001; King et al 2000; see also Heimlich et al, 1998). Though these approaches have almost exclusively been used to estimate the benefits and costs of protecting or restoring wetlands or forests, the methods can be applied to lands that remain in agricultural production as well. Rather than focusing exclusively on the benefits of restoration of natural functions of a non-working landscape, the tools can be used to estimate the economic benefits accruing from practices that mimic ecological functions in a "working" landscape that continues to produce crop and agricultural products. Some of these ecological services include flood water retention, carbon sequestration, or wildlife habitat, depending upon the agricultural system that is utilized. Though these tools depend very heavily on large amounts of data, the greater availability of data through the internet over time can serve to reduce these costs.

The individual farmer need no longer be the sole or primary focus of agricultural policy interventions whereupon, for the purpose of economic analysis, one assumes that he or she decides independently of the actions of other farmers whether or not and how much to pollute. In many drainage districts in agricultural watersheds, particularly in the midwestern cornbelt, where complex drainage systems make possible agricultural production by farmers who independently own or manage the land, farmers must maintain their sections of drains. Failure to maintain a subset of the system affects the functioning of the whole. The most effective techology to reduce pollution from an individual farm would affect the production potential of all the farms because it could affect how much and how quickly the entire system drains. The system can be imagined as a system of quasi-independent factories all of which are connected to the same smokestack. The solution to pollution in these cases may be to affect what comes out of the shared smokestack, such as through construction of artificial or restored wetlands at the terminus of the drainage system before runoff enters surface waters (Kovacic, 2000; Woltemade, 2000).

There are a numerous recent examples of landscape-scale management efforts that provide a variety ecosystem services. In some areas, the service may entail managing the water table to protect water quality and to mitigate floods and droughts (see, for example, Schultz et al, 1995; Isenhart et al, 1997). Wetlands have been integrated into landscapes to capture

---

[6]Well-documented examples of how location of a disturbance within a watershed can affect water quality are given by Bormann and Likens [1985] or Hornbeck et al, 1987].

agricultural runoff. But with few exceptions, there have been few examples of private markets involving the ecological services of the agricultural landscape. A rare illustration of the latter, though narrowly focused on the objective of drinking water quality, is New York City's purchase of services from agricultural landowners in the watershed that supplies the city's drinking water supply (Chichilnisky and Heal, 1998). Entering into agreements to restrict activities that might threaten water quality represented a savings in costs from the alternative of constructing a new filtration plant.

2.2 Putting It All Together: Economic Incentives for Working Landscapes

There is an extensive literature describing the fundamentals of an ecosystem approach to environmental problems. The concept is known by many names, including integrated resource management, ecosystem management, integrated catchment management, and integrated water management (Laane and Lindgarrd-Jorgensen, 1992; Margerum, 1997) and is defined in many ways (Burt, 1999). In nearly all cases, examples have been government-funded projects. Scarce, however, are descriptions of policies for implementing or encouraging an ecosystem approach in managing agroecosystems or for effectively achieving proper valuation of ecosystem services within an agroecosystem (Simpson, 2000). Even rarer are examples of where these policies have been implemented and how they have fared.

Most examinations detail what needs to be done and who should do it, which generally assumes that various levels of government will propose or advance regulatory and voluntary goals and targets and then enact and coordinate regulatory policy to achieve these ends (Laane and Lindgarrd-Jorgensen, 1992). Though technical instruments and policy issues pertinent to an ecosystem approach have been articulated, effectuation of policies to implement the approach for a sustained period of time is either rare or has occurred on a local scale in such a way that hinders its generalizability to the nation as a whole (see Crouzet et al, 2000.).

An alternative policy strategy for implementing an ecosystem approach [Here we use the term in a broad sense that posits the integration of biophysical and social sciences, as defined by Perry and Vanderklein, 1996] uses market-incentives and contracts for services of the "working landscape."[7] The working landscape approach employs these key elements: 1) the use of comprehensive, integrated biophysical and economic models at the farm, watershed, regional, and national scales that can predict and quantify a multitude of benefits of farm and landscape-management practices; 2) a set of policies to encourage the development of markets in ecosystem services; and 3) the coordination or harmonization of government policy at various levels to facilitate these markets. An institutional structure is encouraged at the proper spatial scale in order to encompass the set of stakeholders necessary for coordination, development and implementation of the approach. Furthermore, the concept accepts a principle of the economic theory of federalism that states that economic efficiency in the provision of public goods is

---

[7]Working Landscapes is a term used by a coalition of groups in the Rural U.S. Midwest and includes federal, state, non-profit organizations, and agricultural interests that stresses the multi-functionality of the agricultural landscape. At the same time, it takes into account ecological needs, culture, and economics in a holistic way that conforms to a broad definition of ecosystem approach. The group seeks to wed the production and protection of the working landscape through the implementation of profitable approaches utilizing the functions of the natural landscape to solve both economic and environmental needs (Franz, 2001).

generally best served by delegating responsibility for the provision of the good to the lowest level of government that encompasses all the associated benefits and costs (Shortle, 1995).

The services entail adoption of agricultural practices or landscape-scale management that can be shown by a variety of approaches, including simulation modeling, to reduce the costs of meeting existing services. The analytical tools discussed previously allow expression, over various geographic temporal scales, of the damages of costs avoided in stochastic terms. This provides a means for comparing costs and benefits of policy alternatives using a common metric. The role of government role recedes to that of defining property rights, informational and data management support, and oversight and enforcement of the public interest--ensuring that the proper market signals are transferred to producers and land managers.

An ecosystem approach based upon market-incentives to protect the common pool resource recognizes the capability of agricultural land and the watershed in which it is located to provide cost-effective services to urban interests, downstream users of the common pool resource, interests physically located in floodplains, as well as far-away interests who share in the benefits of the resource or resources (Manale, 2000). The use of economic incentives provided by the market-place allows for decentralized flexibility that can achieve greater economic efficiency, critical in dealing with diffuse sources of pollution (Stewart, 1987;Costanza et al,1997, p.192-203; Baumol and Oates, 1975.).

Though environmental performance contracts exist now, such as the ten-year contracts of the Environmental Quality Incentive Program established by the 1996 farm bill or the green payment programs in the EU (Crouzet, 2000), the buyer of the service is exclusively government and depends upon continuation of the government program to maintain the funding for the service. In the U.S., the effectiveness of the program has been handicapped by annual appropriations which have been set far below demand (USDA, 2000). There is also concern that the success of the program could be hampered by uncertainty regarding the long-term continuation of the program, causing farmers to rethink the value of investing in the necessary technology.

Because efforts to value ecological services are meaningful only in the context of policy actions, valuation of the services occur for the purpose of answering the following question: "What is it worth to us (or how much do we save) in having the land managed in an alternative way?" The valuation of the service is not an exercise in economic estimation of its relative worth to society, but strictly whether or not it is relatively more valuable than an alternative option. Furthermore, the amount of data necessary for the evaluation is that which suffices to make a relative comparison. Hence, a full accounting of services and their costs and benefits may not be necessary if a simple model using data readily available provides sufficient precision to differentiate between options, particularly options involving restoration or rehabilitation of a landscape function versus the status quo. The goal is not to compensate for land manager or managers for all the ecological services that the land provides or the quantified value of the reduction in ecological damage (if that value were even able to be estimated); rather it is to provide the minimum financial inducement such that land managers adopt the practice that provides an ecological benefit equaling or exceeding its costs.

In many cases, the option against which to compare is the investment in structures to minimize damages from deteriorating ecosystems or avoidance of the costs associated with meteorological events. This because the costs to society of floods and droughts tend to be so great--and the structures necessary to reduce these risks so costly--that any agricultural practice that reduces these costs and restores ecological functions will likely compare favorably.

The models are used to identify and define the change from "wholeness," the set or vector of ecological services that were modified in converting the land to agricultural use. The objective of policy is to restore or mimic the extent of the service or services that was lost at the lowest cost by establishing the proper economic signal to landowners or producers to adopt the practices or land management systems that provide the service. Because the objective of this policy approach is to provide local stakeholder involvement, a market in services (or trading) is only likely to be feasible if there is a net gain for the buyers of the service in terms. A regional or national authority may decide to contribution to the cost of the trade so as to increase the price signal to suppliers of the service if public outside of the scope of the immediate trading area also economically benefit.

Graphically visualizing tradeoffs and forcing the explicit statement of key scientific and economic assumptions are critically important since the science of ecology is continually advancing and our ability to predict limited. Establishing what is essentially a policy baseline allows for evaluation of interventions over time to discover what works and what does not, for identification of research gaps and data needs, and for a framework for building social consensus on policy where uncertainty and unknowability predominate.

The focus on services to local and regional interests increases the likelihood of sustained performance. Encouraging the involvement in the market by owners of contiguous lands expands the kinds and values of the services that are provided. Groups of farmers or landowners, for example, either themselves or brokered through third-parties, such as government or even insurance companies, are encouraged to form consortia to provide the services to private entities and quasi-governmental agencies, such as sewage treatment plants, drinking water purveyors, or even energy companies (see, for example, Donnelly, 2000).

Similarly, potential beneficiaries, public and private, could, either individually or collectively in the form of consortia, bid on the rights--or options--to use specified agricultural lands or to specify the system of agricultural practices that are used to produce agricultural goods. The value of the bid depends on the array of array of ecological services that the lands could potentially provide, which in turn depends upon the pool of land put up to bid, its inherent characteristics, and its spatial features, including its "completeness," that is absence of holes in the covered landscape. Beneficiaries potentially supplement other interests' bids to rachet up the set of practices and hence the system of services that the landscape provides. Again, the role of government in facilitating these new markets is to define the commodity (in reality, "service") that is traded, clarify property rights, represent public interests where the interests of society at large are at stake, and to either monitor and enforce agreements or oversee the conduct of the latter through disinterested third parties. In so doing it reduces uncertainty and hence the costs of transactions and thereby the scope of the market.

Identifying the proper unit of organization to supply the service that addresses the problem in a cost-effective manner is key. Rather than the traditional focus on the individual land owner or farmer, a higher organizational unit, such as a drainage or flood district, may more appropriately be targeted as the potential supplier of the service. Nitrogen, for example, can be prevented from reaching water bodies by constructing or restoring wetlands at key junctures where drainage systems reach rivers or streams (Vitousek et al, 1997). Yet the success of constructed wetlands to protect the resource requires cooperation of each member of the drainage district, not just to provide the necessary conditions, but also to address the free-rider problem whereby other farmers cannot be excluded from the benefits provided by farmer or landowner on whose land the wetlands are constructed. Hence, the parties to a contract for water quality would involve the drainage district and a supplier or user of the resource.

The strategy is not meant as a substitute for regulation. It would not work in watersheds or problemsheds where the value of environmental services to local or regional stakeholders either cannot be established or where the monetizable value of the service does not exceed the economic value derived from continuing existing agricultural management practices. In other words, the transformation of natural capital appears to result in society being "better off," in lieu of scientific evidence to the contrary. Instead it serves to reduce the expected costs of natural events, such as floods and droughts; of complying with existing regulation, through such means as preventing the pollution that otherwise must be treated; serving as a sink for pollution that otherwise would have been emitted, such as greenhouse gases; or of providing social amenities, such as wildlife habitat. It also services to direct focus on practices that provide a multitude of environmental benefits, though small individually, rather than a single benefit that, alone justifies implementation. In this way, it complements existing regulatory and incentive-based efforts.

2.2.1 Harmonizing federal policy with ecosystem protection

National policy designed to support farm incomes can impede, if not thwart, environmental policy if it creates incentives to overuse inputs and natural resources in the production of agricultural commodities (Browne et al, 1992). The higher the commodity price support payments, the higher the hurdle for any economic incentive program for managing the landscape in an ecologically sound manner. Recent trends in the European Union and, until recently, in the United States have seen the decoupling of support payments from crop production and shift to a greater portion of farm income for green services (European Environment Agency, 2000). Such trends need to continue if successfully working landscapes that provide ecological services are to become a reality.

Government programs, particularly at the national and state level, should reward good stewardship, encouraging the adoption of practices and management systems that restore ecologic functions and rehabilitate the land and landscape. Rather than encouraging the overuse of soil resources, government programs should reward the provision of ecologic services that benefit the economy locally and the environment globally. Property rights should be clarified to facilitate market-based approaches, where feasible, and provide for greater economic efficiency given the changing structure of agriculture. They should be consistent with or even encourage the institutional structure for collective action to restore the ecologic functions of landscape or to manage animal waste in an environmental sound manner where individual initiative alone is unlikely to achieve the environmental results.

2.2.2 Creating demand

For some services, the role of government may be in creating demand for the service (CRS, 1998). Thi may entail imposing a limit to ambient concentrations of the pollutant in a resource, establishing rights to emit the pollutant, and allowing trading of the rights (see also USEPA. The European Directive on Nitrates is one example in this regard; the United States' Clean Water Action Plan and the setting of TMDLs (Total Maximum Daily Loads, numerical ecological standards for pollutants in watersheds) under its Clean Water Act is another (CRS, 2000b). TMDLs serve to increase demand to reduce total loading of pollutants, such as nitrogen and phosphorus into watersheds. Strict state or national drinking water standards are another. For agriculture, government can also help create demand for its services by recognizing a property right to one or more characteristics of the land, such as soil carbon, and the conditions under which the right can be established, including who monitors and how measurements can be made. In other cases, such as where upland agricultural land is used to store runoff to prevent downstream flooding, it is establishing the limits of liability.

Establishing, for example, a market in carbon by setting a limit on carbon emissions for the purpose of mitigating climate change, and allowing a trade in this new commodity may increase the incentive to invest in this resource (ELI, 1997). To the extent that agricultural practices (such as conservation tillage which increases the carbon content of soil) that assist in the maintenance of this component of the resource provide other ecological services of the larger landscape–such as clean water--, then this expansion of the marketplace can help solve the common pool resource dilemma.

National agricultural farm or commodity-support policy can also serve to create demand for services either directly by purchasing services, providing matching grants for services, or indirectly by reducing their cost through subsidies of flood or crop insurance that are coupled with environmentally sound practices. For example, the Conservation Reserve Enhancement Program established under the Federal Agricultural Income Reform Act (the Fair Act) that provides matching grants to regional or local authorities for targeted conservation improvements could be expanded and broadened in scope to cover a multitude of ecological services that agricultural land can provide (CRS, 1998). Other federal programs, such as federal disaster assistance or federally-subsidized multiperil insurance can be made contingent upon local programs that reduce the likelihood of flood or drought disasters by landscape-scale restoration or management programs.

Federal support for agricultural land preservation programs, such as the Farmland Preservation Program under the 1996 Fair Act, could be expanded to include lands that provide ecological services. Maintaining agricultural land near urban areas subject to flooding could provide flood storage benefits. Alternatively the land could be managed to mitigate the impacts of drought by protecting groundwater recharge zones and being maintained in a manner to encourage percolation (Tourbier, 1994).

2.2.3 Defining the Commodity That Is Traded and Clarifying Property Rights

The new models that can simulate complex relationships between agricultural land management practices, economic costs, and environmental impacts also allow for commoditization of services, and hence the possibility of new markets. Markets, however, require clarity regarding property rights to what is traded. Government needs to define what is it that is traded with the use of these models; help establish who owns the product of the services; as well as who is liable for nonperformance of service and the extent of liability in many cases; and establish conditions for monitoring and enforcement where the primary beneficiary of the service is the public.

Redefining property rights can also help shift the burden of complying with environmental regulations upon the entity within the agricultural production and distribution system better able to pass costs onto consumers. In the case of confined livestock operations, for example, according to Zilberman (2001) "[s]hifting the liability to integrators [The integrator is the livestock processor who contracts with growers to raise the animals for later slaughter.] will lead to more investment in waste management technologies and eventually adoption of alternative technologies."

Defining the commodity or service that is traded is complicated because, unlike the crops that are grown that are readily measurable, the ecological services provided by farmland and the larger landscape cannot be easily quantified. Ecological services are not like bushels of corn, discrete items or set of items that can be clearly and unequivocally measured, given agreement on how measurement will occur. Even the carbon, for instance, in a discrete plot of soil is not likely ever to be actually traded for the purpose of meeting international or even domestic commitments to reduce greenhouse gas emissions. Indeed, what is traded is generally the promise of the performance of a service or services whereby soil quality is increased--which may entail sequestering carbon in soils and maintained there for a given period of time--, the land is managed to reduce the possibility of downstream loadings of pollutants, the landscape is managed to provide wildlife habitat and increase ground water recharge, and the watershed is managed to retain water and reduce the likelihood of downstream floods and droughts. The service involves using agricultural practices that have been scientifically shown to mimic or restore natural ecological functions. The precision of the measurement that is important to trading relates to the ability to predict the quantity of product, or stock, provided by the service at a given point in time and within a given quantity of land.

3. EUROPEAN UNION APPROACHES

The European Union suffers from similar problems as the United States with excessive loadings of nutrients into the environment from agriculture (Manale, 1991; Crouzet, 2000). Since the late 1980s, the European Union has been reforming their farm support policy under the Common Agricultural Policy (CAP), which, similar to that of the United States, provided output-related support, towards area-based payments and payments for environmental services. In 1996, 3% of total farm-support spending under the CAP was spent on encouraging the production of environmental goods or reducing the negative impacts of agricultural production.

In 1991, the European Union promulgated a directive that establishes limits on nitrogen that can be emitted into the environment from agricultural activities (EU, 1997). Member states are required to establish programs to protect waters affected by nitrate pollution. Few have complied (EU, 1997).

Nevertheless, individual member states have implemented programs that show promise. For example, in the United Kingdom, agri-environmental policy has been implemented under a "management agreement" model (Hanley, 1999) whereby society must compensate farmers for any opportunity cost associated with environmental outputs (that is, increased cost or foregone profit associated with reducing a negative impact or with providing an environmental service). Since farmers implicitly have the right to use agricultural production practices that are the most profit-maximizing, society must pay for environmental goods if they entail additional expense. The payments may vary from area to area and target the prevention of environmental degradation or the improvement of environmental quality.

The success of the program has yet to be definitively evaluated. Nevertheless, economic cost-benefit studies suggest that the benefits have exceeded the social cost. The manner in which the benefits were established in most cases, however, is known to result in overestimates. Furthermore, improvements in environmental quality cannot be easily disentangled from reform of the Common Agricultural Policy, the EU-wide system of agricultural support.

4. REFERENCES:

Antle, John. 2000. The New Economics of Agriculture. Am. J. of Ag. Econ. 81(5).

Armentano, T.V. 1980. Drainage of organic soils as a factor in the world carbon cycle. Bioscience 30(12): 825-830.

Babcock, B. A., P. G. Lakshminarayan, J. Wu, and D. Zilberman. 1997. Targeting tools for the purchase of environmental amenities. Land Economics, 73:325_39.

Babcock, Bruce A. and Keith H. Paustian. 2001. Agriculture's Response to the Climate Change Challenge (Report) Council on Agricultural Science and Technology (CAST). Ankeny, Iowa. In press.

Batie, Sandra S. and David E. Ervin, 1997. Flexible incentives for environmental management in agriculture: a typology. Proceedings of the June **8-10**, 1997 conference Flexible Incentives for the Adoption of Environmental Technologies in Agriculture organized by the University of Florida at Gainesville, Michigan State University, and the US Department of Agriculture, Economic Research Service.

Bauer, Kurt. 1996. J. of Soil and Water Conservation 51(1):29-33. Creating integrated rural resource land information systems.

Baumol, W. and W.E. Oates. 1975. The theory of environmental policy. Prentice Hall, Inc. Englewood Cliffs, New Jersey.

Bellamy, Jennifer A., Andrew K. L. Johnson. 1999. Integrated resource management: moving from rhetoric to practice in Australian agriculture. *In* Environmental Management 25 (3): 265-280.

Boehlje, Michael. 1995. Industrialization of Agriculture: What are the Implications? *In* Choices, Fourth Quarter, 1995.

Bormann, F. Herbert and Gene E. Likens. 1985. Air and watershed management and the aquatic ecosystem. *In* Gene E. Likens [ed.] An Ecosystem Approach to the Aquatic Ecology: Mirror Lake and the Environment, Springer Verlag, New York.

Brown, Joel R. and Neil D. MacLeod. 1996. Integrating ecology into natural resource management policy. Environmental Management 20(3):289-296.

Browne, W. J. Skees, L. Swanson, P. Thompson, L.J. Unnevehr. 1992. Sacred cows and hot potatoes: agrarian myths in agricultural policy. Westview Press, 1992.

Burt, T. P. 1999. Towards integrated management of rural drainage basins with particular reference to water quality issues. *In* Stephen T. Trudgill, E. Walling, and Bruce W. Webb [ed.]. Water Quality: Processes Policy and Sustainability. Wiley and Sons, New York.

Camboni. S. M. and T. L. Napier. 1994. Socioeconomic and farm structure factors affecting frequency of use of tillage systems. Paper presented at the Sept 1994 Agrarian Prospects III Symposium in Praque, Czech Republic.

Center for Agricultural and Rural Development (CARD). 1997. RAPS 1997: Agricultural and Environmental Outlook. Iowa State University, Ames, Iowa. Available online at http://www.ag.iastate.edu/card/RAPS.

Center for Compatible Economic Development. 1996. A citizen's guide to achieving a healthy community, economy, and environment. Pamphlet published by The Nature Conservancy. Leesburg, Virginia, 1996.

Chichilnisky, G. and G. Heal. 1998. Economic returns for the biosphere. *In* Nature 391:629-630.

Congressional Research Service (CRS), 1998. Report to Congress 98-946: Clean Water Issues in the 106th Congress, Nov. 24. [Online]. Available at http://www.cnie.org/nle/h2o-33.html. Verified on September 23, 2000.

Congressional Research Service (CRS). 1999. Report to Congress, RL30331: conservation Spending in Agriculture: Trends and Implications [Online]. Available at http:www.cnie.org/nle/ag-89.htm (verified October 6, 1999).

Congressional Research Service (CRS). 2000. Report for Congress: RL 30437: Water Quality Initiatives and Agriculture, Washington, D.C.

Congressional Research Service (CRS). 2000b. IB89102: Water Quality: Implementing the Clean Water Act, July 19. [Online]. Available at http://www.cnie.org/nle/h2o-15.html. Verified December 23, 2000.

Costanza, Robert, John Cumberland, Herman Daly, Robert Goodland, and Richard Norgaard. 1997. An introduction to ecological economics. St. Lucie Press, Boca Raton, Florida.

Crosson, Pierre R. and Sterling Brubahn. 1982. Resource and environmental effects of U.S. agriculture. The Johns Hopkins Press, Baltimore, Maryland.

Crouzet, P., J. Leonard, S. Nixon, Y. Rees, W Parr, L. Laffon, J. Bogestrand, P. Kristensen, C. Lallana, G. Izzo, J. Bak, T. J. Lack. 2000. N. Thyssen [ed.] Environmental Assessment Report #4: Nutrients in European Ecosystems. [Online]. Available at located at http://themes. Eea.eu.int/toc.php/state/water?doc=39216&l=en. Verified on Dec. 23, 2000.

Crumpton, William G. and James L. Baker. 1993. Integrating wetlands into agricultural drainage systems: prediction of nitrate loadings and loss in wetlands receiving agricultural subsurface drainage," p. 118-126. *In* J. Kent Mitchell [ed.] Integrated Resources Management and Landscape Modification for Environmental Protection, Proceedings of the International Symposium 13-14 Dec. 1993. Chicago, Ill. Published by the American Soc. of Ag. Eng., St. Joseph, Missouri.

Daily, Gretchen (ed.). 1997. Nature's Services: Societal Dependence on Natural Ecosystems. Island Press, Washington, D.C.

Daily, Gretchen. 2000. Chapter 1, Introduction: What are Ecosystem Services? [Online]. Available at http://www.islandpress.com/ecocompass/nature/NSintro.html (verified August 15, 2000).

Daily, Gretchen, Susan Alexamder, Paul R. Ehrlich, Larry Goulder, Jane Lubchenco, Pamela A. Matson, Harold A. Mooney, Sandra Postel, Stephen H. Schneider, David Tilman, George W. Woodwell. 2000. Ecosystem Services: Benefits Supplied to Human Societies by Natural Ecosystems [Online]. Available at http://esa.sdsc.edu/daily.htm. (verified August 15, 2000).

Daily, Gretchen and Paul Ehrlich. 1992. Population, sustainability and earth's carrying capacity. Bioscience 42:761-771.

Dasgupta, Partha., 2000. The British Academy Keynes Lecture in Economics: Economics and the Environment [Online]. Available at http://britac3.britac.ac.uk/pubs/keynes95/00index.html, (verified October 24, 2000).

DeLaney, Taylor A. 1995. Benefits to downstream flood attenuation and water quality as a result of constructed wetlands in agricultural landscapes. J of Soil and Water Conservation 50 (6):620-629.

Dinar, Ariel and Edna Tusak Loehman. 1995. p. xxii-xxiii. *In* Ariel Dinar and Edna Tusak Loehman [ed.] Water Quantity/Quality Management and Conflict Resolution: Institutions, Processes, and Economic Analyses. Praeger, Westport, Connecticut.

Donnelly, Aldyen. 2000. January 11, 2000 email from the president of GEMCo, a consortium of Canada energy companies, to Andrew Manale at USEPA describing the background and contractual details of a contract with GEMCo and a group of Iowa farmers to provide carbon sequestration benefits.

Dyke, Paul. 1997. Personal telephone communication. Texas Agricultural Experiment Station, Temple, Texas.

Environmental Law Institute (ELI). 1997. Implementing an emissions cap and allowance trading system for greenhouse gas emissions: lessons from the acid rain program. ELI Project #941722, report produced under a cooperative agreement with the US Environmental Protection Agency, CR-822795-01, Washington, D.C.

European Union (EU). 1991. Directive 91/676/ EEC on nitrates from agricultural sources. Report COM (97) 473. [Online]. Available at http://europa.eu.int/water/water-nitrates/directiv.html. Verified December 20, 2000.

European Environment Agency. 2000. Environmental Signals 2000. [Online]. Available at http://themes.eea.eu.int/toc.php/state/water?doc=39249&1=en, page 2-8,]. Verified on December 27, 2000.

Evans, R.O., J.E. Parsons, K. Stone, and W.B. Wells. 1992. Water table management on a watershed scale. J. of Soil and Water Conservation 47(1):58-64.

Feather, Peter M. and Joseph Cooper. 1995. Voluntary incentives for reducing agricultural nonpoint source water pollution. USDA Economic Research Service Agricultural Information Bulletin, #716.

Franz, William. 2001. Email to Andrew Manale, USEPA, regarding "Working Landscapes." USEPA, Region V. Chicago, Ill.

Gregory, Robin. 2000. Using stakeholder values to make smarter environmental decisions. Environment 42(5):34-44.

Gunderson, Lance, C. S. Hollings, and Stephen S. Light [ed.]. 1995. Barriers and bridges to the renewal of ecosystems and institutions. Columbia University Press, New York.

Gunderson, Lance H., C.S. Holling, and Stephen S. Light (ed). 1995. Barriers and Bridges to the Renewal of Ecosystems and Institutions, Columbia Univ. Press. p.8.: quotation from Holling. (Holling, C. S., 1973. Resilience and stability of ecological systems. Annual Review of Ecology and Systematics 4: 1-23.).

Haapala, J.J. 2000. Global Warming and Agriculture. *I*n Good Tilth [Online]. Available at http://www.tilth.org/>www.tilth.org> (posted November 2000, verified December 21, 2000).

Hall, Millard W. 1998. Extending the resources: integrating water quality considerations into water resources management. *In* Water Resources Update: Water Policies for the Future, Issue 111, Universities Council on Water Resources, Carbondale, Ill.

Halwell, Brian. 2000. Hired Hands on their own Land. Worldwatch Institute. Distributed by the Los Angeles Times Syndicate, Nov. 8, 2000.

Hanley, Nick and David Oglethorpe. 1999. Emerging policies on externalities from agriculture: an analysis for the European Union. American Journal of Agricultural Economics 81(5):1222-1227.

Hardin, G. 1968. The tragedy of the commons. Science 162:1243-48.

Hatfield, J.L. and B. A. Stewart [ed.]. 1998. Animal Waste Utilization: Effective Use of Manure as a Soil Resource. Ann Arbor Press, Chelsea, Michigan.

Heimlich, Ralph E., Keith D. Weibe, Roger Claasen, Dwight Gadsby, and Robert M. House. 1998. Wetlands and agriculture: private interests and public benefits. USDA, Economic Research Service, Agricultural Economic Report No. 765. Washington, D.C.

Herrick, Jeffrey E. 2000. Soil quality: an indicator of sustainable land management? *In* Applied Soil Ecology 15:75-83.

Horan, Richard D., James S. Shortle, and David G. Abler. 1999. "Green Payments for Nonpoint Pollution Control American Journal of Agricultural Economics 81 (5).

Hornbeck, J.W., C.W. Martin, R.S. Pierce, F.H. Borman, G.E. Likens, J. S. Eaton, 1987. The Northern Hardwood Forest Ecosystem: Ten Years of Recovery from Clear-cutting. USDA, Forest Service, NE Forest Experiment Station, NE-RP-596, pamphlet.

Isenhart, Thomas M. Richard C. Schultz, and Joe P. Colletti. 1997. Watershed restoration and agricultural practices in the midwest: Bear Creek of Iowa, p. 313-334. *In* J. E. Williams, C. A. Woody, M. P. Dombeck [ed.] Watershed restoration: principles and practices. American Fisheries Society, Bethesda, MD.

Kemper, W. Doral. 1997. Organic matter--100 years to bring it down. How long to bring it back? National Conservation Tillage Digest. Perryville, Missouri. January 1997.

King, Dennis M., Lisa A. Wainger, Candy C. Bartoldus, and James S. Wakeley. 2000. Expanding wetland Assessment procedures: linking indices of wetland function with services and values. Prepared for the US Army Corps of Engineers Engineer Research and Development Center, ERDC/EL TR-00-17. [Online]. Available at http://www.wes.army.mil/el/wetlands/pds/tre/00-17. Verified December 23, 2000.

Kovacic, David. 2000. The effectiveness of constructed wetlands in reducing nitrogen from agricultural tile drainage. Paper presented at the Global Sustainability 2000 Conference in Springfield, Illinois, March 28-29, 2000. *In* Summary of presentations published by Gallatin County Soil & Water Conservation District.

Laane, W.E. M. and P. Lindgarrd-Jorgensen. 1992. Ecosystem approach to the integrated management of river water quality, p.97-118. *In* P. J. Newman, M.A. Piavaux, and R. A. Sweeting [ed.] Commission of the European Communities, River Water Quality: Ecological Assessment and Control. Commission of the European Communities, Luxembourg.

Lakshminarayan, P.G. and Bruce A. Babcock. 1995. An economic and environmental evaluation of farm bill policy options using the CEEPES_FAPRI modeling system. Iowa State University, Center for Agricultural and Rural Development, report 95_WP 137.

Loucks, Daniel P. 1998. Watershed planning: changing issues, processes, and expectations. *In* Water Resources Update Issue #111, spring. The Universities Council on Water Resources.

Ludwig, D., R. Hilborn, and C. Waters. 1993. Uncertainty, resource exploitation, and conservation: lessons from history. Science 260:17-36.

MacGregor, R.J. and T. McRae. 2000. Driving forces affecting the environmental sustainability of agriculture. In T. McRae, C.A.S. Smith and L.J. Gregorich (ed). Environmental sustainability of canadian agriculture: report of the agri-environmental indicator project. Agriculture and Agri-Food Canada.

Manale, Andrew. 1991. European community programs to control nitrate emissions from agriculture. International Environmental Reporter 14 (12).

Manale, Andrew and Clare Narrod. 1994. Environmental Implications of Industry Structure in Dairy, Swine, and Poultry Industries. *In* Proceedings of the Great Plains Animal Waste Conference on Confined Animal Production and Water Quality: Balancing Animal Production and the Environment, October 19-21, 1994, Denver, Colorado. GPAC publ. no. 151. Printed by National Cattlemen's Association, Englewood, Colorado.

Manale, Andrew. 2000. Managing floods through targeted, selective restoration of landscape functions: paying upland farmers to store or manage runoff during threats of floods. J. of Soil and Water Conservation 55(3):285-295.

Margerum, Richard D. 1997. Advancing integrated resource management: processes and policies. Paper presented at the 2nd National Workshop on Integrated Catchment Management, International Conference held at the Manning Clark Centre, Australian National University, Sept 29-Oct. 1 1997. Available through River Basin Management Society, Inc. Mt. Waverley, Vic., Australia.

Miranowski, John and Mark Cochran. 1993. Economics of land in agriculture. *In* Gerald Carlson, David Zilberman, and John Miranowski Agriculture and environmental resource economics. Oxford University Press, New York.

Napier, Ted L. 1994. Regulating approaches for soil and water conservation. *In* Louis E. Swanson and Frank B. Clearfield [ed.] Agricultural policy and the environment: iron fist or open hand. Soil and Water Conservation Society, Ankeny, Iowa.

Napier, T. L. and D. E. Brown. 1993. Factors affecting attitudes toward ground water pollution among Ohio farmers. J. of Soil and Water Conservation 48(5):432-438.

National Research Council (NRC). 1989. Alternative Agriculture. National Academy Press. Washington, D.C.Natural Resource Conservation Service (NRCS). 1997. Water Quality and Agriculture: Status, Conditions, and Trends. Working Paper #16. USDA. Washington, D.C.

National Research Council (NRC). 1993. Soil and Water Quality: An Agenda for Agriculture. National Academy Press. Washington, D.C.

Nowak, P. J. 1987. The adoption of agricultural conservation technologies: economic and diffusion explanations. Rural Sociology 52(2):208-220.

Ostrom, Elinor, Joanna Burger, Christopher B. Field, Richard B. Norgaard, and David Policansky. 1999. Revisiting the commons: local lessons, global challenges. Science 284(9).

Perry, James and Elizabeth Vanderklein. 1996. Water Quality: Management of a natural resource Blackwell Science, Cambridge, Mass, USA.

Perry, James and Elizabeth Vanderklein. 1996. Water Quality: Management of a Natural Resource. Blackwell Science, Cambridge Mass, USA.

Phipps, Tim T. and Pierre R. Crosson. 1986. Agriculture and the Environment: An Overview, p3-31. *In* Tim T. Phipps, Pierre R. Crosson, and Kent A. Price (ed.) Agriculture and the Environment. Resources for the Future. Washington, D.C.

Ribaudo, Marc O., Richard D. Horan, and Mark E. Smith. 2000. Economics of Water Quality Protection from Nonpoint Sources: Theory and Practice. Resource Economics Division, Economic Research Service, USDA, Agricultural Economic Report No. 782 (AER-782), available at Http://www.ERS.USDA.giv/epubs/pdf/aer782, (Verified October 18, 2000).

Rickson, Roy E.,David Burch and Richard Sanders. 1997. New and Emerging Structures. *In* Agriculture: Challenges and Opportunities for Natural Resource Management. 2nd National Workshop on Integrated Catchment Management: Advancing Integrated Resource Management: Processes and Policies, ISBN 09586668. River Basin Mangement Society, 22 Mt. Waverley, Victoria, Australia.

Saleh, A., J.G. Arnold, P.W. Gassman, L.M. Hauck, W.D. Rosenthal, and A.M.S. McFarland. 2000. Application of SWAT for the Upper North Bosque River Watershed. Transactions of the American Society of Agricultural Engineers 43(5):1077-1087.

Schertz, Lyle P. (ed.) and Lynn M. Daft. 1994. Food and agricultural markets: the quiet revolution. National Policy Association, ISBN 0890681244, Washington, D.C.

Skees, J. R. 1999. Statement before the U.S. Senate Agriculture, Nutrition, and Forestry Committee. U.S. Congress. Hearing on Crop Insurance Reform, March 10, 1999.

Schultz, R.C., J.P. Coletti, T.M. Isenhart, W.W. Simpkins, C.W. Mize, and M.L. Thompson. 1995. Design and Placement of a multi-species riparian buffer strip system. Agroforestry systems 29:201-226.

Segerson, K. 1988. Uncertainty and Incentives for Nonpoint Pollution Control. Journal of Environmental Economics and Management 15: 87-98.

Shortle, James S. 1995. Environmental federalism: the case of U.S. agriculture. *In* J.R. Braden, H. Folmer, and T Ulen (ed); Environmental policy with economic and political integration: the European Union and the United States. Edward Elgar Publisher, Cheltenham, England.

Shortle, James S. and David G. Abler. 1999. Agriculture and the Environment, p.162-3. [*In*] Jeroen C. J. M. van den Bergh [ed.] Handbook of Environmental and Resource Economics. Edward Elgar Publishing, Northhampton, Mass.

Simpson, R. David. 2000. The economic valuation of ecosystem services: a review of literature and issues. Report prepared for the World Bank. Resources for the Future. Washington, D.C

Stewart, Richard B.. 1988. Controlling environmental risks through economic incentives. *In* Columbia Journal of Environmental Law 13:153-169.

The Guardian. 2000. Climate change may be down to farming. [Online]. Available at http://guardian.co.uk./uk_news/story/0,3604,392724,00.html. Verified November 4, 2000.

Tourbier, Toby. 1994. Journal of Soil and Water Conservation 49:1, p. 14.

U.S. Department of Agriculture (USDA). 2000. Agricultural Resources and Environmental Indicators, 2000. Economic Research Service, Resource Economics Division. Washington, D.C.[Online]. Available at http://www.ers.usda.gov/Emphases/Harmony/issues/arei2000/arei2000.htm#conspolicy. Verified on January 26, 2001.

U.S. Geological Survey (USGS). 1996. Using a geographic information system and scanning technology to create high-resolution land-use data sets. Water-Resources Investigations Report 96-4100. Iowa City, Iowa.

U.S. Geological Survey (USGS). 1999. The quality of our nation's waters: nutrients and pesticides. US Geological Survey Circular 1225, Reston, VA.

U.S. Geological Survey (USGS). 2000. Nitrogen in the Mississippi Basin--estimating sources and predicting flux to the Gulf of Mexico. USGS Fact Sheet 135--00. U.S. Department of the Interior. Reston, Virginia. December 2000.

US Geological Survey (USGS). 2000b. Volumes of recent floods and potential for storage in upland watershed areas of Iowa. Fact Sheet FS-097-00.

US Environmental Protection Agency. 1997. Fact Sheet: National Water Quality Inventory__ 1996 Report to Congress. EPA841_F_97_003. Washington, D.C.

US Environmental Protection Agency (USEPA). 2000. Basins 2.0 fact sheet: A powerful tool for managing watersheds. [Online]. Available at http://www.epa.gov/OST/BASINS/basinsv2.htm#model.

US Environmental Protection Agency (USEPA). 2001. Region 10: Lower Boise River effluent trading demonstration project. [Online]. Available at http:yosemite1.epa.gov/r10/oi.nsf...984abd8825696c00793442?OpenDocument. Verified January 22, 2001.

US Environmental Protection Agency (USEPA). 2001b. [Online]. Available at http://www.cleanwater.gov/action/overview.html. Verified January 1, 2001.

Vanderholm, D. H. 1994. Livestock Productions Trends, Water Quality and Economic Impacts. *In* Proceedings of the Great Plains Animal Waste Conference on Confined Animal Production and Water Quality: Balancing Animal Production and the Environment, October 19-21, 1994, Denver, Colorado. GPAC publ. no. 151. Printed by National Cattlemen's Association, Englewood, Colorado.

Virginia Tech. 2001. Summary of presentations at the symposium on integrated decision-making watershed management symposium: processes and tools. Held on January 7-9, 2001 in Chevy Chase, Maryland. [Online]. Available at http://www.conted.edu/watershed/2000/10updates. Verified on January 5, 2001.

Vitousek, Peter M. , John D. Aber, Robert W. Howarth, Gene E. Likens, Pamela A. Matson, David W. Schindler, Willima H. Schlesinger, and David G Tilman. 1997. Human alteration of the global nitrogen cycle: sources and consequences, Ecological Approaches 7(3):737-750.

Wainger, Lisa A., Dennis King, Leonard Shabman, and Laura Zepp. 2001. The benefits and costs of reforesting economically marginal cropland: the case of the Mississippi delta. Prepared for the Delta Land Trust, Jackson, Mississippi.

Weitzman, M. 1974. Prices versus quantity. *In* Review of Economic Studies 41:477.

Welsh, Rick. 1996. The Industrialization of U.S. Agriculture: An Overview and Background Report. Henry A. Wallace Institute, Greenbelt, Maryland.

Whitford, W.G, D.J. Rapport, and A.G. DeSoyza. 1999. Using resistance and resilience measurements for 'fitness' tests in ecosystem health. Journal of Environmental Management 57: 21-29.

Wiche, G.G., S.K. Jensen, J.V. Baglio, and O. Domingue. 1990. Application of digital elevation models to delineate drainage areas and compute hydrologic characteristics for sites in the James River Basin, North Dakota. U.S. Geological Survey Open File Report 90-593 as reported in Preliminary Report of the Scientific Assessment and Strategy Team, Report of the Interagency Floodplain Management Review Committee to the Administration Floodplain Management Task Force: Science for Floodplain Management into the 21st Century, Blueprint for Change Part V, Washington, D.C: 1994.

Wiener, J.G., C.R. Frelling, C.E. Korshgen, K.P. Kenow, E.M. Kirsch, S.J. Rogers, Y. Yin, and J.S. Sauer. 1998. Mississippi River, p. 351-384 *In* M. J. Mac, P.A. Opler, C.E. Puckett Haeker, P.D. Doran [ed.]. Status and trends of the nation's biological resources, vol 1. U.S. Geological Survey, Reston, VA.

Woltemade, C.J. 2000. Ability of restored wetlands to reduce nitrogen and phosphorus concentrations in agricultural drainage water. J. of Soil and Water Conservation 55(3): 303-308.

World Resources Institute. 2000. Ecosystems: What is an ecosystem approach? [Online]. Available at http://www.wri.org/wr2000/ecosys_approach.html. Verified December 10, 2000.

Wu, JunJie and Bruce Babcock. 1999. Metamodeling Potential Nitrate Water Pollution in the Central United States, p. 1916-1922. Journal of Environmental Quality 28:6.

Zilberman, David. 2001. The evolution and situation of the swine industry. Department of Agricultural Resource Economics, University of California, Berkeley. Publication pending.

Appendix

APPENDIX

Abbreviations and Units of Measure

Abbreviations

AWC = available water capacity
BMP = best management practices
CEC = cation exchange capacity
C:N = carbon:nitrogen ratio
CRP = Conservation Reserve Program
DOC = dissolved organic carbon
EC = electrical conductivity
$g/m^2/yr$ = gram per square meter per year
GHGs = greenhouse gases
GT = gigaton = petagram = 1000 MMT
GWP = global warming potential
ha = hectare
IPCC = intergovernmental panel on climate change
kg = kilogram
m = meter
Mha = million hectares
mmhos/cm = millimhos per centimeter
MMT = million metric tons = Tg
MT = metric ton = 1000 kg = 1 Mg
MT/ha = metric ton per hectare
MWD = mean weight diameter
N = nitrogen
NPP = net primary productivity
Pg = petagram = 10 15 g = 1000 MMT = GT
ppbv = parts per billion by volume
ppmv = parts per million by volume
ppptv = parts per trillion by volume
RMP = recommended management practices
SIC = soil inorganic carbon
SOC = soil organic carbon
SON = soil organic nitrogen
SOM = soil organic matter
Tg = teragram = 10 12 g
USDA = United States Department of Agriculture
WRP = Wetland Reserve Program

APPENDIX

Units of Measurement

Metric (SI) multipliers

Prefix	Abbreviation	Value
Exa	E	10^{18}
peta	P	10^{15}
tera	T	10^{12}
giga	G	10^{9}
mega	M	10^{6}
kilo	k	10^{3}
hecto	h	10^{2}
deka	da	10^{1}
deci	d	10^{-1}
centi	c	10^{-2}
milli	m	10^{-3}
micro	m	10^{-6}
nano	n	10^{-9}
pico	p	10^{-12}
femto	f	10^{-15}
atto	a	10^{-18}

SUBJECT INDEX